1997–1998 DIETARY REFERENCE INTAKES (DRI)

Age (yr)	Recommended Dietary Allowances (RDA)								Adequate Intakes (AI)					
	Thiamin (mg)	Riboflavin (mg)	Niacin (mg NE)‡	Vitamin B₆ (mg)	Folate (μg DFE)	Vitamin B₁₂ (μg)	Phosphorus (mg)	Magnesium (mg)	Vitamin D (μg)	Pantothenic acid (mg)	Biotin (μg)	Choline (mg)	Calcium (mg)	Fluoride (mg)
Infants†														
0.0–0.5	0.2	0.3	2§	0.1	65	0.4	100	30	5	1.7	5	125	210	0.01
0.5–1.0	0.3	0.4	4	0.3	80	0.5	275	75	5	1.8	6	150	270	0.5
Children														
1–3	0.5	0.5	6	0.5	150	0.9	460	80	5	2.0	8	200	500	0.7
4–8	0.6	0.6	8	0.6	200	1.2	500	130	5	3.0	12	250	800	1.1
Males														
9–13	0.9	0.9	12	1.0	300	1.8	1250	240	5	4.0	20	375	1300	2.0
14–18	1.2	1.3	16	1.3	400	2.4	1250	410	5	5.0	25	550	1300	3.2
19–30	1.2	1.3	16	1.3	400	2.4	700	400	5	5.0	30	550	1000	3.8
31–50	1.2	1.3	16	1.3	400	2.4	700	420	5	5.0	30	550	1000	3.8
51–70	1.2	1.3	16	1.7	400	2.4	700	420	10	5.0	30	550	1200	3.8
>70	1.2	1.3	16	1.7	400	2.4	700	420	15	5.0	30	550	1200	3.8
Females														
9–13	0.9	0.9	12	1.0	300	1.8	1250	240	5	4.0	20	375	1300	2.0
14–18	1.0	1.0	14	1.2	400	2.4	1250	360	5	5.0	25	400	1300	2.9
19–30	1.1	1.1	14	1.3	400	2.4	700	310	5	5.0	30	425	1000	3.1
31–50	1.1	1.1	14	1.3	400	2.4	700	320	5	5.0	30	425	1000	3.1
51–70	1.1	1.1	14	1.5	400	2.4	700	320	10	5.0	30	425	1200	3.1
>70	1.1	1.1	14	1.5	400	2.4	700	320	15	5.0	30	425	1200	3.1
Pregnancy	1.4	1.4	18	1.9	600	2.6	*	+40	*	6.0	30	450	*	*
Lactation	1.5	1.6	17	2.0	500	2.8	*	*	*	7.0	35	550	*	*

*Values for these nutrients do not change with pregnancy or lactation. Use the value listed for women of comparable age.
†For all nutrients, an AI was established instead of an RDA as the goal for infants; for the B vitamins and choline, the age groupings are 0 through 5 months and 6 through 11 months.
‡Abbreviations: NE = niacin equivalents; DFE = dietary folate equivalents; RE = retinol equivalents; TE = α-tocopherol equivalents.
§The AI for niacin for this age group only is stated as milligrams of preformed niacin instead of niacin equivalents.

Source: Adapted with permission from *Recommended Dietary Allowances,* 10th Edition, and the first two of the *Dietary Reference Intake* series, National Academy Press. Copyright 1989, 1997, and 1998, respectively, by the National Academy of Sciences. Courtesy of the National Academy Press, Washington, D.C.

www.wadsworth.com

wadsworth.com is the World Wide Web site for
Wadsworth Publishing Company and is your direct source
to dozens of on-line resources.

At *wadsworth.com* you can find out about supplements,
demonstration software, and student resources.
You can also send e-mail to many of our authors and
preview new publications and exciting new technologies.

wadsworth.com
Changing the way the world learns®

Community Nutrition in Action

An Entrepreneurial Approach

Second Edition

Marie A. Boyle
The College of Saint Elizabeth

Diane H. Morris
Mainstream Nutrition

West/Wadsworth
I(T)P® **An International Thomson Publishing Company**

Belmont, CA • Albany, NY • Boston • Cincinnati • Johannesburg • London • Madrid • Melbourne
Mexico City • New York • Pacific Grove, CA • Scottsdale, AZ • Singapore • Tokyo • Toronto

Nutrition Publisher: Peter Marshall
Development Editor: Laura Graham
Editorial Assistant: Tangelique Williams
Marketing Manager: Becky Tollerson
Marketing Assistants: Shannon Ryan, Jonathan Larson
Project Editor: Sandra Craig
Print Buyer: Barbara Britton
Permissions Editor: Susan Walters

Production: Martha Emry
Text and Cover Design: Ellen Pettengell
Copyediting: Laura Larson
Illustrations: Jim Atherton
Cover Image: Harvest Scene with Twelve People by R. Mervilus. Private collection / van Hoorick Fine Arts / SuperStock
Composition: Parkwood Composition Service, Inc.
Printer: R. R. Donnelley & Sons

Printed in the United States of America
1 2 3 4 5 6 7 8 9 10

For more information, contact Wadsworth Publishing Company, 10 Davis Drive, Belmont, CA 94002, or electronically at http://www.wadsworth.com

International Thomson Publishing Europe
Berkshire House
168-173 High Holborn
London, WC1V 7AA, United Kingdom

Nelson ITP, Australia
102 Dodds Street
South Melbourne
Victoria 3205 Australia

Nelson Canada
1120 Birchmount Road
Scarborough, Ontario
Canada M1K 5G4

International Thomson Publishing Southern Africa
Building 18, Constantia Square
138 Sixteenth Road, P.O. Box 2459
Halfway House, 1685 South Africa

International Thomson Editores
Seneca, 53
Colonia Polanco
11560 México D.F. México

International Thomson Publishing Asia
60 Albert Street
#15-01 Albert Complex
Singapore 189969

International Thomson Publishing Japan
Hirakawa-cho Kyowa Building, 3F
2-2-1 Hirakawa-cho, Chiyoda-ku
Tokyo 102, Japan

Library of Congress Cataloging-in-Publication Data
Boyle, Marie A. (Marie Ann)
    Community nutrition in action: an entrepreneurial approach/
    Marie A. Boyle, Diane H. Morris.—2nd ed.
        p.   cm.
    Includes bibliographical references and index.
    ISBN 0-534-53829-0 (hc.)
    1. Nutrition policy—United States.   2. Nutrition—United States.
3. Community health services—United States.   I. Morris, Diane H.
II. Title
    TX360.U6B69   1999
    363.8'58'0973—dc21   98-38212

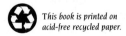

To my beloved Jesse
and to all those who
surround themselves
with lives even more
temporary than their own.

—Marie Boyle

For my parents,
who set a good example,
and my husband,
who never once complained
about his Internet charges.

—Diane Morris

About the Authors

Marie A. Boyle, Ph.D., R.D., received her Ph.D. in nutrition from Florida State University in 1992. She is coauthor of the basic nutrition textbook *Personal Nutrition* and nutrition consultant to the Regal Ware corporation. She presently works as Associate Professor and Director of the Didactic Program in Dietetics at the College of Saint Elizabeth, Morristown, New Jersey. She teaches undergraduate courses in Community Nutrition, Basic and Advanced Nutrition, Medical Nutrition Therapy, and Lifecycle Nutrition. She also teaches Research Methods, Advanced Metabolism of the Micronutrients, Nutrition and Aging, and Alternative Medicine in the Graduate Program in Nutrition at the college. Her other professional activities include teaching a community-based "Culinary Hearts" cooking class for the American Heart Association, acting as legislative chairperson for the local dietetic association, and serving as a member of the Osteoporosis Coalition of New Jersey.

Diane H. Morris, Ph.D., R.D., received her Ph.D. in nutrition in 1982 from the University of Tennessee at Knoxville. She has held faculty or staff positions with the American Medical Association, the Harvard School of Public Health, and the University of Massachusetts Medical School. She serves on the Board of Directors of the Heart and Stroke Foundation of Manitoba, where she is involved in strategic planning and health promotion. She has written more than 40 research, technical, and general articles for consumers and health professionals and has self-published a book, *An Internet Guide for Dietitians,* which is advertised on the Internet and sold globally. She has given numerous presentations on the use of the Internet in dietetics practice. She is president of Mainstream Nutrition and works to educate other health professionals and the general public through her writing, public speaking, and volunteer service.

Contents in Brief

Contents

Section II

Community Nutritionists in Action: Assessing and Planning   147

Section III

Community Nutritionists in Action: Delivering Programs   345

Preface

According to an article in the *New York Times Magazine* (March 1, 1998, p. 62), John Doerr, a Silicon Valley venture capitalist who helped jump start Intuit, @Home, Netscape, and Sun Microsystems, gave a presentation at the Stanford Business School in which he showed a slide that read:

OLD economy	NEW economy
a skill	lifelong learning
managers	entrepreneurs
labor vs management	teams
monopolies	competition
wages	ownership, options
hierarchical	networked
sues	invests
status quo	speed, change

Mr. Doerr was speaking of technology and the new world order wrought by the Internet when he presented this slide, but he could have been speaking about the practice of community nutrition. To succeed in community nutrition today, you must be committed to lifelong learning, because every day brings new research findings, new legislation, new ideas about health promotion, new technologies, all of which affect the ways in which community nutritionists gather information, solve problems, and reach vulnerable populations. You will likely be an entrepreneur—one who uses innovation and creativity to guide individuals and communities to proper nutrition and good health. You work in teams to lobby policy makers, gather information about your community, and design nutrition programs and services. You assess the activities of the competition—the myriad messages about foods, dietary supplements, and research findings that appear in advertising and articles about diet and health on television, radio, and the Internet, and in newspapers and magazines. You take ownership of your career and seek opportunities for growth. You network with colleagues to learn and share ideas. You invest in the future—your own and that of your profession and community.

And you accept change, the one constant in our lives today. We spoke in the first edition of a sea of change—a shift toward globalization of the work force and communications, a shift from clinical dietetics to community-based practice. Since the first edition was published, we have experienced the growth of the Internet—a virtual *tsunami* in communications—and witnessed the collapse of health care reform, the emergence of managed care, the publication of new dietary recommendations, the drive to reform welfare, and the rise of complementary and alternative medicine. What an exciting and challenging time to be a community nutritionist!

In this, the second edition, we continue to discuss the important issues in community nutrition practice and to present the core information needed by students who are interested in solving nutritional and health problems. The book is organized into three sections. Section I shows the community nutritionist in action within the community. Chapter 1 describes the activities and responsibilities of the community nutritionist and introduces the principles of entrepreneurship and the three arenas of community nutrition practice: people, policy, and programs. Chapter 2 makes it perfectly clear that if you're a community nutritionist, you're involved in policy making. Chapter 3 discusses the shift to managed care and the challenges facing federal, state, and municipal governments in providing quality health care to all citizens. Chapter 4 focuses on the nuts and bolts of national nutrition policy, including national nutrition monitoring, nutrient intake standards, and dietary recommendations.

Section II focuses on the tools used by community nutritionists to identify nutritional and health problems in their communities and to design programs to address those problems. Chapter 5 gives a step-by-step analysis of the community needs assessment and describes the types and sources of data collected about the community. Chapter 6 outlines the questions you'll ask in obtaining information about your target population. Chapter 7 describes the program planning process, covering everything from the factors that trigger program planning to the types of evaluations undertaken to improve program design and delivery. Chapter 8 discusses the reasons why people eat what they eat, what research tells us about how to influence behavior, and what program interventions look like. Chapter 9 addresses the heart of the program: the nutrition messages used in community interventions. Chapter 10 introduces you to the principles of marketing, an important endeavor in community nutrition practice. You are more likely to get good results if your program is marketed successfully! Chapter 11 brings the discussion around to important management issues such as how to control costs, manage people, and write grants.

Section III describes current federal and nongovernmental programs designed to meet the food and nutritional needs of vulnerable populations. Chapter 12 reviews some of the issues surrounding poverty and food insecurity in the domestic arena and considers how these contribute to nutritional risk and malnutrition. Chapter 13 focuses on programs for pregnant and lactating women and infants. Chapter 14 describes the programs for children and adolescents. Chapter 15 covers a host of programs for adults, including the elderly. Finally, Chapter 16 closes with a discussion of international issues in community nutrition.

Many of the unique features of the first edition have been retained. These aspects include:

- **Focus on Entrepreneurship.** Successful practitioners in community nutrition have a mind- and skill-set that opens them up to new ideas and ventures. They don't think, "This is how it has always been done." They think, "Let's try this. Let's do something different." We want you to begin thinking of yourself as a "social entrepreneur," as someone who is willing to take risks, try new technologies, and use fresh approaches to improving the public's nutrition and health.
- **Focus on Multiculturalism.** The growing ethnic diversity of our communities poses many challenges for community nutritionists. To increase your awareness and appreciation of cultures beyond your own, we have woven examples and illustrations of various food-related beliefs and practices from different cultures throughout the text.
- **Professional Focus.** The eleven Professional Focus features are designed to help you develop personal skills and attitudes that will boost your effectiveness and confidence in community settings. The topics range from goal setting and time management to writing, public speaking, and leadership. This feature is meant to help build your professional skills.
- **Program Spotlight.** Each chapter in Section III includes a Program Spotlight that describes one assistance program

such as the Food Stamp Program or the National School Lunch Program. The Spotlights cover such topics as the policy issues underlying the program, current legislative issues affecting the program, and the program's effectiveness in reaching the needs of its intended audience.

- **Community Learning Activity.** At the end of every chapter is a Community Learning Activity designed to get you involved in learning about your community and its health and nutritional problems. While most activities can be completed independently, some are meant to be undertaken by teams. The purpose of these activities is to give you experience in designing a program, creating marketing strategies, choosing nutrition messages, and developing goals and objectives.

We have added something new to this edition: Internet resources and activities. Every chapter contains a list of relevant Internet addresses and many chapters contain an Internet activity as part of the Community Learning Activity. You'll use this new technology to obtain data about your community and scout for ideas and educational materials. Moreover, you can link with the Internet addresses presented in this book through the publisher's Nutrition Resource Center on-line (www.wadsworth.com/nutrition). If you aren't using the Internet regularly, this is the time to begin. The Internet promises to become as indispensable to community nutrition practice as the telephone!

Finally, a word is needed about personal pronouns. On the advice of our reviewers and editors, we used gender-neutral terms whenever possible, recognizing that there are male community nutritionists and the involvement of men in community nutrition is likely to increase in the coming years. In some places, however, we used the pronouns "she" and "he" to make the text more personal and engaging and to showcase the image of community nutritionists, particularly women, as leaders, managers, and entrepreneurs. We want you, whether you are a man or woman, to think of yourself as a planner, manager, change agent, thinker, and leader—in short, a nutrition entrepreneur—who has the energy and creativity to open up new vistas for improving the public's health through good nutrition.

Marie Boyle

Diane Morris

November 1998

Acknowledgments

This book was a community effort. Family and friends provided encouragement and support. Colleagues shared their insights and experiences about the practice of community nutrition and the value of focusing on entrepreneurship. We are especially grateful to Beverley Demetrius, WIC Office, Cobb/Douglas Health District (Georgia); Wanda L. Dodson, Mississippi State University; Linda Goodwin, American Dietetic Association; Theresa A. Nicklas, North Dakota State University; Ling C. Patty, Arizona Department of Health Services, Office of Nutrition Services; and Linda Rider, Arizona Department of Education, for sharing information, ideas, and program materials. The text is richer for their contributions.

Special thanks go to our editorial team: Pete Marshall, Publisher; Sandra Craig, Project Editor; Laura Graham, Associate Developmental Editor, and Becky Tollerson, Marketing Manager. They guided us through unfamiliar territory in our new home at Wadsworth, offering both support and assistance. Laura was our constant companion, explaining how things were done, who did what, and what the next step was—all with grace and enthusiasm and despite a close encounter with poison ivy. We appreciate Susan Walters' help finalizing the permissions—in fact, we could not have done them without her. We thank Martha Emry for organizing our production activities, paying attention to details, keeping us focused, counting coincidences, and making us laugh. Martha is a first-rate negotiator and hardly the terror she claims to be. Three other members of our production team also have our thanks: Laura Larson, copyeditor; Jim Atherton, artist; and Marian Selig, proofreader. The fine quality of this product reflects their hard work and diligence. And we give both thanks and congratulations to our cover designer, Ellen Pettengell. We held our breath during the cover design phase, wondering whether the cover for the second edition would match the vibrancy of the first edition's. Ellen came through with style. We are thrilled with our new look.

Last, but not least, we owe much to our colleagues who provided articles and course outlines, their favorite Internet addresses, and expert reviews of the manuscript. Their ideas and suggestions are woven into every chapter. We appreciate their time, energy, and enthusiasm and hope they take as much pride in this book as we do. Thanks to all of you:

Shawna Berenbaum, University of Saskatchewan
Patricia B. Brevard, James Madison University
Marian Campbell, University of Manitoba
Nancy Cohen, University of Massachusetts
Nancy Cotugna, University of Delaware
Marie Dunford, California State University—Fresno
Barbara H. J. Gordon, San Jose State University
Sareen Gropper, Auburn University
Evette Hackman, Seattle Pacific University
Margaret Hedley, University of Guelph
Carolyn J. Hoffman, Central Michigan University
Tanya Horacek, Syracuse University
Jana Kicklighter, Georgia State University
Barbara A. Kirks, California State University—Chico
Bernice Kopel, Oklahoma State University
Kathleen McBurney, California Polytechnic State
    University—San Luis Obispo
Shortie McKinney, Drexel University
Nweze Nnakwe, Illinois State University
Jenice Rankins, Florida State University
Marsha Read, University of Nevada—Reno
Martha L. Rew, Texas Woman's University—Denton
Noreen B. Schvaneveldt, Utah State University
Janet Schwartz, Framingham State College
Padmini Shankar, Georgia Southern University
Chery Smith, University of Minnesota
Edward Weiss, D'Youville College

Community Nutrition in Action

Community Nutritionists in Action: Working in the Community

On Saturday morning, Irene H. opens her kitchen cabinet and takes down six small bottles. She lines them up on the countertop and works their caps off. The process takes a few minutes because her fingers are stiff from arthritis. Let's see, there's cod liver oil, chondroitin sulfate, and glucosamine for arthritis; ginkgo biloba and St. John's wort to relieve anxiety and depression; and DHEA to restore youthful vigor. Irene knows her doctor would be surprised—maybe shocked—to learn that she takes these supplements regularly. She knows, too, that her doctor would not approve of her consultations with a naturopath whose office is just a couple of miles from her home.

At 48, Irene figures she is doing all she can to manage the pain from arthritis and depression after her divorce. The supplements and naturopathic counseling are expensive, but she stretches the income from her job as a checkout clerk at a paint supply store to allow for them. After washing down the pills with orange juice, she pops two frozen waffles in the toaster and pours another cup of coffee. She figures she shouldn't eat the waffles—she was diagnosed with type 2 diabetes just three months ago—but she wants them. After breakfast, she'll enjoy a cigarette with her coffee and then call her oldest daughter. Maybe they can drive out to the mall.

Irene is a typical consumer in many respects. She has chronic health problems for which she has sought traditional medical advice and treatment. Like one in three U.S. adults, she has also sought help from an alternative practitioner. She smokes cigarettes, is overweight, and about the only exercise she gets is browsing the sale stalls out at the mall. She could do more to improve her health, but she isn't motivated to change her diet or quit smoking. She's looking for the quick fix.

Irene and the thousands of other consumers like her are a challenge for the community nutritionist. To help Irene make changes in her lifestyle—changes that will reduce her demands on the health care system and improve her physical well-being—the community nutritionist must be familiar with a broad spectrum of clinical and epidemiologic research, understand the health care system, and draw on the principles of public health and health promotion. The community nutritionist must know where Irene and people like her live and work, what they eat, and what their attitudes and values are. The community nutritionist must know about the community itself and how it delivers health services to people like Irene. And the community nutritionist must know how to influence policy makers. Perhaps now is the time to call for tighter regulation of dietary supplements and greater government support for health promotion and disease prevention programs.

This section describes the work that community nutritionists do in their communities. It outlines the principles of public health, health promotion, and policy making and reviews the current health care environment. It focuses on entrepreneurship—the discipline founded on creativity and innovation—and how its principles can be used to reach Irene and other people in the community with health and nutritional problems. The material in this section sets the stage for all that follows. It lays the groundwork for understanding what community nutritionists do: they focus on people, policies, and programs.

Chapter 1

Opportunities in Community Nutrition

Outline

Learning Objectives

After you have read and studied this chapter, you will be able to:

- Describe the three arenas of community nutrition practice.
- Describe how community nutrition practice fits into the larger realm of public health.
- Describe the three types of prevention efforts and the three levels of intervention.

- List three major health objectives for the nation and explain why each is important.
- Outline the educational requirements, practice settings, and roles and responsibilities of community nutritionists.
- Explain why entrepreneurship is important to the practice of community nutrition.

Something To Think About...

Education and health are the two great keys. We must use all public sector institutions, flawed though they may be, to close the gap between rich and poor. We must work with the political sector to convincingly paint the breadth and depth of the problem and the size of the opportunity as well. . . . Above all, we must not abandon the hope of progress. – *Sir Gustav Nossal, writing on health and the biotechnology revolution in* Public Health Reports, *March/April 1998*

Introduction

Community nutritionists face many challenges in the practice of their science and art. There is the challenge of improving the nutritional status of different kinds of people with different education and income levels and different health and nutritional needs: teenagers with anorexia nervosa, pregnant women living in public housing, the homeless, new immigrants from southeast Asia, elderly women alone at home, middle-class adults with high blood cholesterol, professional athletes, children with disabilities. There is the challenge of forming partnerships with colleagues, business leaders, and the public to advocate for change. There is the challenge of influencing lawmakers and other key citizens to enact laws, regulations, and policies that protect and improve the public's health. There is the challenge of studying the scientific literature for new angles on how to help people make good food choices for good health. And there is the challenge of mastering new technologies like the Internet to help meet the needs of clients and communities.

In addition to these challenges, certain social and economic trends also present challenges for community nutritionists. Immigrants from Mexico, Asia, and the Caribbean, many of whom have poor English skills, have streamed into North America in recent years, searching for jobs and improved living conditions.[1] The North American population is aging rapidly, as baby boomers mature and life expectancy increases.[2] Financial pressures and increased global competition have forced governments, businesses, and organizations to be creative in the face of scarce resources. Indeed, according to one survey of employers undertaken by the American Dietetic Association, the single greatest challenge for the dietetics professional today is "the need to do more and better with less."[3] Community nutritionists in all practice settings face rising costs, changing consumer expectations about health care services, increased competition in the market, and greater cultural diversity among their clients. They are pressured by downsizing, mergers, cross training, and managed health care.

Community nutritionists who succeed in this changing environment are flexible, innovative, and versatile. They are *focused* on recognizing opportunities for improving people's nutritional status and health and on helping society meet its obligation to alleviate hunger and malnutrition. It is an exciting time for community nutritionists. It is a time for learning new skills and moving into new areas of practice. It is a time of great opportunity and incredible need.

Opportunities in Community Nutrition

Community nutrition A discipline that strives to prevent disease and enhance health by improving the public's eating habits.

Founded on the sciences of epidemiology, food, nutrition, and human behavior, **community nutrition** is a discipline that strives to improve the health, nutrition, and well-being of individuals and groups within communities. Its practitioners develop policies and programs that help people improve their eating patterns and health. Indeed, these three arenas—people, policy, and programs—are the focus of community nutrition.

People

The individuals who benefit from community nutrition programs and services range from young single mothers on public assistance to senior business executives, from immigrants with poor English skills to college graduates, from pregnant teenagers with iron-deficiency anemia to grandfathers with Alzheimer's disease. They are found in worksites, schools, community centers, health clinics, churches, apartment buildings—virtually any community setting. Through community nutrition programs and services, these individuals and their families have access to food in times of need or learn skills that improve eating patterns. It is the community nutritionist who identifies a group of people with an unmet nutritional need, gathers information about the group's socioeconomic background, ethnicity, religion, geographical location, and cultural food patterns, and then develops a program or service tailored to the needs of this group.

Policy

Policy A course of action chosen by public authorities to address a given problem.

Policy is a key component of community nutrition practice. **Policy** is a course of action chosen by public authorities to address a given problem.[4] Policy is what governments and organizations *intend* to accomplish through their laws, regulations, and programs.

How does policy apply to the practice of community nutrition? Consider a situation in which a group of community nutritionists are compelled to address food waste in their community. The impetus for their action came from reading the results of a U.S. Department of Agriculture study that found that one-fourth of all food produced in the United States is wasted[5] and from reading about a successful food assistance program called "gleaning." Gleaning began as a project to deliver an abundance of apples from communities with apple orchards to food banks in neighboring states where apples were scarce.[6] The community nutritionists wanted to try gleaning on a small scale using farmers' markets in their community. Unfortunately, there was no city bylaw that allowed surplus foods from farmers' markets to be made available to local food banks and soup kitchens. After getting the owners of the farmers' markets, food banks, and soup kitchens on their side, the community nutritionists lobbied city council to enact a bylaw to allow such transactions. City council voted to pass a bylaw to support gleaning projects. In other words, city council altered its *policy* about recovering and recycling surplus foods.

Community nutritionists are involved in policy when they write letters to their state legislators, lobby Congress to secure Medicare coverage for Medical Nutrition

Therapy, advise their municipal governments about food banks and soup kitchens, and use the results of research to influence policy makers. Many aspects of the community nutritionist's job involve policy issues.

Programs

Programs are the instruments used by community nutritionists to seek behavior changes that improve nutritional status and health. They are wide-ranging and varied. They may target small groups of people—children with developmental disabilities in Nevada schools or teenagers living in a Brooklyn residential home—or they may target large groups such as all adults with high blood cholesterol concentrations. Programs may be as widespread as the U.S. federal Food Stamp Program or as local as a diabetes prevention program for Mohawk people living in the Akwesasne community in northern New York state. They may be tailored to address the specific health and nutritional needs of people with obesity or osteoporosis, or they may be aimed at the general population. Two examples of population-based programs are "Particip*action*," a Canadian program designed to get people moving and fit for health, and "5 a Day for Better Health," a program of the U.S. National Cancer Institute aimed at making Americans more aware of how eating fruits and vegetables can improve their health and may reduce their cancer risk. A new take on the 5 a Day program is "Gimme 5," a school-based nutrition program tested in 12 high schools in the Archdiocese of New Orleans.[7] Regardless of the setting or target audience, community nutrition programs have one desired outcome: behavior change.

Public Health and Community Interventions

Community nutritionists promote good nutrition as one avenue for achieving good health. They develop programs to help people improve their eating habits, and they seek environmental changes (in the form of policy) to support good health habits. But community nutritionists do not work in a vacuum. They work closely with other practitioners, particularly those in public health, to help consumers achieve and maintain behavior change.

Public health can be defined as an effort organized by society to protect, promote, and restore the people's health through the application of science, practical skills, and collective actions. "Public health is what we, as a society, do collectively to assure the conditions in which people can be healthy," wrote the authors of a report for the Institute of Medicine.[8]

In the nineteenth century, the scope of public health was generally restricted to matters of general sanitation, including building municipal sewer systems, purifying the water supply, and controlling food adulteration. Major public health efforts focused on controlling infectious diseases such as tuberculosis, smallpox, yellow fever, cholera, and typhoid. In 1900, the leading causes of death and disability in the United States were pneumonia, tuberculosis, and diarrhea/enteritis,[9] as shown in Figure 1-1. The morbidity and mortality linked with these disease outbreaks shaped public health practice for many years. Such runaway epidemics, which sometimes killed thousands of people in a single outbreak, are uncommon today because of

Public health focuses on protecting and promoting people's health through the actions of society.

large-scale public efforts to improve water quality, control the spread of communicable diseases, and enhance personal hygiene and the sanitation of the environment.

The leading causes of morbidity and mortality in the United States today are chronic diseases such as heart disease, cancer, and stroke (refer to Figure 1-1). Seven of every 10 Americans, or roughly one and a half million people, die from chronic disease each year. Cardiovascular disease (mainly heart disease and stroke) cause about 40 percent of all deaths, killing over 950,000 U.S. adults yearly. Although cardiovascular disease is usually assumed to be primarily a disease of men, over half of all deaths resulting from cardiovascular disease occur among women. Cancer kills more than 550,000 Americans each year—more than 1,500 people every day. Other

FIGURE 1-1 *Leading Causes of Death, United States, 1900 and 1995*

Source: Centers for Disease Control and Prevention, *Unrealized Prevention Opportunities: Reducing the Health and Economic Burden of Chronic Disease* (Atlanta, GA: Department of Health and Human Services, Public Health Service, 1997).

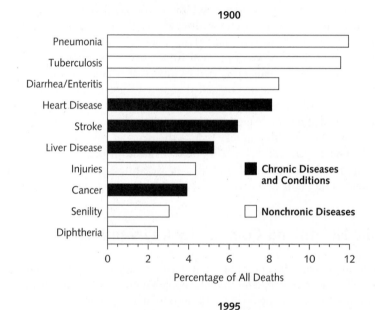

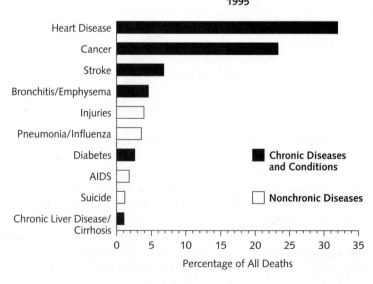

serious chronic diseases that reduce the quality of life, disable, or kill include arthritis, osteoporosis, and Alzheimer's disease.[10]

Infectious diseases remain a problem, however. The AIDS epidemic continues to escalate, particularly among women. About 1 million people in the United States are infected with HIV, the virus that causes AIDS,[11] and about 450,000 people received a diagnosis of AIDS between 1988 and 1995.[12] HIV/AIDS has become the third leading cause of death among women aged 25 to 44 years and the first among African-American women of the same age group.[13] It is the second leading cause of death for American men aged 35 to 64 years.[14] Another infectious disease is tuberculosis, whose incidence had been declining in the general U.S. population for several decades. Tuberculosis has reemerged as a new health threat among recent immigrants.[15] The AIDS epidemic is partly responsible for the recent outbreaks of tuberculosis, although there are other causes, including increases in homelessness, drug abuse, immigration from other countries where tuberculosis is widespread, and crowded housing among the poor.[16]

The leading causes of death in Canada mirror those of the U.S. population in many respects. The top-ranking cause of death among Canadian men and women in 1995 was cardiovascular disease, followed by cancer. Because more women smoke today than smoked 20 years ago, lung cancer is now the leading cause of cancer death among Canadian women.[17]

These changes in disease patterns over the last century have spawned changes in public health actions. Because the goals of public health reflect the values and beliefs of society and existing knowledge about disease and health, public health initiatives change as society's perception of health needs changes. At the close of the twentieth century, public health initiatives have shifted from financing basic population-based measures such as immunization to efforts focused on achieving universal health services, responding rapidly to threats from new infectious diseases such as the Hanta and Ebola viruses, and fostering better collaboration between public health agencies and other organizations involved in promoting the public's health.[18]

The Concept of Health

Most of us equate health with "feeling good," a concept we understand intuitively but cannot define exactly. The term *health* is a derivative of the old English word for "hale," which means whole, hearty, sound of mind and body.[19] Health can be viewed as the absence of disease and pain, or it can be pictured as a continuum along which the total living experience can be placed, with the presence of disease, impairment, or disability at one end of the spectrum and freedom from disease or injury at the other. These extremes in the health continuum are shown in Figure 1-2.[20]

Health is properly defined from an ecological viewpoint—that is, one that focuses on the interaction of humans among themselves and with their environment. In this sense, **health** is a state characterized by "anatomic integrity; ability to perform personally valued family, work, and community roles; ability to deal with physical, biological, and social stress; a feeling of well-being; and freedom from the risk of disease and untimely death."[21] A healthy individual, then, has the physical, mental, and spiritual capacity to live, work, and interact joyfully with other human beings.

Health The World Health Organization defines health as a state of complete physical, mental, and social well-being, not merely the absence of disease.

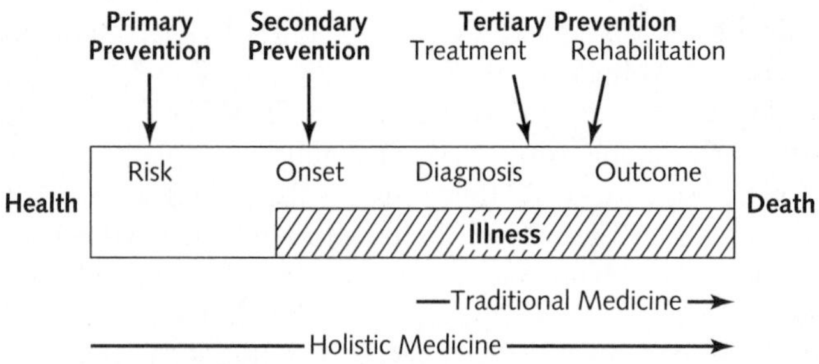

FIGURE 1-2 *The Health Continuum and Types of Prevention to Promote Health and Prevent Disease*

Source: Adapted from M. P. O'Donnell, Definition of health promotion, *American Journal of Health Promotion* 1 (1986): 4; and R. H. Fletcher, S. W. Fletcher, and E. H. Wagner, *Clinical Epidemiology: The Essentials* (Baltimore, MD: Williams & Wilkins, 1982), p. 129.

But how is good health achieved? Why does one child in a family become addicted to cocaine while another never touches illicit drugs? Why do people start smoking? Why do some people overeat? Why do some teenagers consume adequate amounts of iron and calcium whereas others do not? Why is an 80-year-old healthy and vigorous while a 70-year-old is infirm? The answers to these questions still elude epidemiologists and other scientists. We know that a constellation of factors, shown in Table 1-1, influence health. Certain individual factors such as age, sex, and race are fixed, inherited traits that influence an individual's health potential. Other factors, such as lifestyle, housing, working conditions, social networks, community services, and even national health policies, represent layers of influence that can theoretically be changed to improve the health of individuals. In truth, however, less is known about the specific determinants of health than is known about the factors that contribute to disease, injury, and disability. Understanding the causes of disease and ill health does not necessarily lead to an understanding of the causes of good health.

Health Promotion

Some people do things that are not good for their health. They overeat, smoke, refuse to wear a helmet when riding a bicycle, never wear seat belts when driving, fail to take their blood pressure medication—the list is endless. These behaviors reflect personal choices, habits, and customs that are influenced and modified by social forces. We call these lifestyle behaviors, and they can be changed if the individual is so motivated. Educating people about healthy and unhealthy behaviors is one way to help them adopt positive health behaviors.

Health promotion is helping all people achieve their maximum potential for good health.

Intervention is a health promotion activity aimed at changing the behavior of a target audience.

Risk factors Factors associated with an increased probability of acquiring a disease.

Health promotion focuses on changing human behavior, on getting people to eat healthy diets, be active, get regular rest, develop leisure-time hobbies for relaxation, strengthen social networks with family and friends, and achieve a balance among family, work, and play.[22] It is "the science and art of helping people change their lifestyle to move toward a state of optimal health."[23] Behavior change is the desired outcome of a health promotion activity—what we call an **intervention**—aimed at a target audience. Interventions focus on promoting health and preventing disease and are designed to change a preexisting condition related to the target audience's behavior.[24]

There are three types of prevention efforts, shown in Figure 1-2. Primary prevention is aimed at preventing disease by controlling **risk factors** that are related to injury and disease. Low-fat cooking classes, for example, help people change their eating and cooking patterns to reduce their risk of cardiovascular disease. Secondary

Biology	Lifestyle	Living, Working, and Social Conditions	Community Conditions	Background Conditions
Sex	Physical activity	Housing	Climate and	National food and
Race	Diet	Education	geography	nutrition policy
Age	Hobbies	Occupation	Water supply	National minimum
Other hereditary	Leisure time	Income	Type and condition	wage
factors	activities	Social networks	of housing	Cultural beliefs
	Use of drugs:	such as family,	Number and type	Cultural values
	• Cigarettes,	friends, coworkers	of hospitals and	Advertising
	cigars,	Socioeconomic status/	clinics	Media messages
	chewing	class	Health and medical	Food distribution
	tobacco		services	system
	• Alcohol		Social services	
	• Prescription		Leading industries	
	medications		Political/government	
	• Illicit drugs		structure	
	such as		Community health	
	cocaine,		groups and	
	marijuana,		organizations	
	etc.		Number, type, and	
	Religion		location of	
	Safety practices		grocery stores,	
	such as		etc.	
	wearing seatbelts,		Recreation	
	wearing wrist		Transportation	
	guards and		systems	
	knee pads			
	Medical self-care			
	Stress management			

Source: Adapted from G. Pickett and J. J. Hanlon, *Public Health: Administration and Practice* (St. Louis: Times Mirror/Mosby College Publishing, 1990), p. 50; and M. P. O'Donnell, Definition of health promotion, *American Journal of Health Promotion* 1 (1986): 4–5.

prevention focuses on detecting disease early through screening and other forms of risk appraisal. Public screenings for hypertension at a health fair identify people whose blood pressure is high; these individuals are then referred to a physician or other health professional for follow-up and treatment. Tertiary prevention aims to treat and rehabilitate people who have experienced an illness or injury. Education programs for people recently diagnosed with diabetes help prevent further disability and health problems such as blindness and end-stage renal disease arising from the condition and improve overall health.[25] Prevention has become increasingly important, as the medical community moves away from traditional medicine, which focuses on diagnosing and treating diseases, to a holistic approach that encompasses all aspects of the health spectrum.

TABLE 1-1

Determinants of Health

There are three levels of intervention: (1) building awareness of a health problem, (2) changing lifestyles, and (3) creating a supportive environment for behavior change.[26] At any one level, interventions may target individuals in small groups such as families, schools, worksites, and health clinics; people in social networks such as worksites, churches, and bridge clubs; entire organizations; or the community at large, which can be a city, province, state, or nation.[27] An intervention may

A cooking demonstration is an intervention that promotes awareness of the importance of healthful eating and teaches low-fat cooking skills.

be as simple as providing a brochure describing the benefits of breastfeeding to clinic clients or distributing a fact sheet listing the fat content of snack foods to grocery store shoppers. An intervention may be fairly complex, such as a mass media campaign that targets health writers across the nation or adolescents with low calcium intakes. Some interventions are full-service programs, complete with training manual, lesson plans, reproducible handouts, and videos.

Although many questions remain unanswered about why people make the choices they make, the ways to promote good health are widely recognized. Born of decades, if not centuries, of scientific observation and testing, the strategies for promoting good health are outlined in Table 1-2. Although these strategies seem relatively straightforward, putting them into practice is a major challenge for most communities and nations.

Health Objectives

The challenge of improving the nutrition, health, and quality of life for humans is complex. As outlined by the nations of the world in the 1978 conference on primary health care, convened by the World Health Organization (WHO) and the United Nations Children's Fund (UNICEF), the goal of the world community is to "protect and promote the health of all people of the world."[28] When translating this global goal into action at the local level, one challenge is to understand the many physical, biological, social, and behavioral factors that influence the health of individuals and communities. Another challenge is to change human behavior.

• Safe environment	Control physical, chemical, and biological hazards.	**TABLE 1-2** *Ways to Promote Good Health*
• Enhance immunity	Immunize to protect individuals and communities.	
• Sensible behavior	Encourage healthy habits, and discourage harmful habits.	
• Good nutrition	Eat a well-balanced diet, containing neither too much nor too little.	
• Well-born children	Every child should be a wanted child, and every mother fit and healthy.	
• Prudent health care	Cautious skepticism is better than uncritical enthusiasm.	

Source: Adapted, with permission, from J. M. Last, *Public Health and Human Ecology*, 2nd ed. (Stamford, CT: Appleton & Lange, 1998), p. 10.

Nations differ in how they formulate health objectives for their peoples to achieve behavior change, although there are common themes. Working groups in the European Region of the World Health Organization, for example, outlined the following prerequisites for health:[29]

- Freedom from the fear of war—"the most serious of all threats to health"
- Equal opportunity for all peoples
- The satisfaction of basic needs for food, clean water and sanitation, decent housing, and education
- The right to find meaningful work and perform a useful role in society

Achieving these necessities requires both political will and public support, according to the working groups, which translated these prerequisites into specific targets for health. One such target, for example, called for enhancing life expectancy by reducing infant and maternal mortality. Other targets focused on enhancing social networks and promoting healthy behaviors, controlling water and air pollution, and improving the primary health care system.

In Canada, a new vision for promoting health and preventing disease among Canadians was expressed in a document released by Health Canada, which aims to promote a balance between individual and societal responsibilities for health. Titled *Achieving Health for All: A Framework for Health Promotion*, this document cites three challenges to achieving health for all: reducing inequities in access to and use of the health care system; increasing prevention efforts to change unhealthy behaviors; and enhancing the individual's ability to cope with chronic illnesses, disabilities, and mental health problems. A key focus of the proposed implementation strategies is the strengthening of community-based health services, including worksite programs.[30] This new vision for health in Canada is a "window of opportunity" for community nutritionists to promote food and nutrition policies in all Canadian communities.[31]

The health and well-being of individuals and the prosperity of the nation require a well nourished population.
–Joint Steering Committee (Canada), Nutrition for Health: An Agenda for Action

The health objectives for the peoples of the United States differ slightly from those of the European and Canadian communities, reflecting the health needs of the U.S. population. A national strategy for improving the health of the nation was laid out in the publication *Healthy People 2000: National Health Promotion and Disease Prevention Objectives*.[32] This document represents a national health agenda developed by a consortium of national health organizations, state health departments, the Institute of Medicine, and the U.S. Public Health Service. It grew out of a national health strategy that began in 1979 with the publication of *Healthy People: The Surgeon General's Report on Health Promotion and Disease Prevention*[33] and continued in 1980 with the release of the report *Promoting Health/Preventing Disease: Objectives for the Nation*, which established health objectives leading up to the year 1990.[34] The expert working groups of the *Healthy People 2000* consortium developed three broad goals designed to help Americans achieve their full potential:

- Increase the span of healthy life for Americans.
- Reduce health disparities among Americans.
- Achieve access to preventive services for all Americans.

These goals represent the nation's hope for the improved health of its citizens, and they can serve as the foundation for all work toward health promotion and disease prevention. As stated, however, they are too broad to implement. Thus, the working groups also laid out specific, measurable targets or objectives to be achieved by the year 2000. These objectives are grouped into 22 broad categories, shown in Table 1-3. Health promotion priorities include increasing daily physical activity, reducing cigarette smoking and alcohol consumption, and reducing child and spouse abuse, among others. Nutrition-related activities are considered essential, because five of the 10 leading causes of death in the United States are related to dietary imbalance and excess (coronary heart disease, some types of cancer, stroke, diabetes mellitus, and atherosclerosis), and another three are linked with the excessive consumption of alcohol (cirrhosis of the liver, accidents, and suicides). Diet also contributes to the development of other conditions, such as hypertension, osteoporosis, obesity, dental caries, and diseases of the gastrointestinal tract.[35] Some of the *Healthy People 2000* nutrition objectives focus on improving health status. For example, one health status objective calls for reducing coronary heart disease deaths to no more than 100 per 100,000 people. Several objectives focus on risk reduction and specify targets for the intake of nutrients such as fat, saturated fat, and calcium and foods such as fruits, vegetables, and grain products. Other risk reduction objectives set targets for the prevalence of iron deficiency and anemia and the proportion of people who adopt sound dietary practices and reduce their use of salt and sodium in foods and at the table.[36]

Health protection strategies encompass broad environmental measures, such as improving the workplace to reduce work-related injuries and lowering people's exposure to lead, air pollutants, and radon. Preventive services focus on screening, immunization, counseling, and other interventions for individuals in clinical settings. The primary objective for the **surveillance** and data systems is the systematic collection, analysis, interpretation, dissemination, and use of health data to understand a population's health status and plan prevention programs.[37] Some of the health objectives outlined in the *Healthy People 2000* report address the special needs of various age groups, such as children and adolescents, while others focus on special population groups, such as blacks, Hispanics, and Asians.

How are Americans doing in terms of meeting the *Healthy People 2000* goals? Progress has been made in meeting goal 1—increase the span of healthy life.[38] The good news is that people are living longer. When *Healthy People 2000* was released in 1990, life expectancy was 75 years. Babies born in 1992 will live slightly longer, or nearly 76 years. Unfortunately, even though life expectancy has increased, health-

Surveillance An approach to collecting data on a population's health and nutritional status in which data collection occurs regularly and repeatedly.

TABLE 1-3 *Priority Areas of the* Healthy People 2000 *Initiative*

Health promotion

1. Physical activity and fitness
2. Nutrition
3. Tobacco
4. Substance abuse: Alcohol and other drugs
5. Family planning
6. Mental health and mental disorders
7. Violent and abusive behavior
8. Educational and community-based programs

Health protection

9. Unintentional injuries
10. Occupational safety and health
11. Environmental health
12. Food and drug safety
13. Oral health

Preventive services

14. Maternal and infant health
15. Heart disease and stroke
16. Cancer
17. Diabetes and chronic disabling conditions
18. HIV infection
19. Sexually transmitted diseases
20. Immunization and infectious diseases
21. Clinical preventive services

Surveillance and data systems

22. Surveillance and data systems

Age-related objectives

Children
Adolescents and young adults
Adults
Older adults

Source: U.S. Department of Health and Human Services, Public Health Service, *Healthy People 2000: National Health Promotion and Disease Prevention Objectives* (Washington, D.C.: U.S. Government Printing Office, 1990), p. 7.

related quality of life has diminished. Moreover, progress in meeting goals 2 and 3 has not been ideal. Health disparities are evident among Americans, with significant differences between whites and minorities in mortality, morbidity, and the use of health services. And the percentage of Americans with health insurance coverage has fallen. In 1993, about 20 percent of blacks and 32 percent of Hispanics were uninsured.

Progress toward the health status objectives for major populations in the nutrition priority area is summarized in Table 1-4. (Health status objectives for ethnic populations are not shown in the table. Risk reduction and services and protection objectives for major populations are summarized in Appendix A.) Since the 1980s, the number of deaths from cardiovascular disease, stroke, and certain types of cancer has decreased, but the prevalence of overweight has soared. In fact, overweight increased among all ethnic and age subgroups of the population. The widespread problem of

Number	Health Status Objective	Target Population	Baseline	Most Recent Figure*	2000 Target
2.1	Reduce cardiovascular disease deaths	Adults aged 20–74 years	135 per 100,000	108 per 100,000	100 per 100,000
2.2	Reverse the rise in cancer deaths	Adults aged 20–74 years	134 per 100,000	130 per 100,000	130 per 100,000
2.3	Reduce overweight prevalence	Adults aged 20+ years	26%	35%	20%
		Men	24%	34%	20%
		Women	27%	37%	20%
		Adolescents aged 12–19 years	15%	24%	15%
2.4	Reduce growth retardation	Low-income children aged 5 years and younger	11%	8%	10%
2.22	Reduce stroke deaths	Adults	30.4 per 100,000	26.7 per 100,000	20.0 per 100,000
2.23	Reduce colorectal cancer deaths	Adults	14.7 per 100,000	12.8 per 100,000	13.2 per 100,000
2.24	Reduce the incidence and prevalence of diabetes	Adults			
		Incidence (per 1,000)	2.9	NA†	2.5
		Prevalence (per 1,000)	28	31	25

*Most recent figure was obtained from survey data published in 1994 or 1995.
†NA = Data not available.
Source: National Center for Health Statistics (NCHS), *Healthy People 2000 Review, 1997* (Hyattsville, MD: Public Health Service, 1997). Data obtained from the NCHS Web site at www.cdc.gov/nchswww.

TABLE 1-4 *Status of the* Healthy People 2000 N*utrition Objectives*

Development of the *Healthy People 2010* objectives are under way. Refer to the boxed insert for Internet addresses related to *Healthy People* initiatives and other sites of interest to community nutritionists.

overweight has occurred despite a slight decrease in dietary fat intake and a modest increase in consumption of fruits, vegetables, and grains. One contributing factor is that people seem to be taking fewer steps to control their weight by adopting sound dietary patterns and being physically active. Additional health promotion efforts are needed to reduce the prevalence of iron deficiency and anemia and increase the number of women who breastfeed their infants early in the postpartum period.

The ultimate objective of public health is to prevent increased risk of disease and disability and risky behaviors in the first place.[39] Thus, many of the educational programs and services developed by public health practitioners to meet the objectives of *Healthy People 2000* focus on people in groups, whether they are families, workplaces, cities, or nations. Such strategies target people of all ages and segments within the community.

The Concept of Community

"There is no complete agreement as to the nature of community," wrote G. A. Hillery, Jr.[40] Such diverse locales as isolated rural hamlets, mountain villages, prairie towns, state capitals, industrial cities, suburbs or ring cities, resort towns, and major metropolitan areas can all be lumped into a single category called "community."[41] The concept

Internet Resources

Check out these Internet addresses for information related to professional dietetics organizations around the world and issues in community nutrition and public health.

Professional Organizations

American Dietetic Association   **www.eatright.org**

American Public Health Association   **www.apha.org**

Canadian Public Health Association   **www.cpha.ca**

Dietitians of Canada   **www.dietitians.ca**

Health Organizations

Food and Agriculture Organization   **www.fao.org**

Pan American Health Organization   **www.paho.org**

World Health Organization   **www.who.org**

Canadian Federal Government Agencies

Canadian Food Inspection Agency   **www.cfia-acia.agr.ca**

Health Canada   **www.hc-sc.gc.ca**

U.S. Federal Government Agencies and Offices

Center for Food Safety and Applied Nutrition   **vm.cfsan.fda.gov/list.html**

Centers for Disease Control and Prevention   **www.cdc.gov**

Department of Health and Human Services   **www.hhs.gov**

Food and Drug Administration   **www.fda.gov**

Food and Nutrition Information Center   **www.nal.usda.gov/fnic**

Food Safety and Inspection Service   **www.fsis.usda.gov**

National Cancer Institute   **www.nci.nih.gov**

National Center for Chronic Disease Prevention and Health Promotion (CDC*)   **www.cdc.gov/nccdphp**

National Center for HIV, STD, and TB Prevention, Division of HIV/AIDS Prevention (CDC)   **www.cdc.gov/nchstp/hiv_aids/dhap.htm**

National Institutes of Health   **www.nih.gov**

Consumer Health Sites

CancerNet   **cancernet.nci.nih.gov**

Continued

Internet Resources—continued

Consumer Health Sites—continued

CHID[†] On-line   **chid.nih.gov**

Healthfinder   **www.healthfinder.gov**

InteliHealth   **www.intelihealth.com**

National Health Information Center   **nhic-nt.health.org**

NIH[‡] Consumer Health Information   **www.nih.gov/health/consumer/conicd.htm**

Health Promotion Initiatives and Other Sites

Hispanic Customer Service Home Page   **www.dhhs.gov/about/heo/hispanic.html**

NCI's[§] 5 a Day for Better Health Program   **dccps.nci.nih.gov/5aday**

5 a Day Week Community Intervention Kit   **www.dcpc.nci.nih.gov/5aday/week98/CommunityKit98.html**

Healthy People 2000   **www.odphp.osophs.dhhs.gov/pubs/hp2000**

Healthy People 2010   **web.health.gov/healthypeople**

---

*CDC = Centers for Disease Control and Prevention.
[†]CHID = Combined Health Information Database.
[‡]NIH = National Institutes of Health.
[§]NCI = National Cancer Institute.

Community is a group of people who are located in a particular space (including cyberspace), have shared values, and interact within a social system.

of community is not always circumscribed by a city limits sign or zoning laws. Sometimes it describes people who share certain interests, beliefs, or values, even though they live in diverse geographical locations. Thus, we refer to the academic community, the gay community, and the immigrant community. For our purposes, a **community** is a grouping of people who reside in a specific locality and who interact and connect through a definite social structure to fulfill a wide range of daily needs. By this definition, a community has four components: people, a location in space (which can include the realm of cyberspace), social interaction, and shared values.

Communities can be viewed on different scales: global, national, regional, and local. Each of these can be further segmented into specialized communities or groups, such as those individuals who speak Spanish, own computers, or observe Hanukkah. In the health arena, communities tend to be segmented around particular wellness, disease, or risk factors—for example, adults who exercise regularly, children infected with HIV, black men with high blood pressure, or people with peanut allergy.

Community Nutrition Practice

Earlier in the chapter, we defined community nutrition as a discipline that strives to improve the nutrition and health of individuals and groups within communities. How do community nutritionists do this? What skills are needed to accomplish this goal? What job responsibilities do community nutritionists have? This book answers these questions and introduces you to the challenges of working in communities today. Imagine for a moment that *you* are a community nutritionist in one of the following situations:

- An article in the New York Times describes the high rates of substance abuse, teen pregnancy, HIV infection, sexually transmitted diseases, smoking, and eating disorders among U.S. adolescents. Long concerned about this issue, your public health department plans an assessment of the health and nutritional status of teenagers in your county. Your job is to coordinate and lead the community assessment. Where do you start? What is the purpose of your assessment? What types of data do you collect? What information already exists about this population? Should your department work with other agencies to collect data? How will the results of your assessment be used to improve the health of teenagers in your community?

- As the director of health promotion for a large nonprofit health organization, you are responsible for developing and implementing programs to reduce the risk of cardiovascular diseases among people living in your state. Your organization's board of directors has called for an assessment of the effectiveness of all programs in your area. How do you evaluate program effectiveness? What types of data should be collected to show that each program reaches an appropriate number of people at a reasonable cost and helps them make behavioral changes to reduce their risk of heart attack and stroke? How will you present your findings to the board?

- You are attracted to the challenge of building a business and believe your training in nutrition and exercise physiology can help people in your community get fit and improve their lifestyles. What is an attractive name for your business? Where should it be located? What services will you offer and to whom? Who are your competitors? How will you market your services? Can you use the Internet to enhance your business?

- You are employed by the Special Supplemental Nutrition Program for Women, Infants, and Children (WIC) in your state and notice that Spanish is the first language for an increasing number of your clients. You and your colleagues want to offer these clients more materials and services in Spanish. Should you adapt existing English-language materials for these clients or develop new materials from scratch? Are the existing English-language materials culturally appropriate for your Hispanic clients? What are other state WIC programs doing to address this issue?

Common themes are apparent in these situations. All refer to gathering information about the community itself or about people who use or implement community-based programs and services. Although it may not be clear to you now, all involve issues of policy, program management, and cost. All entail making decisions about how to use scarce resources. All are concerned with determining whether nutrition programs and services are reaching the right audience with the right messages and having the desired effect. All describe challenges of a trained professional—the community nutritionist—who identifies a nutritional need in the community and then puts into place a program or service designed to alleviate the need.

Community Versus Public Health Nutrition

Community nutrition and public health nutrition are sometimes considered to be synonymous. In this book, *community nutrition* is the broader of the two terms and encompasses any nutrition program whose target is the community, whether the program is funded by the federal government, as with the WIC program or the National School Lunch program, or sponsored by a private group, such as a work-site weight-management program. *Public health nutrition* refers to those community-based programs conducted by a government agency (federal, state, provincial, territorial, county, or municipal) whose official mandate is the delivery of health services to individuals living in a particular area.

The confusion over these terms stems partly from the traditional practice settings of community dietitians and public health nutritionists, as shown in Figure 1-3. Community dietitians, who are always registered dietitians (RDs) or licensed dietitians (LDs), tend to be situated in hospitals, clinics, health maintenance organizations, voluntary health organizations, worksites, and other nongovernment settings. Some community dietitians work in federal, state, and municipal health agencies. Public health nutritionists, some of whom are RDs or LDs, provide nutrition services through government agencies.

In today's practice environment, these two designations have considerable overlap, and practitioners in both areas share many goals, responsibilities, target groups, and practice settings. The community dietitian plans, coordinates, directs, manages, and evaluates the nutrition component of his organization's programs and services. The public health nutritionist carries out similar activities for her government agency.[42] For our purposes, all nutritionists whose major orientation is community-based programming will be called community nutritionists, whether their official title is community dietitian, public health nutritionist, nutrition education specialist, or some other designation.

Educational Requirements

Community nutritionists have a solid background in the nutrition sciences. They have competencies in such areas as nutritional biochemistry, nutrition requirements, nutritional assessment, nutrition in health and disease, nutrition throughout the life cycle, food composition, and food habits and customs. They are knowledgeable about the theories and principles of health education, epidemiology, community organization, management, and marketing. Marketing skills are especially important, as it is no longer sufficient merely to know *which* nutrition messages to deliver, it is also necessary to know *how* to deliver them effectively in a variety of media formats to a variety of audiences.

The minimum educational requirements for a community nutritionist include a bachelor's degree in community nutrition, foods and nutrition, or dietetics from an accredited college or university. Most community nutrition positions require registration as a dietitian by the American Dietetic Association (ADA). Some positions also require graduate-level training to obtain additional competencies in areas such as quality assurance, biostatistics, research methodology, survey design and analysis, and the behavioral sciences.

Although dietetic technicians registered (DTRs) are most often employed in the food service sector and clinical settings, some do work in the community arena.[43] Community-based DTRs assist the community nutritionist in determining the community's nutritional needs and in delivering community nutrition programs and ser-

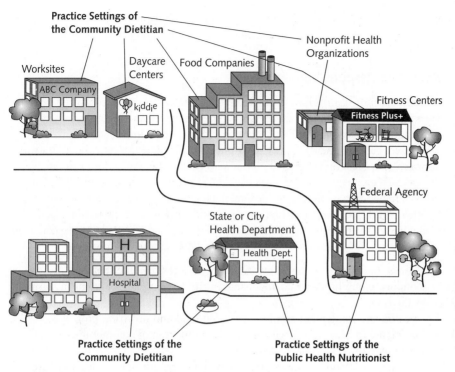

Practice Settings of the Community Dietitian

Worksites — ABC Company

Daycare Centers — kiddie

Food Companies

Nonprofit Health Organizations

Fitness Centers — Fitness Plus+

Federal Agency

Hospital

State or City Health Department — Health Dept.

Practice Settings of the Community Dietitian

Practice Settings of the Public Health Nutritionist

FIGURE 1-3 *Practice Settings of Community Dietitians and Public Health Nutritionists*

The community dietitian is a registered dietitian (RD) or licensed dietitian (LD) who works in day care centers, nonprofit health organizations, hospitals, worksites, food companies, fitness centers, sports clinics, and numerous other community settings. Some community dietitians work in federal, state, and municipal health agencies. The public health nutritionist, who may be an RD or LD, works in a federal, state, or municipal health agency. Both work to improve the health of people in their communities and are called "community nutritionists" in this book.

vices. DTRs must have at least an associate's degree and pass the ADA registration examination.

Practice Settings

The practice settings of community nutritionists include worksites, universities, colleges, medical schools, voluntary and nonprofit health organizations, public health departments, home health care agencies, day care centers, residential facilities, fitness centers, sports clinics, hospital outpatient facilities, food companies, and homes (their own or those of their clients). Some community nutritionists work as consultants, providing nutrition expertise to government agencies, food companies, food service companies, or other groups who are planning community-based services or programs with a nutrition component.

Community nutritionists are also employed by world and regional health organizations. WHO's Division of Family Health, located in the headquarters office in Geneva, Switzerland, includes an office of nutrition. Likewise, the North American regional WHO office in Washington, D.C., which is known officially as the Pan American Sanitary Bureau and coexists with the Pan American Health Organization (PAHO), has a strong nutrition mandate. The PAHO directs its efforts toward solving nutritional problems in Latin America and the Caribbean.[44] Another prominent organization in global community nutrition is the Food and Agriculture Organization (FAO) of the United Nations. The programs of the Food Policy and Nutrition

Division of FAO are directed toward improving the nutritional status of at-risk populations and ensuring access to adequate supplies of safe, good-quality foods.[45]

Roles and Responsibilities

Community nutritionists have many roles: educator, counselor, advocate, coordinator, ideas generator, facilitator, and supervisor. They interpret and incorporate new scientific information into their practice and provide nutrition information to individuals, specialized groups, and the general population. Their focus is normal nutrition, although they sometimes cover the principles of diet therapy and nutritional care in disease for certain groups (for example, HIV-positive children or people with diabetes).[46] In addition to serving the general public, community nutritionists refer clients to other health professionals when necessary and participate in professional activities. Community nutritionists who are registered dietitians are expected to have core competencies and community-based competencies such as those outlined here:[47]

- Manage nutrition care for population groups across the life span.
- Conduct community-based food and nutrition program outcome assessment/ evaluation.
- Develop and implement community-based food and nutrition programs.
- Participate in nutrition surveillance and monitoring of communities.
- Participate in community-based research.
- Participate in food and nutrition policy development and evaluation based on community needs and resources.
- Consult with organizations regarding food access for target populations.
- Develop and implement a health promotion/disease prevention intervention project.
- Participate in screening activities such as measuring hematocrit and cholesterol levels.

Community nutritionists are responsible for planning, evaluating, managing, and marketing nutrition services, programs, and interventions. Nutrition services range from individual counseling for weight management, blood cholesterol reduction, and eating disorders to consulting services provided for food companies and institutions such as residential centers and nursing homes. Nutrition programs may be national in focus, such as the WIC program, or local, such as "Healthy Start for Mom & Me," a Winnipeg prenatal nutrition program for low-income, high-risk pregnant women.[48] Community nutritionists who develop programs identify nutrition problems within the community, obtain screening data on target groups, locate information on community resources, develop education materials, disseminate nutrition information through the media, evaluate the effectiveness of programs and services, negotiate contracts for nutrition programs, train staff and community workers, and document program services.[49]

The job responsibilities of community nutritionists are similar across practice settings. The community nutritionists whose responsibilities are shown in Table 1-5 are all involved in assessing the nutritional status of individuals or identifying a nutritional problem within a community. They all have opportunities for teaching their clients about foods, diet, nutrition, and health and for addressing emerging issues in community nutrition (see the box "Emerging Issues in Community Nutrition").

Position: Child Nutrition Specialist,
State Department of Education

Responsibilities

1. Interprets USDA's regulations, policies, and procedures related to the National School Lunch Program for users of the program
2. Trains program users in such areas as how to count meals and keep records accurately, how to determine that menus meet current nutrition requirements, and which foods can be ordered through the commodities program
3. Revises training manuals
4. Audits program user compliance with USDA's regulations and procedures, including assessing whether student eligibility for participation in the program was determined correctly, ensuring that proper accounting procedures have been followed, and creating an action plan when the program user has not complied with the USDA regulations and procedures

Position: Director of Health Promotion, First-Rate Spa and Health Resort

Responsibilities

1. Develops, implements, and evaluates programs in the areas of nutrition, fitness, weight management, and risk reduction (e.g., blood cholesterol reduction, stress management, smoking cessation)
2. Assists the director in developing marketing strategies for programs
3. Prepares program budgets
4. Tracks program expenses
5. Teaches nutrition/fitness to groups of clients
6. Supervises dietitians and other staff involved in counseling clients and fitness assessments

Position: Nutritionist, Special Supplemental Nutrition Program for Women, Infants, and Children (WIC)

Responsibilities

1. Determines client's eligibility for program, using WIC program criteria (e.g., presence of anemia, underweight/overweight, prior pregnancies, inadequate diet)
2. Assesses the nutritional status of clients
3. Determines the adequacy of the client's diet
4. Provides one-on-one diet counseling
5. Conducts group sessions for clients on basic nutrition topics such as food sources of iron and calcium
6. Helps clients understand how to use the WIC-approved foods in their daily diets
7. Assists in developing educational materials as required
8. Assists in reviewing client records and monitoring health data posted to records as required

Position: President/Owner,
Millennium Nutrition Services

Responsibilities

1. Sets business goals and objectives
2. Manages all aspects of company's programs and services, including developing programs and services, developing educational materials and teaching tools, and evaluating success of programs and services
3. Develops and evaluates a marketing plan
4. Identifies new business opportunities
5. Tracks income (including billings) and expenses for tax purposes
6. Maintains client records
7. Networks with colleagues in business and the community

TABLE 1-5 *Responsibilities of Community Nutritionists in Diverse Practice Settings*

Some are involved in the budget process and in developing marketing strategies, while others are not. The community nutritionist in private practice does it all!

In today's market, community nutritionists are increasingly expected to manage projects, resources, and people. One survey of 350 community dietitians found that

◎ Emerging Issues in Community Nutrition

• The WIC program promotes breastfeeding, even though it also issues vouchers for clients to purchase infant formula. Should WIC and other breastfeeding advocacy groups continue to promote breastfeeding by mothers with HIV/AIDS, especially as some recent studies suggest that HIV-positive mothers who breastfeed are more likely to transmit the virus to their infants?

• Since the welfare law was passed in 1996, many advocacy groups have expressed concern that legal immigrants and their families will lose their food stamp and Social Security benefits. Should your state government restore benefits to legal immigrants residing in the state?

• Health claims for dietary supplements such as chromium picolinate, pyruvate, selenium, and vitamin E are currently regulated under the Nutrition Labeling and Education Act of 1990. Some consumer groups believe that the Food and Drug Administration (FDA) should have more power in regulating health claims made on product labels for dietary supplements. In other words, they think the supplement industry should be required to prove that a product is safe—and not that the FDA be required to prove that a product is unsafe. If you were responding to a proposed rule in the *Federal Register,* would you favor giving the FDA more power in this area?

• Several recent studies have suggested that elevated plasma total homocysteine levels (above 12–15 micromoles per liter) are a strong predictor of mortality in patients with confirmed coronary heart disease (CHD). Homocysteine is a metabolite of methionine, an essential amino acid found mainly in meat and dairy products. High blood levels of homocysteine are associated with low availability of certain B vitamins such as folic acid, vitamin B6, and vitamin B12, although it has not been proven that B vitamin supplements reduce CHD risk. Given the current state of research, should the nutrition education materials used for clients in your organization's heart disease risk reduction program be changed to reflect plasma homocysteine levels as a CHD risk factor?

---

more than two-thirds had mid- to upper-level management responsibilities. Community dietitians in management positions reported spending more time planning, coordinating, and evaluating programs and less time interacting with clients than did those in lower-level positions.[50] A role delineation study conducted by the ADA in 1989 found that RDs working in community settings also had advising and policy-setting roles.[51] The time allocated to these activities varied somewhat by practice level. Community dietitians reported having major responsibility for teaching students, other dietitians, and health professionals. Their roles overlapped significantly with those of RDs in clinical dietetics.

Community nutritionists are also expected to be multiskilled or cross trained. Multiskilled practitioners perform more than one function, often in more than one discipline.[52] The multiskilled community nutritionist knows not only how to conduct a needs assessment and provide dietary guidance but also how to design and conduct a survey, use an Internet Web site for marketing health messages, and obtain funding to support a program's promotional plan. Survey design and analysis, marketing, Internet technology, and grant writing are new and important disciplines for modern community nutritionists. And in today's culturally diverse environment, bilingual community nutritionists are in demand. Being fluent in a language other than English aids in gathering information from at-risk or hard-to-reach populations and developing programs that meet their needs.

Entrepreneurship in Community Nutrition

Entrepreneurship is important in community nutrition. What is entrepreneurship? Who is an entrepreneur? How is entrepreneurship related to community nutrition? In the business world, entrepreneurship is defined as the act of starting a business or the process of creating new "values," be they goods, services, methods of production, technologies, or markets.[53] The essence of entrepreneurship is innovation. Consider the late Ray Kroc of McDonald's. He did not invent the hamburger, but he did develop an entirely new way of marketing and delivering it to his customers. In the process, he revolutionized the food service industry. **Entrepreneurship,** then, is the creation of something of value, be it a product or service, through the creation of organization. In this context, organization refers to the orchestration of the materials, people, and capital required to deliver a product or service. This definition encompasses the myriad actions of individuals—the entrepreneurs—who invent or develop some new product or service that is valued by the community or marketplace.[54]

An **entrepreneur** is an enterpriser, innovator, initiator, promoter, and coordinator. Entrepreneurs are change agents who seek, recognize, and act on opportunities. They ask "What if?" and "Why not?" and translate their ideas into action. They tend to be creative, are able to see an old problem in a new light, and are willing to break new ground in delivering a product or service. When they spot an opportunity to fill a niche in the marketplace, they work to bring together the expertise, materials, labor, and capital necessary to meet the perceived need or want. Two entrepreneurs in community nutrition are Oklahoma dietitians, Kellie Bryant, MS, RD/LD, and Mary S. Callison, MS, RD/LD, who developed a nutrition newsletter for Head Start programs. They observed that Head Start programs, particularly on Indian reservations and in rural areas, lacked practical information pieces on child health and on how to shop for, cook, store, and serve healthy foods. Their "Primarily Nutrition" newsletter is marketed to Head Start programs, which copy it for distribution to clients and their families.[55] The three registered dietitians—Gretchen Forsell, MPH, RD, LD, Jean Nalani Trobaugh, RD, and Kim Ziemer, MS, RD, CNSD—who launched *Dietetics Online*® on the Internet's World Wide Web are also entrepreneurs.[56] They spotted a new technology trend and learned how to harness it for educating and informing dietitians and helping them network and become more efficient on-line.

Entrepreneurs share some common personality traits. They are achievers, setting high goals for themselves. They work hard, are good organizers, enjoy nurturing a project to completion, and accept responsibility for their ventures. They strive for excellence and are optimistic, believing that now is the best of times and anything is possible. Finally, entrepreneurs are reward oriented, seeking recognition and respect for their ventures and ideas. Recognition and respect are often more important than money to entrepreneurs.[57]

These qualities are typically applied to the self-starting, independent entrepreneur in business, but they also describe the **intrapreneur,** the corporate employee

Entrepreneurship Creating something of value through the creation of organization.

Entrepreneur One who undertakes the risk of a business or enterprise.

Intrapreneur A risk taker whose job is located within a corporation, company, or other organization.

who is creative and innovative. Intrapreneurs are seldom solely responsible for the financial risk associated with a new venture, but they share the same entrepreneurial spirit as their more independent counterparts. Like entrepreneurs, intrapreneurs use innovation to exploit change as an opportunity for creating something of value. In other words, intrapreneurs seek to better the existing state of affairs within their organizations through creative problem solving.[58] A good example of intrapreneurship is the action of Dr. Cheryl Sonnenberg at the Economic Opportunity Board of Clark County in Las Vegas, Nevada. Dr. Sonnenberg had been struggling to find ways of delivering WIC services to clients in remote areas of Clark County. She had read of a Texas experiment in which a renovated Cadillac had been used to deliver WIC services in rural areas. When she discovered that the renovation costs far exceeded her budget, she began exploring other innovative delivery systems. Eventually, she arranged to buy a used Winnebago, which was renovated with the help of engineers in the transportation division. Her solution, launched in 1995, became known as "WIC on Wheels," or W.O.W. Her intrapreneurial action helped solve a problem and improve rural service delivery to 450 WIC clients.[59]

What do entrepreneurs do? (In this book, innovators in both the private [corporate] and public [government] sectors who embody the spirit and principles of entrepreneurship will be considered entrepreneurs.) One study of entrepreneurs identified at least 57 separate activities associated with launching a new venture, an indication of how complex entrepreneurial behavior can be. Entrepreneurs have wide-ranging competencies in areas such as planning, marketing, networking, budgeting, and team building, as shown in Table 1-6. They turn their creative vision into intentional decision-making and problem-solving actions to accomplish their goals. They are not just managers, although they typically "manage" themselves well. Their high self-esteem stems from a strong belief in their own personal worth, which strengthens their capacity for self-management. Successfully managing oneself means being in control (that is, having willpower), knowing one's personal strengths and weaknesses, and being willing to change one's own behavior and receive feedback and criticism.

Community nutritionists who are RDs *and* entrepreneurs are expected to have competencies in addition to the core competencies expected of all dietitians. These business competencies, many of which overlap with those shown in Table 1-6, include the following:[60]

- Perform organizational and strategic planning.
- Develop and implement a business or operating plan.
- Supervise procurement of resources.
- Manage the integration of financial, human, physical, and material resources.
- Supervise coordination of services.
- Supervise marketing functions.

What about failure? Do entrepreneurs experience failure, and if they do, how do they handle it? In one study, 80 percent of the 152 entrepreneurs interviewed admitted to making mistakes, especially during the start-up phase of their new ventures. Surprisingly, 20 percent reported making no mistakes![61] This result was unexpected, because entrepreneurial activities are rich in opportunities for making mistakes. Given that key mistakes can be made at any point in the entrepreneurial process—from product concept, design, and marketing to team building and budgeting—it is

Identify an opportunity

Create the solution

Decide to go into business

Analyze individual strengths and weaknesses

Conduct market research

Assess potential share of the market

Establish business objectives

Set up organizational structure

Determine personnel requirements

Select the entrepreneurial team

Determine physical plant requirements

Prepare a financial plan

Locate financial resources

Prepare a production plan

Prepare a management plan

Prepare a marketing plan

Produce and test market the product

Build an organization

Respond to government and society

TABLE 1-6 *Activities of Entrepreneurs and Intrapreneurs*

Source: From *Entrepreneurial Behavior* by Barbara J. Bird. Copyright © 1989 by Barbara J. Bird. Reprinted by permission of Addison Wesley Educational Publishers, Inc.

safe to say that entrepreners do make mistakes and suffer setbacks, just like everyone else. They differ from some of the rest of us, perhaps, in that they learn quickly from their failures and are able to apply the insight gleaned from their mistakes to other problem areas and situations.

What relevance does entrepreneurship have to community nutrition? The answer to this question will become increasingly clear as you read the remaining chapters of this book. Suffice it to say at this point that creativity and innovation—the essence of entrepreneurship—are as important to the discipline of community nutrition as to any other field. Consider the entrepreneurial activities in Table 1-6. Nearly all are relevant for the community nutritionist: recognizing an opportunity to deliver nutrition and health messages, developing an action plan for a target audience, building the team for delivering a nutrition program or service, developing a marketing plan, and evaluating the effectiveness of the nutrition program or service.

Community nutritionists who want to change people's eating habits must be able to see new ways of reaching desired target groups. The strategy that works well with Hmong young people in California will probably not work well with institutionalized elderly women living in Ohio. Community nutritionists must draw on theories and skills from the disciplines of sociology, educational psychology, medicine, communications, health education, technology, and business to develop programs for improving people's eating patterns. The twin stanchions of entrepreneurship—creativity and innovation—assist the community nutritionist in achieving the broad goal of improved health for all.

Leading Indicators of Change

The world has changed! We have to address cultural diversity in our customer service workshops.
  –Beverley Demetrius, MA, RD, LD, director of nutrition services, Cobb/Douglas Health District, Georgia

Recent social and economic trends have important consequences for community nutrition. For example, the demographic profile of many communities is changing rapidly, along with the client mix served by community nutritionists. Analysts predict that by the year 2000 the global workforce will be more ethnically diverse, the result of massive relocations of people, including immigrants, refugees, retirees, temporary workers, and visitors, across national borders. Women are expected to enter the global workforce in unprecedented numbers in the coming decade, a trend that may lead to changes in traditional family norms, structures, and child-rearing practices and thus affect the format and delivery of nutrition programs and services. A worldwide increase in the educational level of the workforce is also anticipated. Consequently, a variety of alternative educational strategies will be needed to reach consumers whose education, training, income, language skills, and economic potential will be highly diverse.[62]

In North America, the aging of the population, coupled with a more ethnically diverse society, will challenge community nutritionists to develop new products and services. In the United States, for example, the fastest-growing segment of the nation's population consists of people over 65 years of age. Indeed, the most rapidly growing segment is the 80+ age group, which tends to be more frail, have more chronic health problems, and use more hospital health services such as inpatient and outpatient care compared with younger groups.[63] The aging population will likely place many demands on health care services, home care services, and food assistance programs—as will the millions of people who lack health insurance under existing insurance programs. In 1995, more than 40 million persons in the United States, including 10.5 million children younger than 19 years of age, were uninsured.[64] Securing their access to nutrition and health services will be challenging.

The so-called baby boomers, those people born during or just after World War II, will reach their peak earning years in the next decade and are expected to be a leading market force. Baby boomers will likely place increased demands on the health care system, as will immigrants. An interesting consequence of the influx of immigrants into Canada and the United States is an anticipated change in consumer marketing strategies due to the family-oriented shopping behaviors of Asian and Hispanic consumers. The cultural values and lifestyles of these ethnic groups tend to reinforce family decision making and collective buying behavior. Marketers will have to change their strategies to appeal to the decision style of the extended family. They will need to market their products and services to groups rather than individuals.[65]

Watchwords for the Future

Several terms have surfaced repeatedly in this chapter: *change, innovation, creativity, community, entrepreneurship.* These watchwords herald the approach of a new century marked by unprecedented global social change. The world is growing smaller, and its peoples seem to be moving toward the birth of a single, global nation. "Citizen of the world," a phrase popular during the mid–twentieth century, takes on added meaning as we move into the twenty-first century. Where once a community was circumscribed by a distant ridge or the next valley, today the "information highway" links us via satellite and optic fibers to unseen faces on the other side of the earth. The growing

By the Year 2000 . . .

- The U.S. population will have grown about 7 percent to nearly 275 million people.
- People in the oldest age groups will increase both in number and as a proportion of the total population.
- The number of people over 65 years of age will grow to about 35 million; the number of people over 85 years of age will grow to more than 4 million.
- The average household size is expected to decline from 2.69 in 1985 to 2.48 in the year 2000, with husband-wife households decreasing from 58 to 53 percent of all households.
- The racial and ethnic composition of the American population will form a different pattern:

    Whites, not including Hispanic Americans, will decline from 76 to 72 percent of the population.

    Hispanics will rise from 10 to 15 percent, to more than 39 million Hispanic people, becoming the largest minority group.

    Blacks will increase their proportion from 12.4 to 13.1 percent.

    Other racial groups, including American Indians and Alaskan Natives and Asians and Pacific Islanders, will increase from 3.5 to 4.3 percent of the total.

- The American population may increase by up to 6 million people through immigration. Certain states and cities, especially those on the East and West coasts, can be expected to receive a disproportionately large number of these immigrants.

Sources: U.S. Department of Health and Human Services, Public Health Service, *Healthy People 2000: National Health Promotion and Disease Prevention Objectives* (Washington, D.C.: U.S. Government Printing Office, 1990), pp. 2–3; U.S. Department of Health and Human Services, Hispanic Customer Service Home Page at www.dhhs.gov/about/heo/hisp.html; U.S. Bureau of the Census, *Population Profile of the United States, 1995*, Current Population Reports, Special Study Series P23-189 (obtained from the Web site at www.census.gov/population/pop-profile/p23-189.pdf); and U.S. Bureau of the Census, *Population Projections of the United States by Age, Sex, Race, and Hispanic Origin: 1995 to 2050*, Current Population Reports, P25-1130, issued 1996 (obtained from the Web site at www.census.gov/prod/1/pop/p25-1130).

connectedness of the human race promises to create new challenges for community nutritionists in their efforts to enhance the nutrition and health of all peoples.

Community Learning Activity

Activity 1

Interview a community nutritionist in your community about his or her job responsibilities, career path, and new ventures. A few questions you might ask this individual are listed here, but do not limit yourself to these questions.

1. What do you do in your job?
2. What skills do you use in your job, and what are the most challenging aspects of your work?

3. What other positions in community nutrition have you held?
4. Describe an activity that you believe is entrepreneurial—that is, one that represents a new venture for your yourself or your department or organization or that was creative or innovative. Did you make mistakes while working on this project, and, if so, what did you learn from them? Was there a point at which you thought about giving up? If so, what kept you going? What would you do differently next time?

Activity 2

Obtain information about your community and its health problems by completing the following tasks:

1. Describe your community in terms of its location, size (number of residents), economic base (for example, agriculture, manufacturing), types of nonprofit organizations with a health focus (such as the American Diabetes Association, Alcoholics Anonymous, and so forth), and number of hospitals and clinics, universities and/or colleges, libraries, and newspapers. This information can usually be obtained from telephone books and the local public library.
2. For a designated period (for instance, one week), clip or photocopy articles on health and nutrition from your local paper and analyze them for their local content—that is, whether they describe local health and nutrition issues or national or even international issues. Summarize your findings, indicating whether you believe the newspaper articles provide an accurate indication of the health and nutrition problems in your community. Share your analysis with the class.

Activity 3

List the communities to which you belong, and explain why you are linked with each one.

Activity 4

Consult the nutrition objectives for minority populations in *Healthy People 2000 Review, 1997*, available on the homepage of the National Center for Health Statistics at www.cdc.gov/nchswww. (The document is located at www.cdc.gov/nchswww/data/hp2k97.pdf and requires Adobe Acrobat Reader® to view on-line.) Summarize the progress made in meeting those nutrition objectives that target ethnic populations. What are the implications of your findings for community nutrition practice?

Activity 5

For this activity, class members are divided into four teams. Each team will take one of the four emerging issues outlined in the boxed insert on page 22. For a particular issue, outline your response to the question, indicating pros and cons of both a yes and no answer. Discuss your conclusion with the class. Does the class support your team's conclusion?

Professional Focus | *Getting Where You Want to Go*

Imagine for a minute that circumstances require you to travel from Kansas City to Chicago. Spreading a map across your lap, you plot your course. You could take an interstate highway all the way, starting with I-35 traveling north to Des Moines and then turning east toward Chicago on I-80. Or you might take I-70 to St. Louis and turn due north onto I-55, a course that would take you right into the Chicago Loop. Or you might decide to bypass the interstate highways altogether and stick to the so-called blue highways, those tiny threads on the map that snake along from town to town. Your decision about which route to take depends on many factors, including the purpose and urgency of your trip and how much time you can allocate for traveling.

In many respects, your decision about what you do in life is much like plotting a journey by car to a distant city. Many choices confront you, and the possibility always exists that circumstances may compel or entice you to change your route along the way. Right now, you might be asking yourself the following questions: How can I get where I want to go? More important, how do I determine where I want to go in the first place? The answers to these questions are unique to each of you, because each of you is unique. As you read through this discussion, write down your thoughts to help clarify your vision.

Square 1: Know Yourself

The first step in determining where you want to go is to know yourself. Hold yourself up to the light, so to speak, so that you can see yourself from every angle. Evaluate your strong points and the areas marked for improvement. (The "To Be Improved" areas are sometimes called weaknesses. Weaknesses are not personality defects or deficits. They are areas of personal development that you have not had the time or inclination to explore and strengthen.) Consider your personality, your view of the world, and what you want out of life. Do you like working with people? Do you enjoy tinkering with gadgets and gizmos? Do you value public service? Are you an optimist or a cynic? Would you describe yourself as impulsive, dependable, funny, unfocused, inquisitive, theatrical, or lazy? Write down the words that describe all aspects of your personality and character. There are no right or wrong answers.

Square 2: Define Your Dreams

Knowing who you are (and who you are not) will help you move to the next tier: defining your dreams. Your vision of your future lies in your dreams, for what you *imagine* yourself doing is what you are ultimately going to do. So, what do you see yourself doing? To help you define your dreams, answer the following questions:[1]

- What would you ideally most like to *be*?
- What would you ideally most like to *do*?
- What kind of experiences help you feel complete?
- In what kind of situations do you most want and tend to share yourself?

Let yourself dream freely and without constraints. Do not be concerned at this point about finances or family obligations. Give your dreams room to grow.

Square 3: Set Goals

Having dreams won't get you very far if you don't put some structure to them. As Thoreau stated so eloquently, "If you have built castles in the air, your work need not be lost; that is where they should be. Now put the foundations under them."[2]

Setting specific goals for your future is one of the most challenging tasks you will undertake. There are many areas in which goal setting is desirable: economic, spiritual, social, physical, mental, emotional, educational, personal, and vocational.

For this exercise, set at least one goal for your personal life. The goal should be achievable but broad enough to accommodate your dreams. Joe D. Batten, author of the book *Tough-Minded Leadership*, wrote his personal goal as follows: "I will make the lives of others richer by the richness of my own."[3] Your personal goal might be entirely different.

Another way to approach this exercise is to write your personal mission statement. A personal mission statement is much like a nation's constitution; it is a set of principles to live by. In his best-selling book *The 7 Habits of Highly Effective People*, Stephen R. Covey cites the personal mission statement developed by a friend, a portion of which is shown on the next page.

Continued

*An Example of a Personal Mission Statement**

1. Succeed at home first.
2. Never compromise with honesty.
3. Be sincere yet decisive.
4. Develop one new proficiency a year.
5. Plan tomorrow's work today.
6. Maintain a positive attitude.
7. Keep a sense of humor.
8. Do not fear mistakes—fear only the absence of creative, constructive, and corrective responses to those mistakes.
9. Help subordinates achieve success.
10. Concentrate all abilities and efforts on the task at hand; do not worry about the next job or promotion.

---

* Adapted from S. R. Covey, *The 7 Habits of Highly Effective People* (New York: Simon & Schuster, 1989), p. 106.

Square 4: Develop an Action Plan

To paraphrase a Chinese proverb, if you don't know where you are going, then any road will take you there. To get where you want to go, you *must* develop an action plan. Action is the essence of achievement. Stephen Covey calls this "beginning with the end in mind." You must begin your journey with a clear picture of your destination. Use the following steps to develop an action plan for your personal life:

- Develop a picture in your mind's eye of what you want to do with your life. You may see yourself having a family and a career position with a major food company, or helping an isolated community in a developing country improve its standard of living, or starting your own business. The technique of mental imaging allows you to fine-tune your picture, so that when opportunities present themselves, you can determine whether they fit your action plan.
- Pretest your mental picture. If your mental picture shows you working with small animals as part of a research project, then find a way to test your decision before you commit yourself to this path. You may discover that you don't like working with rats or hamsters!

Pretesting your decisions saves time and allows you to discard opportunities that are not useful or don't fit your action plan.
- Predetermine your alternatives. Have a backup plan to help you maximize your opportunities and forestall any crises. Explore your alternatives by talking to people who have pursued a similar dream.

Learn to Manage Yourself

Shirley Hufstedler, a lawyer who became the secretary of education, remarked, "When I was very young, the things I wanted to do were not permitted by social dictates. I wanted to do a lot of things that girls weren't supposed to do. So I had to figure out ways to do what I wanted to do and still show up in a pinafore for a piano recital, so as not to blow my cover. You could call it manipulation, but I see it as observation and picking one's way around obstacles. If you think of what you want and examine the possibilities, you can usually figure out a way to accomplish it."[4]

Getting where you want to go is nearly impossible if you don't learn to manage yourself—your goals, your time, your work. Aristotle observed that the hardest victory is the victory over self. Successful people have mastered themselves through discipline. For some people, discipline is a dirty word. In truth, discipline means *training*. Any athlete will attest to the power of training, which builds, molds, and strengthens the body and mind for strong performance. Discipline is as important to life as it is to athletic competition. Without it, little can be accomplished. Acquiring discipline, the mastery of self, is a lifelong process for most of us, and there is no simple pattern by which it can be attained. The process involves developing a vision, setting goals, and following through on an action plan to reach those goals. The first step in acquiring discipline begins at square 1: know yourself.

References

1. J. D. Batten, *Tough-Minded Leadership* (New York: American Management Association, 1989), p. 177.
2. As cited in R. N. Bolles, *The Three Boxes of Life* (Berkeley, CA: Ten Speed Press, 1981), p. 34.
3. As cited in Batten, *Tough-Minded Leadership*, p. 179.
4. As cited in W. Bennis, *On Becoming a Leader* (Reading, MA: Addison-Wesley, 1989), pp. 53–54.

References

1. W. B. Johnston, Global work force 2000: The new world labor market, *Harvard Business Review* 69 (1991): 115–27.

2. P. H. Mirvis, Human resource management: Leaders, laggards, and followers, *Academy of Management Executive* 11 (1997): 43–56.

3. G. I. Balch, Employers' perceptions of the roles of dietetics practitioners: Challenges to survive and opportunities to thrive, *Journal of the American Dietetic Association* 96 (1996): 1301–5.

4. L. A. Pal, *Beyond Policy Analysis: Public Issue Management in Turbulent Times* (Scarborough, Ontario: ITP, 1997), p. 1.

5. Community Nutrition Institute, Looking for solutions: Food recovery, recycling, and education, *Nutrition Week* 27 (July 11, 1997): 4–6.

6. Community Nutrition Institute, USDA to support "human gleaning" with new fund, *Nutrition Week* 27 (March 21, 1997): 3.

7. T. A. Nicklas and coauthors, Development of a school-based nutrition intervention for high school students: *Gimme 5, American Journal of Health Promotion* 11 (1997): 315–22.

8. Institute of Medicine, *The Future of Public Health* (Washington, D.C.: National Academy Press, 1988).

9. The discussion of the leading causes of death was adapted from Centers for Disease Control and Prevention, Department of Health and Human Services, *Unrealized Prevention Opportunities: Reducing the Health and Economic Burden of Chronic Disease* (Atlanta, GA: Department of Health and Human Services, Public Health Service, 1997).

10. S. J. Ventura, K. D. Peters, J. A. Martin, and J. D. Maurer, Births and deaths: United States, 1996, *Monthly Vital Statistics Report* 46 (September 11, 1997): 1–33.

11. T. C. Quinn, Screening for HIV infection—Benefits and costs (editorial), *New England Journal of Medicine* 327 (1992): 486–88.

12. J. J. Neal and coauthors, Trends in heterosexually acquired AIDS in the United States, 1988 through 1995, *Journal of Acquired Immune Deficiency Syndromes and Human Retrovirology* 14 (1997): 465–74.

13. P. Phillips, No plateau for HIV/AIDS epidemic in US women, *Journal of the American Medical Association* 277 (1997): 1747–49.

14. Centers for Disease Control and Prevention, *Unrealized Prevention Opportunities*.

15. P. L. F. Zuber and coauthors, Long-term risk of tuberculosis among foreign-born persons in the United States, *Journal of the American Medical Association* 278 (1997): 304–7.

16. D. E. Snider, Jr., and W. L. Roper, The new tuberculosis (editorial), *New England Journal of Medicine* 326 (1992): 703–5.

17. F. Nault, Narrowing mortality gaps, 1978 to 1995, *Health Reports* 9 (Summer, 1997): 35–41.

18. P. Lee and D. Paxman, Reinventing public health, *Annual Review of Public Health* 18 (1997): 1–35.

19. The discussion of the concepts of health and public health was adapted from J. M. Last, *Public Health and Human Ecology* (East Norwalk, CT: Appleton & Lange, 1987), pp. 1–26; and G. Pickett and J. J. Hanlon, *Public Health: Administration and Practice* (St. Louis: Times Mirror/Mosby College Publishing, 1990), pp. 3–20.

20. M. P. O'Donnell, Definition of health promotion, *American Journal of Health Promotion* 1 (1986): 4–5.

21. J. Stokes III and coauthors, Definition of terms and concepts applicable to clinical preventive medicine, *Journal of Community Health* 8 (1982): 33–41.

22. M. Minkler, Health education, health promotion, and the open society: An historical perspective, *Health Education Quarterly* 16 (1989): 17–30.

23. The quote is from O'Donnell, Definition of health promotion, p. 4.

24. J. P. Elder and coauthors, *Motivating Health Behavior* (Albany, NY: Delmar, 1994), pp. 1–7.

25. The discussion of the types of prevention efforts was adapted from American Dietetic Association, Position of The American Dietetic Association: The role of nutrition in health promotion and disease prevention programs, *Journal of the American Dietetic Association* 98 (1998): 205–8; and R. H. Fletcher, S. W. Fletcher, and E. H. Wagner, *Clinical Epidemiology: The Essentials* (Baltimore, MD: Williams & Wilkins, 1982), pp. 127–31.

26. M. P. O'Donnell, Definition of health promotion: Part II. Levels of programs, *American Journal of Health Promotion* 1 (1986): 6–9.

27. Elder and coauthors, *Motivating Health Behavior*, pp. 195–97.

28. As cited in Last, *Public Health and Human Ecology*, p. 16.

29. World Health Organization, *Targets for Health for All* (Copenhagen: World Health Organization Regional Office for Europe, 1985).

30. H. Nielsen, *Achieving Health for All: A framework for nutrition in health promotion, Journal of the Canadian Dietetic Association* 50 (1989): 77–80.

31. C. R. Connolly, A commentary on *Achieving Health for All: A Framework for Health Promotion, Journal of the Canadian Dietetic Association* 50 (1989): 89–92.

32. U.S. Department of Health and Human Services, Public Health Service, *Healthy People 2000: National Health Promotion and Disease Prevention Objectives* (Washington, D.C.: U.S. Government Printing Office, 1990).

33. U.S. Department of Health and Human Services, Public Health Service, *Healthy People: The Surgeon General's Report*

on *Health Promotion and Disease Prevention*, PHS Pub. No. 79-55071 (Washington, D.C.: U.S. Government Printing Office, 1979).

34. U.S. Department of Health and Human Services, Public Health Service, *Promoting Health/Preventing Disease: Objectives for the Nation* (Washington, D.C.: U.S. Government Printing Office, 1980).

35. U.S. Department of Health and Human Services, Public Health Service, *The Surgeon General's Report on Nutrition and Health*, DHHS Pub. No. 88-50210 (Washington, D.C.: U.S. Government Printing Office, 1988), pp. 1–20.

36. U.S. Department of Health and Human Services, *Healthy People 2000: National Health Promotion and Disease Prevention Objectives*, pp. 117–124.

37. The definition of surveillance was adapted from Federation of American Societies for Experimental Biology, Life Sciences Research Office, *Third Report on Nutrition Monitoring in the United States*, Vol. 1 (Washington, D.C.: U.S. Government Printing Office, 1995), p. I-8.

38. The discussion of the progress toward meeting the *Healthy People 2000* goals was adapted from U.S. Department of Health and Human Services, Public Health Service, *Healthy People 2000 Midcourse Review and 1995 Revisions* (Washington, D.C.: U.S. Government Printing Office, 1995), pp. 27–33 and 164–70; and National Center for Health Statistics, *Healthy People 2000 Review, 1997* (Hyattsville, MD: Public Health Service, 1997), as obtained from the NCHS Web site at www.cdc.gov/nchswww.

39. H. Blackburn, Research and demonstration projects in community cardiovascular disease prevention, *Journal of Public Health Policy* 4 (1983): 398–421.

40. G. A. Hillery, Jr., Definitions of community: Areas of agreement, *Rural Sociology* 20 (1955): 111–23.

41. The discussion of the concept of community was adapted from T. N. Clark, *Community Structure and Decision-Making: Comparative Analyses* (San Francisco: Chandler, 1968), pp. 83–9; A. D. Edwards and D. G. Jones, *Community and Community Development* (The Hague: Mouton, 1976), pp. 11–39; R. M. MacIver, *On Community, Society and Power* (Chicago: University of Chicago Press, 1970), pp. 29–34; and D. E. Poplin, *Communities* (New York: Macmillan, 1979), pp. 3–25.

42. M. Kaufman, Preparing public health nutritionists to meet the future, *Journal of the American Dietetic Association* 86 (1986): 511–14.

43. American Dietetic Association, *Role Delineation for Registered Dietitians and Entry-Level Dietetic Technicians* (Chicago: American Dietetic Association, 1990).

44. Pan American Health Organization, *Food and Nutrition Issues in Latin America and the Caribbean* (Washington, D.C.: Pan American Health Organization/Inter-American Development Bank, 1990).

45. Food and Agriculture Organization of the United Nations white paper on the Food Policy and Nutrition Division, pp. 24–29.

46. L. M. Brown and M. F. Fruin, Management activities in community dietetics practice, *Journal of the American Dietetic Association* 89 (1989): 373–77.

47. C. J. Gilmore and coauthors, Determining educational preparation based on job competencies of entry-level dietetics practitioners, *Journal of the American Dietetic Association* 97 (1997): 306–16.

48. Winnipeg Prenatal Nutrition Initiative, *Healthy Start for Mom & Me*, Winnipeg, Manitoba, Canada R2W 4J5.

49. The discussion of the role delineation study and the responsibilities of dietitians was taken in part from American Dietetic Association, *Role Delineation for Registered Dietitians and Entry-Level Dietetic Technicians*; ADA Reports, President's Page: Beyond the RD, *Journal of the American Dietetic Association* 90 (1990): 1117–21; Of Professional Interest, Commentary on the role delineation study, *Journal of the American Dietetic Association* 90 (1990): 1122–23; and M. T. Kane and coauthors, Role delineation for dietetic practitioners: Empirical results, *Journal of the American Dietetic Association* 90 (1990): 1124–33.

50. Brown and Fruin, Management activities, 373–77.

51. American Dietetic Association, *Role Delineation*.

52. G. Gates and W. Sandoval, Teaching multiskilling in dietetics education, *Journal of the American Dietetic Association* 98 (1998): 278–84.

53. The discussion of entrepreneurship was adapted from B. J. Bird, *Entrepreneurial Behavior* (Glenview, IL: Scott, Foresman, 1989), pp. 1–33, 57–76, and 349–75; and J. G. Burch, *Entrepreneurship* (New York: Wiley, 1986), pp. 4–42.

54. Bird, *Entrepreneurial Behavior*, p. 3.

55. *Health Start*, Pawnee, OK 74058-0462.

56. *Dietetics Online®* can be found on the Internet at www.dietetics.com.

57. Burch, *Entrepreneurship*, pp. 28–29.

58. A. L. F. Foong and coauthors, Entrepreneurs and intrapreneurs: Common blood, different languages, *Entrepreneurship, Innovation, and Change* 6 (1997): 67–72.

59. Community Nutrition Institute, WIC on Wheels Winnebago serves a sprawling Vegas, *Nutrition Week* 27 (August 1, 1997): 6.

60. Gilmore and coauthors, *Journal of the American Dietetic Association*.

61. K. A. Egge and F. J. Simer, An analysis of the advice given by recent entrepreneurs to prospective entrepreneurs, in *Frontiers of Entrepreneurship Research* (Wellesley, MA: Babson College, 1988), pp. 119–33.

62. Johnston, Global work force 2000.

63. M. N. Haan and coauthors, The impact of aging and chronic disease on use of hospital and outpatient services in

a large HMO: 1971–1991, *Journal of the American Geriatric Society* 45 (1997): 667–74.

64. K. E. Thorpe, Incremental strategies for providing health insurance for the uninsured: Projected federal costs and

number of newly insured, *Journal of the American Medical Association* 278 (1997): 329–33.

65. J. L. Zaichkowsky, Consumer behavior: Yesterday, today, and tomorrow, *Business Horizons* 34 (1991): 51–58.

Chapter 2

The Art and Science of Policy Making

Outline

Learning Objectives

After you have read and studied this chapter, you will be able to:

- Describe the policy-making process.
- Describe how laws and regulations are developed.
- Describe the federal budget process.
- Identify a minimum of four emerging policy issues in the food and nutrition arena.
- Prepare a letter addressed to your congressperson.
- Summarize the importance of policy making to nutritionists working in the community.
- Identify three ways in which the community nutritionist can influence policy making.

Something To Think About...

All politics is local. – *Thomas P. "Tip" O'Neill*

Introduction

In bold letters the newspaper headline proclaimed, "Little Ones Doomed: Child Malnutrition a 'Silent Emergency.'"[1] According to Stephen Lewis, deputy executive director of UNICEF, who was quoted in the accompanying article, "The silent emergency of malnutrition is so shocking and simultaneously unnecessary that it must be brought to public attention." His comments coincided with the release of UNICEF's *State of the World's Children 1998* report, which indicated that more than 200 million children in developing countries under the age of 5 years are malnourished. Two weeks after the newspaper article appeared, *Time* magazine featured a two-page column, donated jointly by *Time* and Canon, that called attention to UNICEF's message about the silent emergency of malnutrition.[2]

What relevance do these documents have for community nutritionists? What do they have to do with policy, the topic of this chapter? The answer to both questions is, A great deal. The newspaper article and magazine column are examples of how an organization (UNICEF) used a particular strategy (a press release to the media and partnerships with two companies, *Time* magazine and Canon) to help convince people that a serious problem (global childhood malnutrition) exists. Addressing problems is the core activity of the policy-making process. Whether the issue is regulating experimentation with cloning animals, controlling air and water pollution, or providing quality health care for all citizens, policy making is an ongoing process that affects our lives daily. As a community nutritionist, both local and national policy issues affect the way you work, how you deliver nutrition services, and the dietary messages you give to clients in your community. Consider how you would respond to the following issues in the nutrition policy arena:

- Should all Americans reduce their intake of dietary fat to 30 percent or less of total calories, considering that recent evidence suggests that diets high in fat may be associated with a lower risk of stroke in men?[3]
- Are dietary guidelines developed for adults appropriate for children?
- Should U.S. manufacturers of baby formula be allowed to sell their products in developing countries where the use of such products may undercut breastfeeding practices?
- Should television stations be required to run public service announcements that feature healthful food messages for children to balance current food-related advertising, which tends to promote high-fat, high-sugar foods?

A photograph such as this one can be an effective method of bringing an urgent issue to the public's attention. Convincing people that a problem exists is the first step in the policy process.

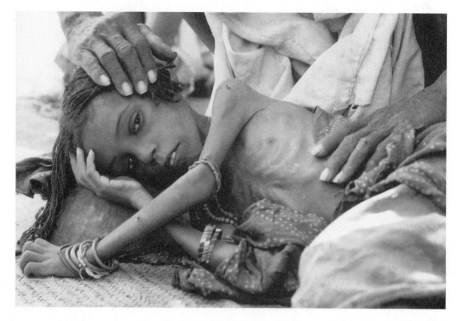

- Should dietary guidelines be changed to state that consuming a minimum of two glasses of red wine daily may reduce the risk of coronary heart disease?
- Is the recommendation to consume moderate amounts of alcohol to help prevent coronary heart disease appropriate, considering that excessive alcohol intake is associated with increased risk of some types of cancer?[4]

There are no simple answers to these difficult policy questions. This is not surprising, for public policy is complex and ever-changing. The purpose of public policy is to fashion strategies for solving public problems. In the nutrition arena, the strategies for solving problems typically include food assistance programs, dietary recommendations, and reimbursement mechanisms for nutrition services. This chapter describes the policy-making process, examines emerging policy issues, and discusses the policy-making activities of community nutritionists.

The Process of Policy Making

You may not think of yourself as a political animal. You may think of politics as being confined to senate hearings, city council meetings, and elections. But if you have ever lobbied a professor to allow you to take an exam at a later date or signed a petition calling for increased funding for a local food bank, then *you* have walked onto the political stage. You have tried to get something you want by presenting compelling reasons why an existing policy should be changed.

Recall that policy was defined in Chapter 1 as the course of action chosen by public authorities to address a given problem. A **problem** is a "substantial discrepancy between what is and what should be."[5] When public authorities state that a

Problem A significant gap between current reality (the way things are) and the desired state of affairs (the way things should be).

problem exists, they are recognizing the gap between current reality and the desired state of affairs. Policies, then, are guides to a range of activities designed to address a problem.[6]

Policy making is the process by which authorities decide which actions to take to address a problem or set of problems, and it can be viewed as a cycle, as shown in Figure 2-1.[7] This diagram has the advantage of simplifying the policy-making process but also makes it a little too simple and neat. In reality, the various stages often overlap as a policy is fine-tuned. Sometimes the stages occur out of sequence, as when agenda setting leads directly to evaluation. For instance, when the issue of hunger became a part of the national policy agenda in the 1960s, existing nutrition and welfare programs were evaluated to discover why they were not reaching hungry children, effectively by-passing the policy design and implementation stages. Nevertheless, viewing the policy-making process as a cycle allows us to see how policies evolve over time.

The discussion that follows focuses on policy making at the national level, because the laws that arise from federal policy may affect some aspects of community nutrition practice. As you study this section, think of ways in which the policy cycle can be applied to lower levels of government, such as your state or municipal government, and to institutions, such as your college or university or place of employment. As a student, for example, your life is affected by your school's policies on course requirements for graduation, residency, use of campus libraries, and many

Policy making The process by which authorities decide which actions to take to address a problem or set of problems.

FIGURE 2-1 *The Policy Cycle*

Source: Adapted from W. Lyons, J. M. Scheb II, and L. E. Richardson, Jr., *American Government: Politics and Political Culture*, p. 468. Copyright © 1995 West Publishing Company. Used by permission of Wadsworth Publishing Co.

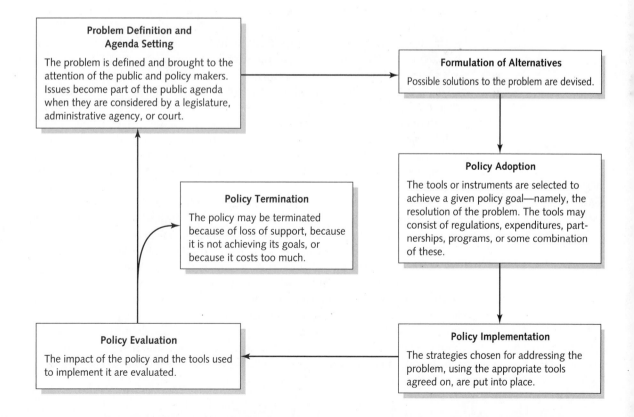

Problem Definition and Agenda Setting

The problem is defined and brought to the attention of the public and policy makers. Issues become part of the public agenda when they are considered by a legislature, administrative agency, or court.

Formulation of Alternatives

Possible solutions to the problem are devised.

Policy Adoption

The tools or instruments are selected to achieve a given policy goal—namely, the resolution of the problem. The tools may consist of regulations, expenditures, partnerships, programs, or some combination of these.

Policy Termination

The policy may be terminated because of loss of support, because it is not achieving its goals, or because it costs too much.

Policy Implementation

The strategies chosen for addressing the problem, using the appropriate tools agreed on, are put into place.

Policy Evaluation

The impact of the policy and the tools used to implement it are evaluated.

other activities. Consider how the policy cycle described here reflects the process by which your school formulated its policies.

1. **Problem definition and agenda setting.** The first step in the policy process is to convince other people that a public problem exists. For example, the fact that approximately 41 million people in the United States do not have medical insurance could be seen as either a public problem or a private (individual) problem.[8] Before a problem can be addressed, then, a majority of people must be convinced that it is a *public* issue. A clear statement of the problem is derived by asking questions: Who is experiencing the problem? Why did this problem develop? How severe is the problem? What actions have been taken in the past to address the problem? What action can be taken *now* to solve the problem? What resources exist to help alleviate the problem?[9] The manner in which the problem is defined will likely determine whether it succeeds in capturing the public's attention and whether action is taken to address it.

Agenda The set of problems to which policy makers give their attention.

Once a problem is defined and gains attention, it is placed on the policy agenda. This **agenda** is not a written document or book but a set of controversial issues that exist within society. Agenda setting is a process in which people concerned about an issue work to bring the issue to the attention of the general public and policy makers. Getting policy makers to place a problem on the official agenda can be difficult. An issue may be so sensitive that policy makers or public attitudes work to keep it from reaching the agenda-setting stage. Consider, for example, the question of whether gay couples should be allowed to marry legally or adopt children. In the absence of widespread public support, some gay couples have taken this issue to the courts as a means of accessing the policy agenda. In other situations, an issue may be perceived as a problem only by a small number of people who lack the political clout required to get the issue onto the agenda. Sometimes a major catastrophe, such as an earthquake, assassination, riot, or an unusual human event, triggers a public outcry and pushes an issue onto the policy agenda. When CBS aired its television report "Hunger in America" in May 1968, the problem of malnutrition and the failure of major government feeding programs for families became vividly real. Millions saw the televised images of starving and dying American children. The public reaction was swift and angry, pushing Congress to form the Senate Select Committee on Nutrition and Human Needs to investigate hunger and malnutrition among America's poor.[10]

Public opinion in this country is everything.
  – Abraham Lincoln

How are issues placed on the policy agenda? The first step is to build widespread public interest for the issue that deserves government attention. One of the most effective ways to build public interest and support for an issue is to work through the media—radio, television, newspapers, and the Internet. Because the media can both create and reflect public issues, they are one of the most powerful tools for setting the public policy agenda. For example, the publication in 1962 of Rachel Carson's book *Silent Spring* opened the public's eyes to environmental dangers and launched the modern environmental movement.[11] UNICEF's success at getting newspapers and magazines to carry the message about childhood malnutrition is another example of using the media to increase public awareness of an issue and seek support for efforts to address it. UNICEF also uses the Internet to keep the issue before the public's eye. Its Web site provides information about its *State of the World's Children 1998* report.

However, it is not enough merely to bring the issue to the attention of policy makers and the public through the media. The issue must get onto the **institutional agenda** defined by each legislative body of the government (for example, Congress, state legislatures, city councils). This is accomplished by winning support for the issue among policy subgovernments. Some political scientists have called **policy subgovernments** the "iron triangles," because they basically control the policy process. Iron triangles refer to three powerful participants in the policy-making process: interest groups, congressional committees or subcommittees, and administrative agencies. The policy subgovernments are not formal, recognized units of government, but they often exert enormous control over the policy-making process. The policy subgovernments consist of anyone interested in policy issues and outcomes, such as government administrators, members of Congress and their staffs, bureau chiefs, interest groups, professionals (for example, dietitians, physicians, bankers, real estate agents), university faculty members, governors, and members of state and local governments, coalitions, and networks.

2. **Formulation of alternatives.** Possible solutions to the problem are devised in this, the most creative phase of the policy-making process. How can the WIC program be modified to meet the needs of working women? How can school food programs be improved to feed more children and teenagers? Should families experiencing food insecurity be given greater access to food banks and food programs or should they receive job training to help them increase their income?

Discussions of possible solutions to public problems—what is sometimes called "policy formulation"—often begin at the grassroots level. Interest groups, coalitions, and networks of experts and people interested in the issue craft a set of possible solutions and bring them forward to policy makers, who continue the discussion of solutions in legislative assemblies, government agencies, other institutions, congressional hearings, town hall meetings, and even focus groups. These forums give the general public an opportunity to express its opinions about possible solutions, potential costs and benefits of alternatives, and the "best" course of action. A key consideration is whether the best proposed solution—in other words, the action that will become policy—is reasonable. During World War II, for example, American and British military officers were stymied in their efforts to stop German submarine attacks on Allied ships. In exasperation, they turned to an operations researcher for a solution. He thought for a moment and then responded, "That's easy—all you have to do is boil the ocean." The military officers replied, "But how do we do that?" The operations researcher answered, "I don't know. I only make policy. It's your job to implement it."[12] In this example, the suggested policy was not realistic or achievable, and the solution proposed by the operations researcher did not consider how the policy would be implemented or even whether it *could* be implemented. The act of making policy was uncoupled from the process of implementing it—an unworkable situation in real life. Policies designed to address food and nutrition-related problems are nearly always workable, although they change as circumstances, information, and priorities change.

UNICEF's Internet address is www.unicef.org.

Institutional agenda The issues that *are* the subject of public policy.

Policy subgovernments Although not formal units of government, these "iron triangles" have a powerful influence on policy and agenda setting.

In the United States, policy is formulated by the legislative, executive, and judicial branches of the government at the national, state, and local levels. (Refer to Appendix B for an organizational chart of the U.S. federal government.) Two examples of national policies formulated by Congress are the Federal Food, Drug, and Cosmetic Act, the legislation that regulates the U.S. food supply,[13] and the Nutrition Labeling and Education Act (NLEA), the legislation that specifies national uniform food labels and mandatory nutrition labeling information on nearly all foods marketed to U.S. consumers.[14]

3. **Policy adoption.** In this step, the tools or instruments for dealing with the problem are chosen. Examples of policy "tools" include regulations, cash grants, loans, tax breaks, certification, fines, price controls, quotas, public promotion, public investment, and government-sponsored programs.[15] The tools are wielded, so to speak, by federal, state, and municipal departments and agencies that are responsible for implementing policy.

At the federal level, two departments are important for our purposes: the Department of Health and Human Services (DHHS) and the U.S. Department of Agriculture (USDA). The mission of the DHHS is to promote, protect, and advance the nation's physical and mental health. Its organizational chart is shown in Figure 2-2. The DHHS includes more than 300 programs, covering a broad spectrum of activities from conducting medical science research and preventing outbreaks of infectious disease to assuring food and drug safety and providing financial assistance for low-income families and older Americans. The DHHS works closely with state and local governments, and many of its services are provided by state or county agencies. The Public Health Service operating division of the DHHS includes the National Institutes of Health, which houses 17 separate health institutes such as the National Library of Medicine and supports about 30,000 research projects worldwide; the Food and Drug Administration, which assures the safety of the food supply and cosmetics; and the Centers for Disease Control and Prevention, which collects national health data and works to prevent and control disease. The Human Resources operating division includes the Health Care Financing Administration, which administers the Medicare and Medicaid programs; the Administration for Children and Families, which administers some 60 programs for needy children and families, including the Head Start program; and the Administration on Aging, which provides services, including some 240 million meals each year, to the elderly.[16]

The USDA is also concerned with some important aspects of public health and policy making. Its overall mission is to enhance the quality of life for all Americans by working to ensure a safe, affordable, nutritious, and accessible food supply; reducing hunger in America and in other parts of the world; and supporting the production of agriculture. Its organization is illustrated in Figure 2-3. The USDA's Food Safety mission strives to assure that the nation's meat and poultry supply is safe for consumption, wholesome, and packaged and labeled properly. The agency responsible for carrying out this mission is the Food Safety and Inspection Service.

The mission of USDA's Food, Nutrition, and Consumer Services is to ensure access to nutritious, wholesome food and healthful diets for all Americans and to provide dietary guidance to help them make healthful food choices. Two agencies

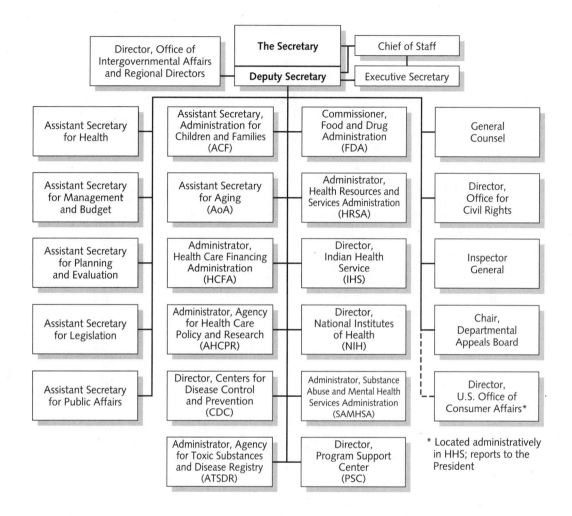

within this division are important to community nutritionists. The Food and Nutrition Service administers 15 food assistance programs such as the Food Stamp Program and the National School Lunch Program (described along with other food assistance programs in Section III of this book). The Center for Nutrition Policy and Promotion coordinates nutrition policy within USDA and provides national leadership in educating consumers about nutrition. This center works with the Department of Health and Human Services to review, revise, and disseminate the *Dietary Guidelines for Americans* (described in Chapter 4).

The mission of USDA's Research, Education, and Economics division is to develop innovative technologies that improve food production and food safety. This area includes three agencies of interest to community nutritionists: The Agricultural Research Service, which works to solve broad agricultural problems and ensure an adequate supply of food for all consumers; the Cooperative State Research, Education, and Extension Service (CSREES), which strives to develop national priorities in research, extension, and higher education; and the Economic

FIGURE 2-2

Organization of the Department of Health and Human Services

Source: The U.S. Department of Health and Human Service's Web site at www.hhs.gov/about/orgchart.html.

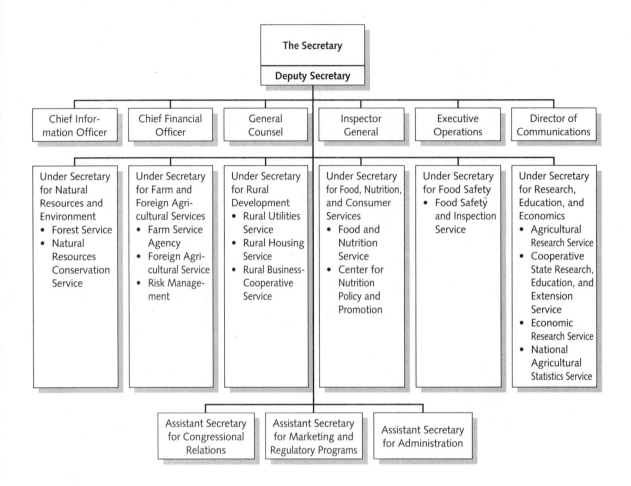

FIGURE 2-3

Organization of the U.S. *Department of Agriculture*

Source: The U.S. Department of Agriculture's Web site at www.usda. gov/agencies/agchart.htm.

Research Service, which produces economic and social science data to help Congress make decisions about the practice of agriculture and rural development. The main research division of the USDA is the Agricultural Research Service. It oversees research related to nutrient needs throughout the life cycle, food trends, the composition of the diet, nutrient interactions, and the bioavailability of nutrients. It compiles data on the nutrient composition of foods through its Nutrient Databank; the food values are published in the Agriculture Handbook No. 8 series, *Composition of Foods*.

Policy adoption also occurs at the state level. States are the basic unit for the delivery of public health services and the implementation of health policies. They determine the form and function of local health agencies, select and appoint local health personnel, identify local health problems, and guarantee a minimum level of essential health services. State and federal public health services have a similar organization. Diagrams of a generic state and municipal health agency appear in Appendix B.

4. **Policy implementation.** After the best solution to the problem has been agreed on and the tools for dealing with the problem have been chosen, the policy is mod-

ified to fit the needs, resources, and wants of the implementing agencies and the intended clientele. Implementation refers to the process of putting a policy into action. In Tacoma, Washington, for example, the Tahoma Food System, a nonprofit organization, implemented a policy to promote a sustainable food system. Its policy supported a community coalition called Bridging Urban Gardens Society (BUGS), which helps create "greening" projects such as organic gardens on unused, littered vacant lots.[17] The implementors of public policy in the United States number literally in the millions and include employees of federal, state, and local governments who work with private organizations, interest groups, and other parties to carry out government policy.

5. **Policy evaluation.** As soon as public policies move into the agenda-setting stage, the evaluation process begins. The purpose of policy evaluation is to determine whether a program is achieving its stated goals and reaching its intended audience, what the program is actually accomplishing, and who is benefiting from it.

From almost the moment of their conception, public policies undergo both formal and informal evaluations by citizens, legislators, administrative agencies, the news media, academicians, research firms, auditors, and interest groups. Ideally, public policies should be evaluated *after* they have been implemented, using the best available research methods and according to a systematic plan. In the real world, policy evaluation seldom works this way. It is often undertaken without a preconceived plan and before the strategy chosen to solve a problem has been fully implemented.

6. **Policy termination.** A policy or program may be terminated for any of several reasons: the public need was met, the nature of the problem changed, government no longer had a mandate in the area, the policy lost political support, private agencies relieved the need, a political system or subgovernment ceased to function, or the policy was too costly. Determining when a policy should be terminated is somewhat subjective. At what point do you decide that the public's need has been met? What measures do you use to conclude that the problem was solved? Typically, policy termination represents a process of adjustment in which the policy makers shift their focus to other policy concerns. While some policy systems go out of existence, others survive and expand, bringing the policy cycle full circle to a redefinition of the public problem.[18]

The People Who Make Policy

The stages of the policy cycle outlined in Figure 2-1 do not correspond directly to the agencies and institutions involved in making government policy. We tend to assume that policy is formulated by legislatures and implemented by administrative agencies, such as the Public Health Service. In fact, administrative agencies sometimes formulate policy and legislatures become involved in policy implementation.

This point leads us to ask, "Who makes policy?" The authorities who "make policy" may be executives, administrators, or committees of an organization or company; elected officials; officers and employees of municipal, state, or federal agencies;

Street-level bureaucrats
Individuals within government who have direct contact with citizens.

members of Congress and state legislatures; and even **street-level bureaucrats:** welfare workers, public health nurses, police officers, schoolteachers, sanitation workers, housing authority managers, judges, and many other people working in government agencies. In the course of carrying out their jobs, these street-level bureaucrats daily make policy decisions by interpreting government laws and regulations for citizens. For example, as a community nutritionist, you will find yourself making policy decisions when you tailor a program to meet a particular client's needs or when you recommend a calcium supplement for a client, based on your interpretation of the new Dietary Reference Intakes (DRIs), described later in this chapter and in Chapter 4.

Legitimating Policy

Once it has been decided that a policy should be put into effect, a choice must be made about *how* it will be implemented. This is not a trivial decision. Consider, for example, the decision by the FDA to allow food product labels to carry health claims. Some consumers, scientists, and food companies objected to this policy, believing that some health claims on food product labels might distort research findings or be extravagant in presenting evidence of a link between food components and health. Others believed it would help the public make healthful food choices. Thus, a policy may be perceived as benefiting some citizens and working to the detriment of others. Because achieving universal agreement on a policy and its effects is impossible, careful attention must be given to the process by which policy decisions are made. This is the point at which legitimating policy is important.

Legitimacy is "the belief on the part of citizens that the current government represents a proper form of government and a willingness on the part of those citizens to accept the decrees of that government as legal and authoritative."[19] In this sense, legitimacy is mainly in the mind, for it depends on a majority of the population accepting that the government has the right to govern. In the case of FDA's health claim policy, the appearance of health claims on food labels indicates that consumers and food companies accept the FDA's *authority* to allow this action. In contrast, the failure of health care reform in President Clinton's first administration was a signal that the American people did not consider the federal government a legitimate provider of comprehensive health care.[20]

Government, then, must somehow legitimate each policy choice. Several mechanisms exist for legitimating policies: the legislative process; the regulatory process; the court system; and various procedures for direct democracy, such as referenda, which put sensitive issues directly before the people. The next section will explore the legislative and regulatory processes in greater detail, as most policies in the areas of health, food, and nutrition arise through these mechanisms.[*]

---

[*]The use of the court system for legitimating policy in the nutrition arena is not discussed in this chapter, although it is important in formulating food and nutrition policy. Regulations *are* challenged through the court system. When the FDA issued a final rule establishing nutrition labeling regulations in 1973, a portion of the regulations dealing with special dietary foods was challenged in the courts by the National Nutritional Foods Association (see *National Nutritional Foods Association v. FDA* and *National Nutritional Foods Association v. Kennedy* as cited in the *Federal Register* 1990 [July 19]: 29476–77).

The Legislative and Regulatory Process

Governments can use any number of instruments to influence the lives of their citizens: taxes and tax incentives, services such as defense and education, price supports for commodities, unemployment benefits, and laws, to name only a few. Laws are a unique tool of government. In the United States, we traditionally associate lawmaking with Congress, the primary legislative body. It is Congress that sets policy and supplies the basic legislation that governs our lives.

Laws and Regulations

The laws passed by Congress tend to be vague. A law defines the broad scope of the policy intended by Congress. For example, the Special Supplemental Nutrition Program for Women, Infants, and Children (WIC) was authorized by Public Law 92-433 and approved on September 26, 1972. This law authorized a two-year, $20 million pilot program for each of the fiscal years 1973 and 1974. It gave the secretary of agriculture the authority to make cash grants to state health departments or comparable agencies to provide supplemental foods to pregnant and lactating women, infants, and children up to 4 years of age who were considered at "nutritional risk" by competent professionals. (The WIC program presently covers children up to 5 years of age.) As written, this law, like most others, was too vague to implement. It did not define or specify which professionals would determine the eligibility of clients. It did not define the concept of "nutritional risk" and other eligibility requirements, the method by which clients would obtain food products, or other aspects of the proposed program. Sorting out these details was left to the USDA.

Thus, once a law is passed, it is up to the administrative bodies such as USDA to interpret the law and provide the detailed regulations or rules that put the policy into effect. These regulations are sometimes called "secondary legislation." The total volume of this activity is enormous, as seen by the size of the *Federal Register*, a weekly publication that contains all regulations and proposed regulations, and the *Code of Federal Regulations* (CFR), a compendium of all regulations currently in force. When the WIC program was started, the details of the regulations were not published in the *Federal Register* until July 11, 1973, nearly 9 months after Congress passed the law. Over time, new laws and amendments to the existing law were enacted to increase the amount of money allocated for the WIC program, specify the means by which the program should be implemented, and authorize the continuation of the program for additional budget years.[21] (A detailed discussion of the WIC program appears in Chapter 13.)

How an Idea Becomes Law

All levels of government pass laws. (At the local level, laws are sometimes called ordinances or bylaws.) The process by which an idea becomes law is complicated. It may take many months, or even years, for an idea or issue to work its way onto the policy agenda. Then, once it reaches the legislative body empowered to act on it, the formal rules and procedures of that body can delay decision making on a proposed bill or scuttle it altogether. Logjams are common, especially at the end of the legislative sessions. Lewis Carroll, writing in *Alice in Wonderland*, could have been speaking of the U.S. Congress when he wrote, "I don't think they play at all fairly, and they quarrel so dreadfully one can't hear oneself speak—and they don't seem to have any rules in particular: at least, if there are, nobody attends to them—and you've no idea how confusing it is."[22]

The general process by which laws are made is outlined in Figure 2-4, which shows the path a bill would take on its way through Congress. The process is much the same for bills introduced into state legislatures. The process begins when a concerned citizen, group of citizens, or organization brings an issue to the attention of a legislative representative at either the local, state, or national level. Typically, the issue is presented to private attorneys or the staff of the legislative counsel, who draft the bill in the proper language and style. A bill is introduced by sending it to the clerk's desk, where it is numbered and printed. It must have a member as its sponsor. Simple bills are designated as either "H.R." or "S." depending on the house of origin. For instance, the Nutrition Labeling and Education Act was designated "H.R. 3562," indicating that the bill was introduced into the House of Representatives. When the bill is introduced, the bill's title is entered in the *Congressional Journal* and printed in the *Congressional Record*.[23]

Markup session A congressional committee session during which a bill is put into its final form before being reported out of committee.

As bills work their way through the House and Senate, they are considered by several committees and subcommittees, which may hold public hearings and seek the testimony of interested persons or experts before deciding whether to move the bill forward. The bill is revised during a **markup session.** If the bill approved by the Senate is identical to the one passed by the House, it is sent to the president to be

FIGURE 2-4 *How a Bill Becomes a Law*

Source: W. Lyons, J. M. Scheb II and L. E. Richardson, Jr., *American Government: Politics and Political Culture*, p. 360. Copyright © 1995 West Publishing Company. Used by permission of Wadsworth Publishing Co.

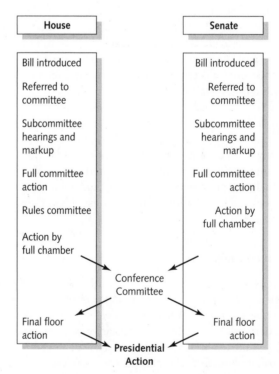

signed. If the two versions differ, a joint House-Senate conference committee is formed to modify the bill by mutual agreement. Once both houses agree on the compromise bill, it is sent to the president, who may sign it into law, allow it to become law without his signature, or veto it. When Congress overrides the president's veto by a two-thirds majority vote in both houses, the bill becomes law without the president's signature.[24] (In state legislatures, of course, the bill is sent to the governor.) When the bill is signed into law by the president, it becomes an *act* and is given the designation "PL," which stands for Public Law, and a number: the first two or three digits indicate the number of the congressional session in which the law was enacted, and the remaining digits represent the number of the bill. Recall that the bill authorizing the WIC supplemental feeding program became Public Law 92-433 (that is, bill number 433, enacted by the 92nd Congress).

Before a law enacted by Congress goes into effect, it is reviewed by the appropriate federal agency, which is responsible for issuing guidelines or regulations that detail how the law will be implemented and any penalties that may be imposed if the law is violated. These regulations are published in the *Federal Register*. Because federal law requires that the public have the opportunity to comment on an agency's proposed guidelines, the agency first issues "proposed regulations" such as the proposed rule shown in Figure 2-5. During the comment period that follows the publication of proposed regulations, the general public, experts, companies, and interested organizations submit their written comments and, in some cases, present their views at public hearings. From 30 to 60 days are allowed for comment, depending on the type and complexity of the regulations. At the end of the comment period, the agency reviews all comments, both positive and negative, before issuing its final regulations, which are incorporated into the CFR. The CFR and the *Federal Register* are available at most local libraries and county courthouses; recent rules and regulations can be found on the Internet. (See the boxed insert for a list of government Internet addresses.)

A list of standing and select congressional committees related to health, food, and nutrition appears in Appendix B.

The Federal Budget Process

Laws and regulations will have no effect unless there are funds to enforce them. Congress must enact bills to fund the programs and services mandated by federal legislation. The federal budget process has been described as "fractured, contentious, and chaotic," mainly because it forces the president and Congress to negotiate and agree on the problems that deserve top priority.[25] Budgets are designed to count and record income and expenditures, demonstrate the government's intention regarding the funding (and, more importantly, the priority) of programs, and control and shape the activities of government agencies. In its simplest form, the budget process has two stages: the president proposes a budget, and then Congress reacts to the president's proposal. The actual budget process is complex and cumbersome, with the final budget reflecting the distribution of power among competing concerns and groups within the current political system.[26]

The Language of the Budget. The budget is the president's financial plan for the federal government. It accounts for how government funds have been raised and spent, and it proposes financial policy choices for the coming fiscal year and some-

FIGURE 2-5 *A Portion of a Proposed Rule Published by the Food and Drug Administration in the Federal Register*
This *Federal Register* notice describes the FDA's proposal to authorize a health claim on labels about the relationship between folate and the risk of neural tube defects. The final rule authorizing this health claim was published on March 5, 1996 (*Federal Register* Vol. 61, No. 44) and became effective April 19, 1996.

Federal Register/Vol. 58, No. 197/

Thursday, October 14, 1993/Proposed Rules

DEPARTMENT OF HEALTH AND HUMAN SERVICES

Food and Drug Administration

21 CFR Part 101

[Docket No. 91N–100H]

RIN 0905–AB67

Food Labeling: Health Claims and Label Statements; Folate and Neural Tube Defects

AGENCY: Food and Drug Administration, DHHS.

ACTION: Proposed rule.

Summary: The Food and Drug Administration (FDA) is proposing to revise its food labeling regulations to authorize the use of a health claim about the relationship between folate and the risk of neural tube birth defects on labels or in labeling of foods in conventional food form or dietary supplements. This rule is being proposed in response to provisions of the Nutrition Labeling and Education Act of 1990 (the 1990 amendments) and the Dietary Supplement Act of 1992 (the DS Act) that bear on health claims. FDA has reviewed the scientific data in conformity with the requirements of the 1990 amendments and has considered recommendations provided by the Folic Acid Subcommittee of its Food Advisory Committee (the Folic Acid Subcommittee) as well as comments received. The agency has tentatively decided to authorize a health claim for folate and neural tube defects (NTD's) and to provide for safe use of folic acid in foods by amending several of its regulations that permit use of folic acid in foods.

DATES: Written comments by December 13, 1993. The agency is proposing that any final rule that may issue based upon this proposal become effective 30 days after the date of publication, except that for foods that are fortified with folic acid to ensure that the folic acid is safely used, any final rule authorizing a health claim will not be effective until the effective date of the amendment to the food additive regulation for folic acid proposed elsewhere in this issue of the **Federal Register.**

ADDRESSES: Submit written comments to the Dockets Management Branch (HFA–305), Food and Drug Administration, rm. 1–23, 12420 Parklawn Dr., Rockville, MD 20857.

FOR FURTHER INFORMATION CONTACT: Jeanne I. Rader, Center for Food Safety and Applied Nutrition (HFS–175), Food and Drug Administration, 200 C St. SW., Washington, DC 20204, 202–205–5375.

Receipts or revenue Amounts the government expects to raise through taxes and fees.

Budget authority Amounts that government agencies are allowed to spend in implementing their programs.

Budget outlays Amounts actually paid out by government agencies.

Entitlements Programs that require the payment of benefits to all eligible people as established by law.

times beyond. Financial or fiscal policy refers to the effect of government taxation and spending on the economy in general.[27] The budget describes the government's **receipts or revenue** (amounts the government expects to raise in individual income taxes, corporation income taxes, payroll taxes, sales and excise taxes, property taxes, and other fees), the **budget authority** (amounts that government agencies are allowed to spend in implementing their programs), and **budget outlays** (amounts actually paid out by government agencies, including funds spent for equipment or property or to pay employees' salaries).

The budget allocates funds to cover two types of spending. *Mandatory spending* is required by law for **entitlements**—that is, programs that require the payment of benefits to any person that meets the eligibility requirements established by law.[28] For these programs, Congress provides whatever money is required from year to year to maintain benefits for eligible people. Many of these programs are indexed to the cost of living or similar measures, resulting in steady increases in benefits with a rise

◎ Internet Resources

Canada

Agriculture and Agri-Food Canada's Electronic Information Service	**aceis.agr.ca/newintre.html**
Government of Canada	**canada.gc.ca**

United States

Information Locators

FedWorld	**www.fedworld.gov**
GPO* Access	**www.access.gpo.gov**
Thomas Legislative Information on the Internet	**thomas.loc.gov**

Federal Agencies

Office of Management and Budget	**www1.whitehouse.gov/WH/EOP/OMB/html/ombhome-plain.html**
Department of Health and Human Services	**www.hhs.gov**
Administration for Children and Families	**www.acf.dhhs.gov**
Administration on Aging	**www.aoa.dhhs.gov**
Centers for Disease Control and Prevention	**www.cdc.gov**
Food and Drug Administration	**www.fda.gov**
Health Care Financing Administration	**www.hcfa.gov**
Health Resources and Services Administration	**www.hrsa.dhhs.gov**
Indian Health Service	**www.tucson.ihs.gov**
National Institutes of Health	**www.nih.gov**
Substance Abuse and Mental Health Services Administration	**www.samhsa.gov**
U.S. Department of Agriculture	**www.usda.gov**
Agricultural Research Service	**www.ars.usda.gov**
Center for Nutrition Policy and Promotion	**www.usda.gov/cnpp**
Cooperative State Research, Education, and Extension Service	**www.reeusda.gov/csrees.htm**
Economic Research Service	**www.econ.ag.gov**
Food and Nutrition Service	**www.usda.gov/fcs**
Food Safety and Inspection Service	**www.fsis.usda.gov**

Continued

Internet Resources—*continued*

United States—continued

The Hill Sites

U.S. House of Representatives   **www.house.gov**

U.S. Senate   **www.senate.gov**

White House   **www.whitehouse.gov**

Political Party Sites

Democratic Party Headquarters   **www.democrats.org/party**

Republican National Committee   **www.rnc.org**

Government Publications Online

Code of Federal Regulations   **www.access.gpo.gov/nara/cfr/index.html**

Federal Register   **www.access.gpo.gov/su_docs/aces/aces140.html**

State and Local Government

Library of Congress State and Local
   Government Information   **lcweb.loc.gov/global/state/stategov.html**

NASIRE[†] State Search   **www.nasire.org/ss/index.html**

U.S. State and Local Government Gateway   **www.health.gov/statelocal**

---

*GPO = Government Printing Office.

†NASIRE = National Association of State Information Resource Executives.

in inflation. Entitlements such as Social Security, Medicare, food stamps, agricultural subsidies, and veterans' benefits—so-called "uncontrollable expenditures"—command a major portion (about two-thirds) of the federal budget. *Discretionary spending* refers to the remainder of the federal budget—that is, the budget choices that can be made in such areas as defense, energy assistance, and education after the mandatory allocations have been made.[29]

Principles of Federal Budgeting. The federal fiscal year begins on October 1 and runs through September 30 of the following year. The fiscal year is named for the year in which it ends; thus, fiscal year 1999, or FY99, begins October 1, 1998, and ends September 30, 1999. (States and municipalities differ in terms of fiscal years, with some following a calendar year [January–December], some starting on July 1, and others having two-year fiscal cycles.[30]) The fiscal year is the year in which money allocated in the budget is actually spent, but important steps must be taken both before and after the fiscal year that can affect the government's or an agency's programming. These steps form the budget cycle and include budget formulation, approval, implementation (sometimes called execution), and audit.

Figure 2-6 shows the federal budget cycle for FY99. The first step, budget formulation, begins 15-18 months before the start of the fiscal year. (For FY99, the budget process began in March, 1997). During this phase, the Office of Management and Budget (OMB), the central budget office at the federal level, works with federal agencies to outline their funding projections for new and ongoing programs. After this consultation process, a single budget document is prepared and released, usually in September. This document is revised by the president in November and December and becomes the basis of his budget message—the most important statement of his priorities and concerns—submitted to Congress in late January or early February. In his FY99 budget proposal, President Clinton asked Congress for $20 million in grants for states and nonprofit organizations to support food recovery and gleaning projects[31] and restoration of food stamps, cut to most legal immigrants in the welfare law signed in August 1996. He also requested an increase in funding for the WIC program.[32] Whether he gets what he asks for will not be known until Congress has reviewed and agreed on the budget many months later.

In general, Congress can approve, disapprove, or modify the president's budget proposal, adding or eliminating programs or altering methods of raising revenue. After the president's budget is submitted to Congress, it is reported to committees and subcommittees that must make decisions about budget authority, taxes, appropriations, and the reconciliation of the budget. In this process, Congress passes revenue bills that specify how funds to support the government's activities will be raised. In the House, the Ways and Means Committee has jurisdiction over revenue bills, making it one of the most powerful committees in Congress. In terms of spending, congressional committees must pass bills to authorize government programs. An **authorization** defines the scope of a program and sets a ceiling on how much money can be spent on it. Before money can be released to a program, however, an **appropriation** bill must be passed. The appropriation for a program may cover a single year, several years, or an indefinite period of time.

Authorization A budget authorization provides agencies and departments with the legal authority to operate.

Appropriation A budget appropriation is the authority to spend money.

FIGURE 2-6 *Federal Budget Cycle for* 1999 (FY99)

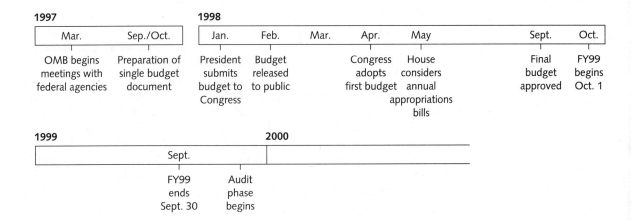

All revenue and appropriations bills passed by the House are forwarded to the Senate for consideration. Differences between the two houses are worked out in conference committee, and ultimately a *reconciliation bill* is passed. The end result of the authorization and appropriations work is that Congress adopts its version of the budget in a *budget resolution*. The first budget resolution is usually passed by May 15 and the second, after all spending bills have been passed, by September 15. If Congress is unable to pass a budget by the beginning of the fiscal year, it may adopt *continuing resolutions*, which authorize expenditures at the same level as in the previous fiscal year, until a budget agreement can be reached.

Fiscal year 1999 began October 1, 1998, marking the implementation stage of the budget cycle when government agencies execute the agreed-on policies and programs. At the end of FY99, the audit phase begins, during which time the agencies' operations are examined and verified. In recent years, the audit phase has come to include performance auditing, or determining whether the agency's goals and objectives were met and whether the agency made the best use of its resources.

The Political Process

The complexities of the legislative and policy-making process present many challenges, and years may be required to reach a critical mass in public support for a policy change. As a case in point, consider that until 1906 no *federal* legislation regulated the nation's food supply and protected consumers against food adulteration, mislabeling, and false advertising. Even then, more than 25 years of persistent pressure from consumers and the agricultural community had been required to achieve this legislative milestone.[33] Check the boxed insert for a description of the imperfect process leading to the passage of the Food and Drugs Act.

A more recent example of the legislative process is the campaign by the American Dietetic Association (ADA) supporting Medical Nutrition Therapy (MNT). MNT is a service provided by a registered dietitian or nutrition professional that includes counseling, nutrition support, and nutrition assessment and screening to improve people's health and quality of life. ADA launched its MNT campaign in 1992 with the publication of a position paper on health care services[34] and a report on the importance of including reimbursable nutrition services as part of any health care reform legislation.[35] It organized a grassroots campaign among its members to raise the visibility of the dietetics profession on Capitol Hill and to lobby Congress to support MNT,[36] and it formed coalitions with other health-oriented organizations to develop a uniform position on health care reform.[37] ADA members wrote letters to legislators, met with Washington lobbyists and key members of Congress,[38] and testified at a hearing held by the Health Subcommittee of the House Ways and Means Committee.[39] Advertisements, such as the one shown in Figure 2-7, helped educate consumers and policy makers about the importance and potential impact of MNT. ADA's efforts resulted in the introduction of the Medical Nutrition Therapy Act (H.R. 2247 and S. 1964) during the 104th Congress. Although the House version had 91 cosponsors and the Senate version had 4 cosponsors, both bills expired when Congress adjourned.[40] Even so, the bills'

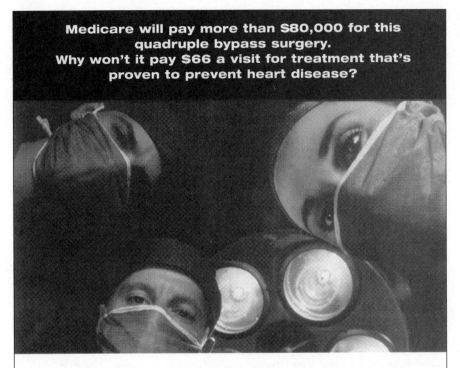

FIGURE 2-7 *An Advertisement for Medical Nutrition Therapy*

Source: Property of American Dietetic Association. Used with permission.

support encouraged ADA to continue its efforts. On April 17, 1997, the Medicare Medical Nutrition Therapy Act of 1997 (H.R. 1375 and S. 597) was introduced in the 105th Congress,[41] and ADA launched a new effort, the Majority by March Campaign, to secure a majority of congressional members as cosponsors of the legislation by March 1998.[42] After 6 years of effort, the ADA continues to work for passage of legislation which, in the words of the Honorable John E. Ensign, who introduced the bill in the House, "will help to save Medicare, and most importantly,

The Legislative Process in Real Life

Between 1880 and 1906, nearly 200 bills designed to protect consumers against food adulteration, misbranding, and false advertising were introduced into Congress without success. One such bill was submitted to the House and passed in 1904; the Senate began debating and eventually passed a similar bill, with amendments, in 1906. During the debates on the bill, Dr. Harvey W. Wiley, who was the chief chemist of the Department of Agriculture, testified about adulterated foods, bringing examples to show before Congress. At his urging, women who were concerned about food safety also lobbied Congress to pass the bill. When the bill went to the conference committee to iron out differences between the two houses, tensions were high. At one point, President Theodore Roosevelt felt compelled to lean on Congress and express his support for the pure foods bill. Finally, the bill passed both the House and Senate on June 27, 1906, and was signed into law by President Roosevelt on June 30, 1906. The Food and Drugs Act became law effective January 1, 1907.

Unfortunately, the law was defective. In the beginning, Congress failed to pass appropriation bills to provide the funds to enforce the law. Congress also failed to authorize the development of standards of food composition and quality, an omission that made it difficult for authorities to prove in court that a food was an imitation and not the genuine food product. Thus, because the Food and Drugs Act lacked teeth, food adulteration remained a threat to public health. Attempts to strengthen the law failed until 1938, when Congress passed the Federal Food, Drug, and Cosmetic Act, which was signed into law by President Franklin D. Roosevelt in 1938. It became effective one year later and is the primary legislation by which our food supply is regulated today.

Source: Adapted from H. W. Schultz, *Food Law Handbook* (Westport, CT: Avi, 1981), pp. 3–21.

to save lives."[43] When the bills pass Congress, the Department of Health and Human Services will craft the regulations that specify how dietitians will be reimbursed by Medicare under the law. Until these regulations are finalized, no nutrition professional will be reimbursed under the Medicare program.

Current Legislation and Emerging Policy Issues

In this chapter, we have seen that policies, and the laws and regulations derived from them, are the means by which public problems are addressed. We have also come to appreciate that policies are not static. Like the discipline of community nutrition itself, policies are dynamic, changing as conditions and circumstances change. In the food and nutrition arena, existing policies are constantly being challenged by market forces, scientific knowledge, and consumer practices and attitudes. Current legislation and emerging issues have the potential to affect the delivery of food and nutrition programs and the way in which community nutritionists work.

Current Legislation

Current legislation at the national and state level may affect the practice of community nutrition. The potential impact of current legislation is described next.

State Licensure Laws. Licensure is a state regulatory action that establishes and enforces minimum competency standards for individuals working in regulated professions such as dietetics. Licensure is designed to help the general public identify individuals qualified by training, experience, and testing to provide nutrition information and medical nutrition therapy.[44] States vary in their regulation of dietitians. The ADA's ongoing effort to achieve licensure in every state means that, eventually, community nutritionists who are dietitians must be licensed.

Consolidation of Food Agencies. The Safe Food Act of 1997 (S. 1465) was introduced to the Senate and referred to the Committee on Governmental Affairs. The corresponding House bill, H.R. 2801, was introduced and referred to the Agriculture and Commerce Committees.[45] The legislation called for the formation of a single, independent agency—the Food Safety Administration—that would be responsible for food safety, inspection, and labeling.[46] The Safe Food Act would consolidate the USDA's Food Safety and Inspection Service and FDA's Center for Food Safety and Applied Nutrition, along with the Commerce Department's National Marine Fisheries Service, the Environmental Protection Agency's Office of Pesticide Programs, and possibly other offices designated by the president.[47] Some education programs may be affected by this legislation.

Implementation of the Results Act. The Government Performance and Results Act was passed in 1993. It requires federal agencies to define their missions, goals, and objectives and to specify the performance measures they plan to use. That is, each federal program must demonstrate that it is doing what it is supposed to do and at an appropriate cost. The legislation is designed to make federal programs more accountable to stakeholders such as Congress and the general public. The act requires that agencies submit an annual performance plan along with their budgets and that this government-wide performance plan be made part of the president's budget. The first government-wide plan was sent to Congress in February 1998 for FY99.[48] This legislation may result in changes to the delivery of some programs, as agencies work to link program goals with program outcomes.

Emerging Issues

National policy in the food and nutrition arena will likely be affected in the coming years by advances in science. New medical and biochemical findings often present challenges for policy makers, as outlined here.

Dietary Reference Intakes. The Dietary Reference Intakes (DRIs) are the new standard for evaluating nutrient intake and diet quality in the United States. (In Canada, the DRIs will be adopted by Health Canada and eventually replace the Recommended Nutrient Intakes [RNIs]). The DRIs represent a major shift in the

More information about
the Dietary Reference
Intakes (DRIs) appears in
Chapter 4.

policy of the last 50 years, because they focus on intakes designed to prevent chronic diseases such as osteoporosis, rather than on recommended intakes to avoid nutrient deficiencies such as pellagra and beriberi.[49] DRIs for calcium, phosphorus, vitamin D, fluoride, and magnesium were released in 1997 and for folate and other B vitamins in 1998. The new reference values will affect nutrition labeling, the nutrient requirements for food programs, and the Dietary Guidelines.

Labeling of Dietary Fats. Under the Nutrition Labeling and Education Act of 1990,[50] every nutrition label must list the amount of total fat and saturated fat per serving. A food manufacturer may list voluntarily the amounts of monounsaturated fat and polyunsaturated fat provided in a serving. The FDA allows no other fatty acids or groups of fatty acids to be listed in the nutrition facts panel, nor does it allow any claims about specific fatty acids or groups of fatty acids on the label. However, several unresolved issues surround fatty acid labeling, including whether and how to label the following: omega-3 and omega-6 fatty acids, *trans* fatty acids, stearic acid and other fatty acids that do not raise blood cholesterol, partially absorbed fats, claims for specific fatty acids such as γ-linolenic acid, and fatty acids found in foods derived from biotechnology. Beyond the issue of whether information about such fatty acids should appear on food product labels, there are complex issues surrounding the definition of a fatty acid and whether the general public can be properly educated about these nutrients.[51]

Biotechnology. Biotechnology uses DNA recombinant technology to custom design protein molecules and other compounds of great purity.[52] Biotechnology is not new; it has been practiced for more than 8,000 years. Early applications of food biotechnology include the production of vinegar, alcoholic beverages, sourdough, and cheese. Today, biotechnology has many applications in the dairy, baking, meat, enzyme, and fermentation industries.[53] In the last decade, plant biotechnology has been applied to producing plants that are resistant to viruses, insects, fungi, and herbicides. From a regulatory standpoint, the development and testing of genetically modified plants are monitored by the FDA, USDA, Environmental Protection Agency (EPA), and most state governments.[54] Both FDA and USDA view biotechnology as no different from any other food manufacturing process.[55] The FDA has approved as safe several genetically engineered crops, including herbicide resistant soybeans, potatoes that resist a damaging beetle, and a virus-resistant squash. One genetically engineered food for consumers approved by FDA is the Flavr-Savr tomato.[56] Future biotechnology goals are to improve the nutritional quality of plants, increase harvest yield, and produce special oils, carbohydrates, and proteins.[57] New advances in this area will continue to challenge existing regulations.

Complementary and Alternative Medicine. Complementary and alternative medicine (CAM) is commonplace in many parts of the world, where it is accepted as appropriate therapy. In North America, CAM has emerged recently as a potential adjunct approach to traditional Western medicine, mainly owing to its adoption by consumers who have embraced it as an alternative to the invasive treatments typical of Western medical practice today. The National Institutes of Health defines CAM as "those treatments and health care practices not taught widely in medical schools, not generally available in hospitals, and not usually reimbursed by medical insurance companies."[58] CAM practices include acupuncture, homeopathy, herbal therapy, manual healing methods such as reflexology and chiropractic, methods of controlling the mind and body such as meditation and biofeedback, pharmacological and biological treatments such as chelation therapy, and dietary therapies such as macrobiotics and nutritional supplements. The main objection to CAM is that few controlled, clinical studies of its safety and efficacy have been conducted.[59] The growth in CAM practices will likely challenge existing policies related to health care delivery and the practice of dietetics.

Nutraceuticals. The term *nutraceutical* is used to describe food products created by new technologies and scientific developments. A proposed definition for nutraceutical is "any substance that may be considered a food or part of a food and provides medical or health benefits, including the prevention and treatment of disease."[60] Under this definition, nutraceuticals would include nutrients, dietary supplements, herbal products, genetically engineered "designer" foods, and some processed foods. Nutraceuticals on display at the Institute of Food Technology's 1997 Food Expo ranged from whey proteins with varying contents of proteins and fats to a green drink mix made from broccoli, asparagus, tomato powder, and chorella rich in lycopene, lutein, phenols, indoles, folic acid, and sulfurophanes. Exotic botanical extracts such as bitter melon, catawba, and guarana were also featured. The wide range of foods and ingredients classified as nutraceuticals suggests some controversy over whether they are "healthy" or "healing" foods.[61]

Medical Foods. Medical foods are foods designed to provide complete or partial nutrition support to individuals who have special physiological or nutritional needs or who cannot ingest food in a conventional form. They refer to enteral formulas used to feed hospitalized patients or foods for people with rare diseases.[62] Medical foods are administered under the supervision of a physician and thus differ from the foods and beverages consumed by healthy adults. Under current law, medical foods come under less scrutiny than virtually all other foods. They are not subject to nutrition labeling requirements under the 1990 NLEA because Congress exempted them. This exemption means that medical foods can be sold without standard nutrition labeling, may bear health claims not approved by FDA, and may not have been manufactured under strict quality controls.[63]

Policies in the food and nutrition arena will continue to evolve as our knowledge of foods and their relationship to health expands and the issues of public concern change. The broad scope of food and nutrition policy provides ample opportunity for you to become involved in the policy process.

The Community Nutritionist in Action

Whether the issue is food safety legislation, health care reform, licensure of registered dietitians, or funding of the School Lunch Program, there are many ways in which you, as a community nutritionist, can influence the policy-making process.

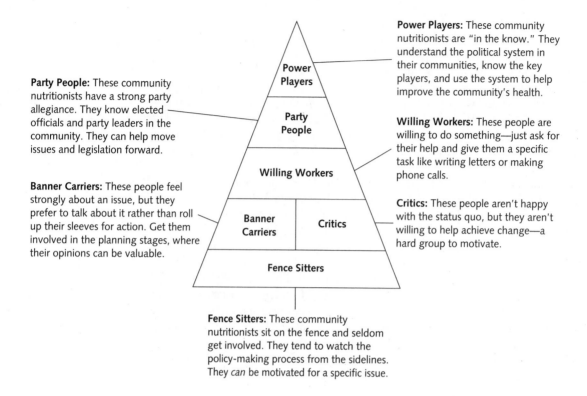

Power Players: These community nutritionists are "in the know." They understand the political system in their communities, know the key players, and use the system to help improve the community's health.

Party People: These community nutritionists have a strong party allegiance. They know elected officials and party leaders in the community. They can help move issues and legislation forward.

Willing Workers: These people are willing to do something—just ask for their help and give them a specific task like writing letters or making phone calls.

Banner Carriers: These people feel strongly about an issue, but they prefer to talk about it rather than roll up their sleeves for action. Get them involved in the planning stages, where their opinions can be valuable.

Critics: These people aren't happy with the status quo, but they aren't willing to help achieve change—a hard group to motivate.

Fence Sitters: These community nutritionists sit on the fence and seldom get involved. They tend to watch the policy-making process from the sidelines. They *can* be motivated for a specific issue.

FIGURE 2-8 *Grassroots Pyramid*

Source: Adapted from the American Dietetic Association's *Grassroots Targeting* (1997).

Whether you are new to this effort or not, consider which category of involvement shown in the grassroots pyramid in Figure 2-8 best describes your current level of involvement.[64] Are you a fence sitter who doesn't think about the big issues in community nutrition? Or are you a "banner carrier" who has strong opinions about issues but would rather talk about them than take action? What kind of player will you be in the future? No matter where you fall on the grassroots pyramid, there are opportunities in community nutrition for you to move up one level of involvement—or even two! Some strategies for influencing and becoming involved in the political process are described next.

Make Your Opinion Known

Expressing your opinion about an issue is one way of influencing the political process. When the issue is important to you, present your ideas and opinions at a public meeting or write a letter to the editor of a newspaper, magazine, or scientific journal. For example, a July 1997 article in the *Globe and Mail* prompted one of us to submit a letter, shown in Figure 2-9, to the editor decrying the article's scare tactics about canola oil and its role in the Canadian diet. Most of the letter's middle paragraph, along with comments from other concerned professionals, was published the following week and helped educate the newspaper's readers about the safety of canola oil. Notice that the letter was written on business stationery, mentioned the date of the original article, and was short.

main~stream NUTRITION

July 11, 1997

Diane H. Morris
Ph.D., R.D.
President

Mr. William Thorsell
Editor-in-Chief
The Globe and Mail
444 Front St. W.
Toronto, ON   M5V 2S9

Dear Mr. Thorsell:

The recent Middle Kingdom article, "Is the hype about canola just snake oil?"
(July 9, 1997), is a good example of bad journalism. The author fails to
present a balanced view of genetic engineering, uses scare tactics to get his
message across and did not appear to interview a trained nutritionist or
registered dietitian about dietary fats and oils or the role of canola oil in a
healthy diet.

The article's alarmist tone is especially unfortunate, as Canada has one of the
safest food supplies in the world. Moreover, there is no evidence that
consumption of genetically engineered foods increases one's risk of cancer,
coronary heart disease or stroke -- the nation's leading causes of death and
disability. Indeed, there is plenty of evidence that these diseases result from
lifestyle choices such as smoking, being inactive and eating a diet high in
calories, fat and saturated fat. The notion that "refined oils" are akin to soft
drinks, while olive oil is akin to orange juice, makes no sense; and trans fatty
acids are not toxic elements of our food supply to be avoided at all costs.
They do raise blood cholesterol, but not to the same extent as saturated fats.

Given the high level of confusion among consumers about what constitutes a
healthy eating pattern, your readers would have been better served by an
article that helped them identify low-fat foods and make food choices for good
health. Unfortunately, as my former boss at the Harvard School of Public
Health used to say, "Common sense doesn't sell."

Sincerely,

Diane H. Morris

Diane H. Morris, Ph.D., R.D.

FIGURE 2-9 *Letter to the Editor of the* Globe and Mail, *July 16, 1997, Toronto, ON. Used with permission.*

Use these principles of letter writing when responding to proposed rules and regulations. When writing the FDA, for example, refer to the proposed rule's publication in the *Federal Register*, outline your concerns, and keep your comments as short
and direct as possible. Use personal, institutional, or company letterhead as appropriate. An example of a letter to the FDA is given in Figure 2-10.

Become Directly Involved

When you become directly involved in policy making, *you* become a political actor
and run for political office, sponsor a referendum, or initiate a campaign to bring an
issue to the attention of the general public or policy makers. For example, you might
seek an elected position within your state or the national dietetic association as one

FIGURE 2-10 *Letter to the Food and Drug Administration*

Marie A. Boyle, Ph.D., R.D.
1412 Ashland Terrace
New York, NY   13904

July 6, 1997

Dockets Management Branch (HFA-305)
Food and Drug Administration
12410 Parklawn Dr., Room 1-23
Rockville, MD   20857

RE: FDA Docket No. 95N-0304

Comments in response to the Food and Drug Administration's request for comments as published in the June 4, 1997, *Federal Register*, Vol. 62. No. 107: Dietary Supplements Containing Ephedrine Alkaloids: Proposed Rule.

Dear Sirs,

I am writing to support the FDA's actions to reduce the risks to consumers of consuming products containing ephedrine alkaloids by proposing limits on the amount of ephedrine alkaloids allowed in dietary supplements and by requiring warning labels related to the recommended length of use of such dietary supplements.

I am a registered dietitian who regularly receives queries from clients about the safety and use of supplements containing ephedrine alkaloids. Even though I discuss my safety concerns with them, many of them retain the impression that if the product is on the market, it must be safe. A suitable warning from the FDA may cause some of my clients to think twice about using these supplements indiscriminately.

I support the agency's actions to regulate such dietary supplements more closely, especially where there is little evidence of safety and efficacy.

Thank you.

Sincerely,

Marie A. Boyle

Marie A. Boyle, Ph.D., R.D.

means of influencing the practice of dietetics, or you might participate on a local advisory board that is in a position to influence the political process in your community. You might organize the collection of signatures for a petition to be sent to your state legislature or city council, or you might work to elect a candidate to political office. Perhaps you help fund-raise for a local politician. Getting involved directly in the political process requires time and energy, but it can be rewarding.

Join an Interest Group

Joining an interest group is another way of becoming involved in policy making. **Interest groups** are pressure groups that try to influence public policy in ways that are favorable to their members. They consist of people who work together in an organized manner to advance a shared political interest.[65] Interest groups may exert pressure by persuading government agencies and elected officials to reach a particular decision, by supporting or opposing certain political candidates or incumbents,

Interest group A body of people acting in an organized manner to advance shared political interests.

through litigation, or by trying to shape public opinion. While interest groups are sometimes accused of being bureaucratic and power-hungry and of not always representing their constituents fairly, they do contribute to the political process. Among other things, they encourage political participation and strengthen the link between the public and the government. In addition, they provide government officials with valuable technical and policy information that may not be readily obtained elsewhere.[66]

There are different types of interest groups. Business, for example, is well represented on Capitol Hill by lobbyists from food companies, such as Kellogg's, Procter & Gamble, and the Coca-Cola Company, and from trade associations, such as the American Meat Institute, the Sugar Association, and the National Soft Drink Association. Professional groups, such as the American Dietetic Association and the American Medical Association, also have certain policy concerns. "Public interest" groups are a special category, in that they work to achieve goals that do not directly benefit their membership and serve to inform, educate, and influence the legislative process.[67] The Sierra Club, Common Cause, and the Food Research and Action Center are well-known public interest groups. Still other interest groups may represent particular segments of the population, such as blacks (for instance, the National Association for the Advancement of Colored People) and women (such as the National Organization for Women). Activities to influence the political process even occur within the government itself. The National Governors' Association and the Hall of the States, for example, have been active in trying to set and direct the policy agenda.

Work to Influence the Political Process

Interest groups and their memberships use a number of tactics to influence the political process. Litigation is increasingly being used to change policy through the court system. Filing a class action suit in court on behalf of all persons who might benefit from a court action is one example. Another tactic is the public relations campaign, in which interest groups try to influence public opinion favorably. Three other common tactics are political action committees, lobbying, and building coalitions.

Political Action Committees (PACs). A political action committee or PAC is the political arm of an interest group. It has the legal authority to raise funds from its members or employees to support candidates or political parties. The purpose of a PAC is to help elect candidates whose views are favorably aligned with the group's mission or goals. PACs also work to keep the lines of communication open between policy makers and the interest group's membership.[68] The PAC of the American Dietetic Association strives to influence the policy-making process and its outcome.

Lobbying. Lobbying is often the method of choice when trying to influence the political system. "It is probably the oldest weapon, and certainly one of the most criticized."[69] Lobbying has acquired a negative connotation, giving rise to images of backdoor, professional power brokers who bend the political process through large campaign contributions. Although this image may be appropriate in some situations, it does not apply to all lobbyists. Remember, **lobbying** means talking to public

Lobbying Providing information to elected officials.

officials and legislators to persuade them to consider the information you provide on an issue you believe is important.[70] Lobbyists' experience, knowledge of the legal system, and political skills make them an important part of the political process. They provide technical information to policy makers, help draft laws, testify before committees, and help speed (or slow) the passage of bills.

When this tactic is used, three issues are important: deciding how to lobby, knowing whom to lobby, and determining when to lobby. One of the first decisions is whether to lobby directly or hire a registered lobbyist. In some cases, you or your organization may not need a registered lobbyist to achieve your goal; in other situations, you may not be successful without the knowledge of the political machine and its players that a registered lobbyist offers.

Knowing whom to lobby and when are also critical decisions. To lobby successfully, identify the politicians or elected officials who are in a position to act on your concern by studying the formal structure of the government and its agencies, reading the newsletters of interest groups, and talking with policy makers who share your concern. Sometimes, reporters who cover certain events can help you identify the proper authority figures. Once you know whom to lobby, choose the right time. You will accomplish little if you lobby a legislator when a bill is up for the final reading instead of when it was being considered in subcommittee. Use a triggering event to bring an issue to a politician's attention. If possible, turn the triggering event into an opportunity for placing the issue on the policy agenda. Above all, remember that you will have to lobby for as long as it takes, even if the process requires years. Establishing rapport and building recognition for your concern among elected officials take time.

Having determined where and when to lobby, you must consider how to do it most effectively. Four points are helpful when trying to influence public officials:

1. Show that you are concerned about the official's image. You want to make it easy for the politician to give you what you want, and *at the same time*, you want to make the politician look good. You must not appear to be applying force directly. Instead, provide compelling reasons why the politician should support your proposal.
2. Accept the constraints under which elected officials work. Politicians must deal daily with many pressing issues and diverse groups, including their constituents and staff, lobbyists, party politics, committee leadership, fund-raising, campaigning, and the media. All of these place demands on the official's time. Effective lobbyists recognize the cross pressures politicians face and, when possible, develop strategies for reducing those pressures.
3. Consider reaching elected officials indirectly. Politicians can be influenced through their staff, campaign workers, former colleagues, financial contributors, business associates, and friends. Approaching someone among these groups may speed your ability to refine your proposal and avoid creating a problem for the politician.
4. Provide information for the official through letters, phone calls, and meetings. Ralph Nader, the consumer advocate, once remarked, "Talking frequently to legislators is the best way to persuade them of your position; the importance of this simple method cannot be overstated."[71]

Building Coalitions. An organization is sometimes too small or isolated to influence the political system effectively. It can better achieve policy changes by working with other organizations toward a common goal. Depending on the scope or depth of the cooperative effort, the joint venture may be a formal coalition or a more informal network or alliance. Formal coalitions tend to arise in geographical areas where problems affect many people across different communities. Coalitions may bring together social service organizations, church groups, professional associations, neighborhood groups, and businesses to develop a long-term, joint commitment to solving problems. The challenge is getting such diverse groups to agree on which problems deserve immediate attention.

Networks tend to arise when different organizations across the country share a variety of problems. A network may support a permanent staff in one location and have a common training and information system, but individual members of the network may pursue different problems. In general, network members share a common philosophy about how to mobilize people for action. Alliances, by comparison, tend to bring together organizations that are dispersed geographically to address one specific problem. The level of participation in alliance activities tends to wax and wane according to the urgency of the issue.

Coalitions, networks, and alliances are usually formed to increase the pressure on the political system. By joining forces, organizations can launch more effective public information and media campaigns to mobilize public support for an issue and bring it to the attention of policy makers. The ADA, for example, has more than 200 partners in government, industry, and health care. One of ADA's new partnerships is with the Human Genome Education Model II Project, a program designed to increase the knowledge and skill levels of dietitians, occupational therapists, psychologists, social workers, and other professionals in providing genetic counseling.[72] These activities enhance the ADA's efforts to educate the public about how food choices prevent and delay progression of chronic disease. In Manitoba, Canada, a new Alliance for the Prevention of Chronic Disease was formed in 1997 to improve the health of Manitobans, prevent chronic disease, and reduce total spending on health care. The alliance consists of six nonprofit health charities that focus on five major chronic diseases: cancer, diabetes, and heart, kidney, and lung diseases. The first organization of its kind in Canada, the alliance plans to work with regional health authorities to help shift the emphasis to disease prevention and health promotion from the traditional medical model.[73]

Take Political Action

Community nutritionists *are* lobbyists at the local, state, and national levels. Here are a few strategies to keep in mind when trying to influence the political process.

Write Effective Letters. A personal letter to an elected official from a constituent can be a powerful instrument for change. Public opinion is important to politicians, and they take note of the number of letters received on a particular issue. Consider the following points when writing to your elected official:[74]

- Get the elected official's name right! Misspelling his or her name detracts from your credibility.

- Limit your letter to one page. Letters should be typed or handwritten legibly.
- Write about a single issue.
- Refer to the legislation by bill number and name. Note the names of sponsors, and refer to hearings that have been held.
- Explain how a legislative issue will affect your work, your organization, or your community.
- Use logical rather than emotional arguments in support of your position. Let the facts speak for themselves.
- Ask direct questions and request a reply.
- Be cooperative. Offer to provide further information. Do not seek confrontation, but don't hesitate to ask for your legislator's position on the issue.
- Follow up. Congratulate your legislator for a positive action or express concern again if the legislator acted contrary to your view on the issue.
- Whenever possible, use an example from your local community to draw attention to the issue.
- Write as an individual rather than as a member of the ADA (although you should identify yourself as a registered dietitian).

Many state dietetic associations prepare sample letters to send to elected officials. The letter in Figure 2-11, for instance, was prepared by the Arkansas Dietetic Association as a sample letter to Senator Dale Bumpers seeking his support for Medical Nutrition Therapy legislation. An identical letter could be sent to a member of the House of Representatives, except that the salutation should be changed to read "Dear Congressman [or "Congresswoman"] _____." In either case, use the sample letter as your guide, but change it to make it more personal. Elected officials do not enjoy getting dozens of form letters in the mail!

Make Effective Telephone Calls. Getting through to a legislator or other elected official can be difficult. When the opportunity presents itself, remember the following points when phoning an elected official:

- Write down the points you wish to make, your arguments supporting them, and the action you want the legislator to take.
- Don't expect to speak directly with the legislator. Contact the staff person responsible for the issue you want to address.
- Request a written response so that you have a record of the legislator's position on the subject.

Work with the Media. Reporters and journalists with radio, television, and newspapers can help build support for your position on an issue. Get to know the key media representatives in your community. When they call about an issue, be prepared to answer their questions about the issue and its impact on the community. When appropriate, prepare a press release to alert them to activities around an issue. Press releases should be short (no longer than two pages, double-spaced), concise, provide one or two quotes by a key spokesperson on the issue, and give a contact's name, address, and phone number.

April 11, 1997

The Honorable Dale Bumpers
U.S. Senate
Washington, D.C.   20510

Dear Senator Bumpers:

As a dietetics professional and voter in your state, I urge you to support legislation to provide Medicare coverage for medical nutrition therapy (MNT) services by registered dietitians. MNT is critical to treating and preventing many diseases that afflict the elderly and saves precious Medicare dollars.

MNT is the assessment of patient nutritional status followed by therapy, ranging from diet modification to the administration of specialized nutrition therapies such as intravenous or tube feedings. Clinical data show that MNT provided by dietitians helps speed recovery and healing for patients with a range of conditions, including cardiovascular disease, diabetes, and bed sores. These three conditions alone cost Medicare billions of dollars every year. My colleagues and I can help greatly reduce the cost of these maladies, but we are prevented from doing so because Medicare will not pay for our professional services.

In the near future, Senators Jeff Bingaman (D-NM) and Larry Craig (R-ID) plan to introduce legislation to allow Medicare coverage of medical nutrition therapy. I urge you to support the legislation as an original cosponsor. I also ask you to encourage your other colleagues in Congress to support this measure with their cosponsorship and votes.

Thank you for your consideration on this issue. If Congress is going to successfully fix the ailing Medicare program, cost effective services like MNT simply must be available to our seniors. Please let me know what action you take on this issue so that I can inform my colleagues and patients of your position.

Sincerely,

Mary Collins

Mary J. Collins
428 South Fourth Street
Benton, AR   72015

FIGURE 2-11 *Sample Letter to a Senator Asking for Support of the Medical Nutrition Therapy Bill. Used with the permission of the Arkansas Dietetic Association.*

Political Realities

Reread the quote at the beginning of this chapter. What does it mean? In the political realm, it means that an issue isn't important on Capitol Hill until it is important back home. It means that constituents can have more influence over elected officials than party officials have. It means that a local example of a problem carries more weight with policy makers than a national statistic. It means that elected officials measure the importance of an issue by the number of messages they receive about it—which explains why your letters and political activities count.

Getting involved in the policy-making process is one way to strengthen your connections with other people and with your community. It can be chaotic at times, but knowing that your effort as an individual improved the health or well-being of people in your community can also provide great personal satisfaction. You *can* make a difference in your community by understanding the policy-making process, taking time to express your opinion, and being persistent and patient.

Community Learning Activity

This Community Learning Activity is designed to help you become acquainted with the policy-making process in foods and nutrition. Activities 2 and 3 ask you to use the Internet to facilitate the learning process.

Activity 1

Go back to the introduction to this chapter and choose one of the issues outlined there (for example, "Are dietary guidelines developed for adults appropriate for children?"). Assume that your organization has been asked to respond to legislation on this topic. Carry out the following activities:

1. Briefly outline your position on this issue, giving at least two reasons in support of it.
2. List the individuals and organizations in your community that you would contact to build a coalition on the issue. Why are these particular people and groups important?
3. Draft a letter to your representative in Congress outlining your concerns about the issue and your position on the policy.
4. How would you go about getting the media involved in this issue?
5. Gather and evaluate the positions of your classmates on the particular issue you selected. Provided that at least one opinion differs from your own, was your position on the issue influenced or changed by your classmate's position? What does this tell you about the process of formulating a position on an issue?

Activity 2

Access the Web site for Thomas Legislative Information on the Internet at thomas.loc.gov. Search for current legislation on a topic of interest to you. You may choose one of the following, if you prefer:

welfare reform	food labeling
dietary supplements	health claims
WIC	child health
food biotechnology	food safety
organic foods	medical foods

Complete the following activities:

1. Prepare a summary of the legislation, showing the sponsor(s) and/or cosponsor(s), current status, and essential points of the legislation.
2. Review the bills from the most recent (for example, the 103rd and 104th) Congresses. Was action taken in this area in the past? If so, what was the outcome of the action?
3. Locate the names of your state's senators and congresspersons.
4. Was the legislation sponsored by a representative from your state? (Refer to the boxed insert for Internet addresses, which will help you locate your state representatives.)

Activity 3

Access the following sites to obtain information about your state government and how it organizes health services:

State and Local Governments   lcweb.loc.gov/global/state/stategov.html
National Association of State
   Information Resource Executives   www.nasire.org

Complete the following activities:

1. Prepare an organizational chart of your state government.
2. Prepare an organizational chart of your state department of health.
3. Prepare a short analysis of the types of information available at these sites (for example, health data on your state's population, the name and address of the health commissioner, and so forth).
4. List and discuss the health and nutrition initiatives undertaken by your state health department.

Professional Focus   *The Art of Negotiating*

Whether we like it or not, we are all negotiators. Every day, we negotiate with family, friends, and coworkers to get something we want. Although each of us negotiates something every day, most of us don't negotiate particularly well. We tend to find ourselves in situations where the negotiations leave us feeling frustrated, taken advantage of, dissatisfied, or just plain worn out. This is unfortunate, because negotiation is the heart of all business deals.

The reason that people sometimes emerge from negotiations feeling this way is that they tend to see only two ways to negotiate: hard or soft.[1] Hard negotiators view the other participants as adversaries. Their goal is victory—their side wins, while the other side loses. Hard negotiators tend to distrust the other party and make threats. They perceive the negotiation process as a contest of wills. Soft negotiators, by comparison, think of the participants as friends. Their goal is agreement among the parties. They tend to be trusting, avoid a battle of wills, change positions easily, and make concessions to maintain the relationship. Many of us use soft negotiation tactics in our dealings with our parents, siblings, friends, or other persons who are important to us.

Are these the only ways of negotiating? There is a better way to negotiate than either the soft or the hard approach, according to Roger Fisher and William Ury of the Harvard Negotiation Project. The method they developed is called *principled negotiation*. Its main precept is that a decision about an issue should be based on its merits, not on what each side says it will or will not do. The method can be boiled down to four basic elements:

- People—Separate the people from the problem.
- Interests—Focus on interests, not positions.
- Opinions—Generate a variety of possibilities before deciding what to do.
- Criteria—Insist that the result be based on some objective standard.

Separate the People from the Problem

When people sit down at the bargaining table, they bring with them certain perceptions about the relationships among the participants and about the problem itself. Whether these perceptions are accurate or false, they pervade the proceedings. There is a tendency to confuse the participants' relationships with the "issue" or "problem."

One of the first steps to take in negotiating is to separate the problem from the people and deal with relationship and problem goals separately. This means thinking about how to get good results from the negotiation and what kind of relationship is likely to produce those results.

Another challenge to negotiating is having to deal with a problem when emotions are running high. People sometimes come to the bargaining table with strong feelings. They may be angry or frightened, feel threatened or misunderstood, or be worried about the outcome. A good way to handle the emotional aspects of the negotiation process is to recognize the emotions and give them legitimacy. The bottom line, write Fisher and Brown, is to "do only those things that are both good for the relationship and good for us."[2]

Focus on Interests, Not Positions

The purpose of negotiating is to serve our interests. Interests motivate people to reach certain decisions. The primary problem in most negotiations is not the difference in positions but the conflicts between the two sides' needs, fears, desires, and concerns—in other words, interests. In addition, most of us tend to think that the other party's interests are similar to our own. This is almost never the case. When negotiating, begin by defining, as precisely as possible, your own interests and allow the other participants to define theirs. Work through the discussion until mutual interests are identified.[3]

Consider a Variety of Options

We sometimes approach a negotiating session with only one outcome in mind. We operate with blinders on and fail to see other dimensions to the problem that may be a source for possible solutions. To get around this barrier, bring the participants together to brainstorm about potential solutions and options. In a good brainstorming session, judgments about possible solutions are suspended, and everyone involved in the negotiation is allowed to contribute ideas. At the end of the session, the parties discuss the various options, picking several that offer the most promise. The parties allow themselves time to evaluate each of these "best and brightest" ideas and consider which of them, if any, would best suit their purpose. Once again, when exploring options, consider the interests of both parties.

Use Objective Criteria

Suppose your roommate wants to buy your car, but you cannot agree on a price. She thinks that your asking price is too high; you think her offer is too low. Where do you go from here? One option is to consult the Blue Book price for your car's make and model. Another is to examine the newspaper listings of used cars for sale to determine the asking price for cars like yours. These options are the criteria or standards that help you reach an agreement. The type of criteria you use will depend on the nature of the issue being negotiated. In this case, the criterion was the fair market value of your car. In other situations, a court decision, tradition, precedent, scientific judgment, cost, or moral standard might serve as well. The important thing is to choose an objective standard that all parties are comfortable with.

Build Good Relationships

The negotiating process is much like a tango—a little give and take on both sides. Regardless of the issue, a "good" negotiation is fueled by a good relationship. When next you enter into a negotiation, take a few minutes to evaluate your relationship with the other person or party. The questions below will help you determine how good your working relationship is and where improvements can be made. With practice you can help build good working relationships.

Does Our Relationship Work?*

Do we want to work together?

In a good relationship, people want to work together. They respect each other and actively pursue strategies for sorting out differences. They work to keep problems to a minimum.

Are we reliable?

Good relationships are built on trust and constancy. All parties have confidence that verbal and written commitments will be kept. The parties work to allay any concerns about trustworthiness.

Do we understand each other?

Even in the best of working relationships, there will be differences of opinion, values, perceptions, and motives. In good relationships, the parties strive to accept each other and work toward understanding their differences.

Do we use our powers of persuasion effectively?

In good relationships, persuasion is used to influence and inform the other party about an issue or proposed action. The parties refrain from using coercive tactics and rely instead on rational, logical discussions of the merits of a particular position or action.

Do we communicate well?

Good communication is based on sound and compassionate reasoning. In good relationships, sensitive issues can be discussed in a supportive environment, one where candor is valued. The parties in a good relationship communicate often, consult each other before making decisions, and practice "active listening," where the parties work to hear each other with an open, flexible mind.

---

*Sources: J. D. Batten, *Tough-Minded Leadership* (New York: AMACOM, 1989), pp. 122–32; M. DePree, *Leadership Is an Art* (New York: Doubleday, 1989), pp. 89–96; and R. Fisher and S. Brown, *Getting Together—Building a Relationship That Gets to Yes* (Boston: Houghton Mifflin, 1988), pp. 178–79.

Work Toward Success

Good negotiating means that all parties leave the bargaining table feeling they have won. The successful negotiation is one in which the outcome meets both parties' interests. It is seen to be fair. The solution was arrived at with an efficient use of everyone's time. Neither party feels that they are at a disadvantage. And the solution will be implemented according to plan. A successful negotiation leaves the parties feeling respect for their counterparts and a desire to work together again.[4]

References

1. The discussion of hard and soft negotiation was adapted from R. Fisher and W. Ury, *Getting to Yes—Negotiating Agreement Without Giving In* (New York: Penguin, 1981), pp. 8–9.
2. The quotation was taken from R. Fisher and S. Brown, *Getting Together—Building a Relationship That Gets You to Yes* (Boston: Houghton Mifflin, 1988), p. 38.
3. K. Albrecht and S. Albrecht, Added value negotiating, *Training* 39 (1993): 26–29.
4. J. Allan, Talking your way to success, *Accountancy* 111 (1993): 62–63.

References

1. *Winnipeg Free Press,* December 17, 1997, page B-1.
2. *Time,* December 29, 1997/January 5, 1998, pp. 6–7.
3. M. W. Gillman and coauthors, Inverse association of dietary fat with development of ischemic stroke in men, *Journal of the American Medical Association* 278 (1997): 2145–50.
4. M. Mezzetti and coauthors, Population attributable risk for breast cancer: Diet, nutrition, and physical exercise, *Journal of the National Cancer Institute* 90 (1998): 389–94; and S. A. Smith-Warner and coauthors, Alcohol and breast cancer in women: A pooled analysis of cohort studies, *Journal of the American Medical Association* 279 (1998): 535–40.
5. As cited in L. A. Pal, *Beyond Policy Analysis* (Scarborough, Ontario: ITP, 1997), p. 72.
6. Ibid., pp. 1–12.
7. The discussion of the policy cycle was adapted from W. Lyons, J. M. Scheb II, and L. E. Richardson, Jr., *American Government: Politics and Political Culture* (St. Paul, MN: West, 1995), pp. 467–75; and D. J. Palumbo, *Public Policy in America—Government in Action* (San Diego: Harcourt Brace Jovanovich, 1988), pp. 1–155.
8. R. B. Reich, When naptime is over, *New York Times Magazine,* January 25, 1998, p. 33.
9. Pal, *Beyond Policy Analysis,* pp. 1–12; and R. B. Denhardt, *Public Administration: An Action Orientation* (Belmont, CA: Wadsworth, 1995), pp. 245–48.
10. K. Schlossberg, Nutrition and government policy in the United States, in *Nutrition and National Policy,* B. Winikoff, ed. (Cambridge, MA: MIT Press, 1978), pp. 334–39.
11. R. Carson, *Silent Spring* (Boston: Houghton Mifflin, 1962).
12. As cited in Palumbo, *Public Policy in America,* p. 23.
13. H. W. Schultz, *Food Law Handbook* (Westport, CT: Avi, 1981), pp. 3–21.
14. Provisions of the Nutrition Labeling and Education Act of 1990 as cited in the *Congressional Record*—House, July 30, 1990, pp. II 5836–40, and Legislative Highlights, Wrap-up of the ADA's issues in the 101st Congress, *Journal of the American Dietetic Association* 90 (1990): 1653–55.
15. Pal, *Beyond Policy Analysis,* pp. 8–9.
16. Information about the organization and activities of the Department of Health and Human Services were obtained from its Web site: www.hhs.gov/progorg/. Information about the U.S. Department of Agriculture was obtained from its Web site: www.ars.usda.gov/nps/programs/.
17. Community Nutrition Institute, Gleaning and greening in the Puget Sound region, *Nutrition Week* 27 (November 21, 1997): 3.
18. C. O. Jones, *An Introduction to the Study of Public Policy* (Belmont, CA: Wadsworth, 1970), pp. 1–15.
19. B. G. Peters, *American Public Policy—Process and Performance* (New York: Franklin Watts, 1982), p. 69.
20. H. H. Schauffler and J. Wilkerson, National health care reform and the 103rd Congress: The activities and influence of public health advocates, *American Journal of Public Health* 87 (1997): 1107–12.
21. J. E. Austin and C. Hitt, *Nutrition Intervention in the United States* (Cambridge, MA: Ballinger, 1979), pp. 37–41.
22. L. Carroll, *Alice in Wonderland* (New York: Washington Square Press, 1951), pp. 74–75.
23. The discussion of how a bill is introduced into Congress was adapted from T. R. Dye, *Politics in States and Communities,* 7th ed. (Upper Saddle River, NJ: Prentice Hall, 1991), pp. 156–75.
24. *How Congress Works,* 2nd ed. (Washington, D.C.: Congressional Quarterly, 1991), p. 140.
25. Society for Nutrition Education, *Influencing Food and Nutrition Policy—A Public Policy Handbook* (Oakland, CA: Society for Nutrition Education, 1987), p. 5.
26. The discussion of the federal budget process was adapted from Lyons and coauthors, *American Government,* pp. 507–20; and G. J. Gordon and M. E. Milakovich, *Public Administration in America,* 5th ed. (New York: St. Martin's, 1995), pp. 314–55.
27. Denhardt, *Public Administration,* pp. 148–60.
28. Gordon and Milakovich, *Public Administration in America,* pp. 315–16.
29. Denhardt, *Public Administration,* pp. 148–60.
30. Ibid.
31. Community Nutrition Institute, Agriculture Department '99 budget will promote food recovery and gleaning, *Nutrition Week* 28 (January 23, 1998): 1.
32. Community Nutrition Institute, President Clinton calls for Food Stamp restorations for legal aliens in FY99, *Nutrition Week* 28 (February 6, 1998): 1–2.
33. Schultz, *Food Law Handbook.*
34. American Dietetic Association, Position of The American Dietetic Association: Affordable and accessible health care services, *Journal of the American Dietetic Association* 92 (1992): 746–48.
35. American Dietetic Association, White paper on health care reform, *Journal of the American Dietetic Association* 92 (1992): 749.
36. American Dietetic Association, ADA's lobbying efforts focus on health care reform, *Journal of the American Dietetic Association* 93 (1993): 754.
37. American Dietetic Association, Health care reform initiatives stress grass-roots lobbying and coalition-building, *Journal of the American Dietetic Association* 93 (1993): 528.
38. American Dietetic Association, ADA continues push for medical nutrition therapy, improved child nutrition programs, and labeling of dietary supplements, *Journal of the American Dietetic Association* 94 (1994): 721.
39. American Dietetic Association, ADA urges Congress to expand Medicare coverage for medical nutrition therapy, *Journal of the American Dietetic Association* 95 (1995): 974.
40. American Dietetic Association, ADA mobilizes grassroots action to secure Medicare coverage for medical nutrition therapy, *Journal of the American Dietetic Association* 96 (1996): 1241.

41. Medicare Medical Nutrition Therapy Act of 1997 introduced in Congress, *ADA Courier* 36 (1997): 1.

42. American Dietetic Association, New ADA campaign seeks more cosponsors for Medicare Medical Nutrition Therapy Act; update on child/elderly bills, *Journal of the American Dietetic Association* 97 (1997): 1372.

43. Remarks made by Hon. John E. Ensign on introducing the Medical Nutrition Therapy Act of 1997 in the House of Representatives, from the *Congressional Record*, April 17, 1997, p. E696 (Thomas Legislative Information on the Internet: thomas.loc.gov).

44. American Dietetic Association, Update on state licensure laws and ADA regulatory remarks, *Journal of the American Dietetic Association* 97 (1997): 1251.

45. Senator R. J. Durbin, The Safe Food Act of 1997, *Food Technology* 52 (1998): 112.

46. Community Nutrition Institute, Legislation for new single independent food safety agency proposed on Hill, *Nutrition Week* 27 (November 7, 1997): 1–2.

47. S. A. Smith, Jumping in with both feet, *Food Technology* 52 (1998): 22.

48. The discussion of the Government Performance and Results Act was adapted from C. Wye, Overview: Views from the starting line, *Public Manager* 26 (Fall 1997): 3–4; J. A. Koskinen, Managing for results, *Public Manager* 26 (Fall 1997): 5–6; and C. DeMaio, Agencies have much work to do—but so do OMB and Congress, *Public Manager* 26 (Fall 1997): 10–12.

49. Food and Nutrition Board, Dietary Reference Intakes (DRIs) for calcium, phosphorus, magnesium, vitamin D, and fluoride, *Nutrition Today* 32 (1997): 182–88; and Food and Nutrition Board, Institute of Medicine, *Dietary Reference Intakes for Thiamin, Riboflavin, Niacin, Vitamin B_6, Folate, Vitamin B_{12}, Pantothenic Acid, Biotin, and Choline* (prepublication copy, 1998, obtained from the National Academy of Sciences Web site at www2.nas.edu/iom).

50. Nutrition Labeling and Education Act of 1990, Public Law 101-535, 104 Stat. 2353, 1990.

51. F. E. Scarbrough, Some Food and Drug Administration perspectives of fat and fatty acids, *American Journal of Clinical Nutrition* 65 (suppl.) (1997): 1578S–80S.

52. S. Harlander, Food biotechnology: Yesterday, today, and tomorrow, *Food Technology* 43 (1989): 196–206.

53. D. Knorr and A. J. Sinskey, Biotechnology in food production and processing, *Science* 229 (1985): 1224–29.

54. J. Q. Wilkinson, Biotech plants: From lab bench to supermarket shelf, *Food Technology* 51 (1997): 37–42.

55. N. A. Higley and J. B. Hallagan, Safety and regulation of ingredients produced by plant cell and tissue culture, *Food Technology* 51 (1997): 72–74.

56. National Science and Technology Council, Committee on Health, Safety, and Food, *Meeting the Challenge: A Research Agenda for America's Health, Safety, and Food* (Washington, D.C.: U.S. Government Printing Office, 1996).

57. Wilkinson, Biotech plants, p. 41.

58. D. Shattuck, Complementary medicine: Finding a balance, *Journal of the American Dietetic Association* 97 (1997): 1367–69.

59. A. Barrocas, Complementary and alternative medicine: Friend, foe, or OWA? *Journal of the American Dietetic Association* 97 (1997): 1373–76.

60. The definition of nutraceuticals was taken from D. E. Pszczola, Highlights of "The nutraceutical initiative: A proposal for economic and regulatory reform," *Food Technology* 46 (1992): 77–79.

61. Nutraceutical ingredients: Ill-defined but growing, *Food Technology* 51 (1997): 62–63.

62. The discussion of medical foods was adapted from the Scientific Status Summary of the Expert Panel on Food Safety and Nutrition, Medical foods, *Food Technology* 46 (1992): 87–96.

63. E. A. Yetley and R. J. Moore, Medical foods: A regulatory paradox, *Food Technology* 51 (1997): 136.

64. Capitol Resources, Grassroots targeting, American Dietetic Association 80th Annual Meeting, October 28, 1997.

65. E. C. Ladd, *The American Polity—The People and Their Government* (New York: Norton, 1985), p. 351.

66. G. Starling, *Understanding American Politics* (Homewood, IL: Dorsey, 1982), pp. 184–85.

67. T. D. Bevels, Public interest groups and the public manager, *Bureaucrat* 25 (Winter 1996–97): 8–12.

68. J. M. Burns, J. W. Peltason, and T. E. Cronin, *Government by the People,* 12th ed. (Upper Saddle River, NJ: Prentice Hall, 1984), pp. 167–68.

69. Ibid., p. 169.

70. H. J. Rubin and I. S. Rubin, *Community Organizing and Development* (New York: Macmillan, 1992), pp. 274–95.

71. Ibid., p. 282.

72. Linda Goodwin, RD, Alliance Program Director, American Dietetic Association, personal communication, July, 1998.

73. Alliance for the Prevention of Chronic Disease (Winnipeg, Manitoba, Canada, 1997).

74. The description of how to communicate effectively with elected officials was taken from How to be heard on Capitol Hill, *Journal of the American Dietetic Association* 92 (1992): 296.

Chapter 3

The Reality of Health Care

Outline

Learning Objectives

After you have read and studied this chapter, you will be able to:

- Describe current trends affecting the cost and delivery of health care.
- Explain why health promotion is a major component of the rhetoric about health care reform at the national level.
- Differentiate between traditional fee-for-service systems of health care and managed forms of health care.
- Describe eligibility requirements and services provided to recipients of Medicare and Medicaid.

- Explain the significance and relevance of complementary medicine to dietetics professionals.
- Describe the benefit and current status of state licensing of nutrition professionals.
- State the purpose and value of using medical nutrition therapy protocols to document client outcomes in various health care settings.

Something To Think About...

The enjoyment of the highest attainable standard of health is one of the fundamental rights of every human being without distinction of race, religion, political belief, economic or social condition. . . . Governments have a responsibility for the health of their peoples which can be fulfilled only by the provision of adequate health and social measures. – *Preamble to the Constitution of the World Health Organization*

Introduction

The U.S. health care system is approaching a breaking point. According to the Health Care Financing Administration (HCFA), health care expenditures in the United States in 1965 totaled $42 billion. Three decades later in 1995, Americans spent more than $988 billion for health care.[1] This hefty sum represents over 14 percent of the gross national product (GNP)—as compared with 9 percent in 1980.[2] By the year 2000, health care costs are expected to reach $1.7 trillion.[3] Although the escalation of health care expenditures has slowed nationally, it still exceeds the average rise in consumer pricing.

A strange paradox exists today in health care as we know it in the United States. As Louis Sullivan, the former secretary of health and human services observed, prevention "must become a national obsession."[4] He went on to say that health promotion and disease prevention comprise perhaps our best opportunity to reduce the ever-increasing portion of our resources that we spend to treat preventable illness and functional impairment.

Yet the Medicaid and Medicare systems and the Health Care Financing Administration provide almost no reimbursement for prevention activities and/or procedures. In addition, major third-party payers offer limited reimbursement for preventive procedures.[5]

In 1974, a landmark Canadian document—*The Lalonde Report: A New Perspective on the Health of Canadians*—was the first modern government document to acknowledge that our emphasis on a biomedical health care system is wrong. It stressed that we must look beyond the traditional health care (sick care) system if we wish to improve the public's health. The report was followed by similar reports in Great Britain, Sweden, and the United States.[6]

Public policy is now attempting to direct our medical system toward health promotion, disease prevention, and the efficient use of scarce resources. The American Dietetic Association (ADA) is among the organizations involved in formulating a reform policy that assures access to appropriate and affordable health care for all.[7] The ADA's position paper states that quality health care should be available, accessible, and affordable to all Americans. *Quality* health care is defined to include nutrition services that are integral to meeting the *preventive* and therapeutic health care needs of all segments of the population.[8]

Many studies show that early detection and intervention, immunization, and behavior change could significantly reduce many of the leading causes of death and disability.[9] By investing in health maintenance through health promotion and disease prevention, we can avoid the much greater economic and social costs of disease and injury. As Louis Sullivan has pointed out: we can preserve good health *and* reduce costs if we concentrate on the "front end" rather than waiting to devote substantial resources to illness and disability after they strike.[10]

This chapter introduces you to the challenges facing health care as we move into the twenty-first century. One question, for example, is how we can balance the physician/medical model of health care with a wellness/preventive medicine model. Other issues include resource allocation and cost containment, social justice and adequate access to health care resources, program accountability and quality in health care, and funding for health promotion and disease prevention. Before we talk about changing our present system of health care, however, we must first understand how that system works.

An Overview of the Health Care Industry

Our pluralistic system of health care includes many parts: private insurance, group insurance, Medicare, Medicaid, workers' compensation, the Veterans Health Administration medical care system, the Department of Defense hospitals and clinics, the Public Health Service's Indian Health Service, state and local public health programs, and the Department of Justice's Federal Bureau of Prisons. One other piece is missing from the system: the uninsured. Currently, the system is structured around the provision of health insurance. There are three general categories of **health insurance** in the United States: traditional private, prepaid, fee-for-service insurance; private, prepaid **group contract** insurance; and public health insurance.[11]

Private Insurance

Private, prepaid, fee-for-service plans can be either individual or group policies. Fee-for-service plans include a billing system in which the provider of care charges a fee for each service rendered. Private insurance is provided by both commercial insurance companies and not-for-profit organizations such as Blue Cross and Blue Shield and independent employee health plans. Approximately 147 million Americans have health insurance coverage through an employer-sponsored plan.[12] The traditional fee-for-service plans of the 1980s account for only 16 percent of insurance coverage today. Critics of fee-for-service plans claim that they encourage physicians to provide more services than are necessary.[13] Proponents of fee-for-service systems prefer the greater flexibility and unrestricted access to physicians, tests, hospitals, and treatments.

Group Contract Insurance

Currently, the health care financing system is undergoing a transition from the traditional unmanaged fee-for-service system to a more **managed-care** system, represented by **health maintenance organizations (HMOs)** and **preferred provider organizations (PPOs).** All are prepaid group practice plans that offer health care services through groups of medical practitioners. The presumed goal of managed care is improved quality of care with decreased costs. Group policies make up over

---

Health insurance Protection against the financial burdens associated with health care services and assurance of access to the health care system.

Group contract A health insurance contract that is made with an employer or other entity and covers a group of persons identified as individuals by reference to their relationship to the entity.

Managed care An approach to paying for health care in which insurers try to limit the use of health services, reduce costs, or both. These health plans are subject to utilization review (UR). That review function aims to prevent unnecessary treatment by requiring enrollees to obtain approval for nonemergency hospital care, denying payment for wasteful treatment, and monitoring severely ill patients to ensure that they get cost-effective care.

Health maintenance organization (HMO) A prepaid plan that both finances and delivers health care. HMOs enroll patients as members, charge a fixed fee per year, and provide all medical services deemed necessary. Enrollees generally must use the plan's providers or face financial penalties.

70 percent of all policies and are usually employer-based and offered as a fringe benefit.

Congress provided an impetus for the growth of HMOs when it passed the Health Maintenance Organization Act of 1973 in an effort to contain utilization, control escalating medical costs, and improve the quality of health care. The law requires employers with 25 or more employees to offer their employees HMO membership as an alternative to traditional health insurance plans.

In HMOs, physicians practice as a group, sharing facilities and medical records. The physicians may either be salaried or provide contractual services. There are four general models of HMOs:

1. **Staff HMO:** a model in which the HMO owns and operates its own facility and is equipped for laboratory, pharmacy, and X-ray services, and hires its own physicians and other health care providers
2. **Group practice HMO:** a model in which the HMO contracts with a multispecialty group practice to provide health care services to its members
3. **Network HMO:** a model in which the HMO contracts with multiple group practices, hospitals, and other providers to provide services to its members
4. **Independent practice association (IPA):** a decentralized model in which physicians group themselves together to share risk and profit while contracting to provide medical care to a plan's enrollees for a discounted fee.

HMOs typically provide comprehensive services across the continuum of care. In some HMO programs, the provider receives a **capitation payment,** usually a specific amount per enrollee per month, to provide a defined group of health care services (see Figure 3-1). Dietitians may be included under specialists or as part of the primary care provider portion, depending on the contractual agreement of the HMO.[14]

The HMO idea—a fixed cost to the consumer, with health care insurer and health care provider as one and the same—is viewed as a more cost-effective way of practicing medicine than the traditional fee-for-service systems. Because HMOs make money by keeping you healthy, they have a greater stake in your wellness than most fee-for-service doctors.[15]

Prepaid group health plans emphasize health promotion, since they provide health care services at a preset cost. By keeping people healthy, HMOs avoid lengthy hospitalizations and costly services. Enrollees of HMOs are hospitalized less frequently than patients of fee-for-service physicians.[16]

Public Insurance

The two major public health insurance plans in the United States are **Medicare** and **Medicaid.** A comparison of their features is provided in Table 3-1. Workers' compensation, which pays benefits to workers who have been injured on the job, is another public-sector health benefit program. Health care services are also provided by the Department of Veterans Affairs (VA), the Public Health Service (including the Indian Health Service), the Department of Defense (including the Civilian Health and Medical Program of the Uniformed Services or CHAMPUS), public hospitals and community health centers, and state and local public health programs.[17]

Preferred provider organization (PPO) A group of providers, usually hospitals and doctors, who contract with private indemnity (fee-for-service) insurance companies to provide medical care for a discounted fee. PPOs are subject to peer review and strict use controls in exchange for a consistent volume of patients and speedy turnaround on claim payments.

Capitation A predetermined fee paid per enrollee per month to the participating health care provider.

Medicare A federally run entitlement program through which people age 65 years or older and persons in certain other eligible categories receive health insurance.

Medicaid A federally aided, state-administered entitlement program that provides medical benefits for certain low-income persons in need of health and medical care.

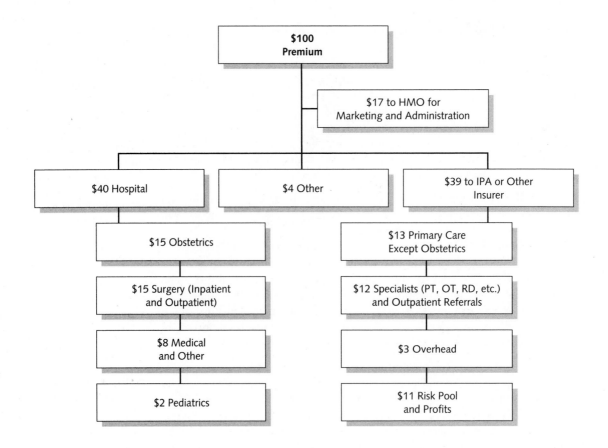

FIGURE 3-1 A *Sample of How a Capitation Payment Is Used*

Source: Adapted from American Dietetic Association, *Medical Nutrition Therapy Across the Continuum of Care* (Chicago: American Dietetic Association, 1996), p. 2.

HCFA (Health Care Financing Administration) (pronounced *hic-fa*) A federal agency that establishes guidelines and monitors Medicare and Medicaid programs.
Coinsurance A cost-

The Medicare Program. Medicare, with over 38 million beneficiaries, is the largest health care insurer in the United States. It was established in 1965 by Title XVIII of the Social Security Act and is designed to assist:

- people 65 years of age or older,
- people of any age with end-stage renal disease,
- people eligible for Social Security disability payment programs for more than 2 years, and
- qualified railroad retirement beneficiaries and merchant seamen.

Medicare is administered by the **Health Care Financing Administration (HCFA)** of the Department of Health and Human Services. The Social Security Administration provides information about the program and handles enrollment.

Medicare consists of two separate parts: hospital insurance (Part A) and medical insurance (Part B). No monthly premium is required for Medicare Part A if a person or his or her spouse is entitled to benefits under either Social Security or the Railroad Retirement System or has worked a sufficient period of time in federal, state, or local government to be insured. Those not meeting these qualifications may purchase Part A coverage if they are at least age 65 and meet certain requirements.

	Medicare	Medicaid
Administration	Social Security Office Federal Insurance Program	Local welfare office Federal-state partnership assistance program
Financing	Trust funds from Social Security; contributions from insured	Taxes from federal, state, and local sources
Eligibility	People 65 years of age and older, people with end-stage renal disease, people eligible for Social Security disability programs for more than 2 years	Needy and low-income people, people 65 or older, the blind, persons with disabilities, all pregnant women and infants with family incomes below 133% of poverty level, possibly others
Benefits*	Same in all states	Varies from state to state
	Hospital insurance (Part A) *helps* pay for inpatient hospital care, skilled nursing facility care, home health care, hospice care.	**Hospital services:** inpatient and outpatient hospital services, other laboratory and X-ray services, physician services, screening, diagnosis, and treatment of children, home health care services
	Medical insurance (Part B) *helps* pay for physicians' services, outpatient hospital services, home health visits, diagnostic X-ray, laboratory, and other tests; necessary ambulance services, other medical services and supplies, outpatient physical or occupational therapy and speech pathology; and partial coverage of mental health treatment.	**Medical services:** many states pay for dental care, health clinic services, eyecare and glasses, prescribed medications, other diagnostic, rehabilitative, and preventive services, including nutrition services
	Exclusions: regular dental care and dentures, routine physical exams and related tests, preventive services, eyeglasses, hearing aids and examinations to prescribe and fit them, prescription drugs, nursing home care (except skilled nursing care), custodial care, immunizations (except for pneumonia, influenza, and hepatitis B), cosmetic surgery	Varies from state to state
Premium Costs	Part A: none if eligible, or $188–$289/month Part B: $43/month	None (federal government contributes 50% to 80% to states to cover eligible persons)

TABLE 3-1 A *Comparison of Medicare and Medicaid Services*
Source: Adapted from U.S. Department of Health and Human Services, *1997 Guide to Health Insurance for People with Medicare* (Washington, D.C.: U.S. Department of Health and Human Services, 1997).

*As a result of the Balanced Budget Act of 1997, Medicare beneficiaries who have both Part A and Part B can choose to get their benefits through a variety of risk-based plans (e.g., HMOs, PPOs), known as Medicare + Choice or Part C of Medicare.

sharing requirement in which the insured assumes a portion of the costs of covered services.

Prospective payment system (PPS) A payment system under which hospitals are paid a fixed sum per case according to a schedule of diagnosis-related groups.

Diagnosis-related groups (DRGs) A method of classifying patients' illnesses according to principal diagnosis and treatment requirements for the purpose of establishing payment rates. Under Medicare each DRG has its own payment rate, which a hospital is paid regardless of the actual cost of treatment.

Deductible The amount of expense that must be incurred by a person who is insured before an insurer will assume any liability for all or part of the remaining cost of covered services. Deductibles may be fixed dollar amounts or the value of specified services (for example, 2 days of hospital care or one visit to a physician).

Supplemental health insurance Health insurance that covers medical expenses, services, and supplies that are covered only partially or not at all by Medicare.

Medicare Part A provides hospital insurance benefits that include up to 90 days of inpatient care annually with a 20 percent **coinsurance** fee for hospital charges. Hospital inpatient charges are reimbursed according to a **prospective payment system** known as **diagnosis related groups (DRGs)**—discussed in detail later in this chapter. Since 1983, the government has shifted a larger portion of health care costs to Medicare beneficiaries through larger **deductibles,** greater use of services with coinsurance, and use of services not covered by Medicare.

Medicare Part B is an optional insurance program financed through premiums paid by enrollees and contributions from federal funds; it provides supplementary medical insurance benefits (see Table 3-1). To date, dietitians are not Medicare-approved practitioners, but they can sometimes achieve reimbursement on a state-by-state basis as determined by the state's private insurance carriers who reimburse for services under Medicare Part B.[18]

The two most notable gaps in Medicare coverage are prescription drugs and long-term institutional care.[19] Prescription drugs are not covered at all under the Medicare program. Only 100 days of long-term care are covered annually. Thereafter, patients or their families must either pay the costs themselves or "spend down" in order to reduce their net worth and to be eligible for Medicaid long-term care coverage. Medicare enrollees have five options to help fill the gaps in Medicare:

1. Purchase Medicare **supplemental insurance,** which may also be called *Medigap insurance*. Such policies help pay the deductible, coinsurance fees, prescription drug costs, and certain services not covered by Medicare.
2. Enroll in a managed care plan such as an HMO that has a Medicare contract. Health care services are then purchased directly for a fixed monthly premium.
3. Continue group coverage through their current or former employer.
4. Purchase nursing home or long-term care policies, which pay cash amounts for each day of covered nursing home or at-home care.
5. Qualify for full Medicaid benefits or at least some state assistance in paying for Medicare costs.

The Medicaid Program. Medicaid was established as a joint state and federal program, with the latter paying 50 percent or more of the costs depending on a state's per capita income. It was established in 1966 by Title XIX of the Social Security Act. Medicaid provides assistance with medical care for:

- eligible, low-income persons;
- the aged, blind, and people with disabilities;
- members of families with dependent children in which one parent is absent, incapacitated, or unemployed.

The individual states define eligibility, benefits, and payment schedules. Typically, one must meet three criteria: income, categorical, and resource. Income must be below—sometimes significantly below—the poverty line. Poverty line monthly income limits in 1997 were $658 for an individual or $885 for a couple. Those eligible for Temporary Assistance for Needy Families (formerly known as Aid to Families with Dependent Children) and Supplemental Security Income (SSI) are automatically eligible for Medicaid.

To meet the categorical requirements, one must be a member of a family with dependent children or be aged, blind, or a person with a disability. The resource test sets a maximum allowable amount for liquid resources and other assets. Income and asset eligibility standards vary widely among the 50 states.

Medicaid covers inpatient and outpatient hospital services, physicians' services, laboratory and X-ray tests, and skilled nursing home services. Some states include other benefits, such as prescription drug coverage, home health services, and dental services, but there is significant variability among states. To date 36 state Medicaid programs cover certain forms of nutrition services provided by dietitians.[20]

Medicaid currently covers less than half of those below the poverty line.[21] The American Medical Association has recommended that Medicaid be expanded to provide acute-care coverage for all persons below the poverty line. Such an expansion would increase the current number of Medicaid recipients from 20 to 40 million people.[22]

The Uninsured

In theory, health care coverage is available to virtually all U.S. citizens through one of four routes: Medicare for the elderly and people with disabilities, Medicaid for low-income women and children and some low-income men and people with certain disabilities, employer-subsidized coverage at the workplace, or self-purchased coverage for those ineligible for the previous three.[23] Efforts to control health care costs, however, have led to an estimated 41 million people under age 65 with no insurance coverage at all and to perhaps an even larger number whose coverage is inadequate for any major illness.[24] The uninsured represent 18 percent of all Americans and an increase of 19.6 percent of uninsured people over the past decade. If this situation continues, the number of people without health insurance will approach 47 million by 2005.[25]

Who, then, are the uninsured? Statistics show that they are not the elderly, who have Medicare, or the very poor, who have Medicaid. Instead, those who lack coverage are primarily people in the middle—for example, the working poor and those who work for small businesses. They include the self-employed, those who work part-time, seasonal workers, the unemployed, full-time workers whose employers offer unaffordable insurance or none at all, and early retirees—aged 55 through 64—who retired from companies that either offered no health insurance or have since dropped it.[26] These persons are classified further as the employed uninsured, the nonworking uninsured, and the medically uninsurable. Included among the uninsured are 11 million children.[27]

The employed uninsured number 15 million and, with their dependents, represent 70 percent of all uninsured persons. The second group, the nonworking uninsured, number about 9 million; they include the homeless, some deinstitutionalized mentally ill patients, and low-income people who do not qualify for Medicaid because they are not categorically eligible or because their income is above the cutoff level for their state. Finally, a small but growing group of 1 million people are unable to obtain insurance because of a preexisting medical condition.[28] For example, some AIDS patients and persons infected with HIV are not able to obtain insurance.

When those without health insurance do get sick, they often wind up using the most expensive treatment available—hospital emergency room care—or they delay getting treatment and later require more expensive and prolonged medical services. These costs are shifted to the people who are insured.

Trends Affecting Health Care

The future of health care in the United States will be shaped by current trends in society at large and in the field of health care, as well as by the choices we make for health care reform. Some of the trends and issues that will shape the future of health care are described in the next sections.

Demographic Trends and Health Care

Between 1946 and 1964, 78 million babies were born in the United States; these individuals—the baby boomers—now make up one-third of the population.[29] By the year 2030, the baby boom will become a senior boom, with 21 percent of the population—30 million more Americans than today—over 65 years of age.[30]

Not only will the elderly be greater in number, but they will potentially require care for a greater number of years, placing a heavier burden on the long-term care system (see Figure 3-2). Since older Americans consume a disproportionate amount

By the year 2000, the number of elderly needing nursing home care will triple.

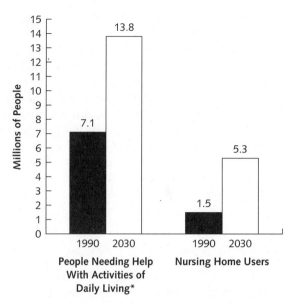

FIGURE 3-2 *Number of Elderly Needing Long-Term Care, 1990 and 2030*

Source: *A Call for Action: Final Report of the Pepper Commission* (Washington, D.C.: U.S. Government Printing Office, 1990).

*Activities of daily living (ADL) include activities such as bathing, dressing, toileting, continence, and feeding.

of medical care, the demand for such care, including pharmaceutical products and services, can be expected to rise.[31]

Racial and geographical factors in the population are also important to the shape of the future. In some parts of the United States, particularly the Southwest, the Hispanic population will dramatically increase. To the extent that such a population may exhibit differing utilization patterns for medical services or pharmaceuticals, such changes may significantly affect the marketplace. Geographical demographics will also be important, especially if the population drift from the Northeast to the Southwest and the Sun Belt continues.[32]

The Paradigm Shift from Sickness to Wellness

For several decades, the dominant paradigm has been the medical model. During the 1970s, people were guided by the philosophy that the health care system would do everything possible in terms of curative and treatment services to make them well. In the 1990s, people viewed wellness as a function of prevention and began to accept responsibility for their own health. The focus on the pursuit of health, marked by an increased interest in nutrition, fitness, and health promotion, was reflected in the growth of corporate "wellness" programs and a widening choice of health care practitioners.

Our health care system still contains a number of barriers to focusing on prevention of poor health habits, however. The biomedical approach to illness underestimates and underemphasizes behavioral or lifestyle influences on disease. Physicians think in terms of treating or correcting conditions rather than preventing them.

To a greater extent than most of us are willing to accept, today's disorders of overweight, heart disease, cancer, blood pressure, and diabetes are by and large preventable. In this light, true health insurance is not what one carries on a plastic card but what one does for oneself.

– L. Power in G. Edlin and E. Golanty, Health and Wellness (3rd ed.)

Also, both formal diagnostic processes and systematic reimbursement for prevention are lacking. Consider, for example, that Medicare Part A will reimburse for enormous hospital bills, but Part B won't reimburse for prevention.

The challenge for the next decade is to change the United States' approach to health care from a system based on treatment of acute conditions to one based on disease prevention and health promotion. Physicians, public health workers, registered dietitians and other health practitioners, health educators, and community health organizations have joined ranks to emphasize health promotion and disease prevention as a more economical route to good health than the more costly procedures necessitated by sickness and disease.[33] According to physician Andrew Weil, "the medical facility of the future should look like a cross between a health spa and a hospital. Not only should it offer a smorgasbord of therapeutic options, its main focus should be on educating patients on how to stay well once they leave."[34]

Complementary Nutrition and Health Therapies

Complementary medicine Using medical techniques, practices, and therapies from both Western and alternative medicine.

Alternative medicine An unrelated group of nonorthodox therapeutic practices, often with explanatory systems that do not follow conventional biomedical explanations. Appendix C includes a glossary of alternative medical practices and related terms.

A new Dietetic Practice Group of the American Dietetic Association— *Nutrition in Complementary Care (NCC)*—was recently formed. Check it out on the ADA Web page: www.eatright.org.

Paralleling the paradigm shift from "sickness" to "wellness" is an expansive interest in and use of **complementary** and **alternative medicine**.[35] Table 3-2 lists the seven major categories of complementary and alternative medicine (CAM). As shown in the table, CAM covers a wide range of healing philosophies, approaches, and therapies. It generally is defined as those treatments and health care practices not taught widely in medical schools, not generally used in hospitals, and not usually reimbursed by medical insurance companies.[36] Many of the therapies are called "holistic," meaning that the health care provider considers the whole person, including physical, mental, emotional, and spiritual health. The therapies shown in Table 3-2 are sometimes used alone (referred to as alternative therapy) or in addition to conventional therapies (referred to as complementary therapy).

A number of important factors have led to the current proliferation of alternative health therapies. These include (1) the failure of conventional health practitioners to acknowledge the human side of medicine and practice preventive medicine, (2) the philosophy that achieving and maintaining health is very different from fighting disease, and (3) the emphasis on individual empowerment in health maintenance.

Worldwide, approximately 70 percent of health care is delivered by alternative practitioners. Alternative therapies have become increasingly popular in recent years—whether it be acupuncture, herbs, or chiropractic.[37] In the past 5 years, an Office of Alternative Medicine was established at the National Institutes of Health, courses on complementary medicine were introduced into medical schools, and herbs and supplements became best-sellers for national pharmaceutical and supermarket chains.[38] The cost of alternative medicine practices to Americans was $13.7 billion in 1990 and included $10.7 billion in out-of-pocket costs and an additional $3 billion paid by third-party payers. Sales for herbal products alone in 1997 were $3.24 billion.[39]

Some managed care organizations and other health insurers now include holistic health providers among the services offered to patients.[40] For example, the Mind/Body Institute in Morristown, New Jersey, in conjunction with a local hospital, provides a 10-week program that demonstrates the healing power of the mind. The program offers classes in stress management, relaxation techniques, guided imaging, nutrition, and meditation to area oncology patients.[41]

Despite the rising popularity of alternative medicine, its domain lacks regulatory control, offers nonstandardized systems of therapy, and needs scientific, outcomes-based research.[42] The implications for the dietetics profession are noteworthy. Clients and others will ask for your opinion on various alternative therapies. The real question to ask about these "alternative" therapies is, Are there sufficient data available from controlled scientific studies to give you confidence that the compound or treatment being considered is safe and effective?

As members of the ADA, dietitians serve the public through the promotion of optimal nutrition, health, and well-being.[43] Dietitians are ethically bound to protect the public's health. To do so, one can follow these tips:

1. Inquire about the use of alternative or unconventional therapies by clients and educate them about the state of scientific knowledge with regard to alternative nutrition and health therapies. Be culturally sensitive to the beliefs and practices of clients; accept and integrate alternative therapies into the overall care plan if they offer comfort without harm. However, recognize the potential hazards associated with the postponement of conventional therapy or the combination of certain alternative therapies with conventional medicine (for example, combining certain herbs with prescription drugs).

2. When considering the use of an alternative therapy, ask, "Are there any double-blind, placebo-controlled randomized clinical trials demonstrating efficacy?" The Office of Alternative Medicine currently funds several specialty research centers to conduct alternative medicine research. Alternative therapies being studied at present include:[44]

   • Acupuncture in the treatment of attention deficit hyperactivity disorder, depression, and the pain associated with osteoarthritis
   • Herbal products to treat Parkinson's disease and other conditions
   • Hypnosis in the treatment of chronic low back pain
   • Biofeedback for the treatment of diabetes and chronic back and jaw pain
   • Imagery to treat breast cancer and asthma

   See Appendix C for a list of alternative medicine research centers and their on-line addresses.

3. Keep an open mind. Enroll in courses to learn more about alternative medicine. Such courses should present both the scientific view of unconventional therapies as well as the potential therapeutic use of these practices.[45]

Licensure of Nutrition Professionals

Licensure protects the public from the fraudulent and unqualified practitioners of various health care therapies that have made inroads into the health care marketplace. Licensure of physicians has been the principal social mechanism for quality control in health care. In the past two decades, increasing numbers of *licensed*, non-physician health care providers, such as nurse practitioners, physician assistants, midwives, physical therapists, dietitians, and dental hygienists, have appeared. They provide at a lower cost many services formerly reserved for doctors.[46]

In medicine, the educational standards for the medical degree (MD) are governed by law. Unfortunately, in some states, the term *nutritionist* is not legally defined, and as a result, the public can be the hapless prey of anyone who wishes to use this title. Some nutritionists obtain their diplomas and titles without the rigorous training

The Professional Focus in Chapter 10 discusses nutrition and health fraud and offers tips on spotting quackery.

TABLE 3-2

Classification and Examples of Complementary and Alternative Medical (CAM) Practices

I. Mind-Body Medicine

Involves behavioral, psychological, social, and spiritual approaches to health. Examples include:

Yoga	Psychotherapy	Art therapy
Tai chi	Hypnosis	Dance therapy
Meditation	Biofeedback	Music therapy
Imagery	Support groups	Humor therapy
		Prayer

II. Alternative Medical Systems

Involves complete systems of theory and practice that have been developed outside the Western biomedical approach. It is divided into four subcategories:

A. Acupuncture and Oriental medicine. Examples include:
   Acupuncture
   Herbal formulas
   Tai chi

B. Traditional indigenous systems. Examples include:
   American Indian medicine
   Ayurvedic medicine
   Traditional African medicine
   Traditional Aboriginal medicine
   Central and South American practices

C. Unconventional Western systems. Examples include:
   Homeopathy
   Cayce-based systems
   Orthomolecular medicine

D. Naturopathy

III. Lifestyle and Disease Prevention

This category involves theories and practices designed to prevent the development of illness, identify and treat risk factors, or support the healing and recovery process. To be classified as CAM, the lifestyle therapies for behavior change, dietary change, exercise, stress management, and addiction control must be based on a nonorthodox system of medicine or be applied in unconventional ways.

IV. Biologically-Based Therapies

Includes natural and biologically-based practices, interventions, and products. Many overlap with conventional medicine's use of dietary supplements. There are four subcategories as shown at the top of the next page.

required for a legitimate nutrition degree. Because of lax state laws, it is even possible for an irresponsible "correspondence school"—a diploma mill—to pass out degrees to anyone who pays a fee.

Licensure of qualified nutritionists protects consumers from unqualified practitioners, particularly those who have no training in nutrition but nevertheless refer to themselves as "nutritionists." Licensure is designed to protect the public, control malpractice, and ensure minimum standards of practice. Its aim is to gain legal recognition of health care professionals with the training and experience to deliver nutrition services.[47]

TABLE 3-2 *Continued*

IV. Biologically-Based Therapies—continued

A. *Individual herbs* (e.g., ginkgo biloba, hypericum, garlic, ginseng, echinacea, saw palmetto, ginger, green tea, psyllium, etc.)

B. *Special diet therapies* (e.g., Pritikin, Ornish, Gerson, vegetarian, fasting, macrobiotic, Mediterranean, Paleolithic, Atkins, natural hygiene, etc.)

C. *Orthomolecular medicine.* This subcategory refers to products used as nutritional supplements for preventive or therapeutic purposes. They are usually used in combinations and at high doses (e.g., ascorbic acid, carotenes, tocopherols, niacinamide, choline, lysine, boron, melatonin, dehydroepiandrosterone [DHEA], amino acids, carnitine, probiotics, glucosamine sulfate, chondroitin sulfate, etc.)

D. *Pharmacological, biological, and instrumental interventions.* Includes products and procedures applied in an unconventional way (e.g., cartilage, enzyme therapies, hyperbaric oxygen, bee pollen, cell therapy, iridology, etc.)

V. Manipulative and Body-Based Systems

Refers to systems that are based on manipulation and/or movement of the body. It has three categories:

A. *Chiropractic medicine*

B. *Massage and body work* (e.g., Swedish massage, applied kinesiology, reflexology, rolfing, polarity, etc.)

C. *Unconventional physical therapies* (e.g., colonics, hydrotherapy, heat and electrotherapies, etc.)

VI. Biofield

Involves systems that use subtle energy fields in and around the body for medical purposes. Examples include:

Therapeutic touch
Healing touch
Reiki

VII. Bioelectromagnetics

Refers to the unconventional use of electromagnetic fields for medical purposes.

Source: Adapted from National Institutes of Health, Office of Alternative Medicine, *Classification of Complementary and Alternative Medical Practices: General Information Package* (Silver Spring, MD: Office of Alternative Medicine Clearinghouse, 1998).

Today, better consumer protection with respect to nutrition services is evident in many states. As of 1998, 38 states, the District of Columbia, and Puerto Rico have enacted some form of licensure law.[48] Several states have passed legislation restricting the fraudulent use of the title "dietitian." In Alabama, self-styled nutritionists with diploma mill degrees are now forbidden to use the nutritionist title. Several other states have passed legislation to prohibit people from calling themselves nutritionists without a license that requires a background in dietetics.

The advantages of licensure are clear. Americans are accustomed to identifying *licensed* health professionals. The initials L.D. (licensed dietitian) after a person's name assure consumers, health professionals, and insurance companies that the person providing the nutrition services meets the specific professional standards established by the state's Department of Professional Regulation.[49]

The Need for Health Care Reform

To determine the rating of a particular health care system, one must examine three crucial variables: cost, quality, and access.[50] At the zero end of the scale is no health care system. As already discussed, millions of Americans cannot afford to buy into or gain meaningful, ongoing access to any health care at all. At the other end of the scale is high-quality, reasonably priced, accessible health care. On such a scale, how does the U.S. health care system rate?

Before you respond, consider the following scenario:[51]

Imagine you are the decision maker in a large corporation and I come in to you and try to sell you a product. I say that I want to sell you a key piece of equipment that meets the following specifications:

- It will cost you $3,200 per employee per year.
- It will consume up to half of each profit dollar and will rise in price by 15 to 30 percent annually.
- There is a tremendous unexplained variation in this product depending on who uses it.
- There is no way to measure its quality in terms of appropriateness, reliability, or outcome.
- And, you'll just have to take my word for it when I tell you that we adhere to the highest professional standards.

Would you buy this product? Many believe the current U.S. health care system fits this description. Not only is it expensive, but we don't necessarily know what we're paying for or whether what we're paying for is worth it.[52]

Health care reform refers to the efforts undertaken to ensure that everyone in the United States has access to quality health care at an affordable price. Among the challenges for health care reform are how to make health care accessible to everyone, contain costs, provide nursing home care to those who need it, and ensure that Medicare and Medicaid can serve all who are eligible.

As you will see, cost, access, and quality are interrelated; manipulating one has an astounding impact on the others. For example, some people argue that we should abandon free enterprise and turn the system over to the government, as has been done in other countries, including Canada. Critics of government-run health care systems say they appear promising at first but soon bog down in bureaucracy, unable to keep pace with advances in medical technology. Some point to the Canadians who come south to the United States to purchase treatment out of their own pockets rather than wait in line.[53] The question arises, How do you extend the scope of the system without sacrificing quality?

Health care policy makers are studying alternative models of delivery and financing in hopes of applying other nations' successes to the United States. The quest for a new health care strategy is particularly pressing because public sentiment for health care reform is at unprecedented levels.[54] The U.S. health care system appears to have both higher costs (see Figure 3-3) and less access than the systems of other industrialized nations. During the last two decades, U.S. health care trends differed from those in other nations in a variety of ways, most notably, rising costs, eroding access, and quality in health care service delivery. The following sections consider each of these in turn.

The High Cost of Health Care

Health care inflation is well established. Figure 3-4 tracks the rise in U.S. health care costs since 1960. Since that year, health care expenditures have increased over 800 percent. The level of health care activity is expected to grow as a result of various factors, including an aging population, increased demand (fostered in part by more consumer awareness of health issues), and continuing advances in medicine, which make it possible to do more for people than ever before.[55]

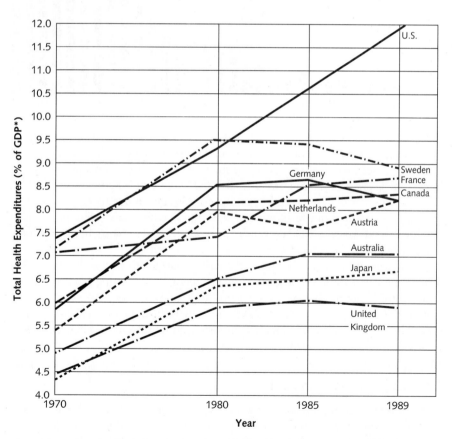

FIGURE 3-3 *Total Health Expenditures as a Percentage of Gross Domestic Product* (GDP), 1970–1989

Source: Organization of Economic Cooperation and Development, Paris, 1991. Reprinted from *Hospitals*, Vol. 65, No. 10, by permission, May 20, 1991. Copyright 1991, American Hospital Publishing, Inc.

*Gross domestic product (GDP) represents the total value of a nation's output, income, or expenditures produced within its borders. GDP is more specific than gross national product (GNP), the total retail market value of all goods and services.

FIGURE 3-4 *National Health Expenditures (billions of dollars), 1960–1995*

Source: Adapted from *Source Book of Health Insurance Data* (Washington, D.C.: Health Insurance Association of America, 1997).

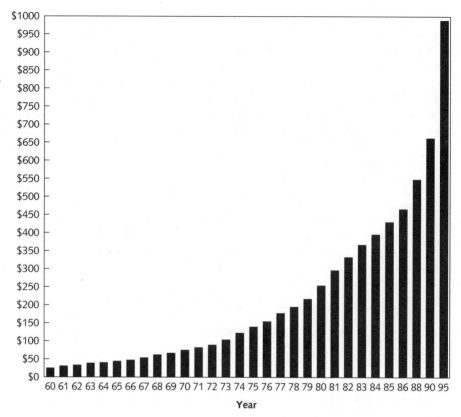

A major contributor to health care expenditures in the United States is the cost of the insurance process itself. Figure 3-5 shows the trends in administrative costs in both Canada and the United States since 1960. Note that 30 years ago, when the U.S. system was still dominated by not-for-profit insurers reimbursing private medical practitioners and not-for-profit hospitals, the overhead costs were proportionally lower—similar to Canada's. Since then, administrative costs have increased, while coverage has decreased.

Yet another factor contributing to the cost of our health care is the practice of defensive medicine and the associated phenomenon of ever-rising professional liability costs. Some say that we have become a litigious society. A neurosurgeon in New York City pays more than $70,000 annually for malpractice insurance.[56] The cost of health care would be dramatically reduced if physicians did not feel forced to practice defensive medicine.

Third-party reimbursement involves three parties in the process of paying for medical services: The *first party* (patient) receives a service from the *second party* (physician, hospital, other health care provider). The *third-party* (insurance company, government) pays the bill.

Efforts at Cost Containment

Efforts to curb soaring health care costs cover a broad spectrum: slowing hospital construction, modifying hospital and physician reimbursement mechanisms, reducing the length of hospital stays, increasing **copayments** and deductibles for insured employees and Medicare recipients, changing eligibility requirements for Medicaid, reducing unnecessary surgery by requiring patients to obtain second opinions,

Copayment The portion of the charge that the patient must pay.

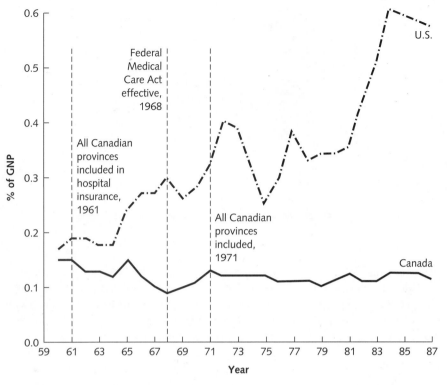

FIGURE 3-5 *Costs of Insurance and Administration as a Share of GNP, Canada and the United States, 1960–1987*

Source: R. G. Evans, Tension, compression, and shear: Directions, stresses, and outcomes of health care cost control, *Journal of Health Politics, Policy, and Law* 15: 1. Copyright Duke University Press, 1990. Reprinted with permission.

restricting the access to new technology, encouraging alternative delivery systems, and emphasizing prevention.[57]

The recent cost containment effort in the United States is actually a fierce competition among *third-party* payers (government, insurance companies, employers) to control their own costs. This effort has been characterized by three trends:[58]

1. There is a movement away from traditional fee-for-service health care to newer models of managed care, evident in the increasing enrollments in HMOs and PPOs.
2. As more and more of their profits are siphoned off into health care coverage, companies are increasingly attempting to manage the health care of their employees themselves to reduce expenditures. In an effort to avoid **cost shifting,** many businesses are moving to **self-insured** health plans, thereby determining which benefits are covered and assuming the risks involved. Sixty-seven percent of employers offering insurance plans are self-insured.[59]
3. The payers (government, insurance, employers) are actively setting **reimbursement** restrictions and limitations.

The largest components of national health care expenditures are hospital care (39 percent) and physician services (21 percent).[60] Therefore, efforts to contain costs have largely been aimed at these providers.

One example of cost containment is the prospective payment system (PPS) that the federal government implemented as a result of the 1983 Social Security Act

Cost shifting A much criticized aspect of the existing health care system in which hospitals and other providers bill indemnity insurers higher rates to recover the costs of charity care and to make up for discounts given to HMOs, PPOs, Medicare, and Medicaid.

Self-insured or self-funded plan A health plan where the risk for medical cost is assumed by the employer rather than an insurance company or managed care plan.

Reimbursement Payment made by a third party (for instance, government or private or commercial insurance).

"It's a get-well card from your hospitalization insurance company."

Principal diagnosis The condition chiefly responsible for the patient's need for health care services. The principal diagnosis determines the payment the hospital receives from Medicare.

Secondary diagnoses are also referred to as **comorbidities.** A comorbid condition is present at the time of admission to the hospital but is not the primary reason for treating that patient. For example, if a patient is admitted for a cholecystectomy and has diabetes mellitus, diabetes mellitus is a comorbid condition.

ICD-9-CM (*International Classification of Diseases—Clinical Modifications,* 9th ed.) Codes used by health care providers on billing forms to classify diseases/diagnoses.

Home health agency An agency that provides home health care. To be certified under Medicare, an agency must provide skilled nursing services and at least one additional therapeutic service (physical, speech, or occupational therapy; medical social services; or home health aide services) in the home.

Amendments. The purpose of the PPS was to change the behavior of health care providers by changing incentives under which care is provided and reimbursed.[61] Prospective payment means knowing the amount of payment in advance. The PPS uses diagnosis related groups (DRGs) as a basis for reimbursement. Patients are classified according to their **principal diagnosis,** secondary diagnosis, sex, age, and surgical procedures.

The DRG approach is based on a system of classifying hospital admissions. The system begins with the ninth edition of *International Classification of Diseases: Clinical Modifications,* abbreviated **ICD-9-CM,** which contains approximately 10,000 possible reasons for a hospital admission, organized into 23 major categories. The 23 categories are subdivided into 490 DRGs. Tables compute average cost per discharge by state, region (rural or urban), hospital bed size, and other factors.[62] All DRGs have been assigned a *relative weight* that reflects the cost of caring for a patient in the particular category. Table 3-3 shows a sample payment based on DRGs. Note that a patient with a complication or comorbidity (for example, with malnutrition) is assigned a higher relative weight, reflecting the need for more intensive services.

One consequence of the PPS has been an increased focus on outpatient services as opposed to more costly inpatient care. For employment of dietitians, the trend will continue to bring increased opportunities for consultation in outpatient settings, such as hospital outpatient clinics, **home health agencies,** private-practice counseling and consulting in physician or other health care provider offices, HMOs, health and fitness facilities, weight loss programs, community health centers and clinics, and group patient education classes.[63]

Equity and Access as Issues in Health Care

Is health care a basic right? Most people in the United States would answer affirmatively. Public opinion polls show that a large majority of people in the United States believe that all citizens are entitled to access to health care.[64] In reality, as deVise has observed, health care may be more of a privilege than a right:

Name of DRG	DRG Number	1991 Medicare Relative Weight		Base Rate		Payment Amount
Respiratory infections and inflammations without complication/comorbid condition	080	1.0404	×	$4,000	=	$4,162
Respiratory infections and inflammations with complication/comorbid condition	079	1.8144	×	$4,000	=	$7,258

Source: D. D'Abate Cicenas, Increasing Medicare reimbursement through improved DRG coding, *Reimbursement and Insurance Coverage for Nutrition Services* (Chicago: American Dietetic Association, 1991), p. 53. Used with permission of Ross Products Division, Abbott Laboratories, Columbus, OH 43216. © 1990 Ross Products Division, Abbott Laboratories.

If you are either very poor, blind, disabled, over 65, male, female, white, or live in a middle- or upper-class neighborhood in a large urban center, you belong to a privileged class of health care recipients, and your chances of survival are good. . . . But, if you are none of these, if you are only average poor, under 65, female, black, or live in a low-income urban neighborhood, small town, or rural area, you are a disenfranchised citizen as far as health care rights go, and your chances of survival are not good.[65]

TABLE 3-3 *Sample Payment Based on DRGs, with and Without Complication/Comorbid Condition*

In 1983, a presidential commission studying ethical issues in medicine reported, "Society has a moral obligation to ensure that everyone has access to adequate [health] care without being subjected to excessive burdens."[66] Proponents of this view argue that just as the federal government provides for defense, postal delivery, and certain other services, it should provide at least a *minimal* amount of basic health care.[67]

View the ADA Web site (www.eatright.org) for an overview of possible career paths for nutrition professionals.

This debate leads to another question: access to what? What *is* an acceptable level of health care? The states that have considered or passed health care plans for their uninsured have aimed at providing "basic" or "minimum" health care benefits unlike the "comprehensive benefits" offered through the national health plans of other industrialized countries.

Comprehensive benefits, of course, do not necessarily mean unlimited care. The right to health care in Britain, Germany, and Canada does not mean the right to all treatments. Although most services provided in these countries are covered, the extent to which services are offered varies substantially across countries. Equity in health care in reality means a commitment to providing some common, adequate level of care. As yet, however, no country has explicitly determined what this level is.[68]

See Appendix D for a comparison of national health care systems in other countries.

In countries with universal access, referral systems tend to restrict access to high-technology services while maintaining comprehensive coverage of services. This is different from the U.S. approach of providing open access to technological services but restricting the type and quantity of services that are covered under the various insurance plans.[69]

Quality and Cost-Effectiveness in Health Care

Unfortunately, spending the most dollars on health care hasn't made us the healthiest. On the basis of crude outcome measures such as life expectancy and access to

Practice guidelines Guidelines to be used by doctors, hospitals, and other health professionals for treating various conditions in order to ensure the most cost-effective care.

Protocol Detailed guidelines for care that are specific to the disease or condition and type of patient.

Outcome An end result of the health care process; a measurable change in the patient's state of health or functioning.

Cost-effectiveness analysis An approach to evaluation that takes into account both costs and outcomes of intervention for a specific purpose. The analysis is especially useful for comparing alternative methods of intervention.

Protocols have been developed for a variety of conditions including: enteral and parenteral feeding support, gestational diabetes, AIDS, hyperlipidemia, hypertension, irritable bowel syndrome, insulin-dependent and non-insulin-dependent diabetes mellitus, oncology, congestive heart failure, pressure ulcer management, pre-end stage renal disease, chronic obstructive pulmonary disease, high-risk prenatal care, weight management, and anorexia and bulimia nervosa. A sample protocol for a client with hyperlipidemia is provided in Appendix E.

care, the United States is below many other industrialized countries. The rate of heart disease in Canada is 20 percent lower than in the United States. Canadians' average life span—78 years—is almost 2 years longer than Americans.[70]

In an effort to enhance the quality, efficiency, and effectiveness of the health care system, policy makers are urging physicians and other health professionals to develop **practice guidelines** or **protocols** that clearly specify appropriate care and acceptable limits of care for each disease state or condition. Care delivered according to a protocol has been linked with positive **outcomes** for the patient or client.[71] Examples of outcomes include measures of control (serum lipid profiles, glycolated hemoglobin), quality of life, dietary intake, or patient satisfaction. The ADA has recently developed 21 client protocols that define the minimum number of office visits and activities required for successful nutrition intervention and the outcomes that can be expected from the dietetics professional implementing the protocol.[72] Nutrition protocols serve as frameworks to help practitioners in the nutrition assessment, development, and evaluation of nutrition interventions. In addition, protocols that produce positive outcomes should include cost-effectiveness information. The ADA encourages all its practitioners to document the **cost-effectiveness** of nutrition services.

Cost-Effectiveness of Nutrition Services

Community nutritionists need to compete successfully for a fair share of the health care dollar. To do so, they must document the demand for nutrition services and their effectiveness so they can market those services to health care officials, providers, payers, and the public.

Obviously, no payer in the health care system wants additional costs. For a new technology or service, including nutrition services, to be a reimbursable benefit, it must prove its cost-effectiveness. Only services that have a proven impact on the quality of patient care will be funded. As Simko and Conklin have said, no expenditure of resources is justified for a service that fails to achieve its intended outcome.[73]

Cost-effectiveness studies compare the costs of providing health care against a desirable change in patient health outcomes (for example, a reduction in serum cholesterol in a patient with hypercholesterolemia).[74] Figure 3-6 shows a model for testing the costs and benefits of nutrition services. As this model shows, effective nutrition therapy can produce economic benefits as a result of altered food habits and risk factors.

Developing standardized protocols of care (practice guidelines) for nutrition intervention is considered a must for achieving payment for nutrition services and expanding current levels of third-party reimbursement.[75] An example of standardized practice guidelines developed by the National Cholesterol Education Program (NCEP) is shown in Figure 3-7. Note the orchestrated steps for identifying persons at risk for heart disease and establishing goals for dietary intervention.

Documentation of specific *outcomes* of nutrition intervention—clinical data, laboratory measures, anthropometric measures, and dietary intake data—is also necessary. Figure 3-8 on page 95 shows examples of outcome measures of nutrition intervention in burn injury, prenatal care, diabetes, and obesity.[76] When determining the outcomes of a given intervention, remember to ask the following questions: (1) Does the nutrition intervention make a difference in terms of disease-specific

Nutrition Intervention
(Screening/Assessment/Counseling)

Increase in knowledge,
skills, and motivation

Changes in food habits

Altered Risk Factors

Positive outcomes
Improved Health and Well-Being

Economic Benefits
(Decreased health costs)

FIGURE 3–6 *Benefits of
Nutrition Intervention*

Source: Adapted from M. Mason
and coauthors, Requisites of
advocacy: Philosophy, research,
documentation. Phase II of the costs
and benefits of nutritional care,
*Journal of the American Dietetic
Association* 80 (1982): 213.

indicators? (2) Is the nutrition intervention worth the cost? Table 3-4 on page 96 offers steps for developing protocols that enhance the effectiveness of nutrition services and make it easier to evaluate their quality and effectiveness.[77]

Health Care Reform, American Style

Practically all industrialized countries except the United States have national health care programs.[78] Coverage is generally universal (everyone is eligible regardless of health status) and uniform (everyone is entitled to the same benefits). Costs are paid entirely from tax revenues or by some combination of individual and employer premiums and government subsidization.

The concept of government-sponsored comprehensive health care is not new to the United States.[79] In 1934, President Franklin D. Roosevelt strongly supported national health insurance (NHI) and almost pushed to have it included with old age and unemployment insurance in the Social Security Act of 1935. Fearing that NHI might jeopardize passage of the Social Security Act, however, he decided to drop the proposal. As a result of World War II and the passage of the Hill-Burton Act of 1946, federal monies were diverted away from NHI and used for construction of new hospitals. Two decades later, the nation shifted its focus from NHI to providing for those without private insurance. Consequently, through the efforts of Presidents John F. Kennedy and Lyndon B. Johnson, Congress enacted the Social Security Amendments of 1965, which created Medicare (Title XVIII) and Medicaid (Title XIX).

The Hill Burton Act
(Hospital Survey and
Construction Act) made
federal funds available for
the first time for hospital
construction, expansion, or
improvement.

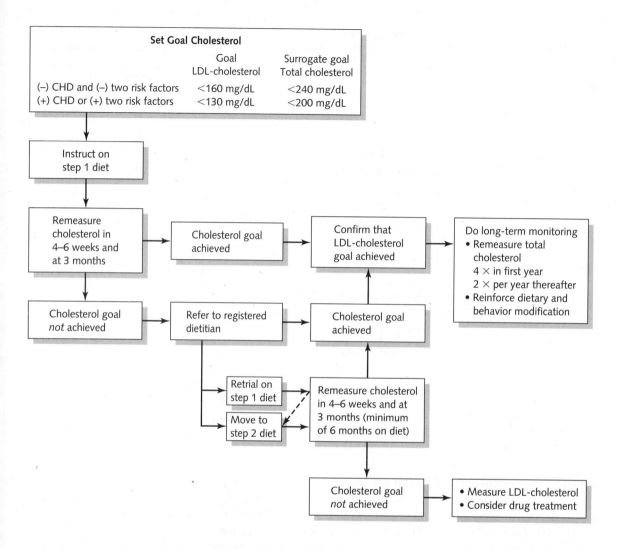

FIGURE 3-7 *National Cholesterol Education Program Guidelines for Dietary Treatment*

Source: National Cholesterol Education Program, *Highlights of the Report of the Expert Panel on Detection, Evaluation, and Treatment of High Blood Cholesterol in Adults* (Washington, D.C.: U.S. Department of Health and Human Services, 1987).

Now, almost 35 years later, increased health care costs, decreased patient satisfaction with available health care, and the current administration's health care reform initiative have together highlighted the need for a new approach to health care. Rather than propose a comprehensive reform of health care, Congress is now considering incremental reforms on broad issues such as health insurance reform, physician malpractice reform, and incentives to businesses for including health promotion initiatives in their insurance plans.[80]

Health care reform for the United States raises a formidable list of issues including overall cost containment, universal access, emphasis on prevention, and reduction in administrative superstructure and costs.[81] These issues require difficult decisions. Consider the following questions:[82]

- Who should be covered?
- How can coverage be increased to reach all people?

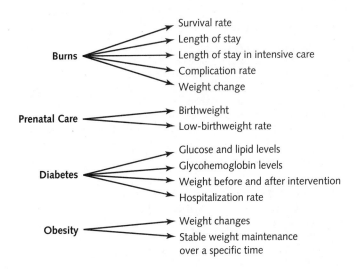

FIGURE 3-8 *Measurable Outcomes of Nutrition Intervention*

Source: R. Gould, The next rung on the ladder: Achieving and expanding reimbursement for nutrition services. © The American Dietetic Association. Reprinted by permission from *Journal of the American Dietetic Association* 91 (1991): 1383.

- What services should be considered in basic health care packages?
- Should health care cover both acute problems and prevention?
- Who should decide what constitutes preventive services?
- Who will pay for this coverage—consumers, employers, government?
- Where will government get the money to pay for it?
- How can health care costs be reduced or contained?
- What are the advantages and disadvantages of managed competition versus single-payer systems?

Although much attention has been focused on the Canadian system, most analysts believe that political and practical considerations will prevent the wholesale adoption of a Canadian or similar plan in the United States.[83] The Canadian plan is government run, offering minimal choice outside the main delivery network. Because costs are sometimes controlled by limiting the availability of technology, long waits to receive care are possible. For these reasons, some predict that the U.S. system will evolve into one similar to the German system, in which the bulk of policy making is left to the private sector. The German government closely regulates this private sector to achieve national goals, however.

While the government remains undecided on what kind of health care system is needed or how to pay for it, health care reform is evolving at an accelerating rate without legislation. The health care industry's determined efforts to curb the growth of costs while increasing access to services has stimulated the growth in managed care. According to one survey, the number of employees enrolled in managed care health plans rose from 52 percent in 1993 to 71 percent in 1995.[84]

Two of the main health reform proposals being considered are the **play-or-pay** and **single-payer** systems.[85] Play-or-pay is an employer-based system in which the employer must either offer health insurance or pay a new federal payroll tax. The single-payer system is a **national health insurance** system that would tax Americans at a rate that would cover all health benefits regardless of the individual's employment status.

Play-or-pay plans A system of paying for health care in which employers must either provide coverage for employees or pay a tax to finance a public insurance plan covering the uninsured.

Single-payer plan A system of paying for health care in which the government or some other entity is the sole payer, financed through a combination of payroll, corporate, and income taxes.

National health insurance A system in which a person is guaranteed coverage merely by being a citizen of the country or resident of the state involved.

TABLE 3-4

Development of Patient Care Protocols

1. Define the patient population:

- Select a prevalent problem (diagnosis).
- Select a population or problem for which evidence (in the literature or your own experience) indicates that nutrition intervention can produce clinical outcomes.
- Define the population as specifically as possible, based on medical diagnosis, state of diagnosis if relevant (e.g., new diagnosis versus long-standing diagnosis of diabetes), and other relevant factors (medication, age, other medical conditions). Example: hypercholesterolemic patients (cholesterol ≥ 200 mg/dL) who are either newly diagnosed or first-time referrals.

2. Define the treatment:

- Base standards for treatment on accepted methods of nutritional management (e.g., the National Cholesterol Education Program guidelines; National Institutes of Health Consensus Panel recommendations on diabetes management).
- Emphasize individualization of specific dietary patterns within limits of treatment standards.
- Specify expected length of treatment (e.g., three-month trial of diet) and expected minimum number of visits.
- Specify expected length of each visit as well as length of time between visits.
- Outline specific activities or topics to be addressed during the course of treatment and provide appropriate educational materials and tools.

3. Identify expected outcomes:

- Include as appropriate:

  Anthropometric measurements (weight, body mass index, skinfolds, etc.)
  Laboratory values (fasting blood sugar, hemoglobin A_{1c}, cholesterol, lipid profile, etc.)
  Clinical data (medication change, blood pressure change, etc.)
  Dietary intake

- Specify appropriate points in time when outcomes should be measured.
- If possible, specify expected magnitude of change and associated timeline.
- Identify expected and measurable intermediate outcomes (e.g., positive changes in knowledge, behavior, decision making, involvement in self-care).

Source: M. K. Fox, Defining appropriate nutrition care. © 1991, The American Dietetic Association. Reprinted by permission from *Reimbursement and Insurance Coverage for Nutrition Services,* 1991, p. 93.

Managed competition The creation of large groups of insurance purchasers to give them the bargaining power needed to create strong incentives for providers to supply the highest- quality service for the lowest possible cost.

Integrated delivery system A combination of two or more traditionally separate medical system components, such as physician, hospital, outpatient services, administration, insurance carrier.

A third proposal for health care reform is a concept that is growing in popularity and has emerged under a variety of names: **managed competition, integrated delivery system/network,** and **accountable health plans (AHPs).** The theory of managed competition holds that the quality and cost of health care delivery will improve if independent provider groups compete for consumers.[86] Embraced favorably by the current administration, managed competition is based on two assumptions:[87]

- The free market approach to health care reform is unworkable because of the imbalance of power between sellers (insurance companies and providers) and purchasers (businesses and individuals).
- Consumers have no incentive to shop among health plans because employer-based insurance is deductible by employers and nontaxable for employees.

Under managed competition, health care is restructured into a vertically integrated, full-service network of seamless health services. Physicians, hospitals, and other health care providers would form regional competing networks to provide a continuum of health care services required under a defined, uniform package of benefits. The success of such community-based networks would depend on their ability to provide quality services at a reasonable cost.[88] These networks would contract with **health insurance purchasing cooperatives (HIPCs)** to provide health care to consumers.[89] Unlike the traditional fee-for-service system, providers would offer standardized packages of health care benefits for fixed per capita rates. Because Medicare may be used as the model for a basic benefits package, the ADA has urged policy makers that Title XVIII of the Social Security Act be amended so that nutrition services will be included in the Medicare package of covered benefits.[90]

On the 200th anniversary of the convening of the nation's first Congress, Howard Nemerov, the poet laureate of the United States, read a poem to a joint session of Congress. It concluded with the following lines:

Praise without end the go-ahead zeal
of whoever it was invented the wheel;
But never a word for the poor soul's sake
that thought ahead and invented the brake.

The poet has clearly identified two essential elements of any future U.S. health care system: "we must think ahead and we must invent (and apply) a brake."[91]

Nutrition as a Component of Health Care Reform

During the summer of 1992, the House Select Committee on Aging held a hearing on "Adequate Nutrition: The Difference Between Sickness and Health for the Elderly." The purpose of the hearing was "to examine the critical role that nutrition plays in the lives of the American elderly population, and to determine how we can optimize the delivery of nutrition services, which include screening, assessment, counseling, and therapy to this population."[92]

As a result of this hearing, its sponsor, Congressman Edward Roybal stated:

Now I am even more convinced that nutrition services play a vital role in maintaining the health, independence, and quality of life of older Americans. Nutrition services must become an integral part of the health care services provided to not only the elderly, but every citizen of the United States. The benefits of proper nutrition have been shown time and time again. Nutrition screening, assessment, and counseling save money. When an older person is malnourished, he/she is at risk for disease and other health problems. Eighty-five percent of all older persons have one or more chronic diseases, such as diabetes, osteoporosis, atherosclerosis, hypertension, and cancer. Nutrition is linked to prevention and treatment of these diseases. . . . We must change the system so that nutrition services are specifically reimbursable and not just included in administrative funds. Nutrition services must be made available to elderly Americans in preventive, acute, long-term care, and home health settings. Nutrition screening, to identify those at risk can be a cost-effective prevention measure. . . . Nutritional care

Accountable health partnerships/plans (AHPs) The plans that current managed care plans (HMOs, PPOs) are encouraged to become under managed competition. AHPs must offer, at minimum, the federally mandated standard benefits package.

Health insurance purchasing cooperatives (HIPCs) The heart of managed competition, HIPCs act as purchasing agents for large groups of consumers. They would offer a range of plans to consumers.

should be considered specialized care and should be reimbursed just as respiratory, occupational, and physical therapies are.[93]

Many believe that nutrition services are the cornerstone of cost-effective prevention and are essential to halting the spiraling cost of health care. The ADA has urged that provision of nutrition services be included in any health care reform legislation.[94]

In addition, health care reform legislation needs to recognize the registered dietitian as the nutrition expert of the health care team with a scope of practice that includes the following:[95]

- *Nutrition assessment* for the purpose of determining individual and community needs and recommendations of appropriate nutrient intake to maintain, recover, or improve health
- *Nutrition counseling and education* of individuals, families, community groups, and health professionals
- *Research and development* of appropriate nutrition practice guidelines
- *Administration* through *management* of time, finances, personnel, protocols, and programs
- *Consultation* with patients, clients, and other health professionals
- *Evaluation* of the effectiveness of nutrition counseling/education and community nutrition programs

The Benefits of Medical Nutrition Therapy

One cannot have good health without proper nutrition. Conversely, poor nutrition contributes substantially to infant mortality, retarded growth and development of children, premature death, illness, and disability in adults, and frailty in the elderly, causing unnecessary pain and suffering, reduced productivity in the workplace, and increased health care costs.[96] Note the large sums of money spent on preventable diseases in the following list:[97]

- $136 billion spent for coronary heart disease in direct health expenditures alone
- More than $11 billion spent for stroke health care
- More than $72 billion spent for cancer treatment including lost productivity
- Between $3.5 and $7.5 billion spent annually on low-birthweight infants. Medicaid paid almost $19,000 per delivery of a low-birthweight infant versus $3,500 per delivery of a normal-weight infant.
- $288 billion, or 36 percent of health care costs, spent for older citizens, while Medicare spent just $97.2 billion on older citizens in 1992
- $20 billion spent annually on diabetes treatment
- Another $33 billion spent annually on illusionary "quick fix" weight-loss solutions by 65 million Americans

The contribution of nutrition to preventing disease, prolonging life, and promoting health is well recognized. Accumulated evidence shows that when nutrition services are integrated into health care, diet and nutritional status change with the following results:[98]

- The birthweight of infants born to high risk mothers improves.
- The prevalence of iron-deficiency anemia is reduced.

- Weight reduction and long-term weight maintenance are achieved.
- The rate of dental caries declines.
- Serum cholesterol and the risk of heart attacks are reduced.
- Glucose tolerance in persons with diabetes improves.
- Blood pressure in hypertensive patients is lowered.

The benefits of providing nutrition services far outweigh the costs of providing those services. The U.S. General Accounting Office review of the Special Supplemental Nutrition Program for Women, Infants, and Children (WIC) indicates that every dollar invested in WIC for pregnant women yields up to $4.21 in Medicaid savings. Nevertheless, in spite of the current evidence linking diet to disease, the United States spends only 4 percent of its health care dollars on disease prevention.[99]

Medical Nutrition Therapy and Medicare Reform

Nutrition services are not routinely covered in public programs (except WIC and, in some states, Medicaid) and only sporadically reimbursed by private health insurance. Some insurance companies will pay for outpatient nutrition counseling when it is ordered by a physician and provided by a registered dietitian.

The ADA believes that reimbursement for nutrition services through both Medicare and Medicaid is wholly inadequate. Often, the elderly choose not to seek appropriate nutrition services because Medicare's coverage is limited and they are unable to pay for the services themselves.[100] The **Nutrition Screening Initiative** recommends that nutrition screening for the elderly be included in the U.S. health care system and that Medicare cover and reimburse nutrition assessment and treatment for those found to be at nutritional risk. Such action is considered crucial to lowering health care costs, since malnourished persons have longer hospital stays and higher hospital costs than persons without malnutrition.[101]

Since 1992, the legislative priority of ADA has been the inclusion of **medical nutrition therapy** as a covered benefit in health care delivery.[102] Since the failure of Congress to pass President Clinton's 1993 Health Security Act and other health care reform bills, the ADA Health Care Reform Team has been focused on securing a mechanism for nutrition reimbursement under existing federal insurance programs. The Medical Nutrition Therapy Bill of 1997 (H.R. 1375) requests that Medicare be amended to cover nutrition therapy as an outpatient benefit under Part B of the Medicare program.

An ADA-financed independent study projected the cost of extending coverage of medical nutrition therapy to all Medicare beneficiaries under Medicare Part B to be less than $370 million over 7 years, when savings are considered. Savings would be greater than costs after the third year of enactment (see Figure 3-9).[103] For example, if coverage began in 1998, in 2001 an additional cost to Medicare Part B of $389 million would be offset by a reduction in cost to Part A of $401 million, resulting in a net savings of $11 million. The savings to the Medicare program comes from fewer hospital admissions and fewer complications requiring a physician's visit. The data used in the study were particularly significant for persons with diabetes and cardiovascular disease. Spending for diabetes and cardiovascular disease comprises about 60 percent of annual Medicare spending.[104] In the long run the program would save more in medical expenses than it costs to operate.

Nutrition Screening Initiative A program of the American Academy of Family Physicians, the ADA, and the National Council on Aging, formed in 1990 as a multifaceted effort to promote nutrition and improved nutritional care for the elderly.

Medical nutrition therapy The range of specific medical nutrition therapies for various conditions is determined following a complete assessment of the client's nutritional status. Medical nutrition therapy includes dietary modifications and nutrition counseling as well as more complex methods of nutrition support using specialized nutrition therapies (for example, nutritional supplements and enteral and parenteral feedings).

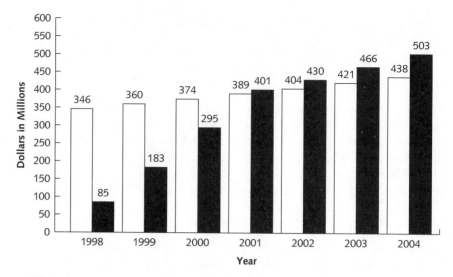

FIGURE 3-9 *Saving Medicare Millions*

Source: Lewin Group, 1997.

☐ Additional costs for expanding coverage.

■ Savings come from reduced hospital admissions and reduced complications requiring a doctor visit.

In summary, medical nutrition therapy is an integral component of cost-effective medical treatment. It can reduce health costs by improving patient outcomes and reducing recovery time. The ADA believes that the coverage of appropriate medical nutrition therapy, when medically necessary, should be included in any basic health care benefit package.

On the Horizon: Changes in Health Care Delivery

The future offers much that is positive for the profession of dietetics. The public's thinking about health and nutrition has matured, interest in positive health is growing steadily, and demand for health promotion products and services is increasing. Yet to be achieved, however, are the effective provision and allocation of resources such as nutrition services as part of preventive care. To accomplish this, a coordinated strategy for health care, political will, and active collaboration of both health care professionals and consumers of health care services will be required.

As we noted earlier, health care reform is a difficult undertaking. It involves more than cost containment and universal access. Community nutritionists need to educate the payers of health care about the inherent value of including nutrition services in their policies.[105] In arguing for reimbursable nutrition services, say, to an HMO benefits coordinator or local legislator, consider highlighting the following list of benefits:[106]

- Nutrition services are attractive, progressive health benefits that are relatively inexpensive compared with other types of benefits.

- Nutrition services benefits enhance the insurance product (employee benefit package).
- Nutrition services have a preventive medicine component (they help keep employees healthy).
- Nutrition services benefits attract healthy subscribers.
- Nutrition services are manageable—they can be easily documented.
- Nutrition services help patients become more self-reliant by helping them fight disease, avoid hospitalization, and reduce the use of other, more expensive medical therapies.
- Nutrition care speeds recovery.

Health care reform for the United States is certain, but the exact nature of the reform will continue to evolve. Undoubtedly, the changes required of the health care system will transform it slowly over time. Nevertheless, the health care reform process stands to have a major impact—for better or for worse—on our profession. It is imperative that you keep up-to-date with both national and local developments in health care reform.

Remember that change at the federal level begins with local advocacy. Remember, too, that the prescription for success is persistence. We need to maintain the pressure to achieve passage of meaningful legislation in health care reform. The Community Learning Activity that follows asks you, as a consumer of health care services, to explore both current legislation concerning health care reform and your own health care coverage benefits.

 ## Internet Resources

Check out these Internet addresses for current information related to nutrition, health care, and health care reform.

Agency for Health Care Policy and Research	www.ahcpr.gov
An arm of the U.S. Dept. of Health and Human Services.	
American Association of Health Plans	www.aahp.org
A trade group for HMOs and PPOs.	
American Association of Retired Persons	www.aarp.org
Offers approval for select managed care plans.	
American Dietetic Association's Washington Report	www.eatright.org
Provides current legislative and political updates from the Government and Legal Affairs Group of ADA.	
Health Care Financing Administration	www.hcfa.gov
Federal agency that administers Medicare and Medicaid.	
Health Insurance Association of America	www.hiaa.org
Trade association for the health insurance industry.	

Continued

Internet Resources—continued

Health Pages   **www.thehealthpages.com**
   Issues report cards on major managed care plans.
Healthfinder   **www.healthfinder.gov**
   Serves to organize the mass of health and nutrition
   information available from federal and state agencies.
Intelihealth   **www.intelihealth.com**
   Offers a wide collection of consumer health information
   from the National Heart, Lung, and Blood Institute, the
   National Institutes of Health, the National Library of
   Medicine, and others.
Joint Commission on Accreditation of Healthcare
Organizations (JCAHO)   **www.jcaho.org**
   The primary accreditation organization that evaluates
   hospitals and outpatient clinics.
National Committee for Quality Assurance (NCQA)   **www.ncqa.org**
   Evaluates managed care plans in terms of patient records,
   complaints, equipment, and personnel.
Office of Alternative Medicine   **altmed.od.nih.gov**
   Evaluates complementary and alternative health therapies.
Social Security Administration   **www.ssa.gov**
   Site for information on the Medicare and Medicaid
   programs.

Community Learning Activity

Activity 1

A good place to start pursuing health care reform is in your own backyard. Does your own health care insurance plan (or that of your family or college) include reimbursement for services rendered by a registered dietitian or for nutritional products? If the answer is no, plan a strategy to have nutrition services become part of the insurer's policy. Write a practice letter to the insurer's benefits coordinator. Start by outlining the potential benefits of nutrition services.

Activity 2

Most state insurance commissioner's offices track the top 10 insurers for their population. Try to obtain a copy of this list for the class. Next, obtain a sample policy from as many of these companies as possible and compare their benefits and exclusions. Are nutrition services covered? If the answer is no, proceed as in step 1.

Activity 3

Grassroots lobbying is an effective way for people to influence legislation that may affect them in one way or another. Lobbying, as discussed in Chapter 2, can take many forms—personal meetings with a legislator or a legislator's aide, a letter, or a phone call. Legislators and their aides are sensitive to public opinion on issues. Get to know your congressperson's position on health care reform. Does he or she support the Medical Nutrition Therapy Act? You can call your congressperson's local district office to obtain copies of bills and testimony. The phone number can be found in the Federal Government section (usually blue pages) of your telephone book, or on-line at www.senate.gov and www.house.gov.

The following House subcommittees have jurisdiction over health issues:

- House Ways and Means Committee/Subcommittee on Health
- Energy and Commerce Committee/Subcommittee on Health

The following Senate committees are also involved with health care reform:

- Committee on Labor and Human Resources
- Committee on Finance

Does your congressperson or senator serve on one of these committees? If you need to learn a legislator's name or find out what committees he or she serves on, you can contact the League of Women Voters in your area, or contact the Web sites mentioned above.

Activity 4

1. Access MEDLINE at www.nlm.nih.gov. The database of greatest interest to alternative medicine research is MEDLINE (MEDLARS On-line). There are 23 main headings in MEDLINE under the term *alternative medicine*. This activity asks you to find an article citation in MEDLINE using either PubMed or Grateful Med for each of the following topics, and print out the abstract to share with your class:
   - Echinacea and the immune system
   - Ginkgo biloba and Alzheimer's disease
   - Soy protein and breast cancer

2. Using the same key words you used in Part A (for example, echinacea, ginkgo biloba, etc.), conduct a literature search using one of the search engines (Infoseek, Altavista, or other). Compare the type of information you found using MEDLINE with the information you found using a search engine.

3. Write down the address for the Herb Research Foundation as found using a search engine.

Professional Focus | *Leading for Success*

Supreme Court Justice Porter Stewart once said about obscenity, "I cannot define it for you, but I know it when I see it."[1] The same might be said about leadership, although the wealth of research in this area has helped define the behaviors and attitudes of good leaders.

Leading is not the same as managing. The difference between the two has been described by Warren Bennis, an internationally recognized consultant, writer, and researcher on leadership: "Leaders are people who do the right thing; managers are people who do things right. Both roles are crucial, but they differ profoundly."[2] In general, managers crunch numbers, orchestrate activities, and control supplies, projects, and data. Leaders fertilize and catalyze. They enhance people, allowing them to stretch and grow. Where managers push and direct, leaders pull and expect.[3]

What exactly is leadership? Leadership is "the process whereby one person influences others to work toward a goal."[4] Leaders understand how an organization works (or doesn't work, as the case may be) and how people work—that is, what motivates them to be top performers. Leaders have the power to project onto other people their vision, their inspiration, their ideas.[5]

The traits of successful leaders number more than one hundred, according to recent studies, although not all experts agree on the importance of every trait.[6] Some of the attributes commonly ascribed to leaders include intelligence, credibility, energy, sociability, discipline, courage, and generosity. Integrity is essential. Integrity has been described as "the most important leadership principle that you will demonstrate to your leadership team and your followers."[7] Integrity means adhering to a high standard of honesty. Leaders with integrity strive to present their values and character honestly in their dealings with other people.

Accountability is also important. Accountability means doing what you say you are going to do. No matter whether the issue is large or small, leaders follow through on their promises and commitments. They know that *not* following through on a promise or commitment reduces their credibility and diminishes the trust other people have placed in them.[8] Leaders hold themselves accountable.

Leaders are forward thinking. They see the big picture and have a vision about where they are going—and where they want their teams to go. Leaders can describe what a team, department, or organization will look and feel like in 6 months, 1 year, or 5 years. Make no mistake, though—

leaders are not ones to complain about the way things are or the way things might be. They are optimistic about the future, seeing many possibilities and positive outcomes. Their enthusiasm is contagious.

Leaders come in all shapes, sizes, and temperaments. They are found in the executive suite and on the factory floor, among support staff and across middle management. In their seminar programs on leadership, James Kouzes and Barry Posner used to say, "Leaders go places. The difference between managing and leading is the difference between what you can do with your hands and what you can do with your feet....You can't lead from behind a desk. You can't lead from a seated position. The only way you can go anyplace is to get up from behind the desk and use your feet."[9] Then, they got a letter from the president of a computer software services company who indicated that about one-third of his employees had disabilities and several were in wheelchairs; yet, they led quite effectively. Kouzes and Posner reformulated their example to reflect the diversity of leaders. You can have a disability and be a leader; you can be a manager and be a leader; you can be a newly minted community nutritionist in your first job and be a leader.

Warren Bennis believes that leaders all have a guiding vision, passion, integrity, curiosity, and daring.[10] They are self-confident and instill self-confidence in others. They are willing to take risks—and responsibility.[11] Following is a description of the principles of leadership outlined by General H. Norman Schwarzkopf, who guided U.S. troops to victory in the Gulf War.*

- **You must have clear goals.** Having specific goals and articulating them clearly makes it easy for everyone involved to understand the mission.
- **Give yourself a clear agenda.** First thing in the morning, write down the five most important things you need to accomplish that day. Whatever else you do, get those five things done first.
- **Let people know where they stand.** You do a great disservice to an employee or student when you give high marks for mediocre work. The grades you give the people who report to you must reflect reality.
- **What's broken, fix now.** If it's a problem, fix it now. Problems that aren't dealt with lead to other problems, and in the meantime, something else breaks down and needs fixing.

- **No repainting the flagpole.** Make sure that all the work your people are doing is essential to the organization.
- **Set high standards.** Too often we don't ask enough of people. People generally won't perform above your expectations, so it is important to expect a lot.
- **Lead, and then get out of the way.** Yes, you must put the right people in the right place to get the job done, but then step back. Allow them to own their work.
- **People come to work to succeed.** Nobody comes to work to fail. Why do so many organizations operate on the principle that if people aren't watched and supervised, they'll bungle the job?
- **Never lie. Ever.** Lying undermines your credibility. Be straightforward in your thinking and actions.
- **When in charge, take command.** Leaders are often called upon to make decisions without adequate information. It is usually a mistake to put off making a decision until all the data are in. The best policy is to decide, monitor the results, and change course if necessary.
- **Do what's right.** "The truth of the matter," said Schwarzkopf, "is that you *always* know the right thing to do. The hard part is doing it."

---

*Source: Reprinted with permission of *Inc.* magazine, Goldhirsh Group, Inc., 38 Commercial Wharf, Boston, MA 02110 (http://www.inc.com). *Schwarzkopf on Leadership, Inc.* 14 (January 1992): 11. Reproduced by permission of the publisher via Copyright Clearance Center, Inc.

References

1. As cited in M. Brown and B. P. McCool, High-performing managers: Leadership attributes for the 1990s, in *Human Resource Management in Health Care*, ed. M. Brown (Gaithersburg, MD: Aspen, 1992), p. 22.
2. W. Bennis, Learning some basic truisms about leadership, *National Forum* 71 (1991): 13.
3. J. D. Batten, *Tough-Minded Leadership* (New York: AMACOM, 1989), p. 2.
4. As cited in D. Hellriegel, J. W. Slocum, Jr., and R. W. Woodman, *Organizational Behavior*, 7th ed. (St. Paul: West, 1995), p. 342.
5. P. J. Palmer, Leading from within, in *Insights on Leadership*, ed. L. C. Spears (New York: Wiley, 1998), pp. 200–1.
6. B. Czarniawska-Joerges and R. Wolff, Leaders, managers, entrepreneurs on and off the organizational stage, *Organization Studies* 12 (1991): 529–46.
7. N. L. Frigon, Sr., and H. K. Jackson, Jr., *The Leader: Developing the Skills and Personal Qualities You Need to Lead Effectively* (New York: AMACOM, 1996), pp. 15–36.
8. R. Carlson, *Don't Worry, Make Money* (New York: Hyperion, 1997), 151–52.
9. As cited in J. M. Kouzes and B. Z. Posner, *Credibility: How Leaders Gain and Lose It, Why People Demand It* (San Francisco: Jossey-Bass, 1993), p. 88.
10. W. Bennis, *On Becoming a Leader* (Reading, MA: Addison-Wesley, 1989), pp. 39–41.
11. R. A. Heifetz and D. L. Laurie, The work of leadership, *Harvard Business Review* 75 (1997): 124–34.

References

1. T. M. O'Neal, Managed Care: Its impact on the delivery of health care in Texas, *Texas Monthly* (July 1995): 44–66.
2. C. Tokarski, 1980s prove uncertainty of instant cures, *Modern Healthcare* 51 (January 8, 1990): 51.
3. O'Neal, Managed care, pp. 44–66.
4. President's Column, A new era for prevention, *The Nation's Health* (January 1990): 2.
5. G. E. A. Dever, *Community Health Analysis*, 2nd ed., (Gaithersburg, MD: Aspen, 1991), p. xiii.
6. T. Hancock, Beyond health care: Creating a healthy future, *The Futurist* 16 (1982): 4.
7. Position of the American Dietetic Association: Affordable and accessible health care services, *Journal of the American Dietetic Association* 92 (1992): 746.
8. Ibid., p. 746.
9. L. W. Sullivan, Healthy People 2000: Promoting health and building a culture of character, *American Journal of Health Promotion* 5 (1990): 5–6.
10. Ibid., p. 6.
11. The definitions provided in the margin throughout this section are from *Reimbursement and Insurance Coverage for Nutrition Services* (Chicago: American Dietetic Association, 1991), pp. 121–25.
12. Health Insurance Institute, *Sourcebook of Health Insurance Data* (New York: Health Insurance Institute, 1989).
13. R. S. Stern, A comparison of length of stay and costs for health maintenance organization and fee-for-service patients, *Archives of Internal Medicine* 149 (1989): 1185–88.
14. American Dietetic Association, *Medical Nutrition Therapy Across the Continuum of Care* (Chicago: American Dietetic Association, 1996), pp. 1–2.
15. S. Wolfe, Handling your health care, *Buyer's Market* 1 (1985): 2.
16. Stern, A comparison of length of stay and costs.
17. J. S. Ahluwalia, Health care in the United States: Our dynamic jigsaw puzzle, *Archives of Internal Medicine* 150

(1990): 256–58; and National Association of Insurance Commissioners and Health Care Financing Administration of the U.S. Department of Health and Human Services, *Guide to Health Insurance for People with Medicare*, Pub. No. HCFA-02110 (Washington, D.C.: U.S. Government Printing Office, 1997).

18. ADA protocols help secure coverage of medical nutrition therapy by Blue Cross Blue Shield of Massachusetts, *Journal of the American Dietetic Association* 98 (1998): 18.

19. Health care reform: A post election guide, *Harvard Health Letter* (January 1993): 11.

20. L. Stollman, ed., *Nutrition Entrepreneur's Guide to Reimbursement Success* (Chicago: American Dietetic Association, 1995), pp. 1–3.

21. Health care reform, *American Dietetic Association's Legislative Newsletter* (October 1991): 1–3.

22. Ahluwalia, Health care in the United States, p. 258.

23. E. Friedman, The uninsured: From dilemma to crisis, *Journal of the American Medical Association* 265 (1991): 2491–95.

24. Consumers Union, Health care today: Who's hurting, *Consumer Reports* 63 (1998): 7.

25. H. Deets, Early retirees, others need health insurance, *Modern Maturity* 41 (1998): 82.

26. B. Harvey, A proposal to provide health insurance to all children and all pregnant women, *New England Journal of Medicine* 323 (1990): 1216–20.

27. Consumers Union, Health care today.

28. Ahluwalia, Heath care in the United States, p. 258.

29. Dever, *Community Health Analysis*, p. 13.

30. Legislative Highlights: Finn offers testimony on nutrition and the elderly, *Journal of the American Dietetic Association* 92 (1992): 1064–65.

31. R. Carlson, Healthy people, *Canadian Journal of Public Health* 76 (1985): 27–32.

32. Ibid., p. 29.

33. L. W. Sullivan, Partners in prevention: A mobilization plan for implementing *Healthy People 2000*, *American Journal of Health Promotion* 5 (1991): 291–97.

34. A. Hawkins, The future of health care: Well Care, *Personal Health Management 97 News* (Summer 1997): 3–55.

35. A. Barrocas, Complementary and alternative medicine: Friend, foe, or OWA? *Journal of the American Dietetic Association* 97 (1997): 1373–76.

36. NIH Office of Alternative Medicine Clearinghouse, *OAM General Information Package: Frequently Asked Questions* (Silver Spring, MD: NIH Office of Alternative Medicine, 1988).

37. D. Eisenberg and coauthors, Unconventional medicine in the United States: Prevalence, costs, and patterns of use, *New England Journal of Medicine* 328 (1993): 246–52.

38. Alternative medicine: Time for a second opinion, *Harvard Health Letter* 23 (1997): 1–2.

39. C. Bartels and S. Miller, Herbal and related remedies, *Nutrition in Clinical Practice* 13 (1998): 5–19.

40. "Alternative" remedies find medicine's middle ground, *Trend Letter: The Global Network* 16 (1997): 16.

41. Atlantic Health System, *Atlantic Currents* (Winter 1998): 7.

42. D. M. Eisenberg, Advising patients who seek alternative medical therapies, *Annuals of Internal Medicine* 127 (1997): 61–69.

43. President's Page: Alternative medicine, *Journal of the American Dietetic Association* 97 (1997): 1431.

44. B. Stehlin, An FDA guide to choosing medical treatments, *FDA Consumer* (June 1995): 10–14.

45. Adapted from the recommendations by the Council on Scientific Affairs at the American Medical Association House of Delegates 1997 Annual Meeting, as cited in Barrocas, Complementary and alternative medicine.

46. C. Bezold, Health care in the U.S.: Four alternative futures, *The Furturist* 16 (1982): 14–18.

47. Pew Commission outlines reforms for health care regulation, *Journal of the American Dietetic Association* 97 (1997): 580.

48. Update on state licensure laws and ADA regulatory remarks, *Journal of the American Dietetic Association* 97 (1997): 1251.

49. S. J. Gillespie, State of cost benefits, in *Reimbursement and Insurance Coverage for Nutrition Services* (Chicago: American Dietetic Association, 1991), p. 107.

50. Wolfe, Handling your health care, p. 2.

51. Mr. Galley, Editorial, *Boston Globe*, September 1991, as quoted by M. K. Fox, Reimbursement practices and trends, presented at the American Dietetic Association's annual meeting in Dallas.

52. Ibid.

53. The discussion of bureaucracy in government-run health care is from E. Brodsky, Government-run health care isn't worth the wait, *Health and You* (Spring 1992): 2.

54. As stated by J. Neel, Healthcare: U.S. looks to German model, *Nature* 351 (1992): 433.

55. K. Hogue, C. Jensen, and K. M. Wiljanen, *The Complete Guide to Health Insurance* (New York: Avon, 1989); S. Finn and G. Martin, The shifting balance of power: A new decade of decision for dietitians, *Dietetic Currents* 18 (1991): 1–6, as cited in M. K. Fox, Overview of third party reimbursement, in *Reimbursement and Insurance Coverage for Nutrition Services* (Chicago: American Dietetic Association, 1991), pp. 3–5.

56. E. Ginzberg, The monetarization of medical care, *New England Journal of Medicine* 310 (1984): 1162–65.

57. The definitions provided in this section are from *Reimbursement and Insurance Coverage for Nutrition Services* (Chicago: American Dietetic Association, 1991), pp. 121–25.

58. The discussion on trends is adapted from M. K. Fox, *Overview of Third-Party Reimbursement* (Chicago: American Dietetic Association, 1991), pp. 3–5.

59. Stollman, *Nutrition Entrepreneur's Guide*.

60. S. W. Letsch, K. R. Levit, and D. R. Waldo, National health expenditures, 1987 health care financing trends, *Health Care Financing Review* 10 (1988): 109–29.

61. P. Stanfill and G. E. Soper, The economic realities of health care, *Health Administration Today* 1 (1988): 8–11.

62. D. Beck, The hospital's financial future, *The Health Care Supervisor* (January 1985): 1–10.

63. Stollman, *Nutrition Entrepreneur's Guide*.

64. Harvey, A proposal to provide health insurance, p. 1216.

65. P. deVise, *Misuses and Misplaced Hospitals and Doctors: A Locational Analysis of the Urban Health Care Crisis,* Resource Paper no. 2 (Commission on College Geography, 1973), p. 1.

66. President's Commission for the Study of Ethical Problems in Medical and Biomedical and Behavioral Research, *Report: The Ethical Implications of Differences in the Availability of Health Services* (Washington, D.C.: U.S. Government Printing Office, 1983), p. 22.

67. Ahluwalia, Health care in the United States, p. 256.

68. C. Grogan, A Comparison of Canada, Britain, Germany, and the United States, *Journal of Health Politics, Policy, and Law* 17 (1992): 213–32.

69. Ibid., p. 226.

70. A. Schmitz, Health assurance, *In Health* (January/February 1991): 39–47.

71. American College of Physicians, Access to health care, *Annals of Internal Medicine* 112 (1990): 641–61.

72. A. Inman-Felton, K. G. Smith, and E. Q. Johnson, eds., *Medical Nutrition Therapy Across the Continuum of Care* (Chicago: American Dietetic Association, 1997).

73. M. D. Simko and M. T. Conklin, Focusing on the effectiveness side of the cost-effectiveness equation, *Journal of the American Dietetic Association* 89 (1989): 485–87.

74. Ibid., p. 486.

75. R. Gould, The next rung on the ladder: Achieving and expanding reimbursement for nutrition services, *Journal of the American Dietetic Association* 91 (1991): 1383–84.

76. P. Splett, Effectiveness and cost effectiveness of nutrition care: A critical analysis with recommendations, *Journal of the American Dietetic Association* 91 (1991): S1–S50.

77. M. K. Fox, Defining appropriate nutrition care, *Reimbursement and Insurance Coverage for Nutrition Services,* pp. 89–95.

78. The discussion of national health care coverage is adapted from American College of Physicians, Access to health care, *Annals of Internal Medicine* 112 (1990): 641–61.

79. The discussion of U.S. history is adapted from Ahluwalia, Health care in the United States, pp. 256–58.

80. R. K. Johnson and A. M. Coulston, Medicare: Reimbursement rules, impediments, and opportunities for dietitians, *Journal of the American Dietetic Association* 95 (1995): 1378–80.

81. G. S. Omenn, Challenges facing public health policy, *Journal of the American Dietetic Association* 93 (1993): 643.

82. Division of Government Affairs, Health care reform, *Legislative Newsletter* (October 1991): 2.

83. This discussion regarding Canada is adapted from M. Hagland, Looking abroad for changes to the U.S. health care system, *Hospitals* 65 (1991): 30–35.

84. A. Foster Higgins & Co., Inc., *Health Benefits Cost Rose 2.1% in 1995* (Washington, D.C.: Foster Higgins, 1996); J. Mandelder, Health care reform, *Business & Health: The state of health care in America* 8 (1995): 30–35.

85. The definitions provided in the margin are from *Insuring the Uninsured: A Guide to the Proposals and the Players* (Washington, D.C.: Faulkner & Gray, 1991).

86. R. Kronick and coauthors, The marketplace in health care reform: The demographic limitations of managed competition, *New England Journal of Medicine* 328 (1993): 148–52.

87. Managed competition: Wonder drug or snake oil? *The Nation's Health* (March 1993): 1, 9–10.

88. C. D. Dauner, Toward the solution, *National Forum* (Summer 1993): 38–44.

89. The discussion on managed competition is adapted in part from J. K. Iglehart, Managed competition, *New England Journal of Medicine* 328 (1993): 1208–12.

90. M. Tate, Health care reform, *Public Health Nutrition Practice Group, The Digest* (Chicago: American Dietetic Association, Spring 1993), p. 3.

91. This quote and comment are from the *New York Times,* March 3, 1989, p. 10, and N. E. Davies and L. H. Felder, Applying brakes to the runaway American health care system, *Journal of the American Medical Association* 263 (1990): 73–76.

92. Legislative Highlights: Finn offers testimony.

93. E. Roybal, Support for nutrition services for the elderly, *Congressional Record* 138 (August 12, 1992): 119.

94. Position of the American Dietetic Association: Cost-effectiveness of medical nutrition therapy, *Journal of the American Dietetic Association* 95 (1995): 88–91.

95. Position of the American Dietetic Association: Nutrition services in health maintenance organizations and alternative delivery systems, *Journal of the American Dietetic Association* 87 (1987): 1391–93.

96. Nutrition Screening Initiative, *Managing Nutrition Care in Health Plans* (Washington, D.C.: Greer, Margolis, Mitchell, Burns, & Associates, 1996), pp. 1–5.

97. The list of diet-related disease is from *Executive Summary of the Legislative Platform of the American Dietetic Association: Economic Benefits of Nutrition Services* (Chicago: American Dietetic Association, 1992).

98. The list of benefits is from Nutrition services in state and local public health agencies, *Public Health Reports* 98 (1983): 7–20.

99. W. O. Weiner, Year 2000 objectives: Altering the medical model, *Medicine and Health Perspectives* (October 23, 1989): 3.

100. Legislative Highlights: Finn offers testimony.

101. G. Robinson, M. Goldstein, and G. M. Levine, Impact of nutritional status on DRG length of stay, *Journal of Parenteral and Enteral Nutrition* 12 (1988): 587–91.

102. A. M. Coulston, Health care reform: ADA's number one priority, *Topics in Clinical Nutrition* 11 (1996): 1–3.

103. Community Nutrition Institute, Dietitians pushing a bill for medical nutrition, *Nutrition Week* 27 (May 2, 1997): 2.

104. E. R. Monsen, From the environment to MNT: Dietitians face key issues, *Journal of the American Dietetic Association* 97 (1997): 360.

105. Public Policy News: Medicare reform offers ADA opportunity to promote MNT, *Journal of the American Dietetic Association* 97 (1997): 378.

106. The list of benefits is from K. Smith, How to argue for nutrition services, *DBC Dimensions* (Fall 1992): 7.

Chapter 4

A National Nutrition Agenda for the Public's Health

Outline

Learning Objectives

**After you have read and studied this
chapter, you will be able to:**

- Describe the relationship of nutrition
  research and nutrition monitoring to
  U.S. national nutrition policy.
- Describe five key components of the
  National Nutrition Monitoring and
  Related Research Program.
- Describe one advantage and one dis-
  advantage of the National Health
  and Nutrition Examination Survey
  (NHANES) series.

- Discuss the Dietary Reference
  Intakes, including the Estimated
  Average Requirement, Recommended
  Dietary Allowance, Adequate Intake,
  and Tolerable Upper Intake Level,
  and explain how they are used to plan
  and assess diets.
- Describe appropriate uses of current
  dietary guidance systems.

Something To Think About...

On my counter is my own huddle of voodoo dolls, starting with the A's, going through Super Ginkgo Biloba, ending on St. John's Wort. There are new weapons every hour. . . . Nutrition is an ever-shifting minefield. I must have the vigilance of a wolverine. – *Richard Woodley, coauthor of* Donnie Brasco *(as cited in the* New York Times Magazine, *February 22, 1998)*

Introduction

Time was when scientists investigating the role diet plays in health zeroed in on the consequences of getting too little of one nutrient or another. Until the end of World War II, in fact, nutrition researchers concentrated on eliminating deficiency diseases such as goiter and pellagra. Today, however, the focus is just the opposite. While an abundant food supply and the practice of fortifying foods with essential nutrients have virtually eliminated deficiency diseases in North America, diseases related to dietary excess and imbalance are widespread. Four of the 10 leading causes of death—namely, heart disease, cancer, stroke, and diabetes—have been linked to diet. Another three are associated with excessive alcohol consumption: liver disease, accidents, and suicides. Together these seven problems account for more than 70 percent of the two million deaths that occur each year, not to mention hospitalizations, time lost on the job, and poor quality of life. Dietary excesses and imbalances contribute to other ills as well, including high blood pressure, dental disease, and osteoporosis.[1]

How do we know that dietary imbalances exist in the United States and that certain population subgroups are at risk of malnutrition? And having determined who is malnourished, what guidelines exist to help community nutritionists and other health professionals address the nutritional needs of malnourished groups? The answers to these questions are found in the policy arena, for nutrition policy dictates the strategies used to determine who is malnourished and outlines the appropriate dietary guidance to improve nutritional intake.

National Nutrition Policy

Let's begin with a simple question: Does the United States have a national nutrition policy? By **national nutrition policy,** we mean a set of nationwide guidelines that specify how the nutritional needs of the American people will be met and how the issues of hunger, malnutrition, food safety, food labeling, food fortification, and nutrition research will be addressed. The answer is both yes and no. The answer is no in the sense that no one federal body or agency has as its sole mandate to establish, implement, and evaluate national nutrition policy. This deficiency in national policy making and planning was recognized more than 20 years ago, when Senator

National nutrition policy A set of nationwide guidelines that specify how the nutritional needs of the population will be met.

George McGovern, the chairman of the Senate Select Committee on Nutrition and Human Needs (the so-called McGovern Committee), called for the formation of such a body:

> We need a Federal Nutrition Office. The White House Conference on Food, Nutrition and Health recommended such an Office more than five years ago. Events since the 1969 meeting strongly reaffirm the importance of institutionalizing responsibility for nutrition policy. . . . We cannot continue to operate on the assumption that the increasingly complex threads affecting nutrition policy will automatically weave themselves together into a coherent plan.[2]

McGovern went on to say that the policy in existence at that time (in 1975) lacked focus, direction, and coordination, all of which contributed to growing conflicts within the administration over program priorities. The Panel on Nutrition and Government had offered a similar recommendation at the National Nutrition Policy Study Hearings in 1974: There should be an independent office, operating outside all existing agencies, whose function would be to coordinate and direct federal nutrition policy. This federal nutrition office would follow through on the commitments made by federal agencies in implementing a national nutrition plan, help develop surveillance systems to monitor the population's overall health and nutritional status, and guarantee that any secondary nutritional implications of major policy decisions would be recognized and published in a "nutrition impact statement."[3]

More than 20 years later, there is no Federal Nutrition Office, and nutrition policy in the United States is still fragmented. The problem with formalizing federal policy decisions in the nutrition arena lies in determining which agency should be the "power center" responsible for final decisions. This task is both complex and politically sensitive, because nutrition policy cuts across several policy areas, including agriculture, exports, imports, commerce, foreign relations, public health, and even national defense. No one federal agency can claim exclusive jurisdiction over nutrition issues. It is interesting that the comments made by McGovern in 1975 are still relevant in today's health care and policy environment:

> Nutrition is treated as a neglected stepchild of income maintenance programs which themselves are woefully inadequate. This narrow conception virtually denies the nutrition dimension in comprehensive health care, or even that nutrition is a health issue. This parochial view ignores disturbing questions about misleading food advertising and other issues totally unrelated to income inequality. It fails to grapple with the reality that even wealthy Americans are often nutritionally illiterate, and that arteriosclerosis and other diseases associated with the aging process affect more than the poor. These and other issues germane to the health and well-being of the American people go far beyond the perils of poverty, and require a much broader Federal conception of the nation's nutritional policy requirements.[4]

Even though no Federal Nutrition Office currently exists, the United States can still be said to have a national nutrition policy, however fragmented and disjointed it may be. D. J. Palumbo remarked that "we can assume that no matter what was intended by government action, what is accomplished *is* policy."[5] Thus, national nutrition policy in the United States manifests itself in food assistance programs, regulations to safeguard the food supply and ensure the proper handling of food products, dietary guidance systems such as the Dietary Guidelines for Americans and the Food Guide Pyramid, monitoring and surveillance programs, food labeling legislation, and other activities in the nutrition arena.

The activities that form the basis of the nation's agenda to improve the public's health are outlined in Figure 4-1.[6] Research results and data obtained from nutrition monitoring provide information that helps in decision making—and hence, policy making—within the two main federal agencies that deal with food and nutrition issues, the U.S. Department of Agriculture (USDA) and the Department of Health and Human Services (DHHS). Some policy decisions affect the types of data collected during nutrition surveys and research related to human nutrient needs. Some aspects of national nutrition policy, such as food assistance programs, are discussed in later chapters; this chapter focuses on two elements of U.S. nutrition policy—namely, national nutrition monitoring and surveillance activities and dietary guidance systems.

FIGURE 4-1 *Relationships Among Nutrition Research, Monitoring, and Policy Making*

Source: Adapted from C. Woteki and M. T. Fanelli-Kuczmarski, The national nutrition monitoring system, in *Present Knowledge in Nutrition*, 6th ed. (Washington, D.C.: International Life Sciences Institute, 1990), p. 416.

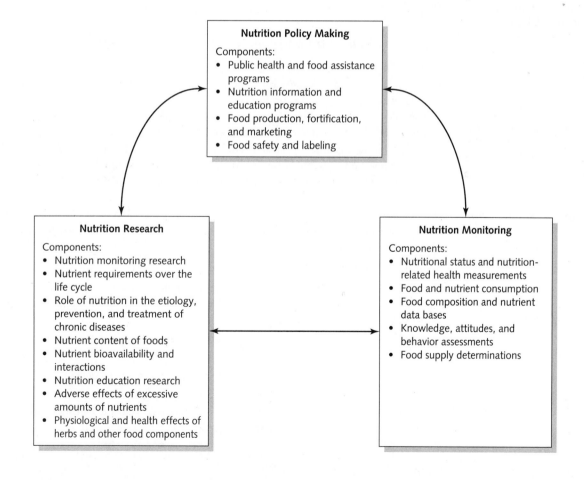

Nutrition Policy Making

Components:
- Public health and food assistance programs
- Nutrition information and education programs
- Food production, fortification, and marketing
- Food safety and labeling

Nutrition Research

Components:
- Nutrition monitoring research
- Nutrient requirements over the life cycle
- Role of nutrition in the etiology, prevention, and treatment of chronic diseases
- Nutrient content of foods
- Nutrient bioavailability and interactions
- Nutrition education research
- Adverse effects of excessive amounts of nutrients
- Physiological and health effects of herbs and other food components

Nutrition Monitoring

Components:
- Nutritional status and nutrition-related health measurements
- Food and nutrient consumption
- Food composition and nutrient data bases
- Knowledge, attitudes, and behavior assessments
- Food supply determinations

Nutrition assessment The measurement of indicators of dietary status and nutrition-related health status to identify the possible occurrence, nature, and extent of impaired nutritional status (ranging from deficiency to toxicity).

Nutrition monitoring The assessment of dietary or nutritional status at intermittent times with the aim of detecting changes in the dietary or nutritional status of a population.

Nutrition surveillance The continuous assessment of nutritional status for the purpose of detecting changes in trends or distributions so that corrective measures can be taken.

Nutrition screening A system that identifies specific individuals for nutrition or public health intervention, often at the community level.

National Nutrition Monitoring

Most nations monitor the health and nutritional status of their populations as a means of deciding how to allocate scarce resources, enhance the quality of life, and improve productivity. National nutrition policies are typically guided by the outcomes of food and health surveys, which are designed to obtain data on the distribution of foodstuffs, the extent to which people consume food of sufficient quality and quantity, the effects of infectious and chronic diseases, and the ways these factors relate to human health. Such information can be derived by any of several methods, including **nutrition assessment, nutrition monitoring, nutrition surveillance,** and **nutrition screening.**[7] These methods are sometimes treated together under the rubric "nutrition monitoring." Such activities provide regular information about nutrition in populations and the factors that influence food consumption and nutritional status. The objectives of any national nutrition monitoring system, as outlined by the United Nations Expert Committee Report, are shown in Table 4-1.[8] The U.S. Congress defined nutrition monitoring and related research as "the set of activities necessary to provide timely information about the role and status of factors that bear on the contribution that nutrition makes to the health of the people of the United States."[9]

Background on Nutrition Monitoring in the United States

The U.S. federal government has been involved in tracking certain elements of the food supply and food consumption for more than eight decades, beginning with the USDA's Food Supply Series undertaken in 1909. In the 1930s, the first USDA Household Food Consumption Survey was conducted (in 1965 this survey became known as the Nationwide Food Consumption Survey). In the late 1960s, concerns about the "shocking" nutritional status of Mississippi schoolchildren and widespread chronic hunger and malnutrition led to the nation's first comprehensive nutrition survey, the Ten-State Nutrition Survey, conducted between 1968 and 1970 in 10

TABLE 4-1 *Objectives of a National Nutrition Monitoring and Surveillance System*

- To describe the health and nutrition status of a population, with particular reference to defined subgroups who may be at risk

- To monitor changes in health and nutrition status over time

- To provide information that will contribute to the analysis of causes and associated factors and permit selection of preventive measures, which may or may not be nutritional (for example, smoking)

- To provide information on the interrelationship of health and nutrition variables within population subgroups

- To estimate the prevalence of diseases, risk factors, and health conditions and of changes over time, which will assist in the formulation of policy

- To monitor nutrition programs and evaluate their effectiveness in order to determine met and unmet needs related to target conditions under study

Source: G. B. Mason and coauthors, *Nutritional Surveillance* (Geneva: World Health Organization, 1984). Used with permission.

states: California, Kentucky, Louisiana, Massachusetts, Michigan, New York, South Carolina, Texas, Washington, and West Virginia.[10] In the 1970s other surveys, such as the National Health and Nutrition Examination Surveys (NHANES I and II) and the Pediatric Nutrition Surveillance System, were added to the roster of methods used to obtain information about the nutritional status of the population.

Beginning in the 1980s, the Joint Nutrition Monitoring Evaluation Committee was set up as a federal advisory committee, jointly sponsored by the USDA and the DHHS, to develop reports on the nutritional status of the U.S. population; these reports were to be submitted to Congress at 3-year intervals. The first report was published in 1986, the second in 1989.[11] Then, in 1990, the U.S. Congress passed legislation (PL 101–445) that established the **National Nutrition Monitoring and Related Research Program (NNMRRP).** The legislation specified that the USDA and the DHHS would jointly implement and coordinate the activities of the NNMRRP to obtain data through surveys, surveillance, and other monitoring activities about the dietary, nutritional, and nutrition-related health status of the U.S. population; the relationship between diet and health; and the factors that influence nutritional and dietary status. The NNMRRP takes a multidisciplinary approach to monitoring the nutritional and health status of Americans in general and high-risk groups such as low-income families, pregnant women, and minorities in particular.[12] Today, the NNMRRP includes more than 50 surveillance activities that monitor and evaluate the health and nutritional status of the U.S. population.[13]

National Nutrition Monitoring and Related Research Program (NNMRRP) The set of activities that provides regular information about the contribution that diet and nutritional status make to the health of the U.S. population and about the factors affecting diet and nutritional status.

The National Nutrition Monitoring and Related Research Program

The NNMRRP includes all data collection and analysis activities of the federal government related to (1) measuring the health and nutritional status, food consumption, dietary knowledge, and attitudes about diet and health of the U.S. population and (2) measuring food consumption and the quality of the food supply.[14] Overall, the NNMRRP has the following goals:[15]

- Provide the scientific foundation for the maintenance and improvement of the nutritional status of the U.S. population and the nutritional quality and healthfulness of the national food supply.
- Collect, analyze, and disseminate timely data on the nutritional and dietary status of the U.S. population, the nutritional quality of the food supply, food consumption patterns, and consumer knowledge and attitudes concerning nutrition.
- Identify high-risk groups and geographical areas, as well as nutrition-related problems and trends, to facilitate prompt implementation of nutrition intervention activities.
- Establish national baseline data and develop and improve uniform standards, methods, criteria, policies, and procedures for nutrition monitoring.
- Provide data for evaluating the implications of changes in agricultural policy related to food production, processing, and distribution that may affect the nutritional quality and healthfulness of the U.S. food supply.

The NNMRRP surveys can be grouped into five areas: nutritional status and nutrition-related health measurements; food and nutrient consumption; knowledge, attitudes, and behavior assessments; food composition and nutrient databases; and food

FIGURE 4-2 A
*Conceptual Model of the
Relationships of Food to
Health, Showing the
Major Components of the
National Nutrition
Monitoring and Related
Research Program*

Source: Federation of American
Societies for Experimental Biology,
Life Sciences Research Office,
prepared for the Interagency Board
for Nutrition Monitoring and
Related Research, *Third Report on
Nutrition Monitoring in the United
States,* Vol. 1 (Washington, D.C.:
U.S. Government Printing Office,
1995), p. 6.

supply determinations. The next sections, which describe the major surveys in each area, are based on the directory of federal nutrition monitoring activities compiled by the Interagency Board for Nutrition Monitoring and Related Research.[16] Consult Figure 4-2 to clarify how the various surveys are used to obtain information about the relationship of food to health. The figure (described in more detail in Chapter 6) indicates that an individual's health and nutritional status is influenced by his or her food intake, which includes the food prepared and eaten at home and away from home. Food intake is influenced by the individual's knowledge about the relationship between diet and health, use of supplements, nutrient requirements, and attitudes about food and dietary and health practices. The composition and types of food available in the food supply also influence food choices. The boxes in the figure represent the five major component areas of the NNMRRP.

Refer to Table 4-2 for a description of the major surveys described in the section that follows. Other major NNMRRP surveys not mentioned in this chapter are described in Appendix F. The boxed insert provides Internet addresses for on-line survey data and other documents.

Nutritional Status and Nutrition-Related Health Measurements. The surveys that form the basis of the health and nutritional status component of the NNMRRP target a variety of specific population groups, including noninstitutionalized civilians over the age of 55 years, children aged 2 to 6 years, women of reproductive age, and individuals residing in nursing homes. The surveys collect data on diverse

Continued on page 121.

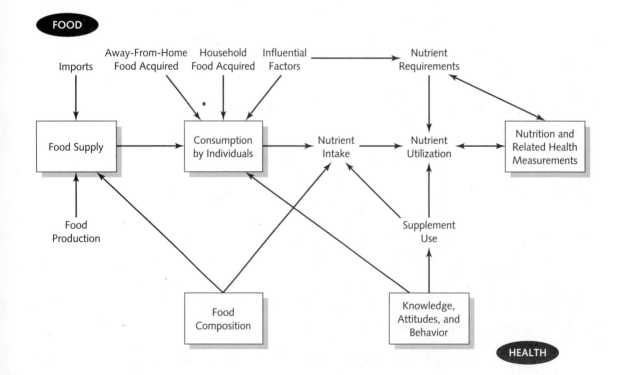

TABLE 4-2 *Sources of Data from the Five Component Areas of the NNMRRP Considered in the Third Report on Nutrition Monitoring*

Component Area and Survey or Study	Sponsoring Agency (Department)	Date	Population	Data Collected
Nutritional Status and Nutrition-Related Health Measurements				
Pediatric Nutrition Surveillance System (PedNSS)	NCCDPHP, CDC (HHS)	1973–92	Low-income, high-risk children, birth–17 years of age, with emphasis on birth–5 years of age	Demographic information; anthropometry (height and weight), birthweight, and hematology (hemoglobin, hematocrit), breastfeeding
Fourth National Health and Nutrition Examination Survey (NHANES IV)	NCHS, CDC (HHS)	1998–01	Civilian, noninstitutionalized population 2 months of age and older. Over-sampling of adolescents, African Americans, and Mexican Americans	Survey elements resembling those of NHANES III and the National Health Interview Survey
Third National Health and Nutrition Examination Survey (NHANES III)	NCHS, CDC (HHS)	1988–94	Civilian, noninstitutionalized population 2 months of age and older. Over-sampling of non-Hispanic blacks and Mexican Americans, children < 6 years of age, and adults aged ≥ 60 years	Dietary intake (one 24-hour recall and food frequency), socioeconomic and demographic information, biochemical analyses of blood and urine, physical examination, body measurements, blood pressure measurements, bone densitometry, dietary and health behaviors, and health conditions. Two additional 24-hour recalls for participants 50 years of age and older
Hispanic Health and Nutrition Examination Survey (HHANES)	NCHS (HHS)	1982–84	Civilian, noninstitutionalized Mexican Americans in five southwestern states, Cuban Americans in Dade County, FL, and Puerto Ricans in New York, New Jersey, and Connecticut; 6 months–74 years of age	Dietary intake (one 24-hour recall), food frequency, socioeconomic and demographic information, dietary and health behaviors, biochemical analyses of blood and urine, physical examination, body measurements, and health conditions
Second National Health and Nutrition Examination Survey (NHANES II)	NCHS (HHS)	1976–80	Civilian, noninstitutionalized population of the United States; 6 months–74 years of age	Dietary intake (one 24-hour recall), food frequency, socioeconomic and demographic information, biochemical analyses of blood and urine, physical examination, and body measurements
First National Health and Nutrition Examination Survey (NHANES I)	NCHS (HHS)	1971–74	Civilian noninstitutionalized population of the conterminous United States; 1–74 years of age	Dietary intake (one 24-hour recall), food frequency, socioeconomic and demographic information, biochemical analyses of blood and urine, physical examination, and body measurements

Continued

Food and Nutrient Consumption

Component Area and Survey or Study	Sponsoring Agency (Department)	Date	Population	Data Collected
5 a Day for Better Health Baseline Survey	NCI (HHS)	1991	Adults 18 years of age and older in the United States	Demographic information; fruit and vegetable intake; knowledge, attitudes, and behaviors regarding fruit and vegetable intake.
Continuing Survey of Food Intakes by Individuals (CSFII)	HNIS† (USDA)	1994–96 1989–91	Individuals in households in the 48 conterminous states. The survey was composed of two separate samples: households with incomes at any level (basic sample) and households with incomes ≤130% of the poverty thresholds (low-income sample).	One-day and 3-day food intakes by individuals of all ages, times of eating occasions, and sources of food eaten away from home. Data collected over 3 consecutive days by use of a 1-day recall and a 2-day record. Data available for 28 food components.
Total Diet Study (TDS)	FDA (HHS)	1982–89	NA	Chemical analysis of nutrients and contaminants in the U.S. food supply. Food composition data are merged with food consumption data to estimate daily intake of nutrients and contaminants.
Nationwide Food Consumption Survey (NFCS)	HNIS (USDA)	1987–88	Households in the 48 conterminous states and individuals residing in those households. The survey was composed of two samples: a basic sample of all households and a low-income sample of households with incomes ≤130% of the poverty threshold.	For households: quantity (pounds), money value (dollars), and nutritive value of food used. For individuals: 1-day and 3-day food intakes by individuals of all ages, times of eating occasions, and sources of food eaten away from home. Data collected over 3 consecutive days using a 1-day recall and a 2-day record. Data available for 28 food components.
Continuing Survey of Food Intakes by Individuals (CSFII)	HNIS (USDA)	1985–86	Individuals in households in the 48 conterminous states. Two samples: households with incomes at any level (basic sample) and households with incomes ≤ 130% of the poverty thresholds (low-income sample), including in 1985, women 19–50 years of age and their children aged 1–5 years and men 19–50 years and women 19–50 years of age in 1986, women 19–50 years of age and their children aged 1–5 years.	Six 1-day food and nutrient intakes by individuals, names and times of eating occasions, and sources of food obtained and eaten away from home. Data collected at about 2-month intervals over a 1-year period.

Component Area and Survey or Study	Sponsoring Agency (Department)	Date	Population	Data Collected
Food and Nutrient Consumption—continued				
Vitamin and Mineral Supplement Use Survey	FDA (HHS)	1980	Civilian, noninstitutionalized adults 16 years of age and older in the United States	Prevalence of use, sociodemographic characteristics of the users, intakes of 24 nutrients (12 vitamins and 12 minerals) and other miscellaneous substances, and supplement use behaviors of the users by telephone interview
Nationwide Food Consumption Survey (NFCS)	ARS (USDA)	1977–78	Private households in the 48 contermi-nous states and the individuals in those households (all income and low income)	For households: quantity (pounds), money value (dollars), and nutritive value of food used. For individuals: 1-day and 3-day food and nutrient intakes by individuals of all ages, names and times of eating occasions, and sources of food obtained and eaten away from home. Data collected over 3 consecutive days using a 1-day recall and a 2-day record. Intakes are available for 15 nutrients and food components.Behavioral
Knowledge, Attitudes, and Behavior Assessments				
Behavioral Risk Factor Surveillance System (BRFSS)	NCCDPHP, CDC (HHS)	1992	Adults 18 years of age and older residing in households with telephones in participating states	Demographic information; height, weight, smoking, alcohol use, weight control practices, diabetes, preventable health problems, mammography, pregnancy, cholesterol-screening practices, awareness, treatment, and modified food frequencies for dietary fat, fruit, and vegetable consumption by telephone interview
Weight Loss Practices Survey (WLPS)	FDA and NHLBI (HHS)	1991	Individuals currently trying to lose weight, 18 years of age and older	Demographic information; body mass index; diet history and other health behaviors; self-perception of overweight by telephone interview
Youth Risk Behavior Survey (YRBS)	NCCDPHP, CDC (HHS)	1990,1991	Youths attending school in grades 9–12 in the 50 states, District of Columbia, Puerto Rico, and the Virgin Islands	Demographic information; smoking, alcohol use, weight control practices, exercise, and eating practices information

Continued

Component Area and Survey or Study	Sponsoring Agency (Department)	Date	Population	Data collected
Knowledge, Attitudes, and Behavior Assessments—continued				
Diet and Health Knowledge Survey (DHKS)	HNIS (USDA)	1989–91	Main meal planner and preparer in households in the 48 conterminous states who participated in CSFII 1989–91	Self-perceptions of relative intake levels, awareness of diet-health relationships, use of food labels, perceived importance of following dietary guidance for specific nutrients and food components, beliefs about food safety, and knowledge about food sources of nutrients. These variables can be linked to data on individuals' food and nutrient intakes from CSFII 1989–91.
Food Composition and Nutrient Data Bases				
National Nutrient Data Bank (NNDB)	ARS (USDA)	NA[‡]	NA[‡]	Nutrient content of foods. Data from the National Nutrient Data Bank are used in the USDA Survey Nutrient Data Base for analysis of national dietary intake surveys and are also made available in published tables of food composition and as computerized databases. Periodic updates to the data are also available on the Nutrient Data Bank Electronic Bulletin Board.
USDA Nutrient Data Base for Standard Reference	ARS (USDA)	NA[§]	NA[§]	A computer file for *Agriculture Handbook No. 8* (USDA, 1992) produced from the National Nutrient Data Bank and the main source of the data for the USDA Survey Nutrient Data Base. This database includes data on food energy, 28 food components, and 18 amino acids for about 5,200 food items.
USDA Survey Nutrient Data Base	ARS (USDA)	NA[‖]	NA[‖]	The database is used for analysis of nationwide dietary intake surveys. It is updated continuously and includes data on food energy and 28 food components for >7,100 food items. The database used for CSFII 1989–91 and NHANES III 1988–91 contained approximately 6,700 items.

Component Area and Survey or Study	Sponsoring Agency (Department)	Date	Population	Data collected
Food Composition and Nutrient Data Bases—continued				
Food Label and Package Survey (FLAPS)	FDA (HHS)	Biennially, 1977–90	NA	Use of nutrition labeling; declaration of selected nutrients and ingredients; nutrition claims; other label statements and descriptors; nutrient analysis of a representative sample of packaged foods with nutrition labels
Food-Supply Determinations				
U.S. Food Supply Series	CNPP and ERS# (USDA)	1970–92	U.S. total population	Quantities of foods available for consumption on a per capita basis; quantities of food energy, nutrients, and food components provided by these foods (calculated)

*Within each component area, entries are listed in reverse chronological order. NCCDPHP, National Center for Chronic Disease Prevention and Health Promotion; CDC, Centers for Disease Control and Prevention; HHS, Department of Health and Human Services; NCHS, National Center for Health Statistics; USDA, U.S. Department of Agriculture; NCI, National Cancer Institute; HNIS, Human Nutrition Information Service; FDA, Food and Drug Administration; ARS, Agricultural Research Service; NHLBI, National Heart, Lung, and Blood Institute; CNPP, Center for Nutrition Policy and Promotion; ERS, Economic Research Service; NA, not available.

†Legislation passed on February 20, 1994, transferred the functions and staff of the USDA's Human Nutrition Information Service (HNIS) to the existing Agricultural Research Service (ARS) of that department.

‡The work leading to the establishment of the NNDB was initiated in 1892 and has been maintained by the USDA since that time. The sponsoring agency is currently the ARS.

§The USDA Nutrient Data Base for Standard Reference was initiated in 1980 and has been maintained by USDA since then. The sponsoring agency is currently ARS.

‖The USDA Survey Nutrient Data Base appropriate for the years of the following surveys was used: NFCS 1977–78, HHANES 1982–84, CSFII 1985–86, NFCS 1987–88, NHANES III 1988–91, CSFII 1988–91, CSFII 1989–91, and Strong Heart Dietary Survey 1989–91. The sponsoring agency was the ARS (USDA) for 1977–81 and the HNIS (USDA) for 1981–91.

#On December 1, 1994, the U.S. Food Supply Series work conducted by the ARS was transferred to the Center for Nutrition Policy and Promotion (CNPP).

Source: Federation of American Societies for Experimental Biology, Life Sciences Research Office, prepared for the Interagency Board for Nutrition Monitoring and Related Research, *Third Report on Nutrition Monitoring in the United States*, Vol. 1 (Washington D.C.: U.S. Government Printing Office, 1995), pp. 7–15.

Internet Resources

Canada

Canada's Food Guide to Healthy Eating	www.hc-sc.gc.ca
The National Plan of Action for Nutrition	www.hc-sc.gc.ca

United States

*Nutrition Monitoring**

Third Report on Nutrition Monitoring in the United States	www.nalusda.gov/fnic/usda/thirdreport.html

Survey Descriptions and Data Tables

1996 BRFSS[†] Summary Prevalence Report	www.cdc.gov/nccdphp/brfss
Pyramid Servings Data: Results from USDA's 1995 and 1996 Continuing Survey of Food Intakes by Individuals[‡]	www.barc.usda.gov/bhnrc/foodsurvey/home.htm
Results from USDA's 1996 Continuing Survey of Food Intakes by Individuals and 1996 Diet and Health Knowledge Survey[‡]	www.barc.usda.gov/bhnrc/foodsurvey/home.htm
Trends in Food and Nutrient Intakes by Adults: NFCS 1977–78, CSFII 1989–91, and CSFII 1994–95[‡]	www.barc.usda.gov/bhnrc/foodsurvey/home.htm
Third National Health and Nutrition Examination Survey (NHANES III) Public-Use Data Files[§]	www.cdc.gov/nchswww/products/catalogs/subject/nhanes3/nhanes3.htm#descriptional
Youth Risk Behavior Surveillance System	www.cdc.gov/nccdphp/youthris.htm

Related Resources

Appendix I: History of the Dietary Guidelines for Americans	www.nal.usda.gov/fnic/Dietary/12dietapp1.htm
U.S. Action Plan on Food Security Framework for the U.S. Action Plan on Food Security	www.fas.usda.gov/icd/summit/framewor.html
Discussion Paper on Domestic Food Security	www.fas.usda.gov/icd/summit/discussi.html
USDA Nutrient Database for Standard Reference, Release 12	www.nal.usda.gov/fnic/foodcomp

*With the exception of the USDA Nutrient Database for Standard Reference, all other reports and data tables mentioned in this table require Adobe Acrobat® Reader to view and download.

[†]BRFSS = Behavioral Risk Factor Surveillance System.

[‡]To locate the report, go to the Internet address shown, click on "Products: Data Sets, Technical Databases, Table Sets, and Reports from CSFII/DHKS 1994–1996 and Earlier Surveys," and click on the appropriate report title.

[§]NHANES III reports can also be downloaded from the *Advanced Data* publication series of Vital and Health Statistics of the Centers for Disease Control and Prevention at www.cdc.gov/nchswww.

issues, such as family structures, community services, risk factors associated with cancer, aspects of family planning and fertility, and the causes of low birthweight among infants. The NHANES series and the Pediatric Nutrition Surveillance System (PedNSS) are included in this component. The PedNSS is a surveillance program that monitors the nutritional status of low-income children from birth to 17 years of age at high risk for nutrition-related problems through the collection of measurements such as height, current weight, weight at birth, hemoglobin, and hematocrit.

The following surveys are among the most important of the health and nutritional status measurements. The NHANES series uses a sample that is representative of the civilian, noninstitutionalized population, and it has a good **response rate.**[17]

Response rate The value obtained by multiplying the participation rates for each survey component.

- **National Health and Nutrition Examination Survey (NHANES I).** Conducted in 1971–1974, the NHANES I was designed to collect and disseminate data that could be obtained best or only by direct physical examination, laboratory and clinical tests, and related measurements. The target population was civilian noninstitutionalized persons aged 1 through 74 years. The measures included dietary intake (one 24-hour recall), body composition, hematologic tests, urine tests, X-rays of the hand and wrist, dental examinations, and other measurements.
- **NHANES II.** Conducted in 1976–1980, this program targeted civilian noninstitutionalized persons aged 6 months through 74 years. It collected the same types of data as the NHANES I.
- **Hispanic Health and Nutrition Examination Survey (HHANES).** This survey, conducted in 1982-1984, was designed to collect and disseminate data obtained from physical examinations, diagnostic tests, anthropometric measurements, laboratory analyses, and personal interviews of Mexican Americans, Puerto Ricans, and Cubans. Dietary intake was assessed by one 24-hour recall and a food frequency questionnaire. The target population consisted of "eligible" Hispanics aged 6 months through 74 years. "Eligible" Hispanics were limited to Mexican Americans living in five southwestern states; Cubans living in Dade County, Florida; and Puerto Ricans living in New York, New Jersey, and Connecticut.
- **NHANES III.** The NHANES III was conducted in 1988–1994 on a nationwide sample of about 34,000 persons aged 2 months and over. The survey was divided into two 3-year surveys (phase 1 and phase 2) so that national estimates could be produced for each 3-year period and for the entire 6-year period.[18] Its target population was civilian noninstitutionalized persons aged 2 months and over. Dietary intake was assessed by one 24-hour recall and a food frequency questionnaire. The hematological and biochemical components of NHANES III are shown in Appendix F. The examination components vary by age. Infants aged 2 to 11 months undergo a physician's exam, body measurements, and an assessment of tympanic impedance; a dietary interview is conducted with the child's caregiver or parent. Between the ages of 1 and 19 years, additional assessments are added, including a dental exam, urine test, cognitive test, allergy skin test, and spirometry (a measurement of lung capacity). Participants over 20 years of age undergo these assessments, plus an oral glucose tolerance test, electrocardiogram, and other tests.[19] NHANES III data were used to develop new, nationally representative equations to predict stature for non-Hispanic white, non-Hispanic black, and Mexican-American adults aged 60 years and older.[20]

Compared with other surveys, the response rate for the NHANES III was good. In NHANES III, 100 percent of households were screened, and 86 percent of those screened were later interviewed. The survey's overall response rate was 73 percent. By comparison, the response rate for the dietary assessment component of the Nationwide Food Consumption Survey, described in the next section, was 25 percent, meaning that only one in four respondents completed the dietary recalls and food records for all 3 days of the assessment period.[21]

- **NHANES IV.** The NHANES IV, which began in 1998, has the same general design as the NHANES III, except that it is linked with the National Health Interview Survey (NHIS). Thus, much of the contents of the NHANES IV questionnaire comes directly from the NHIS survey instrument. Some population groups such as adolescents, African Americans, and Mexican Americans will be oversampled in the NHANES IV. As with other surveys in this series, the NHANES IV collects data at the household, family, and individual levels.[22]

Food and Nutrient Consumption. These surveys collect data on food consumption behavior. The Vitamin and Mineral Supplement Use Survey, for example, assesses quantitatively the nutrient intake from vitamin and mineral supplements and the characteristics of supplement users. The 5 a Day for Better Health Baseline Survey, conducted in 1991, obtained data on the fruit and vegetable intake and knowledge, attitudes, and behaviors regarding fruit and vegetable intake of U.S. adults. The following are the other major surveys of this NNMRRP component:

- **Nationwide Food Consumption Survey (NFCS).** Last conducted in 1987–1988 by the Human Nutrition Information Service of the USDA, the NFCS collects data that are used to describe food consumption behavior and to evaluate the nutritional content of diets for their implications for food policies, marketing, food safety, food assistance, and nutrition education. The target population consists of private households of all incomes. The measures include the quantity (in pounds) of food used from home food supplies during one week by the entire household and the food ingested by individual household members both at home and away from home over a 3-day period. The money value (in dollars) of food used by households is also calculated. This survey has been discontinued.[23]
- **Continuing Survey of Food Intakes by Individuals (CSFII), 1985 and 1986.** The CSFII was designed to provide timely data on U.S. diets in general, the diets of high-risk population subgroups, and changes in dietary patterns over time. The target population was persons of selected sex and age in private households with incomes at any level (basic survey) and incomes at or below 130 percent of the poverty guidelines (low-income survey). The 1985 survey focused on women and men aged 19 to 50 years and children aged 1 to 5 years. The 1986 survey focused on women and children and did not include men. Dietary measures included food intakes from multiple 24-hour recalls collected by interview.
- **Continuing Survey of Food Intakes by Individuals (CSFII), 1989–1991 and 1994–1996.** Popularly known as the "What We Eat in America Survey," these surveys resemble the CSFII series conducted in 1985 and 1986. The target population is men, women, and children of all ages. The dietary component includes the kinds and amounts of food ingested both at home and away from home by individual household members for three consecutive days as measured by a 24-hour recall (obtained by personal interview) and a 2-day food diary.
- **Total Diet Study (TDS).** This survey, conducted annually by the Food and Drug Administration (FDA), is designed to assess the levels of various nutritional components and organic and elemental contaminants of the U.S. food supply. The Selected Minerals in Food Survey, a component of the TDS, estimates the level of 11 essential minerals in representative diets. The target population is eight age-sex groups: infants, young children, male and female teenagers, male and female adults, and male and female older persons. The design includes collecting 234 foods from retail markets in urban areas,

Consult Appendix F for other sources of data on nutrients and food constituents obtained from other NNMRRP surveys.

preparing them for consumption, and analyzing them for nutritional elements and contaminants four times a year.

Knowledge, Attitudes, and Behavior Assessments. Surveys in this component of the NNMRRP gather data on weight loss practices; the general public's knowledge about the relationship of diet to health problems such as hypertension, coronary heart disease, and cancer; awareness among the general public, physicians, nurses, and dietitians of the risk factors of high blood cholesterol and coronary heart disease; and knowledge and attitudes about cancer prevention and lifestyle risk factors. For example, the Diet and Health Knowledge Survey (DHKS) was initiated in 1989 by the USDA as a follow-up to the CSFII to measure consumers' awareness of diet-health relationships and dietary guidance, knowledge of food sources of nutrients, use of food labels, and beliefs about food safety. The Behavioral Risk Factor Surveillance System (BRFSS) is a unique system active in all 50 states. It is the main source of information on risk behaviors among adult populations. The BRFSS collects data related to health status, access to health care, tobacco and alcohol use, injury control (for example, use of seat belts), use of prevention services such as immunization and breast cancer screening, HIV and AIDS, weight control practices, treatment for high blood cholesterol, and frequency of intake of dietary fat, fruits, and vegetables. The BRFSS surveys adults by telephone interview.[24]

The Weight Loss Practices Survey (WLPS) also uses telephone interviews to obtain information about dieting history, health behaviors, and self-perception of overweight from adults trying to lose weight. The high-risk behaviors of youths such as smoking, alcohol use, eating practices, and weight control practices are assessed in the Youth Risk Behavior Survey.

Food Composition and Nutrient Data Bases. Information about the nutrient content of foods is provided by four different activities. The Food Label and Package Survey (FLAPS), undertaken biennially by the FDA, is designed to monitor the labeling practices of U.S. food manufacturers; it analyzes about 300 foods to check the accuracy of nutrient values on food labels. Other activities include the following:

• **National Nutrient Data Bank.** This continuous activity of the Agricultural Research Service of the USDA compiles and disseminates data on the nutrient composition of foods. Sources of nutrient data include government-funded university research, scientific publications, food processors and trade groups, and the results of food analyses by the Nutrient Composition Laboratory.

• **Nutrient Data Base for Standard Reference.** This database was initiated in 1980 and is sponsored by the Agricultural Research Service. It consists of computerized data of the nutrient composition of foods and is published as *Agriculture Handbook 8*.

• **Survey Nutrient Data Base.** This database is updated continuously and is used to analyze nationwide dietary intake surveys. It includes data on food energy and 28 food components for more than 7,100 food items.

Food-Supply Determinations. Food available for consumption by the U.S. civilian population is determined by the USDA's Center for Nutrition Policy and Promotion through its Food Supply Series surveys. These food supply or **disappearance data** have been available annually since 1909.[25] The nutrient content of the available food

Food disappearance data The amount of food remaining after subtracting nonfood uses such as exports and industrial uses from the total available food supply. These data represent the food that "disappears" into the marketing system and is available for human consumption.

TABLE 4-3 *Uses of Data from the National Nutrition Monitoring and Related Research Program*

Assessment of Dietary Intake

- Provide detailed benchmark data on food and nutrient intakes of the population
- Monitor the nutritional quality of diets
- Determine the nature of populations at risk of having diets low or high in certain nutrients
- Identify socioeconomic and attitudinal factors associated with diets

Economics of Food Consumption

- Predict demand for agricultural products and marketing facilities
- Determine the effects of socioeconomic factors on the demand and expenditure for food
- Determine the importance of home food production
- Determine the demand for food away from home and its effects on the nutritional quality of diets

Food Programs and Guidance

- Develop food guides and dietary guidance materials that target nutritional problems in the U.S. population
- Identify educational strategies to increase the knowledge of nutrition and to improve the eating habits of Americans
- Identify factors affecting participation in some food programs and estimate the effect of participation on food expenditures and diet quality
- Estimate the effect of food programs on demand for food
- Identify populations that might benefit from intervention programs
- Identify changes in food and nutrient consumption that would reduce health risks
- Develop food guides and plans that reflect food consumption practices and meet nutritional and cost criteria
- Determine the amounts of foods that are suitable to offer in food distribution programs

Food Safety Considerations

- Estimate intake of incidental contaminants, food additives, and naturally occurring toxic substances
- Identify extreme and unusual patterns of intakes of foods or food ingredients
- Predict food items in which a food additive can safely be permitted in specified amounts
- Determine the need to modify regulations in response to changes in consumption
- Identify size and nature of population at risk from use of particular foods and food products

Historical Trends

- Correlate food consumption and dietary status with incidence of disease over time
- Follow food consumption through the life cycle
- Predict changes in food consumption and dietary status as they may be influenced by economic, technological, and other developments
- Track use and understanding of food labels and their effect on dietary intakes

Source: Food Surveys Research Group, Beltsville Human Nutrition Research Center, Agricultural Research Service, U.S. Department of Agriculture Web site at sun.ars-grin.gov/ars/Beltsville/barc/bhnrc/foodsurvey/uses.htm.

supply is determined using food composition data and then used to estimate the nutrient content of the food supply on a per capita basis.

Uses of National Nutrition Monitoring Data

The primary purpose of national nutrition monitoring activities is to obtain the information needed to ensure a population's adequate nutrition. The collected data are used in health planning, program management and evaluation, and timely warning and intervention efforts to prevent acute food shortages.[26] Data related to the population's nutritional status and dietary practices, obtained through national nutrition monitoring activities, are then used to direct research activities and make a variety of policy decisions involving food assistance programs, nutrition labeling, and education. For instance, the BRFSS allows for comparisons between states and between individual states and the nation. BRFSS data are also used to help states set priorities among health issues, develop strategic plans, monitor the effectiveness of public health interventions, measure the achievement of program goals, and create reports, fact sheets, press releases, and other publications to help educate the public, health professionals, and policy makers about disease prevention and health promotion. National policy makers use BRFSS data to monitor the nation's progress toward the *Healthy People 2000* objectives.[27] Specific uses of NNMRRP surveys are listed in Table 4-3. Congress, in particular, needs the data from nutrition monitoring activities to formulate nutrition and health policies and programs, assess the consequences of such policies, oversee the efficacy of federal food and nutrition assistance programs, and evaluate the extent to which federal programs result in a consistent and coordinated effort (a significant activity considering that no Federal Nutrition Office exists at the present time).[28]

Nutrient Intake Standards

Merely collecting data on a population's nutrient intake and eating habits is not enough. Such data are meaningless on their own; to be valuable, they must be compared with some national standard related to nutrient needs. In the United States, two national committees are engaged in defining the recommended nutrient intakes that best support health: the Committee on Diet and Health and the Committee on Dietary Allowances. Both are funded by the federal government and come under the auspices of the National Research Council of the National Academy of Sciences (NRC/NAS). The Committee on Diet and Health pays particular attention to reducing the risk of chronic disease and makes recommendations regarding certain dietary inadequacies and excesses, as shown in Table 4-4. These recommendations are intended to be used together with the Recommended Dietary Allowances (RDAs) and, where applicable, the Dietary Reference Intakes (DRIs) in planning optimal diets.

Recommended Dietary Allowances (RDAs)

The **RDAs** are recommendations for nutrient intakes published by the Committee on Dietary Allowances, a panel of the NRC's Food and Nutrition Board. The RDAs focus

RDAs The amounts of energy, protein, and other selected nutrients considered adequate to meet the needs of practically all healthy people.

on energy and nutrient needs and the maintenance of health (refer to the inside front cover for a table of the RDAs). The first edition of the RDAs was published in 1943 to provide "standards to serve as a goal for good nutrition."[29] The tenth and most recent edition was published in 1989. Each revision reflects new evidence regarding the nutrient needs of Americans. The following factors offer a perspective on the RDAs:

- The RDAs are based on available scientific research.
- The RDAs are recommendations, not requirements, and certainly not minimal requirements. They include a substantial margin of safety.
- The RDAs take into account differences among individuals and define a range within which most healthy people's intakes of nutrients probably should fall. Individuals whose nutrient needs are higher than the average are included within this range.
- Individuals with special nutritional needs are not covered by the RDAs.

TABLE 4-4 *The National Research Council's Nutrition Recommendations*

- Reduce total *fat* intake to 30% or less of calories. Reduce saturated fatty acid intake to less than 10% of calories, and the intake of cholesterol to less than 300 mg daily.[*]

- Increase intake of starches and other *complex carbohydrates*.[†]

- Maintain *protein* intake at moderate levels.[‡]

- Balance food intake and physical activity to maintain appropriate *body weight*.

- For those who drink *alcoholic beverages,* the committee recommends limiting consumption to the equivalent of less than 1 oz of pure alcohol in a single day.[§] Pregnant women should avoid alcoholic beverages.

- Limit total daily intake of *salt* (sodium chloride) to 6 g or less.[||]

- Maintain adequate *calcium* intake.

- Avoid taking dietary *supplements* in excess of the RDA in any one day.

- Maintain an optimal intake of *fluoride,* particularly during the years of primary and secondary tooth formation and growth.

---

[*]The intake of fat and cholesterol can be reduced by substituting fish, poultry without skin, lean meats, and low- or nonfat dairy products for fatty meats and whole-milk products, by choosing more vegetables, fruits, cereals, legumes, and by limiting oils, fats, egg yolks, and fried and other fatty foods.

[†]Every day eat five or more servings of a combination of vegetables and fruits, especially green and yellow vegetables and citrus fruits, and six or more daily servings of a combination of breads, cereals, and legumes.

[‡]Meet at least the RDA for protein; do not exceed twice the RDA.

[§]The committee does not recommend alcohol consumption. One ounce of pure alcohol is the equivalent of two cans of beer, two small glasses of wine, or two average cocktails.

[||]Limit the use of salt in cooking and avoid adding it to food at the table. Salty, highly processed salty, salt-preserved, and salt-pickled foods should be consumed sparingly.

Source: Adapted with permission from the National Academy of Sciences report, *Diet and Health: Implications for Reducing Chronic Disease Risk,* which was produced by the Committee on Diet and Health of the Food and Nutrition Board of the National Research Council. Copyright 1989 by the National Academy of Sciences. Courtesy of the National Academy Press, Washington, D.C.

The RDAs are nutrient goals to be achieved over time. They can be used to set standards for food assistance programs and for licensing group facilities such as daycare centers and nursing homes, to design nutrition education programs, and to develop new food products. They are used for estimating an individual's nutritional status and his or her risk of nutrient deficiency. Ideally, when comparing an individual's dietary intake to the RDAs, the individual's typical intake should be determined as an average derived from several days' intakes.[30]

Dietary Reference Intakes (DRIs)

The Dietary Reference Intakes (DRIs) consist of reference values developed by Health Canada and the Food and Nutrition Board of the U.S. National Research Council to be used in planning and assessing the diets of individuals and groups. There are three new values—the Estimated Average Requirement, Adequate Intake, and Tolerable Upper Intake Level—and one familiar value, the Recommended Dietary Allowance. These terms are defined as follows:[31]

- The **Estimated Average Requirement (EAR)** is the intake value that is estimated to meet the requirement defined by a specific indicator of adequacy in 50 percent of a group of a certain age and sex. At this level of intake, 50 percent of the specified group would not have its nutrient needs met. The EAR includes an adjustment for nutrient bioavailability and is expressed as a daily value averaged over time (at least 1 week for most nutrients). The EAR is used in setting the RDA and is one reference value for assessing the adequacy of dietary intake of groups and for planning nutritionally adequate diets for groups.

- The **Recommended Dietary Allowance (RDA)** is the daily dietary intake level that is sufficient to meet the nutrient requirements of nearly all (97 to 98 percent) individuals in the specified life stage and gender group. The RDA is based on the EAR, such that if the **standard deviation** of the EAR is available, then the RDA is set at two standard deviations above the EAR (that is, $RDA = EAR + 2\ SD_{EAR}$). The RDA applies to individuals, not groups, and serves as a goal for dietary intake by individuals. The RDA is not intended to be used for assessing the dietary intakes of either individuals or groups or planning diets for groups.

> **Standard deviation** The value of the difference of individual values from the mean.

- The **Adequate Intake (AI)** is not a standard, but it can be used as a guide for the nutrient intake of individuals and groups in situations where no RDA for a nutrient has been established. An AI is established for a nutrient when there is not sufficient evidence to calculate an EAR. The AI is based on observed or experimentally determined estimates of the average nutrient intake that appears to sustain a defined nutritional state such as growth or normal circulating levels of a vitamin in a particular population. The AI exceeds the EAR and possibly the RDA.

- The **Tolerable Upper Intake Level (UL)** is the maximum level of daily nutrient intake that is unlikely to pose any risk of adverse health effects to nearly all of the individuals in the age- and sex-specific group. The UL is *not* a recommended intake level. The need for ULs grew out of the widespread use of dietary supplements by the general population and the increasing practice of food fortification.[32]

The Food and Nutrition Board established subcommittees to focus on developing the Tolerable Upper Intake Level for nutrients and defining the uses and interpretation of DRIs. It also convened seven panels to develop DRIs for the nutrients shown at the top of the next page.

- Calcium, vitamin D, phosphorus, magnesium, and fluoride
- Folate and other B vitamins
- Antioxidants (for example, vitamins C and E, selenium)
- Macronutrients (such as protein, fat, carbohydrates)
- Trace elements (for instance, iron, zinc)
- Electrolytes and water
- Other food components (including fiber, phytoestrogens)

The first report on calcium, vitamin D, phosphorus, magnesium, and fluoride was released in 1997. The DRIs for thiamin, riboflavin, niacin, vitamin B_6, folate, vitamin B_{12}, pantothenic acid, biotin, and choline were released in 1998.[33] The DRIs are shown in the table on the inside front cover. The plan is to release the DRIs for the other nutrients and food components by the year 2000.

The DRIs are a good example of policy making in action. Compared with the RDAs, the DRIs represent a major shift in thinking about nutrient requirements for humans from prevention of nutrient deficiencies to prevention of chronic disease.[34] They also herald new thinking about the role of dietary supplements in achieving good health. A preliminary publication related to the DRIs states that "for some individuals at higher risk, use of nutrient supplements may be desirable in order to meet reference intakes."[35] In other words, older women who are at high risk of osteoporosis may benefit from dietary calcium supplements to help maintain bone mineral mass. This view represents a change in philosophy from that of the RDAs, which supported consuming diets composed of a variety of foods, rather than taking supplements or relying on fortified foods, as the best approach to achieving nutritional adequacy. In addition, the development of the UL signals the widespread recognition that there can be a degree of risk with high intakes of nutrients. For this reason, separate recommendations—and possibly, separate upper limits—may be required for individual supplements of minerals and trace elements.[36]

Overall, the DRIs represent a new approach to dietary guidance and have the potential to influence the dietary messages provided to clients and communities, affect food fortification policy, and stimulate the development of new food products by industry. Consult Table 4-5 for a summary of the appropriate uses of the DRI values.

Dietary Recommendations of Other Countries and Groups

Various nations and international groups have published different sets of standards similar to the RDAs. The Canadian recommendations—the Recommended Nutrient Intakes (RNIs)—differ from the RDAs in some respects, reflecting the needs and requirements of the Canadian population. The Canadian recommendations, for example, include RNIs for the omega-3 and omega-6 fatty acids.[37] The RNIs are intended to guide Canadians in selecting a dietary pattern that supplies the recommended amounts of all essential nutrients and reduces the risk of chronic disease. They can be used to plan diets for individuals and groups and evaluate the need for public health interventions such as fortification.[37]

Among the most widely used recommendations are those of the Food and Agriculture Organization (FAO) and the World Health Organization (WHO). The FAO/WHO recommendations are considered sufficient for the maintenance of health in nearly all people. They differ from the RDAs because they are based on

The Canadian RNIs, Food Guide, and Guidelines for Healthy Eating and the WHO dietary recommendations are given in Appendix G.

Activity	Appropriate to Use	Not Appropriate to Use
Diets of individuals		
Assessment	AI,[†] UL[†]	EAR, RDA
Planning	RDA, AI,[‡] UL[§]	EAR
Diets of groups		
Assessment	EAR,[‖] AI, UL	RDA
Planning	EAR,[‖] AI,[#] UL	RDA

TABLE 4-5 *Uses of the Dietary Reference Intake Values for Planning and Assessing Diets*[*]

[*]Abbreviations are as follows: EAR = Estimated Average Requirement; RDA = Recommended Dietary Allowance; AI = Adequate Intake; UL = Tolerable Upper Intake Level.

[†]May be used for assessing the diets of individuals, depending on the characteristics of population groups, sociocultural settings, and chronic disease risk factors.

[‡]Use as a goal for nutrient intake in the absence of an EAR or RDA.

[§]Use as a goal to indicate the maximal level that will likely not pose a health risk; higher intakes may increase risk of adverse health effects.

[‖]Use in conjunction with data on the group's distribution of intake.

[#]May be used for setting tentative dietary goals.

Source: Adapted from Food and Nutrition Board, Dietary Reference Intakes (DRIs) for calcium, phosphorus, magnesium, vitamin D, and fluoride, *Nutrition Today* 32 (1997): 182–90; and videotapes of the Dietary Reference Intakes Conference, November 24, 1997, Ottawa, Ontario.

slightly different judgment factors and serve different purposes. The FAO/WHO recommendations, for example, assume a protein quality lower than that commonly consumed in the United States and so set a higher intake of this nutrient. They also take into consideration that people worldwide are generally smaller and more physically active than the U.S. population. For the most part, the various recommendations fall within the same general range.

Nutrition Survey Results: How Well Do We Eat?

What do the results of national surveys tell us about the nutritional status and dietary patterns of Americans? How well do we eat? The answers to these questions are mixed. Although we are well nourished, we are also generally overfat, underexercised, and beset to some extent with nutrient deficiencies.

According to results of the 1996 Continuing Survey of Food Intakes by Individuals, fully one-third of the U.S. population over 20 years of age is overweight. (Overweight was defined in this survey as a body mass index equal to or greater than 27.8 for men and 27.3 for women, excluding pregnant women.) Nearly half (45 percent) of men aged 50 to 59 years, more than 35 percent of women over 40 years of age, and about 5 million children—14 percent of children aged 6 to 11 years and 12 percent of adolescents aged 12–17 years—are overweight. This represents a significant increase in overweight in all age groups since NHANES II was conducted in 1976–1980.[38]

The problem of overweight in all age groups may arise from our increasingly sedentary lifestyles, lack of regular physical activity, and dietary patterns. Many young people, for example, are not involved in even moderate physical activity three times per week. About one-third of all adults report exercising rarely, and only one in four reports exercising vigorously two to four times per week as recommended.

The increase in the prevalence of overweight among children, teenagers, and adults indicates that energy balance is a problem for many people. Only about one-third of children under 5 years of age and between 25 and 43 percent of adults consumes a diet that meets current recommendations for total fat intake (that is, a diet providing 30 percent or less of energy from fat). Adults surveyed in NHANES III consumed about 34 percent of their total food energy intake as total fat and about 12 percent as saturated fat.[39] Even though these values represent a decrease since the NHANES II survey, they are still above the national health objective for the year 2000 to reduce average total fat intake to less than 30 percent and saturated fat intake to less than 10 percent of total food energy intake.

Caution must be exercised in interpreting the NHANES data. On the one hand, when average nutrient intakes are examined, severe deficiencies in individuals can be missed. On the other hand, findings based on a single day's intake—as the NHANES findings were—can overestimate the extent of undernutrition. An analysis of NHANES III data found that about 18 percent of men and 28 percent of women underreported their food consumption, resulting in an underreporting of total energy intake.[40] The survey's principal usefulness lies in its ability to identify the population subgroups most at risk of deficiency and the nutrients most in need of attention.

What do survey results indicate about U.S. dietary patterns? The overall dietary pattern for the U.S. population in 1988–1991, based on NHANES III data, was 50 percent of energy from carbohydrate, 15 percent of energy from protein, 34 percent of energy from fat, and 2 percent of energy from alcohol.[41] In 1996, Americans ate fewer fruits, dairy products, and servings of meat than the recommended levels, while their consumption of high-fat, high-calorie foods exceeded the recommendations.[42]

Whereas sodium intakes from food are higher than recommended levels for most Americans over 6 years of age, the intakes of other nutrients are below the RDA. Calcium intakes from food by adolescents, women, and the elderly are below recommended values.[43] And only about half of women of childbearing age (20 to 45 years) consume the recommended amount of folate. Moreover, only about one in four teenagers and young women consume the RDA for iron. Women in general have lower zinc intakes than men. For some populations, eating sufficient food throughout the day is a problem. Table 4-6 lists the nutrients about which there is some current concern, along with those that require further study and those that are not presently considered a concern.

Information is still accumulating from ongoing analyses of the data from the most recent surveys, but a general trend has emerged. Since the 1977–1978 CSFII was conducted, whole milk consumption has decreased, while consumption of grain products; bananas; meat, poultry, and fish mixtures; beer and ale; fruits drinks and ades; and soft drinks has increased. In fact, since the 1989–1991 CSFII, the amount of soft drinks consumed by both women and men has surpassed their milk intakes. Nutrients such as magnesium and zinc that were problem nutrients in 1977–1978— meaning that people's intakes of these nutrients failed to meet the RDA—were also problem nutrients in 1994–1995.[44] In short, people are overconsuming calories, underexpending energy, or both. The majority are adequately nourished with

Current Public Health Issues	Potential Public Health Issues Requiring Further Study	Not Currently a Public Health Issue
Food energy	Total carbohydrate	Thiamin
Total fat	Dietary fiber	Riboflavin
Saturated fatty acids	Sugars*	Niacin
Cholesterol	Monounsaturated fatty acids*	Iodine*
Alcohol	Polyunsaturated fatty acids	
Calcium	Trans fatty acids*	
Iron	Fat substitutes*	
Sodium	Protein	
	Vitamin A	

	Antioxidant vitamins
	    Vitamin C
	    Vitamin E
	    Carotenes	

	Folate	
	Vitamin B$_6$	
	Vitamin B$_{12}$	
	Copper	
	Fluoride	
	Magnesium	
	Phosphorus	
	Potassium	
	Selenium*	
	Zinc	

TABLE 4-6 *Classification of Food Components as Public Health Issues*

*These nutrients are being evaluated for the first time for the National Nutrition Monitoring and Related Research Program.

Source: Federation of American Societies for Experimental Biology, Life Sciences Research Office, prepared for the Interagency Board for Nutrition Monitoring and Related Research, *Third Report on Nutrition Monitoring in the United States*, Vol. 1 (Washington, D.C.: U.S. Government Printing Office, 1995), p. ES-25.

respect to most individual nutrients, but deficiencies and marginal intakes of nutrients do exist.[45]

Dietary Guidance Systems

One approach to improving the public's knowledge about healthful eating involves the dissemination of dietary guidance in the form of food group plans and dietary guidelines. In his report to the American Society for Clinical Nutrition on "The Evidence Relating Six Dietary Factors to the Nation's Health," former assistant secretary for health and surgeon general Julius Richmond commented, "Individuals have the right to make informed choices and the government has the responsibility to provide the best data for making good dietary decisions."[46] Dietary guidance systems are methods by which the federal government helps the American people make prudent dietary decisions.

Food Group Plans

Food group plan A diet-planning tool that sorts foods of similar origin and nutrient content into groups and then specifies that a person eat a certain number of foods from each group.

Dietary guidance in the form of **food group plans** has been provided to the U.S. population since the turn of the century. The early forms of dietary recommendations tended to focus on the consumption of adequate amounts of the foods needed to provide nutrients and energy intake for good health.[47] The USDA, for example, published its first dietary guidance plan in 1916. Between 1916 and the 1940s, a variety of food group plans, featuring anywhere from 5 to 12 food groups, were published by federal agencies and voluntary health organizations. In 1943, the Bureau of Home Economics published a food guide that promoted seven food groups as part of the USDA's National Wartime Nutrition Program. This guide served as the basis for nearly all nutrition education programs for more than a decade. Then, in 1955, the Department of Nutrition at the Harvard School of Public Health published a recommendation that the food groups be collapsed into four food groups.[48] This format was adopted in 1956 by USDA and promoted as the Four Food Group plan. The USDA added a fifth group (fats, sweets, and alcohol) to the Basic Four plan in 1979 to draw attention to foods targeted for moderation, but it did not offer specific limitations for the fifth group.

Since the publication of the Basic Four plan, however, the emphasis of nutritional guidance has expanded from just meeting nutrient needs to also eating a diet low in saturated fat and cholesterol and generous in fiber. With the publication of the *Surgeon General's Report on Nutrition and Health* in 1988, the overconsumption of certain dietary constituents became a major concern. For this reason, the Food Guide Pyramid was introduced to reinforce our understanding of the nutrient composition of foods, human nutrient needs, and the relationship of diet to health. The Food Guide Pyramid, shown on the inside back cover, conveys five of the essential components of a healthful daily diet: adequacy, balance, moderation, energy control, and variety.[49] It can also be used to implement the Dietary Guidelines (described in the next section).

Dietary Guidelines

Food group plans focus on recommended levels of food consumption. Dietary guidelines are typically broader than food group plans and focus on goal statements related to overall nutrient intake and daily eating patterns. The first attempt to formulate national dietary guidelines was undertaken in 1977, with the publication of the "Dietary Goals for the United States," a report of the U.S. Senate Select Committee on Nutrition and Human Needs.[50] In the 1980s, so many dietary guidelines were published that consumers were overwhelmed by the dietary advice, not knowing which group to listen to. In this section, we describe the major dietary guidelines published by government agencies and nongovernment organizations.

"OUR TURNIPS LOOK LIKE CARROTS, OUR POTATOES TASTE LIKE APPLES, OUR GREEN PEAS ARE BLUE... STUFF LIKE THAT."

Government Guidelines. Although the federal government's decision to promote a set of dietary goals or guidelines to improve the public's health would seem to be a straightforward matter of setting national nutrition policy, in fact, there has been much discussion about whether the government *should* establish such guidelines for the entire population. Concerns have focused on whether scientists could achieve a consensus on research outcomes and the relationship of diet to disease processes. Other questions also arose: What obligation do nutrition scientists and educators have to the general public to explain study outcomes and the differences that sometimes arise in interpreting study results? Who decides when research results are firm enough to warrant incorporating them into dietary guidance systems for the public? Is it appropriate to make general dietary recommendations for the *entire* population, when only a portion may be at risk for a specific disease condition? In what ways will the availability of national dietary guidelines—or the lack of them—affect the delivery of health care and educational efforts in the public health arena?[51]

Consider the comments made by Marc LaLonde, Canada's former minister of National Health and Welfare, in 1974:

> Even such a simple question as whether one should severely limit his consumption of butter and eggs can be a subject of endless scientific debate.

Faced with conflicting scientific opinions of this kind, it would be easy for health educators and promoters to sit on their hands; it certainly makes it easy for those who abuse their health to find a real "scientific" excuse.

But many of Canada's health problems are sufficiently pressing that action has to be taken even if all scientific evidence is not in.[52]

These words are relevant today in both Canada and the United States, as scientists debate the merits of butter over margarine, the health effects of *trans* fatty acids, and the implications of biotechnology. Dietary guidelines are a good example of the changeable interface between science and policy.

• **Dietary Guidelines for Americans.** The fourth edition of the Dietary Guidelines for Americans was published jointly by the USDA and the DHHS in 1995 (earlier editions had been published in 1980, 1985, and 1990). The principles of healthful eating promoted by the 1995 edition of the Dietary Guidelines, illustrated on the inside front cover, grew out of reports such as the *Surgeon General's Report on Nutrition and Health*, the National Research Council's report on *Diet and Health*, and the National Cholesterol Education Program's Report of the *Expert Panel on Detection, Evaluation, and Treatment of High Blood Cholesterol in Adults*.[53] These reports all pointed to the need to change and improve the eating patterns of the U.S. population.

Thus, the Dietary Guidelines were designed to be consistent with these reports and the recommendations of other government and voluntary health organizations. They emphasize moderation, the avoidance of over- and underconsuming foods, and enjoyment of eating—"one of life's greatest pleasures." The guidelines apply to the diet as consumed over several days, not to a single meal or food. The key messages are to consume a variety of foods providing adequate amounts of nutrients and balance food intake with physical activity to maintain or improve body weight.[54] By law (PL 101-445), the Dietary Guidelines for Americans must be revised every 5 years, beginning in 1995.[55] A short history of the development of this policy tool can be found at the Internet address shown in the boxed insert of Internet Resources.

Consumers who follow the Dietary Guidelines for Americans eat plenty of fruits, vegetables, and whole grains and balance their food intake with physical activity.

• **Other government dietary guidelines.** A variety of U.S. government agencies, including the Surgeon General's office, the National Cancer Institute, and the National Heart, Lung, and Blood Institute, issued dietary guidelines during the 1980s. Some of these recommendations are quite general, whereas others contain very specific recommended intakes of certain dietary components, such as saturated fat, polyunsaturated fat, and fiber. The nutrition recommendations for Canadians are given in Appendix G.

Nongovernment Dietary Recommendations. During the 1980s, dietary recommendations were also issued by a variety of nonprofit health organizations, such as the American Heart Association, American Cancer Society, and American Institute for Cancer Research. These groups were motivated to provide dietary guidance for the public by the growing scientific evidence linking certain dietary patterns with increased risk for heart disease and some types of cancer. These recommendations represent attempts to create broad, noncontroversial recommendations for dietary patterns in the United States and around the world.

Implementing the Recommendations: From Guidelines to Groceries

The challenge today is to help consumers put the wide assortment of dietary recommendations into practice. This requires translating the recommendations into food-specific guides that consumers can implement in their homes, at the grocery store, or in restaurants.[56] For example, consider this dietary guideline: *Choose a diet low in fat, saturated fat, and cholesterol.* This broad guideline must first be translated into food-specific behaviors for consumers, who must acquire the knowledge and skills to change their eating patterns. Table 4-7 lists three food-specific behaviors that can be derived from the guideline and the knowledge and skill set required for each. Consumers who wish to adopt a dietary pattern low in fat, saturated fat, and

Dietary Guideline	Food-Specific Behavior	Knowledge and/or Skills Required to Support the Behavior
Choose a diet low in fat, saturated fat and cholesterol.	• Choose low-fat foods more often than high-fat foods.	• Know major sources of fat, saturated fat, and cholesterol. • Be able to read food product labels. • Know low-fat cooking techniques. • Know how to adapt recipes. • Know which foods to substitute for higher-fat foods. • Know how to read restaurant menus.
	• Choose lean meats, fish, chicken, and turkey.	• Know lean cuts of meat. • Know low-fat cooking methods. • Remove skin from chicken and turkey before cooking.
	• Choose low-fat dairy products.	• Know which dairy products are low in fat. • Be able to read food product labels.

TABLE 4-7 *Translating a Dietary Guideline into Food-Specific Behaviors*

cholesterol must realize that doing so will require them to choose low-fat foods more often than high-fat foods. Successfully choosing low-fat foods means having some knowledge of food composition, recognizing the major food sources of fat and cholesterol in the diet, knowing which cooking techniques reduce fat intake, knowing how to adapt recipes, and being able to read a restaurant menu. The process of breaking down a dietary guideline into specific behaviors can be carried a step further. The guideline can be translated into an actual eating pattern, as shown in Table 4-8. The eating pattern should reflect the basic principles of the Dietary Guidelines and focus on variety—eating a selection of foods from all of the food groups; proportionality—eating appropriate amounts of foods to meet nutritional needs; moderation—enjoying all foods but avoiding eating patterns that are associated with chronic disease; and usability—being flexible and practical enough to accommodate individual food preferences and meet nutrient needs.[57]

Policy Making in Action

Students often have difficulty visualizing the connection between policy making and their work as community nutritionists. Let's describe an example of this connection, drawing on the job responsibilities of the director of health promotion for the FirstRate Spa and Health Resort (described in Table 1-5 on page 21). The director decides to develop a risk reduction program that will help spa clients make healthier eating choices. The program will address eating strategies to reduce the risk of coronary heart disease, stroke, cancer, and osteoporosis. In organizing the content of the program, the director realizes that clients need to be given information about healthy eating patterns; recommendations related to fat, calcium, and antioxidant vitamin intake; and instructions on how to read food product labels, among other topics. Table 4-9 shows six decisions the director must make in developing the program's content, the policy "tool" she will use for each decision, and the source of each tool. Her first decision is to choose a program instructor who is a registered dietitian. This decision enhances the image of the program and ensures that accurate and timely information will be offered to clients. Decisions to use the Dietary Guidelines for Americans and the Food Guide Pyramid are fairly obvious ones, as these are widely accepted policy tools of the federal government. The company's requirement that all visual aids show the logo of the FirstRate Spa and Health Resort is also a policy. Every organization, agency, and institution has its own policies that affect the practice of community nutritionists. The decision about how much material to present to clients on the essential fatty acid, alpha-linolenic acid, and other omega-3 fatty acids—hot topics for consumers today—is more complex. The United States has no RDA or DRI for omega-3 fatty acids, whereas Canada has developed an RNI for them. Should she use the Canadian policy tool, the RNI, in this situation? Ah, this is where we see policy making in action, for the director is making a policy decision when she chooses to use the RNI for omega-3 fatty acids in program materials. Many decisions made by community nutritionists—the people called "street-level bureaucrats" in Chapter 2—are opportunities for making and implementing policy.

Foods		Energy (kcal)	Fat (g)
Breakfast			
Grapefruit	1/2 medium	39	—†
Sugar	1 tsp.	16	0
Whole-wheat cereal, hot, cooked	1 cup	150	1
Milk (1% low-fat)	1 cup	102	3
Coffee	1 cup	—	0
Milk (1%) in coffee	1 tbsp.	6	—
Lunch			
Hotdog, regular, with bun	1	365	17
Baked beans with tomato sauce and pork	1/2 cup	124	2
Coleslaw made with:			
Cabbage, shredded	1/2 cup	8	—
Carrot, grated	1 small	22	—
Mayonnaise, low-fat	1 tbsp.	35	3
Dill pickle	1 medium	12	—
Diet soft drink	1	2	0
Snack			
Whole-wheat crackers	4	72	3
Swiss cheese	1 oz.	106	8
Dinner			
Roast beef, lean only	4 oz.	245	10
Corn, boiled	1/2 cup	66	—
Salad:			
Spinach, raw	1 cup	7	—
Tomato, raw	1/2 medium	13	—
Italian dressing, low-fat	1 tbsp.	16	2
Rice, herbed	1/2 cup	102	—
Dinner rolls	2	170	4
Margarine, soft	1 tsp.	34	4
Peach melba:			
Peach halves	1 large peach	68	—
Raspberries	1/4 cup, fresh	15	—
Ice milk, vanilla	1/2 cup	164	6
Iced tea, unsweetened	8 oz.	—	—
Total		**1,959**	**63**

TABLE 4-8 *Eating Pattern for a 2,000-kcal Diet with about 30 Percent of Calories from Fat**

*The eating pattern was adapted from U.S. Department of Agriculture, Human Nutrition Information Service, *Preparing Foods and Planning Menus Using the Dietary Guidelines,* Home and Garden Bulletin No. 232-8 (Washington, D.C.: U.S. Government Printing Office); the energy and fat values were taken from U.S. Department of Agriculture, Agricultural Research Service, *USDA Nutrient Database for Standard Reference,* Release 12, 1998, available on the Nutrient Data Laboratory Home Page at www.nal.usda.gov/fnic/foodcomp.
†— = Negligible.

TABLE 4-9 *Connection Between Policy Tools and Policy Decisions*

Decision	Policy Tool	Source of Policy Tool
Instructor will be a registered dietitian (RD)	Credentials	Commission on Dietetic Registration
Program should include a discussion of healthy eating patterns	Dietary Guidelines for Americans, Food Guide Pyramid	U.S. Department of Agriculture
Program segment on osteoporosis should include recommendations for obtaining calcium from foods, including fortified foods, and dietary supplements	Dietary Reference Intakes (DRIs)	Food and Nutrition Board of the National Academy of Sciences
Program should include a segment on reading food product labels	Labeling regulations	Food and Drug Administration
All overheads must show the company logo	Company policy	FirstRate Spa and Health Resort
Program segment on dietary fats should include information about omega-3 fatty acids and alpha-linolenic acid	No current U.S. policy; there is a Canadian RNI* for omega-3 fatty acids	Health Canada

*RNI = Recommended Nutrient Intake.

Policy Making Does Not Stand Still

This is an exciting time in community nutrition. New legislation related to welfare reform and health care, new dietary reference intake values (the DRIs), and market forces in the health care field and the food industry promise many opportunities for community nutritionists to serve as liaisons between policy makers and the general public. The one guarantee is that nutrition policy will change in the next few years.

The task of formulating a national nutrition agenda is daunting. Government must make a commitment to develop and promote a coordinated plan to improve the nation's nutritional and health status, scientists must reach a consensus on the interpretation of scientific findings and appropriate dietary advice for all Americans, and sufficient financial resources must be allocated to implement the policy. Consider the policy-making machinery that has geared up to ensure that all Americans are properly fed. An outgrowth of the 1996 World Food Summit in Rome, the United States developed a blueprint, titled "U.S. Action Plan for Food Security," to strengthen the commitment of the U.S. government and civil society to reducing hunger and malnutrition both at home and abroad by the year 2015. Hundreds of people around the country helped develop the plan, which is scheduled to be adopted and funded by Congress in 1998. By the year 2000, the first steps toward achieving the plan will occur with the launch of the Millennium Food Security Initiative.[58] (Check the boxed insert for Internet addresses related to the U.S. and Canadian action plans.) The actions of the U.S., Canadian, and other governments regarding the issues of hunger and malnutrition are good examples of the importance of policy making in community nutrition.

Community Learning Activity

Activity 1

Choose one of the Dietary Guidelines (other than the one related to choosing a diet low in fat, saturated fat, and cholesterol), and develop a table showing the food-specific behaviors and skill set required to adopt it.

Activity 2

Go back and read the quote at the beginning of this chapter. How would you translate current nutrition recommendations for a client who regularly uses dietary supplements such as ginkgo biloba and St. John's wort? If you believe that you are missing important information about these products, what steps would you take to influence government policy on dietary supplements?

Activity 3

In this Internet activity, access the following page of the American Dietetic Association's Web site: http://www.eatright.org/adap1197.html. There you will find the ADA's position statement on vegetarian diets. Review the on-line material and answer the following questions:

1. Is this position paper an "ADA policy" on vegetarian diets? Why?
2. If the ADA had published no position paper on vegetarian diets and you had discovered the vegetarian Food Guide Pyramid on-line or elsewhere, would you use it as a nutrition education tool for teaching your clients about nutrition? Why?
3. Considering your answer to the last question, what factors contributed to your decision?

Activity 4

Review the 1996 BRFSS Summary Prevalence Report on the Internet at www.cdc.gov/nccdphp/brfss. (The file can be found at www.cdc.gov/nccdphp/brfss/96prvrpt.pdf.) Then, complete the following activities:

1. Using the BRFSS data, summarize the findings about the general health status of your state's population. Include statements about self-reported health status, quality of life, no insurance, diabetes, smoking, overweight, no leisure time physical activity, and consumption of fruits and vegetables.
2. Include definitions of each risk factor. (Refer to the definitions that begin on page xvii of the on-line report.)
3. Describe the progress your state has made in meeting the *Healthy People 2000* objectives for nutrition and health status.

Professional Focus   *Ethics and You*

Life is full of paradoxes. Whereas health promotion and disease prevention are paramount to halting the escalating cost of health care in this country, the United States spends less than 1 percent of all dollars directed toward health care on public health and disease prevention.[1] Even though one goal of nutrition is to apply scientific knowledge to feed all people adequately, every fifth child in the United States is vulnerable to hunger. Although the United States spends more on health care than other nations, certain health disparities between racial and ethnic groups, particularly in pregnancy outcome, infant mortality rate, nutritional status, life expectancy, and food insecurity exist. And finally, one issue confronting developing countries is whether it is fundamentally wrong that so much preventable sickness and death occur in the world.

As a community nutritionist, how do you address such issues? This Professional Focus reviews some of the ethical questions in the field of health promotion that relate specifically to community nutritionists. Its intent is not to arrive at a conclusion or present solutions to ethical dilemmas; rather, it seeks to present the issues for your consideration and stresses the need for moral sensitivity in the planning and implementation of community nutrition programs. As Aristotle once said, "We are what we repeatedly do." Moral sensitivity and characteristics such as honesty, integrity, loyalty, and candor are developed by practice.[2]

What Is Ethics?

Philosophers throughout history have struggled with questions of how to live and work ethically. Ethics is a philosophical discipline dealing with what is morally good and bad, right and wrong. Ethics helps decision makers search for criteria to evaluate different moral stances.[3]

As a community nutritionist, you may wonder what ethics has to do with your professional activities. Certainly, as a community nutritionist, you will not often confront such media issues as euthanasia, abortion, capital punishment, insider trading, maternal surrogacy, infanticide, the withdrawal of nutrition support for terminally ill patients, or the right to die. Nevertheless, situations arise in community settings that will force you to make ethical decisions.

Community nutritionists working with the media or food industry must consider the accuracy of product descriptions and claims as well as words and images that can mislead the public.[4] As a manager, the community nutritionist may face ethical dilemmas in allocating resources. In setting priorities, she may have to make some decisions that call on her ethics. If she believes that all eligible clients have the right to receive optimal nutritional care, then how should she decide which clients will actually receive the care? (In this case, optimal nutritional care may mean the receipt of home-delivered meals by a homebound elderly person.) The community nutritionist involved in research emphasizes honesty, accuracy, and integrity in conducting studies and publishing the results. Consider the impact of using falsified data in determining nutrition policy for funding a new or existing nutrition program.

Codes of Ethics

Simple answers to ethical questions are elusive, but many health care organizations and professional associations have established codes of ethics to provide guidance in resolving ethical dilemmas.[5] Codes of ethics are written to guide decision making in areas of moral conflict; outline the obligations of the practitioner to self, client, society, and the profession; and "assist in protecting the nutritional health, safety, and welfare of the public."[6] The American Dietetic Association (ADA) published its first code of ethics in 1942.[7] The most recent code (presented in the next section) became effective in 1989 and applies to all ADA members and credentialed practitioners.

A code of ethics for nutritionists and other professionals working in international situations is likewise critical, as C. E. Taylor notes: "Needs are so obvious that the temptation is great to rush in with programs that seem reasonable; but international work is full of surprises. Each new activity needs to be carefully tested."[8] Consider the story of the monkey and the fish:

> After a dam burst, a flood raged through an African countryside. A monkey, standing in safety on the riverbank, watched a fish swim into its view. "I will save this poor fish from drowning," thought the monkey. And, swinging from a tree branch, he scooped up the fish and carried it, gasping, to land. "Throw me back," pleaded the fish. Reluctantly, the monkey agreed, scratching his head in bewilderment at the fish's lack of appreciation for the aid he had so selflessly offered.[9]

Code of Ethics for the Profession of Dietetics*

The dietetic practitioner:

1. Provides professional services with objectivity and with respect for the unique needs and values of individuals.
2. Avoids discrimination against other individuals on the basis of race, creed, religion, sex, age, and national origin.
3. Fulfills professional commitments in good faith.
4. Conducts himself/herself with honesty, integrity, and fairness.
5. Remains free of conflict of interest while fulfilling the objectives and maintaining the integrity of the dietetic profession.
6. Maintains confidentiality of information.
7. Practices dietetics based on scientific principles and current information.
8. Assumes responsibility and accountability for personal competence in practice.
9. Recognizes and exercises professional judgment within the limits of his or her qualifications and seeks counsel or makes referrals as appropriate.
10. Provides sufficient information to enable clients to make their own informed decisions.
11. Who wishes to inform the public and colleagues of his or her service does so by using factual information; does not advertise in a false or misleading manner.
12. Promotes or endorses products in a manner that is neither false nor misleading.
13. Permits use of his or her name for the purpose of certifying that dietetic services have been rendered only if he or she has provided or supervised the provision of those services.
14. Accurately presents professional qualifications and credentials:
    - Uses "RD" or "registered dietitian" and "DTR" or "dietetic technician registered" only when registration is current and authorized by the Commission on Dietetic Registration.
    - Provides accurate information and complies with all requirements of the Commission on Dietetic Registration program in which he or she is seeking initial or continued credentials from the Commission on Dietetic Registration.
    - Is subject to disciplinary action for aiding another person in violating any Commission on Dietetic Registration requirements or aiding another person

in representing himself/herself as an RD or DTR when he or she is not.

15. Presents substantiated information and interprets controversial information without personal bias, recognizing that legitimate differences of opinion exist.
16. Makes all reasonable effort to avoid bias in any kind of professional evaluation; provides objective evaluation of candidates for professional association memberships, awards, scholarships, or job advancements.
17. Voluntarily withdraws from professional practice under the following circumstances:
    - Has engaged in any substance abuse that could affect his or her practice;
    - Has been adjudged by a court to be mentally incompetent;
    - Has an emotional or mental disability that affects his or her practice in a manner that could harm the client.
18. Complies with all applicable laws and regulations concerning the profession; is subject to disciplinary action under the following circumstances:
    - Has been convicted of a crime under the laws of the United States which is a felony or a misdemeanor, an essential element of which is dishonesty and which is related to the practice of the profession.
    - Has been disciplined by a state and at least one of the grounds for the discipline is the same or substantially equivalent to these principles.
    - Has committed an act of misfeasance of malfeasance that is directly related to the practice of the profession as determined by a court of competent jurisdiction, a licensing board, or an agency of a governmental body.
19. Accepts the obligation to protect society and the profession by upholding the Code of Ethics for the Profession of Dietetics and by reporting alleged violations of the Code through the defined review process of The American Dietetic Association and its credentialing agency, the Commission on Dietetic Registration.

---

*Source: From the American Dietetic Association, Code of Ethics for the Profession of Dietetics, *Journal of the American Dietetic Association* 88 (1988): 1592–93. Copyright © 1988 The American Dietetic Association, Chicago, Illinois. Reprinted by permission.

Continued

Guiding Principles

Three basic principles are used in ethical decision making and in developing guidelines for professional practice:[10] (1) autonomy—respecting the individual's rights of self-determination, independence, and privacy; (2) beneficence—protecting clients from harm and maximizing possible benefits; and (3) justice—striving for fairness in one's actions and equality in the allocation of resources.

To determine whether an issue in the community setting raises an ethical question, consider these ethical principles expressed as questions:[11]

1. Does the nutritional program, message, product, or service foster or deter the individual's ability to act freely? (The ADA code of ethics items 6, 10, and 15 address this principle of autonomy.)
2. Does the nutritional program, message, product, or service help people or harm them? (Items 7, 8, 9, 17, and 18 in the ADA code of ethics address this principle of beneficence.)
3. Does the nutritional program, message, product, or service unfairly or arbitrarily discriminate among persons or groups? (Items 1, 2, and 16 in the ADA code of ethics address this principle of justice.)

Consumers are eager to know about nutrition. The principle of beneficence moves us, as nutrition educators, to provide consumers with truthful and convincing information based on current scientific knowledge. In this way, we protect them from fraudulent misinformation and also motivate them to change their diet accordingly.[12] The principle of autonomy motivates us to provide consumers with factual information that includes both the weaknesses and the strengths of the scientific data supporting a given behavior, service, or product; with this information, individuals can exercise their right to make an informed choice or decision.

Heath Promotion and Ethics

The purpose of health promotion is to motivate people to adopt and maintain healthful practices in order to prevent illness and functional impairment. Many hold that by investing in health promotion and disease prevention activities, we can avoid the much greater economic and social costs of disease and disability. The challenge today is to provide the public with the opportunity to benefit from appropriate nutrition knowledge and services. However, this challenge raises a hidden moral issue worthy of consideration. At what point is scientific knowledge sufficiently documented to warrant translating it into dietary messages to the public? What responsibilities and rights do we have to alter individual lifestyles in our effort to promote public health? As health promoters, we are sometimes criticized for taking a paternalistic approach with a "We know better than you" attitude. Moral sensitivity demands that we respect the dignity of persons—their right to make their own choices. For example, we may carefully and creatively design messages for older women at risk of osteoporosis, encouraging them to use dairy products in their daily diet, but our target audience has the right to resist our efforts and choose not to do so. An adequate calcium intake is a good thing, but life offers many other good things as well.

Health promotion for a number of issues (for example, cigarette smoking and drinking and driving) necessitates a paternalistic approach that both restricts private liberties and promotes group virtues like beneficence and concern for the common good. Such paternalism is for the most part considered legitimate and reflects the view that the good of each of us is not the same thing as the good of all of us together.[13]

Community nutritionists, as health promoters, call attention to other health risks—a diet high in saturated fat or low in fiber, commercial advertising of empty-calorie foods to children, and nutrition fraud in the marketplace, among others. Should governments, therefore, move from taxing cigarettes and alcohol to taxing companies that manufacture high-fat confections or cereals high in sugar? Since obesity and a sedentary lifestyle are associated with a number of chronic diseases (for instance, hypertension, coronary artery disease, diabetes) and increased health care costs, should persons who eat too many calories or too much fat or those who fail to exercise regularly be taxed to discourage these lifestyles and raise revenues for health care? Should we fine pregnant women who smoke or drink? In other words, to what extent should society tolerate and bear the burden for the health risks that individuals choose to take? Such are the ethical dilemmas facing those working in health promotion. The ethical conflict is how to achieve the goal of protecting and promoting public health while ensuring an individual's freedom of choice.

Ethical Decision Making

Analytical skills are necessary to resolve ethical dilemmas. One must objectively evaluate the individual circumstances of each situation, gather relevant data, consider possible alternatives, consult with experts as necessary, and take appropriate actions to accomplish the greatest good for the greatest number. The particular action chosen must adhere to the general ethical principles of autonomy, beneficence, and justice.

As community nutritionists, you are certain to face ethical dilemmas in both your professional and personal lives. In closing this section, we leave you with a set of questions.[14] In the months to come, consider your responses to these questions based on your moral sensitivity regarding these situations, your understanding of ethical principles, and your discussions with other professionals experienced in making ethical decisions:

- Is it right to save lives by immunization, nutrition, oral rehydration therapy, or chemotherapy when those who are saved face a life of despair?
- Do the United States, Canada, and other developed countries have a moral obligation toward the less developed countries?
- When setting priorities in program planning in the face of limited resources, who should receive benefits— infants? Children? Pregnant women? Working people? Elderly persons?
- What are the ethical limits to promotional activities of multinational corporations? Is it acceptable to market infant formula or soft drinks in developing countries? Should we ban television advertising of empty-calorie foods to children?
- In setting program priorities, are there situations in which one ethnic group should be favored over another?
- Do we have the right to ask individuals to adjust and, in some cases, to abandon their ethnic and cultural customs, traditions, or cuisines for the goal of improved health?

References

1. D. R. Smith, Public health and the winds of change, *Public Health Reports* 113 (1998): 160–61.
2. S. L. Anderson, Dietitians' practices and attitudes regarding the Code of Ethics for the Profession of Dietetics, *Journal of the American Dietetic Association* 93 (1993): 88–91.
3. S. Tamborini-Martin and K. V. Hanley, The importance of being ethical, *Health Progress* 70 (1989): 24.
4. J. N. Neville and R. Chernoff, Professional ethics: Everyone's issues, *Journal of the American Dietetic Association* 88 (1988): 1286.
5. J. Sobal, Research ethics in nutrition education, *Journal of Nutrition Education* 24 (1992): 234–38.
6. American Dietetic Association and Commission on Dietetic Registration, *Code of Ethics for the Profession of Dietetics* (Chicago: American Dietetic Association, 1988).
7. Neville and Chernoff, Professional ethics, p. 1287.
8. C. E. Taylor, Ethics for an international health profession, *Science* 153 (1966): 716–20, as cited by P. F. Basch, *Textbook of International Health* (New York: Oxford University Press, 1990), pp. 407–8.
9. C. Levine, Ethics, justice, and international health, *Hastings Center Report* 4 (1977): 5–6.
10. M. Barry, Ethical considerations of human investigations in developing countries: The AIDS dilemma, *New England Journal of Medicine* 319 (1988): 1083–85.
11. M. W. Kreuter, M. J. Parsons, and M. P. McMurry, Moral sensitivity in health promotion, *Health Education* (November/December 1982): 11–13.
12. K. McNutt, Ethics: A cop or a counselor? *Nutrition Today* 26 (1991): 36–39.
13. D. Beauchamp, Lifestyle, public health and paternalism, in *Ethical Dilemmas in Health Promotion*, ed. S. Doxiadis (New York: Wiley, 1987), pp. 69–81.
14. Adapted from Basch, *Textbook of International Health*.

References

1. U.S. Department of Health and Human Services, Public Health Service, *The Surgeon General's Report on Nutrition and Health: Summary and Recommendations* (Washington, D.C.: U.S. Government Printing Office, 1988), pp. 2–4.
2. As cited in J. E. Austin and C. Hitt, *Nutrition Intervention in the United States* (Cambridge, MA: Ballinger, 1979), p. 355.
3. Ibid., pp. 357–85.
4. Quoted in ibid., p. 356.
5. D. J. Palumbo, *Public Policy in America—Government in Action* (San Diego: Harcourt Brace Jovanovich, 1988), p. 17.
6. C. E. Woteki and M. T. Fanelli-Kuczmarski, The National Nutrition Monitoring System, in *Present Knowledge in Nutrition*, 6th ed. (Washington, D.C.: International Life Sciences Institute, 1990), pp. 415–29.

7. U.S. Department of Health and Human Services, U.S. Department of Agriculture, *Nutrition Monitoring in the United States: An Update Report on Nutrition Monitoring*, DHHS (PHS) Pub. No. 89-1255 (Washington, D.C.: U.S. Government Printing Office, 1989); J. B. Mason and coauthors, *Nutrition Surveillance* (Geneva: World Health Organization, 1984); the discussion of the five types of data collection and end-use activities was adapted from E. Yetley, A. Beloian, and C. Lewis, Dietary methodologies for food and nutrition monitoring, in U.S. Department of Health and Human Services, *Vital and Health Statistics: Dietary Methodology Workshop for the Third National Health and Nutrition Examination Survey* (Washington, D.C.: U.S. Government Printing Office, 1992), pp. 58–67.

8. Mason and coauthors, *Nutrition Surveillance;* and R. R. Briefel and C. T. Sempos, Introduction, in U.S. Department of Health and Human Services, *Vital and Health Statistics*, pp. 1–2.

9. The definition of nutrition monitoring and related research was taken from Federation of American Societies for Experimental Biology, Life Sciences Research Office, prepared for the Interagency Board for Nutrition Monitoring and Related Research, *Third Report on Nutrition Monitoring in the United States*, Vol. 1 (Washington, D.C.: U.S. Government Printing Office, 1995), p. xxiii.

10. G. Ostenso, National Nutrition Monitoring System: A historical perspective, *Journal of the American Dietetic Association* 84 (1984): 1181–85.

11. U.S. Department of Health and Human Services, U.S. Department of Agriculture, *Nutrition Monitoring in the United States* (Washington, D.C.: U.S. Government Printing Office, 1986); and *Nutrition Monitoring in the United States: An Update Report*.

12. Federation of American Societies for Experimental Biology, *Third Report on Nutrition Monitoring in the United States*, Vol. 1, pp. 1–17.

13. N. W. Jerome and J. A. Ricci, Food and nutrition surveillance: An international overview, *American Journal of Clinical Nutrition* 65 (suppl.) (1997): 1198S–202S.

14. The margin definition of the NNMRRP was adapted from Woteki and Fanelli-Kuczmarski, *Present Knowledge in Nutrition*, pp. 415–29.

15. *Nutrition Monitoring in the United States: An Update Report*, pp. 2–3.

16. The section that follows was taken from Department of Health and Human Services, U.S. Department of Agriculture, *Nutrition Monitoring in the United States: The Directory of Federal Nutrition Monitoring Activities* (Washington, D.C.: U.S. Government Printing Office, 1989), pp. 1–64; and Federation of American Societies for Experimental Biology, *Third Report on Nutrition Monitoring in the United States*, Vols. 1 and 2.

17. Federation of American Societies for Experimental Biology, *Third Report on Nutrition Monitoring in the United States*, Vol. 1, p. 25.

18. G. M. McQuillan and coauthors, Update on the seroepidemiology of human immunodeficiency virus in the United States household population: NHANES III, 1988–1994, *Journal of Acquired Immune Deficiency Syndromes and Human Retrovirology* 14 (1997): 355–60.

19. C. E. Woteki and coauthors, National Health and Nutrition Survey—NHANES: Plans for NHANES III, *Nutrition Today* (January/February 1988): 26.

20. W. C. Chumlea and coauthors, Stature prediction equations for elderly non-Hispanic white, non-Hispanic black, and Mexican-American persons developed from NHANES III data, *Journal of the American Dietetic Association* 98 (1998): 137–42.

21. Federation of American Societies for Experimental Biology, *Third Report on Nutrition Monitoring in the United States*, Vol. 1, pp. 24–25.

22. N. Dupree, NHANES—What's available from NHANES III and Plans for NHANES IV, available on the Web site for the Centers for Disease Control and Prevention at www.cdc.gov/nchswww/data/PDF/Dupree.pdf.

23. Food Surveys Research Group, Agricultural Research Service, About the Food Surveys Research Group (FSRG), available on the Agricultural Research Service's Web site at sun.arsgrin.gov/ars/Beltsville/barc/bhnrc/foodsurvey/ fsrg.htm.

24. Information about the BRFSS was obtained from the Web site of the Centers for Disease Control and Prevention at www.cdc.gov/nccdphp/brfss/about.htm.

25. The margin definition of disappearance data was taken from Federation of American Societies for Experimental Biology, *Third Report on Nutrition Monitoring in the United States*, Vol. 1, p. I-3.

26. Mason and coauthors, *Nutrition Surveillance*, p. 12.

27. The description of uses of the BRFSS was adapted from the Web site of the Centers for Disease Control and Prevention at www.cdc.gov/nccdphp/brfss/about.htm.

28. G. E. Brown, Jr., National Nutrition Monitoring System: A congressional perspective, *Journal of the American Dietetic Association* 84 (1984): 1185–89.

29. Food and Nutrition Board, National Research Council, *Recommended Dietary Allowances*, 10th ed. (Washington, D.C.: National Academy Press, 1989), p. 1.

30. Ibid., pp. 8–9.

31. Food and Nutrition Board, Dietary Reference Intakes (DRIs) for calcium, phosphorus, magnesium, vitamin D, and fluoride, *Nutrition Today* 32 (1997): 182–88.

32. Ibid.

33. Information about the DRIs was obtained from the Institute of Medicine on the National Academy of Sciences Web site at www2.nas.edu/iom; and A. A. Yates and coauthors, Dietary Reference Intakes: The new basis for recommendations for calcium and related nutrients, B vitamins, and choline, *Journal of the American Dietetic Association* 98 (1998): 699–706.

34. Food and Nutrition Board, How should the Recommended Dietary Allowances be revised? A concept paper from the Food and Nutrition Board, *Nutrition Reviews* 52 (1994): 216–19.

35. Food and Nutrition Board, Dietary Reference Intakes (DRIs), p. 186.

36. W. Mertz, A perspective on mineral standards, *Journal of Nutrition* 128 (1998): 375S–78S.

37. T. K. Murray and J. L. Beare-Rogers, Nutrition recommendations, 1990, *Journal of the Canadian Dietetic Association* 50 (1990): 391–95.

38. The figures on the prevalence of overweight, levels of physical activity, and dietary fat intake were taken from U.S. Department of Agriculture, Agricultural Research Service, *Data tables: Results from USDA's 1996 Continuing Survey of Food Intakes by Individuals and 1996 Diet and Health Knowledge Survey* (on-line), 1997, available on the Web site of the Food Surveys Research Group (under "Releases") at www.barc.usda.gov/bhnrc/foodsurvey/home.htm; and U.S. Department of Health and Human Services, Public Health Service, Guidelines for school health programs to promote lifelong healthy eating, *Morbidity and Mortality Weekly Report* 45 (No. RR-9) (June 14, 1996): 3–4.

39. C. Lenfant and N. Ernst, Daily dietary fat and total food-energy intakes—Third National Health and Nutrition Examination Survey, phase 1, 1988–91, *Morbidity and Mortality Weekly Report* 43 (1994): 116–17, 123–25.

40. R. R. Briefel and coauthors, Dietary methods research in the Third National Health and Nutrition Examination Survey: Underreporting of energy intake, *American Journal of Clinical Nutrition* 65 (suppl.) (1997): 1203S–9S.

41. M. A. McDowell and coauthors, Energy and macronutrient intakes of persons ages 2 months and over in the United States: Third National Health and Nutrition Examination Survey, Phase 1, 1988–91, *Vital and Health Statistics Advance Data*, No. 255 (October 24, 1994), available on-line at www.cdc.gov/nchswww/data/ad255.pdf.

42. U.S. Department of Agriculture, Agricultural Research Service, *Pyramid servings data: Results from USDA's 1995 and 1996 Continuing Survey of Food Intakes by Individuals* (on-line), 1997. ARS Food Surveys Research Group, available (under "Releases") at www.barc.usda.gov/bhnrc/foodsurvey/home.htm.

43. *Third Report on Nutrition Monitoring in the United States*, Vol. 1, pp. 99–158; and K. Alaimo and coauthors, Dietary intake of vitamins, minerals, and fiber of persons ages 2 months and over in the United States: Third National Health and Nutrition Examination Survey, Phase 1, 1988–91, *Vital and Health Statistics Advance Data*, No. 258 (November 3, 1994), available on-line at www.cdc.gov/nchswww/data/ad258.pdf.

44. C. W. Enns, J. D. Goldman, and A. Cook, Trends in food and nutrient intakes by adults: NFCS 1977–78, CSFII 1989–91, and CSFII 1994-95, *Family Economics and Nutrition Review* 10 (1997): 1–15, available on-line at www.barc.usda.gov/bhnrc/foodsurvey/trends.pdf.

45. A. M. Stephan and N. J. Wald, Trends in individual consumption of dietary fat in the United States, 1920–1984, *American Journal of Clinical Nutrition* 52 (1990): 457–69; G. Block, W. F. Rosenberger, and B. H. Patterson, Calories, fat, and cholesterol: Intake patterns in the US population by race, sex, and age, *American Journal of Public Health* 78 (1988): 1150–55.

46. J. B. Richmond, Forward, *American Journal of Clinical Nutrition* 32 (1979): 2621–22.

47. P. M. Behlen and F. J. Cronin, Dietary recommendations for healthy Americans summarized, *Family Economics Review* 3 (1985): 17–24.

48. O. Hayes, M. F. Trulson, and F. J. Stare, Suggested revision of the basic 7, *Journal of the American Dietetic Association* 31 (1955): 1103–7.

49. U.S. Department of Agriculture, *The Food Guide Pyramid*, Home and Garden Bulletin No. 252 (Hyattsville, MD: Human Nutrition Information Service, 1992).

50. The Report of the U.S. Senate Select Committee on Nutrition and Human Needs, Dietary goals for the United States, *Nutrition Today* (September/October 1977): 20–30.

51. K. W. McNutt, An analysis of the Dietary Goals for the United States, second edition, *Journal of Nutrition Education* 10 (1978): 61–62.

52. The Report of the U.S. Senate Select Committee on Nutrition and Human Needs, Dietary goals, p. 22.

53. U.S. Department of Health and Human Services, Public Health Service, *The Surgeon General's Report;* Committee on Diet and Health, National Research Council, *Diet and Health: Implications for Reducing Chronic Disease Risk—Executive Summary* (Washington, D.C.: U.S. Government Printing Office, 1989); The Expert Panel, Report of the Expert Panel on detection, evaluation, and treatment of high blood cholesterol in adults, *Archives of Internal Medicine* 148 (1988): 36–69.

54. U.S. Department of Agriculture, U.S. Department of Health and Human Services, *Nutrition and Your Health: Dietary Guidelines for Americans*, 4th ed. (Washington, D.C.: U.S. Government Printing Office, 1995).

55. U.S. Department of Agriculture, History of Dietary Guidelines for Americans, as cited on-line at www.nal.usda.gov/fnic/ Dietary/12dietapp1.htm.

56. R. M. Mullis and coauthors, Developing nutrient criteria for food-specific dietary guidelines for the general public, *Journal of the American Dietetic Association* 90 (1990): 847–51.

57. S. Welsh, C. Davis, and A. Shaw, Development of the Food Guide Pyramid, *Nutrition Today* 27 (1992): 16.

58. Information about the U.S. Action Plan on Food Security was taken from two Internet documents: *Framework for the U.S. Action Plan on Food Security* and *Discussion Paper on Domestic Food Security*. Both are available at www.fas.usda.gov/icd/summit.

Community Nutritionists in Action: Assessing and Planning

Serena M. sits on the edge of her bed and surveys the baby things spread around the room. A pearly white crib stands near the window on her left, and an old chest of drawers, newly painted yellow, fills the space on her right. On the floor next to her husband's crumpled jeans and T-shirt is a stuffed rabbit, a stack of folded diapers, two rubber squeeze toys, a box of Q-tips, and an airplane mobile, still in its box. She wonders whether she has everything she needs for the new baby, her first, due in just a few weeks.

Serena is excited about this change in her life and a little scared, too. Money is tight. She and her husband, Todd, live with his parents because they cannot afford their own apartment. They can't even afford a car, which is why Todd works at the auto body shop just four blocks over. She tries not to let her fear show, even though she frets daily about life with their in-laws and the prospects of being a mother. "I'm nearly 17," Serena says to herself. "My mother was about this age when she had my brother." As she gets up to take clothes out of the dryer, she wonders whether her mother ever felt scared, anxious, and alone.

Here is a young woman, in many respects a child herself, who is expecting her first baby. Does she know how to care for an infant? Is the family's income sufficient to cover the costs of providing for another child? Does this family need food assistance? Is there a way for this young woman to have her baby and continue in school? Is Serena eating properly during her pregnancy? Does she plan to breastfeed the baby? How can the health and nutritional outcome for this baby and its mother be improved?

This real-life scenario reflects the many activities undertaken by community nutritionists: identifying a nutritional problem in the community (a significant number of pregnant teenagers give birth to low-birthweight infants), selecting a target population (low-income pregnant teenagers), asking questions about why the problem developed (What do pregnant teenagers know about nutrition during pregnancy? Do

pregnant teenagers use health services for prenatal counseling?), and figuring out how best to address the problem (programs that promote prenatal counseling and breastfeeding and enhance parenting skills among teenagers). The desired outcome is ultimately to reduce the number of low-birthweight babies born to teenagers and to improve infant health.

This section describes how to conduct a similar needs assessment in your community. It introduces important questions you might ask when surveying your community: *Who* has a nutritional problem that is not being met? *How* did this problem develop? *What* programs and services exist to alleviate this problem? *Why* do existing services fail to help the people who experience this problem? The answers to these and other questions help community nutritionists understand the many factors that influence the health and nutritional status of a particular group.

This section describes how you might collect information about the group that is experiencing the nutritional problem—the target population. It outlines some of the tools you might use to assess the nutritional status of the target population: health risk appraisals, screening, focus groups, and direct assessment methods. It also lays out a plan for designing a program to help alleviate the nutritional need or problem within the community. You'll learn how to write goals and objectives, design an intervention, choose nutrition messages, market your program, and evaluate its impact. As in Section I, the thread running through these discussions is entrepreneurship and how its main principles—creativity and innovation—help community nutritionists "do more with less." The ultimate goal of community nutrition is to design programs that improve people's health and nutritional status.

Assessing Community Resources

Outline

Learning Objectives

After you have read and studied this chapter, you will be able to:

- Describe seven steps in conducting a community needs assessment.
- Develop a statement that defines the nutritional problem within the community.
- Discuss the contribution of the target population to community needs assessment planning and priority-setting.
- Define incidence and prevalence and how these concepts describe a population's and the community's health.
- Describe three types of data about the community that can be collected and where these data are found.

Something To Think About...

Plans are everything. – *General Dwight D. Eisenhower*

Introduction

Imagine that you have been asked to take a photograph of your city that will be used for the cover of a tourist brochure. In thinking about a photograph that best represents your city, you are immediately beset by choices. Should you photograph your city during a particular season and, if so, which one? Should you photograph your city's downtown area, showing its architectural and business diversity, or should you choose a popular city park? Should the photograph capture a historical landmark such as a statue or fountain, or should it focus on the *people* who live in your city and show a family picnicking at the fairgrounds or a baseball team at play? Your choice will likely be influenced by the expectations of the brochure sponsor, the time available to you for photographing various aspects of city life, and the budget for producing the brochure.

In many respects, the process of community needs assessment is much like the challenge of producing the "best" photograph of your city. It involves making choices to capture a picture of a nutritional problem or need within your community. It, too, is a process influenced by the expectations of the people and organizations involved in the assessment, the time available for collecting and analyzing data, and the budget allocated for the assessment.

Consider the challenge of capturing a picture of hunger among Hispanics living in the United States. The Hispanic population is composed of about 28 million people, or nearly 11 percent of the U.S. population.[1] By the year 2007, this population will become the nation's largest minority group. Hunger in Hispanic communities is believed to derive from poverty, racism, and high unemployment,[2] factors that limit access directly or indirectly to nutritious foods. Community nutritionists who aim to improve the food intake of Hispanics begin by working with local community leaders, state and federal agencies, and other groups to determine why the Hispanic population experiences hunger. They ask many questions: How many Hispanics in our community experience hunger? Do Hispanics have a higher unemployment rate than other ethnic groups in the community? Are they more likely than other groups to have low-paying jobs? What is the mean income of Hispanic families? How many Hispanic families participate in food assistance programs? What are the barriers to their participation in these programs? What factors contribute to hunger in this population? How does the Hispanic population perceive this problem? Are existing community programs and services reaching Hispanics? If they are not, why not? The answers to these and other questions provide pieces of the puzzle about why some Hispanics experience hunger. The answers help community nutritionists observe the extent of the problem in their community; identify resources available at the national, state, and local levels to alleviate it; determine where existing services and programs can be improved; and suggest areas where new programs and services are needed. These activities are part of the process of community needs assessment, described in this chapter.

Community Needs Assessment

Community needs assessment Evaluating the community in terms of its health and nutritional status, its needs, and the resources available to address those needs.

Health status The condition of a population's or individual's health, including estimates of its quality of life and physical and psychosocial functioning.

Nutritional status The condition of a population's or individual's health as influenced by the intake and utilization of nutrients and nonnutrients.

Community needs assessment is the process of evaluating the health and nutritional status of the community, determining what the community's health and nutritional needs are, and identifying places where those needs are not being met.[3] It involves systematically collecting, analyzing, and making available information about the health status and nutritional status of the community or some subgroup of it. The term **health status** refers to the condition of a population's or individual's health, including estimates of its quality of life and physical and psychosocial functioning.[4] **Nutritional status** is defined as the condition of a population's or individual's health as affected by the intake and utilization of nutrients and nonnutrients.[5] The assessment is undertaken to find answers to basic questions: *Who* has a health or nutritional problem that is not being addressed? *How* did this problem develop? *What* programs and services exist to alleviate the problem? *Why* do existing programs and services fail to help the people who experience this problem? *What* can be done to improve the health and nutritional status of the population?

The assessment process is sometimes called "community analysis and diagnosis," "health education planning," and "mapping."[6] Its overall purpose is to get a better understanding of how the community functions and how it addresses the public health and nutritional needs of its citizens. In some respects, the process is much like the clinical assessment of an individual patient's health, except that the community is the patient. This snapshot of the community identifies areas where the community performs well (for example, local hospitals and clinics have good data on infant morbidity and mortality) and the areas where it does not (although two food banks and several food assistance programs are available in the community, some families go without food).

What triggers a community needs assessment? Any number of factors may compel a city health department, state or federal agency, nonprofit organization, or other group to seek information about a community's health and nutritional status. There may be a need for new data on the community's health, either because existing data are several years old and may no longer accurately reflect a population's health or because data have never been collected on some segment of the population. Often, a government agency at the state or federal level has a mandate to carry out such activities. Sometimes, research findings are the impetus for taking action. For example, an article published in the *Journal of the American Dietetic Association* reported that the infants of teenage mothers who participated in the Higgins Nutrition Intervention Program weighed more and were less likely to have low birthweights than infants whose mothers did not participate in the program.[7] These study findings may prompt municipal health clinics to determine the number of low-birthweight infants born to adolescents and study the feasibility of implementing the Higgins program in their communities. In other cases, a community leader or community action group may raise awareness about a health or nutritional issue and prompt action to undertake a community needs assessment. Sometimes the availability of funding serves as the impetus for collecting data on the community's health and nutritional problems. For whatever reason, a decision is made to gather information about a nutritional problem that is not being addressed adequately in the community.

Organizations approach the community needs assessment by first determining its purpose and then planning how it will proceed. The amount of time available to conduct the needs assessment, the staff members responsible for conducting it, and the scope of the assessment must all be specified. The scope deserves special mention. In some cases, the assessment is designed to identify the health and nutritional problems of a large population such as all residents of the community, which might be the nation, state, province, or city. The assessment cuts across all income, educational, and geographical sectors, and it aims to identify the major causes of disease, disability, and death among the community's residents. The Third National Health and Nutrition Examination Survey (NHANES III), for example, identified overweight among all age groups and food insufficiency among Mexican Americans as two nutrition-related problems in the U.S. population.[8] An assessment of New York state's population found that the top-priority conditions requiring attention were lung cancer among blacks, tuberculosis among blacks and Hispanics, HIV/AIDS, iron deficiency among blacks, prostate cancer among white males, and stroke. Other "second-priority" conditions included lung cancer among whites, anemia among blacks and whites, and diabetes mellitus, emphysema, and breast cancer among women.[9]

Large-scale assessments tend to be expensive and time-consuming and may require the work of hundreds of people with expertise in public health, nutrition, epidemiology, statistics, management, and survey design and analysis. The team is pulled together from various departments within several agencies, all acting under the leadership of a single agency. Because they are costly and labor-intensive, such assessments may be undertaken only once in 5 or 10 years.

More often than not, the community needs assessment is limited in scope, focusing on a particular subgroup of the community. The small-scale assessment is relatively inexpensive to conduct and can be undertaken by a small team of community nutritionists and other professionals from several organizations and agencies. For example, an assessment of the nutritional status of schoolchildren may involve experts from the municipal departments of public health and education, a local dietitian who directs the National School Lunch Program, faculty members of the hospital's community health department, and graduate students from the local university. In today's fiscal environment, where money for personnel, equipment, and other resources is scarce, the small-scale assessment is often the better—and sometimes, the only—choice. It focuses on a high-risk group about which community nutritionists and other health professionals have some knowledge and concern.

Target population The population that is the focus of an assessment, a study, or an intervention.

Regardless of the scope, the purpose of the community needs assessment is to obtain information about the health and nutritional status of a particular group—namely, the **target population.** In the case of the large-scale assessment, the target population may be all residents of the nation. Because it is not practical or feasible to obtain information about every resident, large-scale assessments use statistical methods to select a sample of people whose age, sex, race, and other characteristics reflect the demographic profile of the entire population. NHANES III, for instance, was conducted on a nationally representative sample of about 33,994 persons 2 months of age and older.[10] In a small-scale assessment, the target population may be low-birthweight infants born between December 1 and March 31 to mothers living in the state of Missouri, lactating women with type 1 diabetes living in Chicago, or adults with lactose intolerance who present to three city hospitals.

Certain basic principles apply, no matter the scope of the needs assessment. In this and the next few chapters, we describe these principles and the process of conducting a community needs assessment. In this book, the community nutritionist is given primary responsibility for conducting it. She begins by defining the problem, setting the parameters of the assessment, and determining the types of data that must be collected to paint a picture of the target population's nutritional problems. The steps are diagrammed in Figure 5-1 and described in the sections that follow.

Step 1: Define the Nutritional Problem

In the course of living and working within the community and networking with colleagues, the community nutritionist likely has a notion about the target population's nutritional problems. She may not have a folder full of statistics at this point, but she can make a general statement about its main nutritional problem. What she must do now is develop a concise statement of the problem that concerns her. This statement is used to help plan the assessment and motivate other

FIGURE 5-1 *Steps in Conducting a Community Needs Assessment*

Step 1	Define the nutritional problem.
Step 2	Set the parameters of the assessment.
Step 3	Collect data: about the community. about the background conditions. about individuals who represent the target population.
Step 4	Analyze and interpret the data.
Step 5	Share the findings of the assessment.
Step 6	Set priorities.
Step 7	Choose a plan of action.

agencies to join the assessment team. Thus, the first step in the assessment process is to answer the question "What is the nutritional problem?" Consider the "problems" listed here:

• A study of withdrawal rates from the Maryland Special Supplemental Nutrition Program for Women, Infants, and Children (WIC) found an overall withdrawal rate of 47 percent, with older infants (7 to 12 months of age) having higher withdrawal rates than 6-month-old infants.[11] Infants and children who are withdrawn from the WIC program are more likely to be at increased nutritional risk than those who participate fully in the program.[12] Withdrawal rates from WIC are believed to be high in this state, but there are no data that describe the problem or the factors that contribute to it among the state's WIC participants.
• The prevalence of overweight increased 5 percent for men and women aged 18 to 70+ years between 1987 and 1993, according to data from the Behavioral Risk Factor Surveillance System, a telephone survey of health behaviors conducted among adults in 33 states.[13] Overweight contributes to increased morbidity, including increased risk of hypertension and diabetes mellitus. People who are successful at losing weight and maintaining the weight loss report using a combination of diet and exercise to achieve their goals.[14] The prevalence of overweight in Johnson County is not known, nor is information available about the strategies used by overweight people in Johnson County to lose weight successfully.
• Based on national survey data, the percentage of low-birthweight infants increased from 7.3 in 1995 to 7.4 in 1996, the highest level reported in more than two decades.[15] Low-birthweight infants have a higher risk of mortality and morbidity than normal-weight infants. The incidence of low-birthweight infants across all hospitals and clinics in this city is 8.2 percent, a figure higher than the national one. The reasons for this are not known.
• Results of the 1990 Ontario Health Survey indicate that only one-fifth of all immigrants consumed at least 75 percent of the minimum recommended number of servings of each food group in *Canada's Food Guide to Healthy Eating*.[16] Routine consumption of less than the minimum recommended number of servings of each food group may result in inadequate intakes of essential nutrients. No information is available on the food group intake of immigrants in this province.
• A survey conducted by the Heart and Stroke Foundation of Manitoba found that one-third of students in 11 of the city's secondary schools eats mainly french fries for lunch, and only 1 in 20 students consumes a lunch containing foods from all four food groups.[17] It is not known whether students in other secondary schools have similar eating patterns, and no information about the factors that influence the food choices of secondary school students in the province is available.

These statements of nutritional "problems" cover a range of nutritional issues, target populations, and age groups. Each statement indicates who is affected by the nutritional problem and how many people experience it. In most cases, information about the number of people who experience the problem is derived from published results of epidemiologic studies. (Refer to the boxed insert for a description of the principles of epidemiology.) Each statement indicates the impact of the problem on general health or nutritional status. All indicate areas where there are gaps in the community's knowledge of a nutritional problem. These statements serve as the starting point for undertaking an assessment of a nutritional problem within the community, whether the community consists of a city, county, state, province, or nation.

A problem that is inadequately defined is not likely to be solved.
– P. M. Kettner, R. M. Moroney, and L. L. Martin in Designing and Managing Programs, p. 36

Principles of Epidemiology

pidemiology can be defined as "the study of the distribution and determinants of health-related states and events in specified populations and the application of this study to the control of health problems."[1] Its focus is on the health problems of populations rather than individuals.

The term *distribution* in the definition refers to the relationship between the health problem or disease and the population in which it exists. The distribution includes the persons affected and the place and time of the occurrence of a health problem or condition. It also encompasses such population parameters as age, sex, race, occupational type, income, and educational levels, exposure to certain agents, and other social and environmental features. Distribution is concerned with population trends or patterns of disease among groups of people.

The term *determinants* refers to the causes and factors that affect the risk of disease. Some conditions such as rubella or German measles have a single, necessary cause. Others such as osteoporosis have multiple determinants.

Epidemiologists work to identify the causes of disease and to propose strategies for controlling or preventing health problems.[2] They are concerned with estimating **risk**—the likelihood that people who are without a disease but exposed to certain **risk factors** will acquire the disease at some point in their lives.[3] The basic operation of the epidemiologist is to "count **cases** and measure the population in which they arise" in order to calculate rates of occurrence of a health problem and compare the rates in different groups of people.[4] One expression of how frequently a disease occurs in a population is **incidence,** the fraction or proportion of a group initially free of a disease or condition that develops the disease or condition over a period of time. Another common measure of occurrence of an event is **prevalence,** or the fraction or proportion of a group possessing a disease or condition at a specific time. Incidence and prevalence rates describe the frequency with which particular events occur. By calculating and comparing rates, the strength of the association between risk factors and the health problem being studied can be determined.

The principles of epidemiology are used to describe a community's particular health problems and to determine whether a community's overall health is improving or worsening—much like "diagnosing" a patient. Potential health problems may also be spotted through this ongoing examination of a community's health status. For example, in the early 1980s, an unusual illness was observed among mostly gay men living in San Francisco, Los Angeles, and New York. The finding of an unusually high number of cases of Kaposi's sarcoma, a rare cancer affecting primarily middle-aged men of Jewish or Italian ancestry, led to the discovery of a new, fatal disease, acquired immunodeficiency syndrome (AIDS). The primary reasons for investigating and analyzing health problems is to work toward controlling and preventing them.

Risk The probability or likelihood of an event occurring—in this case, the probability that people will acquire a disease.

Risk factor Clinically important signs associated with an increased likelihood of acquiring a disease.

Case A particular instance of a disease or outcome of interest.

Incidence New events in a given period.

Prevalence All events in a given period.

References

1. C. W. Tyler, Jr., and J. M. Last, Epidemiology, in *Public Health and Preventive Medicine*, 13th ed., ed. J. M. Last and R. B. Wallace (Norwalk, CT: Appleton & Lange, 1992), pp. 11–39.
2. R. I. Glass, New prospects for epidemiologic investigations, *Science* 234 (1986): 951–55.
3. R. H. Fletcher, S. W. Fletcher, and E. H. Wagner, *Clinical Epidemiology—The Essentials* (Baltimore, MD: Williams & Wilkins, 1982).
4. A. D. Langmuir, The territory of epidemiology: Pentimento, *Journal of Infectious Diseases* 155 (1987): 349–58.

Step 2: Set the Parameters of the Assessment

Before the community needs assessment is undertaken, certain parameters or elements must be determined. These parameters, described here, set the direction for the assessment. As you read the following material, consult Table 5-1, which describes the parameters for two assessments in the fictional city of Jeffers (population = 612,000). One assessment (Case Study 1) focuses on the issue of women and coronary heart disease (CHD). The other (Case Study 2) focuses on issues surrounding the nutritional status of elderly persons living at home. These case studies, which form the basis for some of the discussion in this and the next chapters, outline two assessments that differ in scope and complexity.

Define "Community." The scope of the "community" must be specified. The community might include the people who represent the target population and live within the city limits or within the greater metropolitan area bounded by certain suburbs. Sometimes the community is a geographical region, state, nation, province, or several countries. In the case studies described in Table 5-1, the "community" is a typical municipal unit such as the city of Jeffers.

Determine the Purpose of the Needs Assessment. A needs assessment is undertaken to gather information about the social, political, economic, environmental, and personal factors that influence the nutritional problem and the population at risk. The community needs assessment may have one or more of the purposes listed here:[18]

- Identify groups within the community who are at risk nutritionally.
- Identify the community's or target population's most critical nutritional needs and set priorities among them.
- Identify the factors that contribute to a nutritional problem within the community or target population.
- Determine whether existing resources and programs meet the community's or target population's nutritional needs.
- Provide baseline information for developing action plans to address the community's or target population's nutritional needs and for evaluating the program.
- Plan actions to improve the community's or target population's nutritional status, using methods that are feasible and focus on established health priorities.
- Tailor a program to a specific population.

Define the Target Population. The focus of the community needs assessment is the target population, whose health and nutritional status is affected by many community, environmental, and personal factors. The choice of the target population is influenced by the initial perception about the nutritional problem. Sometimes the needs assessment begins with one target population (for example, all women with infants under one year of age) and concludes with a more refined focus (for instance, teenagers with infants under one year of age). Usually, however, the target population remains constant over the course of the needs assessment (such as people with cancer, black men with hypertension, teenagers with eating disorders, persons with alcoholism, people recovering from stroke).

Parameter	Case Study 1: Women and Heart Disease	Case Study 2: Nutritional Status of the Elderly
Lead organization	State affiliate of the American Heart Association or, in Canada, provincial Heart & Stroke Foundation	City of Jeffers Health Department
Statement of nutritional problem	Coronary heart disease (CHD) is the leading cause of death among U.S. and Canadian women. Most women apparently do not realize that they are more likely to die from CHD than from cancer. There are no data on knowledge/awareness of CHD risk factors among women living in Jeffers or about existing CHD prevention programs and services in Jeffers.	The number of independent, noninstitutionalized elderly persons (75+ years) has increased nationally. In Jeffers, this population has increased 12% since the 1995 assessment. Several community-based social service agencies have perceived an increased demand for nutrition services by this population, but there are no data on the availability of such services or on the nutritional status of these persons.
Definition of community	The metropolitan area of Jeffers, including the adjoining municipalities of Oakdale, Chambers, Kastle, and Morgan	City of Jeffers, bounded by the city limits as of July 31, 1998
Purpose of the assessment	To obtain information about women and heart health to help determine whether a program or other intervention should be developed	To obtain information about the nutritional status of independent elderly persons > 75 years of age and their needs for nutrition services
Target population	Females over 18 years of age	Elderly persons > 75 years of age living independently
Overall goal of assessment	Women's knowledge, attitudes, and practices related to CHD risk and existing programs and services designed to reduce CHD risk will be identified	The nutritional status of independent elderly (> 75 years) and their use of nutritional services available through community-based agencies will be determined
Objectives of assessment	On a sample of 250 women over the age of 18 years, within 3 months:	On a sample of 150 elderly persons aged 75 years and older, within 1 year:

TABLE 5-1 *Parameters for Two Community Needs Assessments in the Fictional City of Jeffers, Population 612,000*

Goals Broad statements of what the activity or program is expected to accomplish.

Objectives Statements of outcomes and activities needed to reach a goal.

Set Goals and Objectives for the Needs Assessment. This is an essential step, as the goals and objectives determine the types of data collected and how they will be used. **Goals** are broad statements that indicate what the assessment is expected to accomplish. **Objectives** are statements of outcomes and activities needed to reach a goal. Statements of objectives use a strong verb such as *increase, reduce, begin,* or *identify* that describes a measurable outcome. Each objective should state a single purpose.[19] The assessments described in Table 5-1 specify one overall goal and several objectives. Assessments may have two, three, or more goals and 10 to 15 objectives. The needs assessment cannot proceed without clearly defined goals and objectives.

Specify the Types of Data Needed. The types of data required in a needs assessment depend on the purpose, goals, and objectives of the assessment. Some data related to the target population may already exist in the literature or be available from government sources. Other data must be collected directly from the target population. In general, data are used to identify a high-risk population, describe the nutritional

Parameter	Case Study 1: Women and Heart Disease	Case Study 2: Nutritional Status of the Elderly
Objectives of assessment —continued	• Assess women's knowledge and awareness of the leading causes of death and CHD risk factors. • Assess their practices to reduce CHD risk. • Identify existing services offered to women to help reduce their CHD risk. • Assess women's use of existing services designed to reduce risk. • Identify gaps in the delivery of such programs and services.	• Assess the nutritional status of independent elderly persons (> 75 years) • Determine which community services are available for this population. • Identify the existing community services used by this population. • Identify gaps in program and service delivery.

Types of data needed: Community conditions	• Mortality data for women (50+ years)
	• Morbidity data for women (50+ years)
	• Existing programs and services:
	  Hospitals, medical clinics
	  Fitness/sports centers
	  Offered by health professionals in private practice
	• Educational materials available from:
	  doctors' offices
	  health professionals in private practice
	  food/pharmaceutical companies
	  bookstores, other
		• Types of services offered by community organizations, including personal care services, homemaker services, adult day care, home-delivered meals, hospice, and home health care services
		• Number of elderly persons who use these services.
		• Number of elderly persons who participate in federal/state assistance programs, such as Food Stamps, Social Security, Medicare, Medicaid, Supplemental Security Income, Veteran's Benefits, assistance for housing and home heating, and home-delivered meals.
		• Types of medical and social services offered by health professionals.

Types of data needed: Back-ground conditions	• Advertising related to smoking • Health messages about CHD in magazines, newspapers, etc.	• Changes in eligibility for Medicare, Medicaid. • Current funding of Older Americans Act.
Types of data needed: Target population	See Table 6-1 on page 187 in Chapter 6.	See Table 6-10 on pages 204–5 in Chapter 6.

problem within the target population, define areas where nutritional needs are not being met, identify duplication of services, or develop directories of services. Table 5-1 specifies the main types of data needed for the two assessments in the city of Jeffers. These lists are not meant to be comprehensive, but rather they represent the "first pass" in estimating the types of data needed about the target population and its perceived nutritional problem.

TABLE 5-1 *Continued*

Step 3: Collect Data

When the parameters of the assessment have been set, the next step is to begin collecting data. When approaching the process of data collection, remember that the assessment's overall purpose is to paint a picture of the environment in which the target population lives to understand how the nutritional problem developed and what might be done to address it. Begin collecting data about the "big picture"—namely, the environment or *community* in which the target population lives and works. Then,

FIGURE 5-2 *Types of Data to Collect About the Community*

The focus of the community needs assessment is the target population, whose health and nutritional status are affected by many community and background conditions, as well as by individual characteristics such as lifestyle, living and working conditions, and social networks.

Source: Adapted from M. Whitehead, Tackling inequalities: A review of policy initiatives, in *Tackling Inequalities in Health: An Agenda for Action*, eds. M. Benzeval, K. Judge, and M. Whitehead (London: King's Fund, 1995), p. 23.

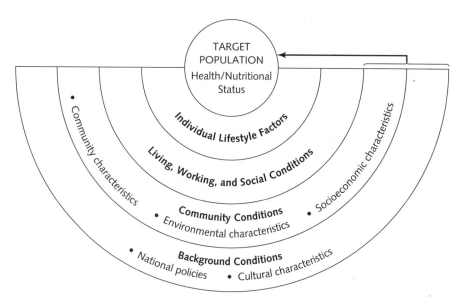

when you have begun to get a sense of the big picture, collect data about *individuals* who make up the target population. Refer to Figure 5-2 as you read this section. The figure shows the general categories of data that might be collected over the course of the assessment and represents another way of thinking about the determinants of health shown in Table 1-1 on page 9. In this chapter, the discussion focuses on the outer layers of the figure, on collecting data about certain community and background conditions that affect the target population's health and nutritional status. Chapter 6 describes issues related to collecting data about the individual lifestyle factors and living, working, and social conditions that influence the target population.

Collect Data About the Community

Qualitative data Data such as opinions that describe or explain, are considered subjective, and can be categorized or ranked but not quantified.

Key informants People who are "in the know" about the community and whose opinions and insights can help direct the needs assessment.

Stakeholders People who have a vested interest in identifying and addressing the nutritional problem.

A good place to begin the community needs assessment is with the community itself. Both qualitative and quantitative data help describe the community, its values, health problems, and needs. **Qualitative data** such as opinions and insights may be derived from interviews with **key informants**—that is, people who are knowledgeable about the community, its history, and its past efforts to address a nutritional or health problem—and with **stakeholders,** the people and organizations who have a vested interest in identifying and addressing the nutritional problem. Members of the target population itself can also provide information about the community. **Quantitative data** may be derived from a variety of databases, including registries of **vital statistics** (for instance, age at death, cause of death), published research studies, hospital records, and local health surveys.[20]

Types of Data to Collect About the Community. This section describes the types of data that can be gathered about the community. These data are outlined in Table 5-2 on pages 160 and 161.

Community Characteristics. Information is needed about how the community operates, how its population is distributed, and how healthy it is. Information about existing health services provides clues about the community's perception of its leading health and nutritional problems.

- **Community organizational power and structures.** Some qualitative data help the community nutritionist understand how the community operates politically and who wields power within the system. The organization charts for government agencies and "city hall" provide information about how the community delivers health services and develops health policy. Knowing the key players in local health organizations and community, business, and media groups helps the community nutritionist identify the concerns of community leaders.
- **Demographic data and trends.** Demographic data help define the people who live in the community by sex, age, race, marital status, and living arrangements. Changes in the demographic profile of a community often serve as an early warning signal about potential gaps in services or undetected nutritional problems. For example, in an assessment of the nutritional and health status of elderly persons aged 65 years and older in a rural community, the assessment team learned that the number of persons older than 85 years who were living alone at home had increased significantly since the 1990 assessment. This trend suggested a need to increase home care services for the 85+ group.
- **Community health.** A variety of health statistics are used to paint a picture of the community's health. Some health data describe the causes and rates of disease, disability, and death within the community; others focus on key life stages or events. Community health data help the community nutritionist describe the population's health and nutritional problems and identify persons who are malnourished. Some common health measures and related terms are as follows:[21]

Rate	The number of events per population at risk.
Crude death rate	The crude death rate (CDR) is expressed as deaths per thousand population. It is calculated using the following formula:

$$\text{crude death rate} = \frac{\text{all deaths during a calendar year}}{\text{population at midyear}} \times 1{,}000$$

Infant mortality rate	Infant death is defined as the death of a liveborn infant under one year of age. It is calculated using the following formula:

$$\text{infant mortality rate} = \frac{\text{number of infant deaths during 1 year}}{\text{number of live births}} \times 1{,}000$$

Live birth rate	The live birth rate is based on the following formula:

$$\text{live birth rate} = \frac{\text{number of live births during 1 year}}{\text{population of area}} \times 1{,}000$$

The infant mortality rate is an important measure of a nation's health and is used worldwide as an indicator of health status. The infant mortality rate in the United States has declined steadily over the past decades and, in 1996, achieved a record low of 7.2 per 1,000 live births. Even so, the United States ranks 24th in infant mortality compared with other

Type of Data	Source of Data
Community Organizational Power and Structures	
• Organization of government (city, state, etc.)	Directory of municipal, state, etc., government
• Organization of health department (city, state, etc.)	Directory of municipal, state, etc., health department
• Local, state, and national organizations with a health mandate (e.g., American Heart Association, American Cancer Society, American Dietetic Association, American Public Health Association)	Yellow pages of the phone book, national or state directories
• Community groups and their leaders (e.g., United Way)	Yellow pages of the phone book, key informants
• Reporters and other people with the media	Local/national newspapers and magazines, television and radio stations
• Members of the Chamber of Commerce	Local Chamber of Commerce
Demographic Data and Trends	
• Total population by age, sex, race, marital status, etc.	Census Bureau, FedStats,* state data centers, data archives, libraries
• Distribution of population subgroups (e.g., percentage of population that is black, Hispanic, Asian, etc.; percentage that is foreign born)	Census Bureau, FedStats, state data centers, data archives
• Size and composition of households (e.g., number of family members in households, number of children in households, percentage of all households consisting of husband-wife families, percentage of single-parent families)	Census Bureau, FedStats, state data centers, data archives
Community Health	
• Mortality statistics (e.g., death rates according to age, sex, cause, location, etc.)	Census Bureau, NCHS,[†] FedStats, Public Health Service, state and municipal health departments
• Morbidity statistics (e.g., frequency of symptoms and disabilities, distribution of disease conditions)	Census Bureau, NCHS, FedStats, Public Health Service, state and municipal health departments
• Fertility and natality statistics (e.g., age and parity of mother, duration of pregnancy, percentage of mothers who get prenatal care, number of unmarried mothers, infant's birthweight, type of birth [i.e., single, twin], fertility rate, infant mortality)	Census Bureau, FedStats, state data centers, data archives
• Communicable diseases (e.g., incidence, distribution)	FedStats, CDC, published studies
• Occupational diseases (e.g., incidence, distribution)	FedStats, National Institute for Occupational Safety and Health, published studies

*The Internet address for FedStats is www.fedstats.gov.
[†]The following abbreviations appear in this table: AOA = Administration on Aging; BRFSS = Behavioral Risk Factor Surveillance System; CDC = Centers for Disease Control and Prevention; NCHS = National Center for Health Statistics; NHANES = National Health and Nutrition Examination Survey; NNMRRP = National Nutrition Monitoring and Related Research Program; RDA = Recommended Dietary Allowance; YMCA/YWCA = Young Men's/Women's Christian Association.

TABLE 5-2 *Types and Sources of Data About the Community* industrialized countries. The infant mortality rate for blacks was 14.2 per 1,000 in 1996, double that of the U.S. national average. American Indians, Alaska Natives, and Hispanics all had infant mortality rates higher than the national average, signaling an urgent need to address this basic health issue.[22]

Type of Data	Source of Data
Community Health—continued	
• Leading causes of death	FedStats, NCHS, published studies
• Life expectancy	FedStats, NCHS, published studies
• Determinants and measures of health (e.g., vaccinations, disability days, cigarette smoking, use of selected substances [i.e., alcohol, marijuana, cocaine], hypertension, obesity, serum cholesterol concentrations, exposure to lead, occupational injuries, incidence of food-borne disease)	FedStats, CDC's *Vital and Health Statistics* series and the *Morbidity and Mortality Weekly Report,* NNMRRP surveys (e.g., NHANES III, BRFSS)
• Food and nutrient intake (e.g., food group intake, nutrient and energy intakes, nutritional adequacy of diets compared with the RDA)	Agriculture Research Service, Elderly Nutrition Program (AOA)
• Use of health resources (e.g., frequency of patient contact with physicians, number of office visits to physicians and dentists)	State Department of Community Health Services
• Health care resources (e.g., persons employed in service, number of active physicians and other health personnel)	FedStats, Health Resources and Services Administration
• Inpatient care (e.g., days of care and average length of stay in hospitals, number and types of operations, number of nursing home residents)	Hospitals, nursing homes, etc.
• Facilities (e.g., short-stay and long-term hospitals, community hospital beds)	Yellow pages of phone book
Existing Community Services and Programs	
• Government-funded food assistance programs (e.g., number of referrals)	Related government agency
• Health and nutrition services and programs offered by hospitals, clinics, community health centers, sports/fitness centers, YMCA/YWCAs, the public health department, voluntary health organizations, schools, universities, colleges, civic groups	Hospitals, clinics, sports/fitness centers, directory of nutrition services (if available from state associations)
• Primary care services (e.g., location, accessibility)	Local hospitals, clinics, etc.
• Soup kitchens, food pantries	Yellow pages of phone book, municipal community services department
• Programs and services offered by nutritionists, dietitians, and other health professionals	Key informants, state associations

• **Existing community services and programs.** Obtaining data on the community's existing health and nutrition services helps pinpoint gaps where services are needed. An inventory of the community's nutrition services and programs can be built by (1) identifying the nutrition services and programs available through government agencies, health organizations, and civic groups; (2) cataloging the educational services and materials offered by voluntary health organizations such as the American Red Cross, National Council on Alcoholism, American Heart Association, and American Cancer Society in the United States and, in Canada, the Canadian Diabetes Association, Canadian Arthritis and Rheumatism Society, and Heart and Stroke Foundation of Canada, among others; and (3) identifying the programs and services delivered by local nutritionists, dietitians, and other health professionals. The United Way of America is a

TABLE 5-2 *Continued*

network of volunteers and local charities that also maintains directories of local community services and programs.[23] In addition, national information centers such as the U.S. National Health Information Clearinghouse (listed together with other clearinghouses and information centers in Appendix H) can be contacted for general information about the availability of educational materials, programs, and referral services.

Environmental Characteristics. Each target population lives and works within a particular physical environment. Certain aspects of this environment, described in Table 5-3, affect the target population's health and nutritional status. Access to medical clinics and ambulatory care services, which provide screening, diagnosis, counseling, follow-up, and therapy, influences the target population's health and

TABLE 5-3 *Types and Sources of Data About the Environment*

Type of Data	Source of Data
Food systems	
• National, regional, and local food distribution networks; extent of emergency and supplemental feeding systems; food wholesale and retail systems; and amount of food grown locally	Census Bureau, data archives, crop reports by county or state agencies (can be accessed through FedStats*)
• Location, type, and number of grocery stores, supermarkets, convenience stores, and farmers' markets	Yellow pages of phone book
• Location, type, and number of restaurants (e.g., family style, fast food, etc.)	Yellow pages of phone book
• Location and number of health food stores	Yellow pages of phone book
Geography and climate	Observation, state department of agriculture
Health systems	
• Location, type and number of hospitals, clinics, health maintenance organizations, long-term care facilities	Yellow pages of phone book
• Types of ambulatory care	Annual reports of hospitals, clinics
Housing	
• Type of housing (e.g., percentage of year-round housing that are single dwelling units; housing characteristics, such as units in structure, year structure was built, number of rooms and bedrooms, plumbing, heating, and kitchen facilities)	Census Bureau, FedStats
• Condition of housing (e.g., percentage of standard housing with an exterior frame made of brick, wood, or concrete block)	Census Bureau, FedStats
Recreation	
• Location and number of fitness clinics, sports facilities	Yellow pages of phone book
• Types of recreation available (swimming, golf, tennis, cross-country skiing, walking trails, etc.)	Observation, yellow pages of phone book
Transportation systems	Municipal/state department of transportation or transit
Water supply (e.g., source of water, distance from residence, water quality)	Municipal/state department of water works and water quality

*The Internet address for FedStats is www.fedstats.gov.

nutritional status, as does access to nutritious foods. Food availability is influenced by the community's geography and climate, which affect the length of the food-growing season; by the type of foods grown in commercial and family gardens; by the type of food storage systems needed to keep foods fresh; and by the location and type of grocery stores, supermarkets, convenience stores, and farmers' markets. Ready access to transportation, whether in the form of personal car, bus, or commuter train, improves the target population's access to medical services and supermarkets.

Socioeconomic Characteristics. Certain economic and sociocultural data related to the community, as shown in Table 5-4, are also useful. Information about the income of families and the number of families receiving public assistance provide a benchmark for comparing the target population's income with the community's mean or median family income. Information about the community's educational level, literacy rate, and major industries and occupations helps identify barriers to improving the health and nutritional status of the target population.

TABLE 5-4 *Types and Sources of Socioeconomic Data Related to the Community*

Type of Data	Source of Data
Sociocultural Data and Trends	
• Labor force characteristics (e.g., occupation, industry, class of workers, hours worked)	Census Bureau, Bureau of Labor Statistics
• Language spoken at home	FedStats*
• Education (e.g., median school years completed by individuals 25 years of age and older; individuals who completed high school)	FedStats, Census Bureau
• Literacy levels of school children and adults	FedStats, Census Bureau
Economic Data and Trends	
• Income of families and unrelated persons living in households	FedStats, Census Bureau
• Median incomes of families	FedStats, Census Bureau
• Percentage of families with incomes below the poverty level	FedStats, Census Bureau
• Number of families receiving Temporary Assistance for Needy Families	Administration for Children and Families (DHHS[†])
• Number of participants in the WIC[‡], National School Lunch, and National School Breakfast programs	USDA,[§] FedStats
• Number of individuals receiving food stamps	National Food Stamp Program
• Number of individuals receiving public assistance	Welfare office
• Unemployment statistics (e.g., percentage of households with one or more unemployed members)	FedStats, Census Bureau
• Tangible wealth (e.g., ownership of land and livestock, ownership of items such as personal computers, telephones, and microwave ovens)	Municipal, county, state records

*The Internet address for FedStats is www.fedstats.gov.
[†]DHHS = Department of Health and Human Services.
[‡]WIC = Special Supplemental Nutrition Program for Women, Infants, and Children.
[§]USDA = U.S. Department of Agriculture.

It is seldom necessary, expedient, or possible to collect and use all data available about the community. Consider a situation where the local health department is aware of the high prevalence of type 2 diabetes in American Indians.[24] The department's health and wellness office wants to increase its initiatives to reduce type 2 diabetes risk among American Indians, especially children. It perceives a need to develop a wellness program specifically for American Indian children, but it needs information about type 2 diabetes in this population. The department's community nutritionist reviews the spectrum of community data that could be collected and decides to collect data about the structure of the tribal council, the distribution of type 2 diabetes among American Indians by sex and age, the types of health care and medical services available to the population, its use of health care and medical services, the availability of diabetes education programs and its participation in these programs, food patterns of American Indians, mean family income and education level, and literacy. Data are not needed on housing, the water supply, recreation facilities, labor force characteristics, tangible wealth, or transportation systems.

In this example, the focus of the assessment is fairly narrow, dealing only with one nutritional problem (type 2 diabetes) experienced by a particular population (American Indian children and their caregivers living on a reservation). In contrast, consider the types of data required to evaluate the health and nutritional status of homeless people living in your community. Because homelessness cuts across all age and ethnic groups, a broad spectrum of data on homeless people and the community's resources for dealing with their health and nutritional problems is needed. In the next section, we describe how to locate information about the community and target population.

Sources of Data About the Community. In real estate, the key to success is location, location, location. In community needs assessment, the key is to observe, observe, observe—and do a little legwork and networking. Consult Tables 5-2, 5-3, and 5-4 for information about sources of data related to the community.

Where does one start to collect data about the community? There is no one right way to begin data collection, but observing the target population in its community setting—doing a walkabout—provides essential information. If your target population is the institutionalized elderly, then visit local nursing and residential homes, senior centers, and other facilities where the elderly live. If it is Muslim, visit local grocery stores where this group shops and find out about Halal food. If it is athletes, visit fitness centers, sports facilities, and schools with sports programs. Observe the target population in the community and ask questions: Where do these people shop for food? What kind of transportation is required for them to reach the supermarket or grocery store where they shop? What are the main occupations of this group? Do they live near a hospital or clinic? Which medical services do they use? Is a food bank, soup kitchen, or other emergency food assistance facility located near where they live? Do they grow some of their own food? What kinds of nutritional problems has this group experienced in the past? Take time to ask members of the target population about how they perceive the nutritional problem. Indeed, you may be surprised to learn that some of them are not aware of a problem! The important thing

Visit grocery stores and supermarkets where your target population shops to learn about its food consumption and shopping practices. Your observation that few members of your target population drive cars, and most walk to the grocery store, is important information about their lifestyle and needs.

is to *listen* to what the target population has to say about the problem, how it arose, and what might be done to address it.

Networking with colleagues also provides information about how the problem is perceived. Colleagues may know of newly published documents that provide recent health statistics about the target population, or they may help you locate unpublished data that are available in preliminary form. They may also be aware of similar needs assessments done in other cities, states, provinces, or regions.

Interviews with key informants are valuable. Key informants may be formal leaders of the community, such as the mayor, who have a broad knowledge of the community, or they may be informal leaders, such as the owner of a community center, whose opinions and connections provide valuable information about the community. Religious leaders, physicians, teachers, volunteers with nonprofit agencies, heads of social services, and members of the media are among those who can provide insights into how the community operates and how the target population perceives the nutritional problem. When arranging interviews with key informants, develop a short list of open-ended questions (that is, questions that require more than a simple yes or no answer); identify a few key informants whose opinions would be most useful to you; contact the informants for permission to interview them; conduct the interview in person, by mail, or by telephone; and summarize the results for other members of the assessment team.[25] It is always appropriate to thank the key informants for their time.

The next step is to turn to the Internet or the library to search the medical, nutrition, and public health databases for literature on the nutritional problem. Table 5-5 gives the Internet addresses for key data sources on-line. Also check the

TABLE 5-5

Internet Resources

The following Internet addresses provide data on vital statistics, income, unemployment, demographic characteristics, and other variables related to U.S. population groups. The Adobe Acrobat™ reader is required to download and view some or all files at each website. Each address provides information on how to download and configure the Adobe Acrobat™ reader.

FedStats

• www.fedstats.gov

More than 70 agencies provide data for this site. Search the site by keyword or alphabetically by topic. Key features at the site include:

• Births	• Marriages
• Breastfeeding	• Personal income
• Chronic diseases	• Population
• Education	• Programs
• Health	• Unemployment
• Government	• Vital statistics

National Agricultural Statistics Service

• www.usda.gov/nass

This site offers a few databases on food consumption. Search under the "Food" keyword for the following topics:

• Changes in Food Consumption and Expenditures
• Expenditures of Food, Beverages, and Tobacco
• Food Away from Home
• Food Consumption
• Food Spending in American Households
• U.S. Food Expenditures

State and Local Governments

• lcweb.loc.gov/global/state/stategov.html

Use this homepage to access information about your state and local government. Some information on major cities is also available at this website.

government documents section of public and university libraries for compilations of vital health statistics and reports on public health issues that relate to the community. Local historical records can also be useful. Information available in newspaper archives, old maps, parish or county records, and other documents can provide a history of a local public health problem and the public attention given to it.

Many demographic and socioeconomic data can be located in publications of the U.S. Bureau of the Census, Bureau of Labor Statistics, Department of Agriculture, and Department of Health and Human Services. The decennial Census of the Population and Housing, for example, which is conducted by the Bureau of the Census, provides data on states, counties, local units of government, school districts, and congressional districts.[26] Census data typically describe age and sex distributions, births and deaths, labor force characteristics (for example, occupation, industry, hours worked), income, housing characteristics (for instance, year built, number of rooms, plumbing, heating, kitchen facilities), and other demographic variables. Many libraries are repositories for census data, and some databases are available on CD-ROM.

TABLE 5-5 *Continued*

U.S. Census Bureau

• www.census.gov

This site provides national, state, and some city/county statistics. The site can be searched by word, place name, ZIP code, or map (geographical location). Browse the alphabetical listing of topics. Key features of the site include:

- Census State Data Centers (Click on your state's name to access information about print-outs, tapes, software, CD-ROM products and services, on-line data service, newsletters/technical journals, or maps. Link with your state's Census Data Center Web site.)
- Health statistics
- Household statistics (e.g., American Housing Survey data)
- Household economic statistics
    Disability
    Income
    Labor force
    Occupation
    Poverty (e.g., Poverty in the United States, 1996; poverty highlights; state poverty rates)
    Small area income and poverty estimates
- Population topics (e.g., estimates and projections for the nation, states, and households and families)
- State profiles (e.g., census tables showing number of persons, number of families, households, races, etc., in your state)

MEDLINE

• www.nlm.nih.gov

Search MEDLINE and other health databases from this Web site sponsored by the National Library of Medicine.

National Center for Health Statistics

• www.cdc.gov/nchswww

This site offers a list of the latest publications and electronic products available from NCHS. Users can check out the new releases, fact sheets, and publications.

Additional legwork helps locate health statistics and related health reports from local, county, and state health departments; social welfare agencies; birth, death, marriage, and divorce registries; and courts. The annual reports of local hospitals, clinics, and health centers provide information on the types of health problems within the community, the existence of screening programs for detecting health and nutritional problems, and the resources available to deal with them.[27]

It is sometimes necessary to resort to secondary data sources such as data archives, which serve as repositories for thousands of surveys conducted over the last two decades. Data archive services have collected and catalogued the surveys of many communities. The services usually charge a fee for data tapes and supporting documentation. Examples of data archive services include the University of Michigan's Institute for Social Research, the largest university-based center for interdisciplinary research in the United States,[28] and the University of North Carolina's Institute for Research in Social Science, which is a source of nonproprietary opinion data such as the Louis Harris public opinion data, national census data, and data from the World Fertility Surveys and the Demographic and Health Surveys.[29]

Collect Data About Background Conditions

Collect information about the broader environment in which the community is positioned. Many political, socioeconomic, cultural, and environmental factors at the national level operate in the background but have the potential to affect how the target population lives, the food choices it makes, and where it obtains medical services. The types and sources of background information are summarized in Table 5-6.

National policy, for example, affects eligibility for food assistance programs, minimum wage levels, and the distribution of commodity foods, all factors that may influence the target population's health and nutritional status. If the "community" consists of tribes of American Indians, the assessment team may review the health care policy of the Bureau of Indian Affairs, the U.S. federal agency charged with providing personal and public health services to American Indians and Alaska Natives. The assessment team may learn that the agency's policies have inadvertently created competition among tribes for money for health care services. This unexpected situation may result in less money being distributed to tribal communities to pay for expensive medical services,[30] an outcome that may affect the target population's health and nutritional status.

The broad "culture" also influences food intake and nutritional status. By **culture,** we are referring to the interconnected web of human knowledge, beliefs, and behav-

Culture The knowledge, beliefs, customs, laws, morals, art, and any other habits and skills acquired by humans as members of society.

TABLE 5-6 *Types and Sources of Information About Background Conditions*

Type of Information	Source of Information
National Policy	
General information	Articles published in journals, magazines, newspapers; commentary on TV, radio, Internet
• Agriculture	Department of Agriculture
• Economics	Department of Commerce
• Education	Department of Education
• Health	DHHS*
• Housing	Department of Housing and Urban Development
• Labor	Department of Labor
• Nutrition	DHHS (FDA, CDC, NIH, etc.)
• Social Security	DHHS
Cultural Conditions	
• Advertising	Television, radio, printed matter, Internet
• Health messages	Television, radio, printed matter, Internet, educational materials available from government, food companies, nonprofit groups, etc.
• Roles of women	Television, radio, printed matter, Internet
• Belief systems	Television, radio, printed matter, Internet, family, friends, other social contacts

*The following abbreviations appear in this table: CDC = Centers for Disease Control and Prevention; DHHS = Department of Health and Human Services; FDA = Food and Drug Administration; NIH = National Institutes of Health.

iors that are learned and transmitted to succeeding generations.[31] Many of our food habits, attitudes, and practices arise from the traditions, customs, belief systems, technologies, values, and norms of the culture in which we live. For instance, in an assessment of the nutritional status of bulimic students, certain background conditions such as advertising, society's emphasis on leanness, and cultural expectations about weight and body size likely influence the students' food patterns and body image. This background information should be evaluated as part of the assessment.

Background information on the community or region's health status is also important and can be obtained from international agencies such as the Food and Agriculture Organization (FAO) and the World Health Organization (WHO), which have regional offices that can furnish relevant population health data. Since 1954, for example, the Pan American Health Organization has published a series of quadrennial reports that document the health progress attained by its members. Entitled *Health Conditions in the Americas*, the reports provide general information on the region's social and political climate, primary demographic characteristics, mortality data, and health conditions, focusing on women, children, and the elderly.[32]

Collect Data About the Target Population

When the community and background data have been gathered, the community nutritionist begins the process of collecting information about the target population. Although details of this process are described in Chapter 6, some general comments are needed here. When gathering information about the target population, the community nutritionist has two choices: (1) use existing data related to the target population, or (2) collect data related to the target population because the needed data do not exist. Both options are described in this section.

Existing Data. The most expedient and cost-effective course for obtaining health statistics and behavioral information may be to use existing data. Data related to the target population can be gathered from large-scale population surveys, such as those conducted by the National Nutrition Monitoring and Related Research Program (NNMRRP), or from small surveys of special populations conducted in the immediate community or region. It is usually desirable to have both types of data on hand. NNMRRP survey data provide a national perspective on the target population and the factors that contribute to its nutritional needs in other regions of the country. In most cases, NNMRRP survey data can be obtained from the National Technical Information Service either as a printed publication or on tape. Some data tapes and publications are available from the sponsoring agency (for instance, U.S. Department of Agriculture, National Institute of Child Health and Human Development) or the U.S. Government Printing Office. In addition, many health and nutrition data are published in periodicals such as the *Journal of the American Dietetic Association, Public Health Reports,* and the *New England Journal of Medicine*. Consult the Public Health Service's publications *Nutrition Monitoring in the United States: The Directory of Federal Monitoring Activities* and the *Third Report on Nutrition Monitoring in the United States*, Volumes 1 and 2, for a list of NNMRRP surveys and pertinent journal publications of survey findings and a description of where NNMRRP survey data can be obtained.[33] Information about the availability of some nutrition monitoring and survey data is provided on the Center for Disease Control and Prevention's Web site.[34]

The decision to use NNMRRP survey data will be influenced by the level of detail needed about the target population, the personnel and facilities available for sorting and analyzing the data, and budget constraints. The cost of purchasing data ranges from $100 to $1,500.

National survey data do not always reflect the nutritional status or food intake of the target population in a particular community. Consider this scenario: national data from the 1987 and 1992 National Health Interview Surveys found that the mean calcium intake of white women fell about 5 percent from the 1987 to 1992 survey;[35] however, a 1997 survey of white women living in your community found that their mean intake of calcium fell nearly 10 percent between 1992 and 1997. In this situation, the national data do not reflect your community particularly well, but they may be a good benchmark against which to compare your target population. The question for you as a community nutritionist is why some white women in your community have a calcium intake lower than the national average. Thus, small surveys carried out locally provide insights into *your* community's needs and values. They tend to be more relevant to defining your community's nutritional problems and needs than large, national surveys.

New Data. In some situations, data related to your target population are not readily available, in which case you must choose a method for obtaining them. The methods for assessing nutritional needs must be simple, cost-effective, and doable within a reasonable time frame. Chapter 6 describes some of the methods used to collect data about target populations, including nutrition surveys, health risk appraisals, screening tools, focus groups, interviews with key informants, and direct assessment techniques.

Step 4: Analyze and Interpret the Data

Data collected about the community's nutritional problem must be analyzed and examined. This requires first collating the data collected about the community itself and any background conditions that may have influenced the target population or the nutritional problem. Then, data collected about the target population (described in Chapter 6) are merged with the community and background data to form a comprehensive report.

Data derived from the assessment are then used to "diagnose" the community. Four steps are involved in making the community diagnosis: (1) interpret the state of health of the target population within the community, (2) interpret the pattern of health care services and programs designed to reach the target population, (3) interpret the relationship between the target population's health status and health care in the community, and (4) summarize the evidence linking the target population's major nutritional problems to their environment. The summary describes the dimensions of the nutritional problem, including its severity, extent, and frequency; its distribution across the urban, rural, or regional setting and across age groups; its causes; and the mortality and morbidity associated with it. The summary should specify the major strengths of existing community resources and health care services as they relate to the target population, the areas where health problems seem to be concentrated, and the areas where health care delivery for the target population can be improved.[36] The summary may also provide information about the cost of treating versus preventing the nutritional problem and the social consequences of not intervening in the target population.[37]

The final step is to prepare an executive summary that captures the three or four key points that emerged from the assessment. The executive summary can be given to stakeholders and other interested parties who request information about the assessment outcomes. It can also be reformatted as a press release for the media.

Step 5: Share the Findings of the Assessment

The results of the community needs assessment are often useful to agencies and organizations that were not directly involved in it. Sharing the assessment results with these other groups and stakeholders is cost-effective, prevents duplication of effort, and promotes cooperation among organizations and agencies. It also enlarges the sphere of awareness about the nutritional problem and increases the likelihood that more than one community group will choose to address the problem. However, releasing the results of an assessment to the community at large without seeking the support and approval of key stakeholders can create ill will. When in doubt, go back to the stakeholders and ask for permission to release sensitive material about the target population.

Step 6: Set Priorities

Setting priorities involves deciding *who* is to get *what* at *whose expense*.[38] When several nutritional problems are identified by the assessment—as often happens—the question asked by the community nutritionist is "Which health outcome is most important?" The term **health outcome** refers to the effect of an intervention on the health and well-being of an individual or population.[39] A health outcome may reflect a change, which can be either positive or negative, or even no change in health status. It may be distinct, such as a drop in blood pressure or an increase in fiber intake, or it may be somewhat subjective, such as an increase in awareness of a health risk.[40] In other words, the community nutritionist is asking, "Given the several nutritional problems and needs of the target population, where should my organization direct its efforts to achieve the *best* health outcome?" In most cases, the best health outcome is an improvement in the nutritional status of the target population.

The challenge for the community nutritionist is deciding which of several nutritional problems or needs of the target population deserves immediate attention. Considering the fierce competition for scarce resources, how does the community nutritionist, the assessment team, and community leaders decide where to put their efforts and money? No one method exists for ranking problems or needs, although various scoring systems that rank risk factors by relative importance have been proposed. A few principles that provide guidance in identifying problems of the highest priority are listed in Table 5-7.

Health outcome The effect of an intervention on the health and well-being of an individual or population.

TABLE 5-7 *Principles Involved in Setting Priorities*

- Community priorities, preferences, and concerns should be given priority.
- Higher priority should be given to common problems than to rare ones.
- Higher priority should be given to serious problems than to less serious ones.
- The health problems of mothers and children that can easily be prevented should have a higher priority than those that are more difficult to prevent.
- Higher priority should be given to health problems whose frequencies are increasing over time than to those whose frequencies are declining or remaining static.

Source: Adapted from D. B. Jelliffe and E. F. P. Jelliffe, *Community Nutritional Assessment* (Oxford: Oxford University Press, 1989), p. 452. By permission of Oxford University Press.

The priority-setting process begins with a review of the summary prepared in Step 4 (described previously). The findings of the community assessment are compared with the nutrition, physical activity and fitness, maternal and infant health, and other *Healthy People 2000* objectives to determine where improvements in health services should be made. Perhaps only 22 percent of worksites with 50 or more employees offer health promotion programs, whereas the *Healthy People 2000* objective is 85 percent, indicating an area that deserves attention. The *Healthy Communities 2000: Model Standards* guidebook is consulted to help the community translate the *Healthy People 2000* objectives into action plans.[41] The seriousness of each problem relative to other nutritional and health problems within the target population is considered. Members of the assessment team rank existing health and nutritional problems and make recommendations about where the community's resources should be directed. Key stakeholders or community leaders help determine which needs of the target population deserve immediate attention.

A description of *Healthy Communities 2000 Model Standards* appears in Appendix I.

In a perfect world, ample personnel, money, and other resources would be available to spend on each of the target population's problems. Priority setting would be unnecessary. In reality, there are never enough resources to address all public health problems, and the decisions about which problems receive attention are not always rational, right, or fair. The process of setting priorities is influenced by the community's political power base, federal and state public health priorities, public opinion, and the beliefs of key stakeholders. The final priority areas generally reflect the community's ranking of the importance of public health problems and its assessment of the probable impact of its interventions.

Step 7: Choose a Plan of Action

The community nutritionist is now ready to make a decision. He has on hand a definition of the nutritional problem, the results of the needs assessment, and a ranking of the nutritional problems and needs that most deserve attention. Now what? As shown in Figure 5-3, he has any number of options, the most important thing being to do *something*. After all, the assessment required planning, team effort, and precious resources. Its findings are too valuable to ignore.

But what should be done? At the very least, the key findings of the assessment should be shared with community leaders and other people who are interested in

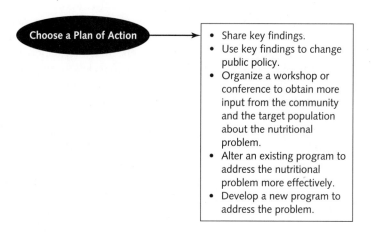

FIGURE 5-3 *Choosing a Plan of Action*

the health and well-being of the target population. These people may use the findings within their own organizations to support interventions aimed at improving the target population's health and nutritional status. For example, the results of an assessment of the dietary changes made by pregnant teenagers can be shared with physicians, nurses, and other health care providers who need to know how and why teenagers change their food patterns during pregnancy.

Another action is to use the assessment's findings to advocate for a change in legislation or public policy that will ultimately improve the health potential of the target population. The term **advocacy** means to build support for an idea, cause, or change. (Advocacy was discussed in Chapter 2.) Releasing the assessment's findings to the media is one way of increasing awareness of the problem and building support for policy changes that address the problem.

The community nutritionist and other team members may elect to organize a workshop or conference to obtain additional data on the problem or pull together community leaders and stakeholders to explore future actions. Or he might decide to alter an existing program by developing new educational materials, enlarging a marketing campaign, or changing the mechanism for delivering the program. He may develop a new program to address the nutritional problem of the target population, in which case he may write a grant proposal to apply for money to pilot test his idea. (Grant writing is discussed in Chapter 11.) In reality, he will likely pursue several actions simultaneously. Regardless, one or more actions should be taken to improve the target population's health and nutritional status through program planning and other activities. (The process of planning, marketing, and evaluating programs is described in Chapters 7 through 11.)

Advocacy To build support for an idea, cause, or change.

Entrepreneurship in Community Needs Assessment

Recall from Chapter 1 that entrepreneurship is the creation of something of value through the creation of organization.[42] In the case of the community needs assessment, the "something of value" is the snapshot of the nutritional and health problems of the community. Obtaining a picture of this valuable commodity requires organization, vision, and new ideas—all aspects of the entrepreneurial process. Community nutritionists can apply the principles of entrepreneurship to community needs assessment by developing new strategies for collecting information about hard-to-reach populations such as new immigrants and the homeless; by forging new partnerships with food producers, retailers, distributors, and marketers to collect information about dietary patterns and beliefs at the local level; and by developing new methods of assessing nutritional needs and problems.

The COMPASS® tool kit developed by United Way of America is an example of entrepreneurship in community needs assessment. COMPASS® helps local United Ways strengthen communities by teaching volunteers how to forge partnerships across the community.[43] Another example is the President's Initiative on Race, a national strategy to eliminate racial and ethnic disparities in health by the year 2010—a goal that will require the concerted efforts of thousands of federal, state, and municipal staff, health professionals, and volunteers. Once again, partnerships among the Department of Health and Human Services, state and local governments, regional minority health and tribal organizations, and community groups are the key to achieving this goal.[44] Look at your community for examples of organizations or people who recognized an opportunity and took the initiative to improve the community's quality of life.

Community Learning Activity

These Community Learning Activities are designed to help you learn about your community and the process of conducting a nutritional needs assessment. Both involve teamwork.

Activity 1

Divide yourselves into five groups. Each group focuses on one of the nutritional problems described in Step 1 of this chapter. For your group's nutritional problem, define the purpose of the assessment, develop one or more goal statements and several objectives, and specify the types of community and background data needed for the assessment. Prepare a summary report for the class that outlines these aspects of the assessment (that is, the purpose, goals, objectives, and so forth), and include answers to the following questions:

1. Which organizations, agencies, and community groups should be involved in the assessment?
2. How does the target population perceive this problem?
3. Does the target population have other health and nutritional problems? If yes, which ones?

Activity 2

Working as a class, begin by defining the community in which you live (for example, the city of Portland, the town of New Bern, and so forth). Divide yourselves into three groups, each responsible for locating data about your community in one of the areas shown in Tables 5-2, 5-3, and 5-4—that is, data about the community, the environment, and socioeconomic information. When the data have been collected and organized, each group presents its findings to the class. Finally, the class answers the following questions:

1. Did your perception of your community change? If yes, what was the basis for the change?
2. What were the five leading health and nutritional problems that emerged from your community study? Present the incidence and/or prevalence of these problems.
3. How would you prioritize these health and nutritional problems?
4. Were you able to obtain all types of data shown in Tables 5-2, 5-3, and 5-4? If not, why not? What does this suggest to you?
5. How would you apply your knowledge of the community and any background conditions to the outcome of your assessment in Activity 1?

Professional Focus   *Teamwork Gets Results*

At the end of the day, you bet on people, not on strategies.
— Larry Bossidy, CEO, AlliedSignal, as cited in
N. M. Tichy's *The Leadership Engine*

Few things are accomplished in community nutrition without teamwork. Community nutritionists often work in teams to lobby state legislators, plan mass media campaigns, identify perceived needs in the community, design programs, and raise awareness about a nutritional problem. When the Arizona Department of Health Services designed a program to help the staff at child care facilities learn how to meet the needs of children with special health care needs, they used a team approach. The course instructors for Project CHANCE included a parent of a child with special health care needs, a nutritionist, an occupational therapist, and a director of a child care facility.[1] Each member of the teaching team brought their personal insights and experiences to the discussion of how to meet the nutritional needs of children with cerebral palsy, cystic fibrosis, epilepsy, cleft palate, and other special needs.

A team is a *small number of people* with *complementary skills* who are committed to a *common purpose* for which they hold themselves *mutually accountable*.[2] Each aspect of this definition is important. The number of people on a team can range from 2 members to a maximum of about 16 members. Twelve members is a size that allows team members to interact easily with one another. Large groups can be unwieldy to manage and direct.[3]

The people chosen to be on the team should be selected carefully, so that the team has balance, variety, and essential expertise.[4] When pulling a team together, prepare a list of potential team members. Think about their similarities and differences. A strong team consists of people with different problem-solving styles and methods of organizing information. Consider diversity of age, sex, educational training, and the like among team members. Diversity ensures a wealth of ideas and views.

Select team members who offer the expertise you believe will be needed to help achieve a goal or solve a problem. If you are developing a restaurant program, for example, you might want team members with expertise in nutrition, food service, marketing, and communications. If you are planning an intervention to address a nutrition problem such as hypertension, include someone from the target population, someone who can speak personally about hypertension, on the team.

"The heart of any team is a shared commitment by the members to their collective performance."[5] Achieving the team's goals requires the cooperation of each team member. In other words, a team probably won't be successful in meeting its goals if one or more members don't participate fully. Successful teams are ones whose team members decide how they will work together, and they know what goal they are working toward.

Team members are mutually accountable for results. They are willing as individuals to assume responsibility for a portion of the work assigned to the team. They are active participants, not passive bench sitters. They prepare in advance for meetings, contribute their ideas, and evaluate themselves.

Most teams have a single leader or captain, but some teams have a "rotating leader"—that is, the team leadership shifts among team members. The team leader is responsible for preparing the meeting agenda, raising important issues, asking questions, keeping the discussion on friendly terms, maintaining order during the meeting, steering the team toward its goal, and recording the team's accomplishments. Diplomacy and humor are two tools the team leader can use to keep the team on track. Much like the football coach or symphony conductor, the team leader's main role is to help team members work together effectively.

Because teams are composed of individuals with different backgrounds, perspectives, ideas, and opinions, conflicts can arise. Thus, team leaders must be able to deal with conflict, a by-product of how people interact in groups. Conflict often has a negative connotation and typically is viewed as a signal that something is wrong. Most people, in fact, try to avoid conflict at all costs. Yet, conflict is inevitable in all relationships. It can arise from competition among workers for scarce resources, questions about authority relationships, pressures arising outside the organization, or even from solutions developed for dealing with previous conflicts. The problem in most organizations and work groups is that conflicts are avoided and left unresolved. When this occurs, team members are likely to be defensive and hostile and to feel that they can't trust one another.[6]

To handle conflict positively, team leaders should remember that conflict usually emerges because of differences in attitudes, values, beliefs, and feelings. Appreciating and being sensitive to such differences helps establish a climate of cooperation. Next, team leaders work to pinpoint the underlying cause of the conflict. Finally, they resolve the differences through open negotiation. If necessary, rules and procedures are changed to address the underlying problem. At all times, leaders work to ensure that team members feel their complaints and concerns are perceived as legitimate. The trick to managing a team on which there are strong differences of opinions is to work toward finding a common ground.[7]

References

1. L. Rider, MS, RD, Arizona Department of Education, personal communication, April 6, 1998.
2. J. R. Katzenbach, The myth of the top management team, *Harvard Business Review* 75 (1997): 83–91.
3. D. Hellriegel, J. W. Slocum, Jr., and R. W. Woodman, *Organizational Behavior,* 7th ed. (St. Paul: West, 1995), p. 281.
4. N. L. Frigon, Sr., and H. K. Jackson, Jr., *The Leader: Developing the Skills and Personal Qualities You Need to Lead Effectively* (New York: AMACOM, 1996), pp. 63–64.
5. Hellriegel and coauthors, *Organizational Behavior,* p. 272.
6. M. S. Corey and G. Corey, *Groups—Process and Practice* (Belmont, CA: Wadsworth, 1992), pp. 150–53.
7. J. G. Liebler and coauthors, *Management Principles for Health Professionals,* 2nd ed. (Gaithersburg, MD: Aspen, 1992), pp. 246–66.

References

1. Data related to the current Hispanic population were taken from "Current Population Reports: The Hispanic Population in the United States: March 1996 (Update)," available on the Census Bureau's Web site at www.census.gov/prod/www/abs/msp20502.html.
2. A. L. Martinez, *Hunger in Latino Communities* (Washington, D.C.: Congressional Hunger Center and Congressional Hispanic Caucus Institute, 1995).
3. G. Christakis, Community assessment of nutritional status, in H. S. Wright and L. S. Sims, *Community Nutrition—People, Policies, and Programs* (Belmont, CA: Wadsworth, 1981), pp. 83–97.
4. E. B. Perrin, Information systems for health outcome analysis, in *Oxford Textbook of Public Health,* 3rd ed., Vol. 2, eds. R. Detels, W. W. Holland, J. McEwen, and G. S. Omenn (New York: Oxford University Press, 1997), pp. 491–97.
5. Department of Health and Human Services and Department of Agriculture, *Nutrition Monitoring in the United States: The Directory of Federal Nutrition Monitoring Activities,* DHHS Pub. No. 89-1255-1 (Hyattsville, MD: U.S. Government Printing Office, 1989).
6. B. Haglund, R. R. Weisbrod, and N. Bracht, Assessing the community: Its services, needs, leadership, and readiness, in *Health Promotion at the Community Level,* ed. N. Bracht (Newbury Park, CA: Sage, 1990), pp. 91–108.
7. S. Dubois and coauthors, Ability of the Higgins Nutrition Intervention Program to improve adolescent pregnancy outcome, *Journal of the American Dietetic Association* 97 (1997): 871–78.
8. K. Alaimo and coauthors, Food insufficiency exists in the United States: Results from the Third National Health and Nutrition Examination Survey (NHANES III), *American Journal of Public Health* 88 (1998): 419–26.
9. W. Carr, *Measuring Avoidable Deaths and Diseases in New York State,* Paper Series 8 (New York: United Hospital Fund, 1988), p. 48; and U.S. Department of Health and Human Services, Public Health Service, *Health United States and Prevention Profile, 1991* (Hyattsville, MD: Public Health Service, 1991), pp. 158–60.
10. National Center for Health Statistics, Third National Health and Nutrition Examination Survey (NHANES III) Public-Use Data Files, available at the center's Web site (under "Catalog of Electronic Products") at www.cdc.gov/nchswww.
11. T. A. Hammad and coauthors, Withdrawal rates for infants and children participating in WIC in Maryland, *Journal of the American Dietetic Association* 97 (1997): 893–95.
12. A. L. Owen and G. M. Owen, Twenty years of WIC: A review of some effects of the program, *Journal of the American Dietetic Association* 97 (1997): 777–82.
13. D. A. Galuska and coauthors, Trends in overweight among US adults from 1987 to 1993: A multistate telephone survey, *American Journal of Public Health* 86 (1996): 1729–35.
14. M. L. Klem and coauthors, A descriptive study of individuals successful at long-term maintenance of substantial weight loss, *American Journal of Clinical Nutrition* 66 (1997): 239–46.
15. Community Nutrition Institute, Teen birth rate declines again; infant mortality rate drops to new record low, *Nutrition Week* 27 (September 12, 1997): 1, 6.

16. J. Pomerleau and coauthors, Food intake of immigrants and non-immigrants in Ontario: Food group comparison with the recommendations of the 1992 Canada's Food Guide to Healthy Eating, *Journal of the Canadian Dietetic Association* 58 (1997): 68–76.

17. The reports on the evaluation of the FitzIn Program are available from the Heart & Stroke Foundation of Manitoba, Winnipeg, Manitoba, Canada R3B 2H8.

18. The description of the purposes of community needs assessment was adapted from Christakis, Community assessment of nutritional status; D. B. Jelliffe and E. F. P. Jelliffe, *Community Nutritional Assessment* (Oxford: Oxford University Press, 1989), pp. 142–55; and Office of Disease Prevention and Health Promotion, Public Health Service and American Dietetic Association, *Worksite Nutrition—A Guide to Planning, Implementation, and Evaluation* (Chicago: American Dietetic Association, 1993), pp. 14–15.

19. T. G. Rundall, Health planning and evaluation, in *Public Health and Preventive Medicine*, ed. J. M. Last et al. (Norwalk, CT: Appleton & Lange, 1992), pp. 1079–94.

20. C. W. Tyler, Jr., and J. M. Last, Epidemiology, in *Public Health and Preventive Medicine*, ed. J. M. Last et al., pp. 11–39; C. S. Reichardt and T. D. Cook, Beyond qualitative versus quantitative methods, in *Qualitative and Quantitative Methods in Evaluation Research*, T. D. Cook and C. S. Reichardt, eds. (Beverly Hills, CA: Sage, 1979), pp. 7–27;

and R. Walker, An introduction to applied qualitative research, in *Applied Qualitative Research*, R. Walker, ed. (Aldershot, England: Gower, 1985), pp. 1–24.

21. R. H. Fletcher, S. W. Fletcher, and E. H. Wagner, *Clinical Epidemiology—The Essentials* (Baltimore, MD: Williams & Wilkins, 1982), pp. 75–90.

22. Infant mortality data were taken from "Eliminating Racial and Ethnic Disparities in Health," available on the U.S. Department of Health and Human Service's Web site at raceandhealth.hhs.gov.

23. Information about the United Way of America was obtained from its Web site at www.unitedway.org.

24. J. Harvey-Berino and coauthors, Food preferences predict eating behavior of very young Mohawk children, *Journal of the American Dietetic Association* 97 (1997): 750–53.

25. M. B. Dignan and P. A. Carr, *Program Planning for Health Education and Promotion*, 2nd ed. (Philadelphia, PA: Lea & Febiger, 1992), pp. 17–58.

26. P. R. Voss and coauthors, Role of secondary data, in *Needs Assessment: Theory and Methods*, ed. D. E. Johnson et al. (Ames: Iowa State University Press, 1987), pp. 156–70.

27. Jelliffe and Jelliffe, *Community Nutritional Assessment*, pp. 355–83.

28. Information about the University of Michigan data archive services was obtained from a booklet published by the Institute for Social Research, University of Michigan (Ann

Arbor); Catalog of Data Collections (Ann Arbor, MI: Inter-university Consortium for Political and Social Research, 1992); and C. Campbell, 1990 Census public use microdata samples (PUMS), *ICPSR Bulletin* 13 (1993): 1–7.

29. Information about the Institute for Research in Social Science was obtained from a booklet published by the institute at the University of North Carolina (Chapel Hill, 1991).

30. S. J. Kunitz, The history and politics of US health care policy for American Indians and Alaskan Natives, *American Journal of Public Health* 86 (1996): 1464–73.

31. The definition of culture was taken from *Webster's Ninth New Collegiate Dictionary* (Springfield, MA: Merriam-Webster, 1988), p. 314.

32. Pan American Health Organization, *Health Conditions in the Americas*, Vol. 1 (Washington, D.C.: Pan American Health Organization, 1990).

33. Department of Health and Human Services and Department of Agriculture, *Nutrition Monitoring in the United States;* and Federation of American Societies for Experimental Biology, Life Sciences Research Office, prepared for the Interagency Board for Nutrition Monitoring and Related Research, *Third Report on Nutrition Monitoring in the United States*, Vols. 1 and 2 (Washington, D.C.: U.S. Government Printing Office, 1995).

34. The Centers for Disease Control and Prevention's Web site is www.cdc.gov.

35. J. Norris and coauthors, US trends in nutrient intake: The 1987 and 1992 National Health Interview Surveys, *American Journal of Public Health* 87 (1997): 740–46.

36. Dignan, and Carr, *Program Planning for Health Education and Promotion*, pp. 51–57.

37. Jelliffe and Jelliffe, *Community Nutritional Assessment*, pp. 452–64.

38. A. D. Spiegel and H. H. Hyman, *Strategic Health Planning: Methods and Techniques Applied to Marketing and Management* (Norwood, NJ: Ablex, 1991), p. 203.

39. Perrin, Information systems, p. 492.

40. M. Samuelson, Commentary: Changing unhealthy lifestyle: Who's ready...Who's not? An argument in support of the stages of change component of the transtheoretical model, *American Journal of Health Promotion* 12 (1997): 13–14.

41. U.S. Department of Health and Human Services, Public Health Service, *Healthy People 2000 Midcourse Review and 1995 Revisions* (Washington, D.C.: U.S. Government Printing Office, 1995), p. 199.

42. B. J. Bird, *Entrepreneurial Behavior* (Glenview, IL: Scott, Foresman, 1989), p. 3.

43. The United Way of America Internet address is www.unitedway.org.

44. The Internet address for the Initiative on Race is raceandhealth.hhs.gov.

Chapter 6

Assessing the Target Population's Nutritional Status

Outline

Learning Objectives

After you have read and studied this chapter, you will be able to:

- Describe the types of data that might be collected about the target population specified in the community needs assessment.
- Describe a minimum of eight methods for obtaining data about the target population.

- Discuss the issues of validity, reliability, sensitivity, and specificity as they apply to data collection.
- Discuss the practical, scientific, and cultural issues that must be considered when choosing a method for obtaining data about the target population.

Something To Think About...

I think it could be plausibly argued that changes of diet are more important than changes of dynasty or even of religion. – *George Orwell*

Introduction

The purpose of the community needs assessment is to obtain answers to basic questions: *What* is the nutritional problem experienced by the target population? How does the target population perceive the problem? *Which* factors contribute to it? *Where* does this group live, work, seek medical care, buy their groceries? *Why* do existing services fail to help this group? *How* can their health and nutritional status be improved? Answers to some of these questions were found during the community phase of the assessment (described in Chapter 5). Consider a target population consisting of teenagers aged 13 to 19 years who live in Carlson County. The community assessment might have determined that the mortality rate for this group is similar to the national average, their level of alcohol and substance abuse exceeds the state average, they access the health care system through pediatric practice and family practice settings, and most live in households earning $22,500 to 35,000 annually. National data on energy and nutrient intakes for this age group are available from the 1996 Continuing Survey of Food Intakes by Individuals. Even so, questions about the health and nutritional status of Carlson County teenagers remain. How many teenagers living in Carlson County are overweight? How many have eating disorders? What are their attitudes about seeking medical attention for these conditions? Are their usual dietary patterns nutritionally adequate? Do Carlson County teenagers take vitamin, mineral, herbal, or other dietary supplements routinely?

When key questions about the target population's health and nutritional status are unanswered, the community nutritionist must identify those data elements that are still needed and choose one or more methods for obtaining them. This chapter describes a plan for collecting data, the types of data that might be gathered about the target population, methods used commonly to obtain such data, and several issues to consider when choosing a method of data collection.

A Plan for Collecting Data

Decisions about which data to collect about the target population are not made willy-nilly. They are made carefully and methodically, following a plan laid out before the first data element is ever collected. In fact, the decision to obtain information directly from members of the target population is a decision to enter the realm of research, and all principles and methods associated with good research apply to the process.

The plan helps determine which questions to ask—and hence which data to collect—about the target population. Some aspects of the plan such as personnel and budget management are activities that fall within the realm of project management and are discussed in Chapter 11. Other aspects such as sample size, data management, quality control, and statistical analysis are beyond the scope of this book. For our purposes, the following planning activities are most important and should be completed before data collection begins:

Step 1: Review the purpose, goals, and objectives of the needs assessment.
Step 2: Develop a set of questions related to the target population's nutritional problem, how it developed, and/or the factors that influence it.
Step 3: Choose a method for obtaining answers to these questions.

Recall that at this point in the assessment process, the community nutritionist has already obtained information about the community in which the target population lives and works and about the broad background issues that influence its nutritional and health status. Now, she must decide which questions about the target population are most important, which methods can be used to obtain answers to these questions, and whether the answers are measurable.

Types of Data to Collect About the Target Population

One key activity in community needs assessment is asking questions. In Chapter 5, questions were asked about the community and environmental factors that affect the health and nutritional status of the target population—the outer layers of influence shown in Figure 5-2 on page 158. In this chapter, we pose questions about the lifestyle choices, dietary patterns, working conditions, and social networks that affect the health and nutritional status of the target population—the inner layers shown in Figure 5-2. A brief review of how these factors influence the target population's health and nutritional status and which questions might be asked in these areas is given in the sections that follow.

Individual Lifestyle Factors

Most community needs assessments are concerned with some aspect of the target population's lifestyle and diet. Many questions can be asked about how the target population's food intake, nutrient intake, and dietary patterns influence nutritional status.

Lifestyle. Lifestyle factors such as physical activity level, choice of leisure time activities, ability to handle stress, smoking status, and use of alcohol or drugs influence health and nutritional status. People who are physically active have lower blood cholesterol concentrations and a lower risk of coronary heart disease (CHD) than sedentary people.[1] Women who become pregnant unexpectedly are less likely to breastfeed their infants than women whose pregnancies are planned.[2] People who are convinced that eating fruits and vegetables protects against cancer are more likely to actually consume these foods than people who see no positive consequences to eating them.[3] People who smoke have higher body levels of oxidized products, including oxidized ascorbic acid, than nonsmokers.[4] The effects of these and other lifestyle choices on health and nutritional status are complex.

When conducting an assessment, the community nutritionist can ask broad questions about lifestyle: *What* effect does a particular lifestyle choice have on the target population's nutritional status? *Why* does the target population behave this way? *How* can this behavior be changed? Or, she can ask specific questions: Do people who begin an exercise program also make other positive lifestyle choices? Do people who routinely have their blood pressure checked also consume low-sodium diets? Do people who consume excessive amounts of alcohol also smoke and eat high-fat and salty foods? The point of these questions is to determine the extent to which a target population's nutritional status is affected by its lifestyle.

Diet. For the community nutritionist, diet is a key individual factor to be analyzed, as nutritional status is affected directly by nutrient intake and utilization and indirectly by the food supply and a host of other factors. The relationships among these factors are shown in Figure 6-1. Along the top of the conceptual model are various stages (for example, food supply, food distribution, and so forth) in which the effects of food and nutrient intake on nutritional status or other health outcomes can be measured. This model is not comprehensive, and some factors and interrelationships that influence nutritional and health status are not shown.[5] Even so, the figure serves as a good overview of the areas where the community nutritionist is likely to ask questions about the target population's food-related behavior and nutritional status.

The community nutritionist may ask questions about the food supply and its effects on the target population's nutritional status. The food supply determines which foods are available to the target population. It is a product of the geographical area, climate, soil conditions, labor, and capital available for building the agricultural base. When assessing the food intakes of low-income families, the community nutritionist may ask questions about how foods are distributed throughout the region or how government regulates the food supply. In Zimbabwe, Africa, for example, an assessment found that low-income families lacked good-quality, affordable foods because the government's agricultural policy allocated land to crops grown for export. This policy reduced the amount of land available for families to grow their own food, forcing them to buy expensive foods in commercial markets.[6] In this example, the question was, What effect does government agriculture policy have on food availability in this region?

Food intake is influenced by a constellation of biological, psychosocial, cultural, and lifestyle factors, as well as by our personal food preferences, **cognitions, attitudes,** and health beliefs and practices. This stage presents many opportunities for asking questions about the target population. Consider an assessment designed to determine the nutrient intakes of pregnant teenagers and to identify what they perceive as their motivations for making dietary changes during pregnancy.[7] The questions that might be asked about this target population include these: Are the diets of pregnant teenagers similar to those of nonpregnant teenagers? What dietary changes do pregnant teenagers make? Do they experience food cravings and aversions and, if so, to which foods? Do they use dietary supplements? Why do pregnant teenagers change their dietary habits? What are their motivations for increasing or decreasing their food and nutrient intakes? Do they follow the health advice given by their health care providers? If not, why not? What are their health concerns during pregnancy? These questions reflect the many factors shown in Figure 6-1 that influence food choices and food intake.

Cognitions The knowledge and awareness we have of our environment and the judgments we make related to it.

Attitude The positive or negative evaluation of performing a behavior.

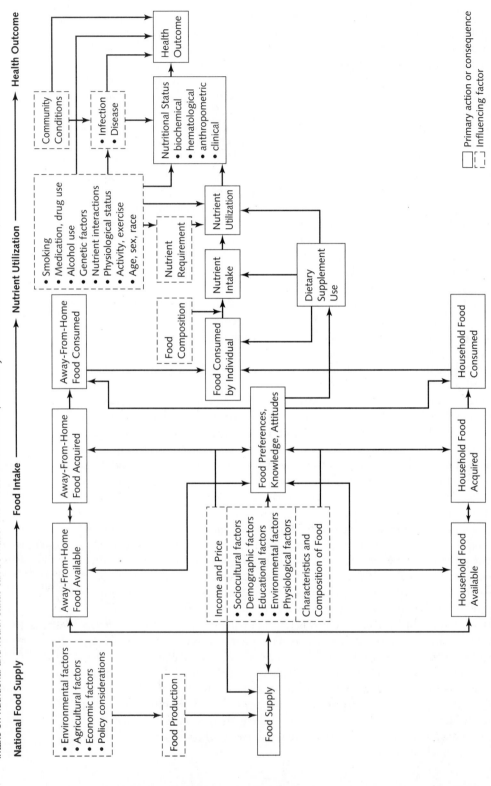

FIGURE 6-1 A Conceptual Model of the Relationships of Food to Health   This conceptual model shows the relationships among food choices, food and nutrient intake, nutritional and health status, and the factors that influence them. It suggests the major areas where measurements of the effects of food and nutrient intake on nutritional and health status can be made. The model is not inclusive, and many factors that affect health status are not shown here.

Source: Adapted from Federation of American Societies for Experimental Biology, Life Sciences Research Office, prepared for the Interagency Board for Nutrition Monitoring and Related Research, *Third Report on Nutrition Monitoring in the United States*, Vol. 1 (Washington, D.C.: U.S. Government Printing Office, 1995), p. 4.

Nutrient utilization is affected by activity levels, smoking status, dietary supplement use, drug-nutrient interactions, and physiological status (for example, growth, pregnancy). A community nutritionist involved in an assessment of the dietary practices of male cigarette smokers may have many questions about nutrient utilization in this target population:[8] How does cigarette smoking affect serum retinol, β-carotene, and α-tocopherol concentrations in this group? How does alcohol intake or the use of vitamin supplements affect these variables? Is nutrient utilization affected by the number of cigarettes smoked daily? Other questions that might arise in this assessment are linked with health outcome, the final stage shown in Figure 6-1. The questions asked here might focus on nutritional status or some measure of health status: What is the nutritional status of male cigarette smokers? Do male cigarette smokers who take vitamin supplements have a lower incidence of cancer than those who don't take supplements? Do male cigarette smokers with high dietary intakes of vitamins A and E have lower mortality rates than those with low intakes of these vitamins? Regardless of the assessment's goals and objectives, many questions can be asked about the important health outcomes for the target population.

Living, Working, and Social Conditions

The manner in which the target population lives and works affects their health and nutritional status. Education, occupation, and income all have powerful effects on health. Individuals who have few or no job skills or who are poor and uneducated tend to have more health problems than those with job training and education.[9] Low socioeconomic status is linked with high prevalence rates of chronic conditions, high stress levels, reduced access to medical care for the diagnosis and treatment of diseases, and poor outcomes following treatment.[10] Among children, poverty—even more than family structure and race—has the strongest association with health. Children who live in families headed by a single mother, black children, and those living below the poverty index are more likely to be in poor health than those living in two-parent families.[11] Low literacy is also a predictor of poor health.[12]

What questions might be asked about socioeconomic status during an assessment? A good example comes from a recent assessment of the food and nutrition situation of low-income Hispanic children living in Hartford, Connecticut.[13] The assessment team asked questions about the children's family situation, including the education and employment status of their caretaker, the number of people in the household, and the family's access to a car. A question was posed about the family's use of various household appliances such as a dishwasher, refrigerator, stove, toaster, television, and microwave. Other questions focused on whether caretakers used emergency food assistance or received food stamps. These questions were designed to obtain information about how these low-income Hispanic children and their caregivers live and also about how the family's living conditions affected the food intake of these children.

Primary **social groups** such as families, friends, and work groups also influence health and nutritional status. The family, not surprisingly, exerts the most influence,[14] as it is the first social group to which an individual belongs, and it is the group to which an individual usually belongs for the longest period of time. The family is a paramount source of values for its members, and its values, attitudes, and

Social group A group of people who are interdependent and share a set of norms, beliefs, values, or behaviors.

traditions can have lasting effects on their food choices and health. For example, Nigerian women feed their newborn infants water from household water pots—a potential source of bacteria that causes infant diarrhea. This practice is passed down from grandmothers and other respected older women to new mothers.[15] Here, the community nutritionist might ask questions about how these practices developed and why women do not perceive the benefits of exclusive breastfeeding to prevent infant diarrhea. Questions involving social networks should be designed to help understand why particular rituals and customs are important and how they influence health and nutritional status.

Case Study 1: Women and Coronary Heart Disease

In this and a later section, we return to our case studies, originally described in Chapter 5. We place the first one here as an example of how the questions asked about the target population are tied to the assessment's objectives. We begin with this case study because it is fairly simple.

Recall that Case Study 1 (summarized in Table 5-1 on pages 156 and 157) described an assessment aimed at obtaining information about women and coronary heart disease (CHD). The "community" consisted of the city of Jeffers and four adjoining municipalities. The target population was women over 18 years of age. In the assessment's community phase, data were collected about community-wide morbidity and mortality associated with CHD, types of educational materials available from nonprofit organizations, doctors' offices, and other sources, and existing programs and services designed to reduce CHD risk. The community nutritionist now reviews the assessment's objectives and, for each objective, develops a list of questions about this population's knowledge, attitudes, and practices related to CHD risk. These questions are shown in Table 6-1.

Two things should be considered when studying this table. First, not all of the data that might be collected about the target population are shown. Some data, particularly demographic data such as age, education level, and income, are collected as a matter of course. These data allow the community nutritionist to compare the findings of this assessment with those of other assessments or studies. Decisions about which demographic data to collect are made with the help of a statistician. Second, the questions posed in column 2 of the table are asked about the target population as a whole—in this case, *all* women over 18 years of age who live in the fictional city of Jeffers and four adjoining municipalities. The answers, however, are obtained from individuals who represent the target population—what is called the **sample.** Once again, the advice of a statistician is required to ensure that the individuals who are sampled represent the target population.

Sample A group of individuals whose beliefs, biological characteristics, or other features represent a larger population.

In this case study, the community nutritionist has developed a list of questions, each one tied to an objective. She then considers the types of data that might be collected to answer the question (for example, knowledge/awareness, health practices) and chooses a method for obtaining them. Her choices include a survey and a 24-hour recall. What is the purpose for placing this information in a table? The main purpose is to help organize data collection. She sees, for example, that a survey can be used to obtain answers to seven questions. In other words, a single sur-

Objective (refer to Table 5-1)	Question(s) Asked*	Types of Data (refer to Figure 6-2)	Method of Obtaining the Answer
• Assess women's knowledge and awareness of the leading causes of death and CHD risk factors.	1. Do women know that CHD is the leading cause of death among women in the United States?	Knowledge/awareness (cognitions)	Survey
	2. Can women identify four major risk factors for CHD?	Knowledge/awareness (cognitions)	Survey
	3. Where do women obtain information about health? About CHD?	Health practices	Survey
• Assess women's practices related to reducing CHD risk.	1. Are women eating a low-fat diet?	Health practices	24-hour recall
	2. Do women exercise regularly?	Health practices	Survey
	3. Do women smoke? If yes, how many packs/day?	Health practices	Survey
• Assess women's use of existing services designed to reduce CHD risk.	1. Which community services related to CHD do women use?	Community conditions	Survey
	2. What aspects of these services are most important to women?	Attitudes	Survey

TABLE 6-1 *Case Study 1: Women and Coronary Heart Disease* (CHD)

*The population in this case study is women over 18 years old living in the fictional city of Jeffers and four adjoining municipalities.

vey instrument or questionnaire can be designed to answer all seven questions. The 24-hour recall method should be a separate tool. This approach is part of the planning process, and it helps streamline data collection. The methods of obtaining answers to questions about the target population are described in the next section.

Methods of Obtaining Data About the Target Population

A variety of methods exist for collecting data related to the target population. The methods range from simple surveys and screening tools to interviews with key informants.

Survey

A **survey** is a systematic study of a cross-section of individuals who represent the target population. It is a relatively inexpensive method of collecting information from a large group of people. Surveys can be used to collect qualitative or quantitative data in formal, structured interviews, by phone or mail, or from individuals or groups. Some survey instruments (for example, questionnaires) are self-administered, while others are administered by a trained interviewer.[16]

Survey A systematic study of a cross-section of individuals who represent the target population.

Designing a questionnaire and conducting a survey involves more than heading out with a clipboard and a list of questions to interview people as they come into your clinic, office, or community center. Survey design and analysis is a discipline in itself, and the process of conducting a survey usually requires a team of experts with knowledge of survey research, statistics, epidemiology, public health, and nutrition. Although a detailed discussion of survey methodology is beyond the scope of this book, a few comments are in order about the issues to consider when designing a **nutrition survey** or adapting an existing survey for your own purposes.

Nutrition survey An instrument designed to collect data on the nutrition status and dietary intake of a population group.

"Planning a survey consists of making a series of scientific and practical decisions."[17] The first step is to determine the purpose of the survey. Most nutrition surveys are carried out to assess the food consumption of households or individuals, evaluate eating patterns, estimate the adequacy of the food supply, assess the nutritional quality of the food supply, measure the nutrient intake of a certain population group, study the relationship of diet and nutritional status to health, or determine the effectiveness of an education program.[18] A nutrition survey does not have to be complex and gargantuan to be meaningful, but it must have a well-defined purpose.

Next, decisions must be made about who will design the survey, who will conduct it, and how it will be carried out. The survey instrument must be designed and pretested, and the sample must be chosen. The personnel responsible for conducting the survey and analyzing data must be trained. Numerous other decisions must be made about the feasibility of the survey, the quality of data obtained by the survey, the costs of carrying it out, and the manner in which data will be analyzed and used. At every step in the planning process, there are practical constraints related to time and money.[19]

Surveys are important tools in assessing the health and nutrition status of individuals, but they must be designed and carried out carefully to provide valid and reliable information. Consult Table 6-2 for a list of questions to ask when designing a survey for your community needs assessment.

Health Risk Appraisal

The health risk appraisal (HRA) is a type of survey instrument used to characterize a population's general health status. We mention it here in a separate category because it is widely used in worksites, government agencies, universities, and other organizations as a health education or screening tool. The HRA is a kind of "health hazard chart" that asks questions about the lifestyle factors that influence disease risk,[20] and it has been used successfully to improve health behaviors.[21]

The HRA instrument consists of three components: a questionnaire, certain calculations that predict risk of disease, and an educational message or report to the participant.[22] A typical HRA asks questions about age, sex, height, weight, marital status, body frame size, exercise habits, consumption of certain foods (for example, fruits and vegetables) and ingredients (such as sodium), intake of alcohol, job satisfaction, hours of sleep, smoking habits, and medical checkups or hospitalizations. An HRA has been developed and tested for people aged 55 years and older.[23] A portion of the Healthier People Network's questionnaire is shown in Figure 6-2 on page 190.

- **Is the survey valid and reliable?** Will it measure what it is intended to measure, and, assuming that nothing changed in the interim, will it produce the same estimate of this measurement on separate occasions?

- **Are norms available?** That is, are reference data or population standards available against which the data from your target group can be compared?

- **Is the survey suitable for the target population?** A survey designed to obtain health and nutrition data on free-living elderly people may not be appropriate for the institutionalized elderly.

- **Are the survey questions easy to read and understand?** Survey questions must be geared to the target population and its level of literacy, reading comprehension, and fluency in the primary language. Having a readable survey is especially important if it is to be self-administered.

- **Is the format of the questionnaire clear?** If the questionnaire is not laid out carefully, respondents may become confused and inadvertently skip questions or sections.

- **Are the responses clear?** A variety of scales and responses may be used in designing a survey. Some questions may require filling in blanks or providing simple yes/no or true/false answers. Others may ask respondents to rank-order their responses from "seldom/never use" to "use often/always." The trick when selecting such scales is to choose one that allows you to discriminate between responses but doesn't provide so many categories that respondents are overwhelmed.

- **Is the survey comprehensive but brief?** Often the length of the survey must be limited to ensure that respondents complete it within a reasonable time frame. With long questionnaires, respondents are likely to answer questions hurriedly and mark the same answer to most questions.

- **Does the survey ask "socially loaded" questions?** Each survey question should be evaluated for how it is likely to be interpreted. Questions that imply certain value judgments or socially desirable responses should be rewritten. This is especially important when dealing with respondents from cultures other than your own.

Source: Adapted from L. Fallowfield, *The Quality of Life* (London: Souvenir, 1990), pp. 40–45.

TABLE 6-2 *Questions to Ask When Designing a Survey*

HRAs are used to alert individuals about their risky health behaviors and how such behaviors might be modified, usually through a lifestyle modification program.[24] For example, a health risk questionnaire was used at a trucker trade show to assess the health status of truck drivers. Truckers who stopped at the trade show enjoyed free food, music, raffles, and other events. At one booth, truckers received free blood pressure measurements and health education materials and were asked to complete a survey of their health risk factors, health status, and driving patterns. An analysis of the survey results showed that truck drivers tend to smoke cigarettes, be sedentary and overweight, and be unaware that they had high blood pressure.[25]

Screening

Screening is an important preventive health activity designed to reverse, retard, or halt the progress of a disease by detecting it as soon as possible. Screening occurs in both clinical practice and community settings, using procedures that are safe, simple, and cheap. Table 6-3 lists some common screening procedures. In community screening programs, people from the community are invited to have an assessment made of a health risk or behavior. Their screening value is then compared with a predetermined cutpoint or risk level. For example, during Heart Month a popular

FIGURE 6-2 A Portion
of a Health Risk
Appraisal Form

Source: Used with permission of The
Healthier People Network, Inc.

The HEALTHIER PEOPLE NETWORK, Inc.

. . . linking science, technology, & education to serve the public interest . . .

IDENTIFICATION NUMBER

The health risk appraisal is an educational tool, showing you choices you can make to keep good health and avoid the most common causes of death (for a person of your age and sex). This health risk appraisal is **not** a substitute for a check-up or physical exam that you get from a doctor or nurse; however, it does provide some ideas for lowering your risk of getting sick or injured in the future. It is NOT designed for people who already have HEART DISEASE, CANCER, KIDNEY DISEASE, OR OTHER SERIOUS CONDITIONS; if you have any of these problems, please ask your health care provider to interpret the report for you.

DIRECTIONS:
To get the most accurate results, **answer as many questions as you can.** If you do not know the answer leave it blank.

The following questions must be completed or the computer program cannot process your questionnaire:

| 1. SEX | 2. AGE | 3. HEIGHT | 4. WEIGHT | 15. CIGARETTE SMOKING |

Please write your answers in the boxes provided. ☞ (Examples: ☒ or [98])

1.	SEX		1 ☐ Male 2 ☐ Female
2.	AGE		[ ] Years
3.	HEIGHT	(Without shoes) (No fractions)	[ ] Feet [ ] Inches
4.	WEIGHT	(Without shoes) (No fractions)	[ ] Pounds
5.	Body frame size		1 ☐ Small 2 ☐ Medium 3 ☐ Large
6.	Have you ever been told that you have diabetes (or sugar diabetes)?		1 ☐ Yes 2 ☐ No
7.	Are you now taking medicine for high blood pressure?		1 ☐ Yes 2 ☐ No
8.	What is your blood pressure now?		
★38.	Do you eat some food every day that is high in fiber, such as whole grain bread, cereal, fresh fruits or vegetables?		1 ☐ Yes 2 ☐ No
★39.	Do you eat foods every day that are high in cholesterol or fat, such as fatty meat, cheese, fried foods, or eggs?		1 ☐ Yes 2 ☐ No
★40.	In general, how satisfied are you with your life?		1 ☐ Mostly satisfied 2 ☐ Partly satisfied 3 ☐ Not satisfied

HPN Health Risk Appraisal (V 6.0)   **Thank You for Participating**
© 1992 The Healthier People Network, Inc.

shopping mall sponsors a health fair that offers screening for high blood cholesterol concentrations. People who come through the screening booth give a finger-stick blood sample, which is analyzed on site by a machine designed for this purpose. Their blood cholesterol concentrations are classified as high, borderline-high, or desirable, based on the classification scheme developed by the National Cholesterol Education Program. Individuals whose screening value suggests an elevated risk are referred for medical diagnosis and treatment. Screening programs are not meant to substitute for a health care visit or routine medical monitoring for people already receiving treatment, but they do have educational value and serve to identify high-risk persons.[26]

Screening Procedure	Target Population
• **Clinical practice**	
Taking a medical history	All ages, both sexes
Height and weight	All ages, both sexes
Phenylketonuria (PKU)	Newborn infants
Posture (for detection of scoliosis)	Children over 3 years of age
Vision	Children over 3 years of age
Hearing	Children over 3 years of age
Tuberculin test	Children over 1 year of age
• **Community settings**	
Health risk appraisal	Primarily adults over 18 years of age
Blood pressure	Adults over 18 years of age
Blood cholesterol level	Adults over 18 years of age

TABLE 6-3 *Common Screening Procedures in Clinical Practice and Community Settings*

Sources: Adapted from J. M. Last, *Public Health and Human Ecology* (East Norwalk, CT: Appleton & Lange, 1987), p. 13; and *The Merck Manual,* ed. R. Berkow, 15th ed. (Rahway, NJ: Merck Sharp & Dohme Research Laboratories, 1987), pp. 1814–20.

Examples of screening programs abound. One example is the Nutrition Screening Initiative (NSI), a project designed to promote nutrition screening and improved nutritional care in the United States, especially among older adults. The NSI uses an educational tool, the one-page DETERMINE Checklist, to identify persons at increased risk of poor nutritional status. The checklist consists of a set of basic questions, written at the fourth- to sixth-grade reading level, that addresses the categories of nutritional risk specified by the NSI. The acronym DETERMINE is used to help remind respondents of the warning signs for nutritional risk: Disease, Eating poorly, Tooth loss/mouth pain, Economic hardship, Reduced social contact,

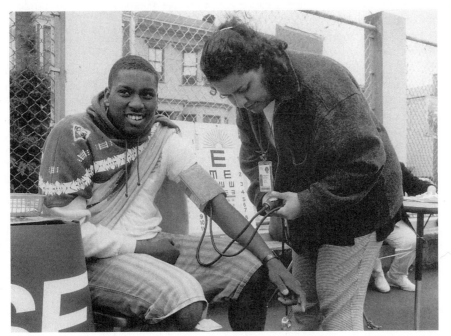

Screening is one method of identifying people with high blood pressure. People whose screening value indicates high blood pressure are referred for medical diagnosis and treatment, which may include nutrition counseling.

More information about the Nutrition Screening Initiative is provided in Chapter 15.

Multiple medicines, Involuntary weight loss/gain, Needs assistance in self-care, and Elder years above age 80. The checklist can be self-administered or completed by anyone who interacts with older family members or friends. It helps increase awareness of the factors that influence nutritional health.[27]

Focus Groups

Focus group An informal group of about 5 to 12 people who are asked to share their concerns, experiences, beliefs, opinions, or problems.

One method of obtaining information about the target population is to hold focus group interviews. **Focus groups** usually consist of 5 to 12 people who meet in sessions lasting about 1 to 3 hours. The group members are brought together to talk about their concerns, experiences, beliefs, or problems. Focus groups are used to obtain advice and insights about new products and services, research data and information about key variables used in quantitative studies, and opinions about products or creative concepts such as advertising campaigns or program logos.[28]

Focus group sessions are led by a trained moderator who is skilled at putting people at ease and promoting group interaction. Listening is the most important skill used during focus group interviews. Like a good teacher, the focus group leader must be able to listen on several levels, concentrating on what participants say—and do not say—and on the progress being made during the interview. A good session leader explores a topic without making participants feel guilty or defensive, avoids asking leading questions, doesn't interrupt participants when they are talking, pays attention to nonverbal cues such as body language, and keeps participants focused on the session's topic. Asking open-ended questions—for example, "How do you make decisions about buying milk, cheese, and other dairy products?"—allows participants to take any direction he or she wants and reconstruct a particular experience.[29] The information obtained from focus group interviews is then used to provide direction for the needs assessment or change a marketing strategy, product, or existing program.[30]

Focus group interviews provide qualitative information that help community nutritionists understand how the nutritional problem developed or even whether the target population perceives a problem. They are less expensive to conduct than face-to-face interviews, and they help community nutritionists obtain information of a sensitive nature or that might otherwise be difficult or costly to get.[31] The Johns Hopkins Hospital staff, for example, used focus groups to help them understand why some clients were not complying with cardiovascular health promotion programs to control hypercholesterolemia. Participants were asked questions about health, high blood cholesterol, current diet, food preferences, grocery shopping, and the types of education programs and materials they usually used. The staff learned that clients preferred to be taught by professionals in small groups and liked to hear from individuals who had successfully lowered their blood cholesterol. The focus group interviews helped the assessment team understand why a certain behavior developed and suggested ways to address the problem.[32]

Interviews with Key Informants

Key informants—the people "in the know" about the community—can also provide information about the target population. Key informants may have worked with the target population in the community or in some other part of the world, or they may have conducted research related to the population's health status, beliefs, or practices and be familiar with its attitudes and opinions about the nutritional problem. Informant interviews can be used to complete a cultural assessment of the target population, described later in this chapter. Interviews with key informants also provide insights about whether the target population perceives a nutritional problem and which actions for addressing the problem are culturally appropriate. For example, if the target population is obese persons, then interviews with physicians in family practice, internal medicine, and endocrinology will provide information about how this group is managed medically, what advice it is given about weight control, and whether it adheres to advice received from physicians.[33] If the target population is teenagers with a high risk of HIV/AIDS, then interviews with teenagers who work in peer education programs will provide valuable information about how these high-risk teenagers perceive risky behavior, what they know about preventing and transmitting HIV/AIDS, how susceptible they are to peer pressure, and what their overall health beliefs are.[34]

Direct Assessment of Nutritional Status

Another method of determining nutritional needs is to conduct a direct assessment of individuals. Direct assessment methods use dietary, laboratory, anthropometric, and clinical measurements of individuals to identify those with malnutrition or a nutritional deficiency state. These methods may be used alone or in combination.

Dietary Assessment Methods. In her comments at the Second International Conference on Dietary Assessment Methods, held in Boston in 1995, Elizabeth Helsing of the World Health Organization Regional Office for Europe remarked that "we still have a lot to learn from one another in the area of dietary assessment about the slow and arduous process of collecting vast amounts of data, the battle to mold them into some shape, and finally the wrestling of meaning from them."[35] Her comments allude to the challenge of collecting meaningful information about what people eat and why they make the food choices they do.

Dietary methods are used to determine an individual's or population's usual dietary intake and to identify potential dietary inadequacies. Dietary inadequacies represent stage 1 of the nutrient depletion scheme shown in Table 6-4. The primary methods of measuring the food consumption of individuals include the 24-hour recall method, food records, diet histories, and the food frequency questionnaire. At the household level, food records and inventory methods are used to estimate food consumption; population dietary intakes are estimated using food balance sheets and market databases.[36] The method chosen depends on many factors, some of which are shown in Table 6-5. All have strong and weak points, advantages and disadvantages. The trick is to choose the most valid method, given the financial resources and personnel available to collect and analyze the dietary intake information.[37]

Food Consumption at the National Level. The primary method of assessing the available food supply at the national level is based on **food balance sheets.** Food balance

Food balance sheets
National accounts of the annual production of food, changes in stocks, imports, and exports, and distribution of food over various uses within the country.

TABLE 6-4 *General Scheme for the Development of a Nutritional Deficiency*

Stage	Depletion Stage	Method(s) Used to Identify
1	Dietary inadequacy	Dietary
2	Decreased level in reserve tissue store	Biochemical
3	Decreased level in body fluids	Biochemical
4	Decreased functional level in tissues	Anthropometric/biochemical
5	Decreased activity in nutrient-dependent enzyme	Biochemical
6	Functional change	Behavioral/psychological
7	Clinical symptoms	Clinical
8	Anatomical sign	Clinical

Source: With permission from Sahn DE, Lockwood R, Scrimshaw NS, *Methods for the Evaluation of the Impact of Food and Nutrition Programmes.* Tokyo: United Nations University, 1984.

sheets do not measure the food actually ingested by a population. Rather, they measure the food *available* for consumption from imports and domestic food production, less the food "lost" through exports, waste, or spoilage, on a per capita basis. The per capita figures are obtained from the population estimate for the country.

Food balance sheets tend to be affected by errors that arise in calculating production, waste, and consumption. Hence, they are not used to describe the nutritional inadequacies of countries. They can be used to formulate agricultural policies concerned with food production and consumption.[38]

Household food consumption The total amount of food available for consumption in the household, generally excluding food eaten away from home unless taken from home.

Food Consumption at the Household Level. Methods of assessing **household food consumption** consider the per capita food consumption of the household, taking into account the age and sex of persons in the household (or institution), the number of meals eaten at home or away from home, income, shopping practices, and other factors. In most cases, no record is made of food obtained outside the household food supply or of food wasted, spoiled, or fed to pets. An exception is the U.S. Nationwide Food Consumption Survey, where a trained interviewer asked the householder to recall all foods used from the home food supplies during the preceding 7 days, including all food that had been eaten, discarded, and fed to pets.

Food Consumption by Individuals. Several methods exist for assessing the food, food group, or nutrient intake of individuals. The oldest of these is the diet history method and the newest, the food image processing method. Each has advantages and disadvantages. They may be used in combination. For example, the National Health and Nutrition Examination Survey (NHANES) III, conducted in 1988–1994, used a 24-hour dietary recall and a 62-item food frequency questionnaire to assess the usual eating patterns and nutrient intake of about 32,000 people.[39]

• **Diet history method.** One of the earliest descriptions of diet analysis was reported by Bertha Burke with the Department of Child Hygiene of the Harvard School of Public Health. In studies of pregnant women and their infants and children, conducted during the 1930s, Burke and her colleagues developed a diet history questionnaire to assess usual dietary intake. Their studies showed statistically significant relationships between a mother's diet during pregnancy and the condition of her infant at birth. In addition, there was a correlation between

TABLE 6-5 *Issues to Consider When Selecting a Dietary Intake Method*

Program Objectives

- Degree of accuracy needed
- Type of intake data needed
  - Food intake
  - Nutrient intake
  - Dietary pattern
  - Food pattern

Study Population

- Sample size needed
- Ability of respondents
  - Age
  - Literacy level
  - Language skill
- Willingness to cooperate
- Time constraints

Financial Issues

- Cost of analysis software
- Cost of training staff to use software
- Cost of intake forms (e.g., printed versus computer-generated)
- Cost of analyzing records (e.g., data entry by humans versus computer scanned data)
- Number of interviewers and coders to process data

Implementation Requirements

- Time required for respondents to complete intake form
- Ease of completion of intake form
- Need for support materials (e.g., instruction forms, food records, scales for weighing food, measuring cups, etc.)
- Training and skill level of interviewer

Analysis Requirements

- Quality of nutrient database
- Quality of analysis software
- Training and skill level of food coder
- Level of quality control

Source: Adapted from J. M. Karkeck, Improving the use of dietary survey methodology, *Journal of the American Dietetic Association* 87 (1987): 869–871.

the dietary ratings of the children's diets and objective measures of nutrition status, such as hemoglobin concentration.[40]

The diet history method has the advantage of being easy to administer, although it is time-consuming and requires a trained interviewer. For these reasons, it is not practical for large population studies. A true validation of the diet history method is probably not possible, although the method allows for reasonable confidence in classifying respondents according to some preset dietary criteria (for example, the Recommended Dietary Allowances).[41] When undertaken repeatedly at different times, this method is fairly reliable. It is perhaps best used to provide qualitative, not quantitative, data.

- **Twenty-four-hour recall method.** The 24-hour recall method is one of the most widely used diet assessment methods. It is easy to administer, can be administered in person or by phone, and lends itself to large population studies, mainly because it requires little time from either the respondent or the interviewer. Its appropriateness for assessing the intake of individuals—what is called "validity" or the ability of the instrument to measure what it is intended to measure—has repeatedly been questioned, however.[42] There are indications that 24-hour recall data are subject to recall bias; that is, respondents cannot accurately recall the foods they ate during the previous 24-hour period and either over- or underestimate their dietary intakes. Gender differences in recalling dietary intakes have also been reported. Thus, a single 24-hour recall does not provide an accurate estimate of an individual's usual dietary intake.

The validity of the 24-hour recall can be improved by administering repeated recalls. Seven or eight 24-hour recalls of an individual's dietary intake over a period of 2 or 3 weeks are more likely to provide a reasonable estimate of that person's usual intake than a single recall. Even so, 24-hour recall data are best suited to describing the intakes of populations, not individuals.[43]

• **Diet record method.** Diet records or food records have been considered the "gold standard" of diet assessment methods. Completed over a period of 3, 4, or 7 days, or even as long as 1 year, they have the advantage of providing detailed information about food products, including brand names, and methods of food preparation, and they eliminate the uncertainty that goes with trying to recall the foods eaten. The amount of food consumed can be estimated or calculated by weighing. If the respondents are properly trained, diet records give a reasonably accurate picture of usual dietary intake. However, the diet record method replaces errors in recall with errors in recording, and the possibility always exists that the act of recording food intake changes the actual foods chosen for recording. Accurate food records can be obtained if the respondents are highly motivated, literate, and well trained.

• **Food frequency method.** One of the first large-scale uses of the food frequency questionnaire was in the Nurses' Health Study, a study of a cohort of more than 95,000 female registered nurses being followed for the occurrence of coronary heart disease and cancer. A semiquantitative food frequency questionnaire—sometimes called "Willett's FFQ," named after the study's principal investigator, Dr. Walter C. Willett—was developed to categorize individuals by their intake of selected nutrients (for instance, vitamin A, vitamin C, animal fat). The validity and precision of this instrument were evaluated using four sets of 7-day food records and a 1-year diet record.[44]

Other food frequency questionnaires besides Willett's are available. Most have been tested and used with large groups of people participating in epidemiological studies, where the use of more labor-intensive instruments such as food records is not practical. A few food frequency questionnaires have been used with mixed success among minority populations in the United States.[45]

The food frequency questionnaire offers several advantages: it is self-administered, requires only about 15 to 30 minutes to complete, and can be analyzed at a reasonable cost. Thus, it has been used in a variety of population studies where no other instrument could have been administered practically. It suffers, however, from the same limitations as any other recall method, in that the accurate reporting of intake depends on memory. In addition, the food list must necessarily be short and thus may overlook some foods commonly consumed by the population being surveyed. Controversy over the appropriate uses of the food frequency questionnaire continues.[46]

- **Other diet assessment methods.** One recent innovation in the diet assessment arena is the use of photography to record dietary intake. Photography reportedly has provided valid and precise results compared with food records.[47] Another method of estimating food intake is the picture-sort approach, in which participants sort into categories various cards on which pictures or drawings of foods appear. The method was developed for use with a diverse population of older adults with low education or literacy levels. It can be self- or interviewer administered. When tested among participants in the Cardiovascular Health Study, the method was an easy and quick way to obtain data on food patterns.[48] Computers, not surprisingly, are also used to conduct personalized interviews and collect data about dietary intake.[49]

Laboratory Methods. Laboratory methods can be used to identify individuals at risk of a nutrient deficiency (stages 2 to 5 in Table 6-4), since tissue stores of nutrients gradually become depleted over time. The depletion may result in alterations in the level or activity of some nutrient-dependent enzymes or the levels of metabolic products. Thus, laboratory methods are used to detect subclinical deficiencies. Static biochemical tests measure a nutrient in biological tissues or fluids or the urinary excretion rate of the nutrient. Examples of static tests include the platelet concentration of α-tocopherol and urinary 3-hydroxyproline excretion. Functional biochemical tests measure the biological importance of a nutrient and the consequences of the nutritional deficiency. Functional biochemical tests include taste acuity, a measure of zinc status; dark adaptation, a measure of vitamin A status; and capillary fragility, a measure of vitamin C deficiency. Functional tests are generally too invasive and expensive to employ in most field surveys of nutrition status.[50]

Anthropometric Methods. Measurements of the body's physical dimensions and composition are used to detect moderate and severe degrees of malnutrition and chronic imbalances in energy and protein intakes. The most common growth indices include measurements of stature (height or length), weight, and circumference of the head. Measurements of skinfolds, mid-upper-arm circumference, and the waist-hip circumference ratio are used to derive equations that predict muscle and fat mass. Body mass index (BMI = weight [kg] / height [m]^2) is a commonly used indirect measure of obesity in adults.[51] Anthropometric measurements are useful in large-scale community assessment programs, because they involve simple, safe, non-invasive procedures; require inexpensive and portable equipment; and produce accurate and precise data when obtained by trained personnel. Such measurements are used to estimate an individual's long-term nutritional history. They do not provide data on short-term nutrition status, nor can they provide information about specific nutritional deficiencies.[52]

Clinical Methods. Clinical assessment of health status consists of a medical history and a physical examination to detect physical signs and symptoms associated with malnutrition. The medical history includes a description of the individual and his or her living situation (for example, married or single, children, employment). It typically obtains information about existing clinical conditions, previous bouts of illness, presence of congenital conditions, smoking status, existence of food allergies and intolerances, use of medications, and usual levels of physical activity. In the physical examination, the clinician evaluates the major organic systems: skin, muscular and skeletal, cardiovascular, gastrointestinal, and nervous. The hair, face, eyes, lips, tongue, teeth and gums, and nails are also examined for signs associated with malnutrition.[53]

Issues in Data Collection

The choice of method for obtaining information about the target populations is influenced by practical, scientific, and cultural considerations. These issues are described in this section.

Practical Issues

The choice of assessment method is influenced by practical issues such as the number of staff available to collect and analyze data, the cost of administering the test, and the amount of time needed to identify and interview or sample members of the target population. For example, it may be impractical to use food diaries to collect data about the food patterns of low-income, immigrant women and their children who live in a city's public housing. This particular population may read English poorly, be uncomfortable with record keeping, and lack transportation to the clinic for training on how to keep food records. An interviewer-administered 24-hour dietary recall is a better choice of diet assessment with this group.

The assessment method chosen should be simple to administer, take only a few minutes to complete, be inexpensive, and be safe. Blood pressure measurements are ideal screening tests, for example, because they are cheap, quick, need little advance preparation and setup, and are not uncomfortable or unsafe for participants.

Scientific Issues

Scientific issues such as the validity, reliability, and sensitivity of the assessment method and the nature of dietary variation also influence the choice of assessment method. Key scientific issues to consider when choosing an assessment method are described here.

Sensitivity The proportion of individuals in the sample with the disease or condition who have a positive test for it.

Sensitivity Versus Specificity. Two issues to consider when choosing an assessment method are sensitivity and specificity. **Sensitivity** is defined as the proportion of subjects with the disease or condition who have a positive test for the disease or condition. A sensitive test rarely misses people with the disease or condition, and it is often used in screening situations in which the purpose of the test is to detect a disease or condition in people who appear to be asymptomatic.[54] For example, a screening test that uses phlebotomy to detect high blood cholesterol concentrations is more sensitive—that is, it is more likely to identify an individual's true blood cholesterol concentration and to yield a positive result in the presence of a high blood cholesterol concentration—than a finger-prick cholesterol test.

Specificity The proportion of individuals in the sample without the disease or condition who have a negative test.

Specificity is defined as the proportion of subjects without the disease or condition who have a negative test. Specific tests are used to confirm a diagnosis, and they are important tests to administer when the possibility of not properly identifying a disease or condition would harm the patient or subject. The oral glucose tolerance test is a highly specific test for diagnosing diabetes mellitus.

Ideally, it is desirable to have an assessment method that is both highly sensitive and highly specific, but this is seldom possible in the real world. The trick is to choose a method that correctly classifies subjects into a particular group and misclassifies few of them.

Validity and Reliability. Two questions arise when evaluating assessment methods: one concerns the instrument's validity, and the other its reliability or reproducibility. **Validity** refers to the ability of the assessment instrument to measure what it is intended to measure. Another word for validity is *accuracy*. In the case of a diet assessment tool, a valid instrument accurately measures an individual's usual or customary dietary intake over a period of time—1 day, 3 days, 7 days, 1 month, 1 year. An instrument's validity can be affected by many factors:[55]

> **Validity** The accuracy of the assessment instrument.

- Characteristics of the respondent—literacy level, education level, conscientiousness in completing the instrument, ability to follow instructions
- Questionnaire design—difficulty of instructions, ease of recording intake, number and types of foods listed, portion sizes given (if any)
- Adequacy of reference data—a sufficient number of days of intake data were obtained to estimate the "true" nutrient intake; completeness of the nutrient database used to calculate nutrient intake
- Accuracy of data input and management—quality control of keypunching or coding of food items

Consider the validity of a survey instrument used during a telephone interview or mailed to respondents. One threat to the survey's validity is ambiguous language in the survey questions.[56] Writing a survey question that is easy for respondents to understand *and* yields the health or nutrition information you want is no simple task. Examine the questions in Table 6-6. The first question is fairly straightforward. Most consumers know what fruits are, and they can estimate their fruit intake with

Survey Question	Response
• Not counting juices, how many fruits do you usually eat per day or week?	_____ fruits per _____ day, week
• How often do you use fat or oil in cooking?	_____ times per _____ day, week, month
• How many servings, on average, do you eat of mayonnaise or salad dressing (serving size = 2 tablespoons)?	_____ never or < 1 per month _____ 1–3 per month _____ 1–4 per week _____ 5–7 per week _____ 2 or more per day

TABLE 6-6 *Examples of Survey Questions Designed to Obtain Information About Fiber and Fat Intake*

Source: The first two questions were taken from the National Cancer Institute's Health Habits and History questionnaire (Bethesda, MD: National Cancer Institute, 1987), p. 4. The third question was adapted from the University of Minnesota Healthy Worker Project questionnaire (Minneapolis: University of Minnesota, 1987), p. 3.

a reasonable degree of accuracy. The second question is more difficult to answer. Respondents must know what a fat is and recognize butter, stick margarine, tub margarine, lard, fatback, bacon fat, and vegetable oils and shortening as fats used in cooking. They must then estimate how frequently they use one or more of these fats in cooking, a complex calculation for most consumers. The third question likewise presents a problem. Respondents must decide what the survey question means by "mayonnaise" and "salad dressing." Is the question asking about all such products, be they low-fat, full-fat, or low-cholesterol, or is it only asking about regular, full-fat mayonnaise and salad dressings? Thus, some survey instruments used to assess dietary intake are not valid because only full-fat foods are indicated on the survey form and respondents cannot indicate that they usually consume reduced- or low-fat foods. These survey instruments tend to overestimate fat intake.

An individual's true usual diet cannot be known with certainty. It is not possible to follow respondents around all day and night and surreptitiously record every morsel they consume. And the very act of recording food intake can have a subtle influence, as the respondents may make choices they might not have made otherwise. (Although people's food and beverage intake can be monitored with precision on a metabolic ward, their dietary intake cannot be considered "usual" under these circumstances.) Attention to an instrument's validity ensures that respondents can be placed with a high degree of accuracy along a distribution of intake, from low to high consumption.

Reliability The repeatability or precision of an instrument.

The second concern is the **reliability** of an assessment instrument—that is, its ability to produce the same estimate of dietary intake, for example, on two separate occasions, assuming the diet did not change in the interim. (Other words for reliability are *precision*, *repeatability*, and *reproducibility*.) This issue is different from validity. It is possible for an instrument to give reproducible results that are also incorrect! An instrument's reliability can be affected by the respondents' ability to estimate their dietary intake reliably, by real dietary changes that occurred between the two assessment periods, and by inaccuracies in coding dietary data.

Sources of Variation. The foods we consume each day are complex mixtures of chemicals, some of which are known to be important to human health, while others have not even been identified or measured. The chemicals found in or on foods include essential nutrients, structural compounds (for example, cellulose), additives, microbes, pesticides, inorganic compounds such as heavy metals, natural toxins (for instance, nicotine), and other natural compounds such as DNA and RNA. The sheer diversity of the chemicals found in foodstuffs creates problems for investigators studying the relationship of diet to disease processes. When assessing the relationship of vitamin A intake to the development of lung cancer, for example, the population's intake of compounds with vitamin A activity—that is, preformed vitamin A (retinol), retinal, retinoic acid, and the carotenoids—must be calculated. And not only must the intake of these compounds from foods be considered but also the intake from vitamin supplements and other sources if they exist. Moreover, in some cases, it is the long-term dietary intake of foodstuffs that is important. In the case of a disease such as lung cancer, which may take 10 to 20 years to develop, the lifelong intake of vitamin A must be estimated.

Another difficulty is that people don't eat the same foods everyday. Work and school schedules, illnesses, holidays, the seasons of the year, weekends, personal preferences, availability of foods, social and cultural norms, and numerous other factors influence our daily food choices. As a result, our nutrient intake varies from day to day. The magnitude of the variation differs according to the nutrient.[57] Scientists with the Human Nutrition Information Service of the U.S. Department of Agriculture conducted a study to determine the number of days of food intake records needed to estimate the "true" average nutrient intake of a small number of adults.[58] The 29 men and women participating in the study completed detailed food records for 365 consecutive days. The ranges and average number of days of food records required to estimate the "true" average intake for individuals are shown in Table 6-7. Note the gender differences in the average values. For example, more food records were required to estimate the true average intake of calories by women than men. Note, too, that there were differences between nutrients. Compare the average values for calories with those for vitamins A and C, sodium, and fat. Finally, note the wide ranges in food records required to estimate true intakes. To estimate vitamin A intake, some men needed to complete 115 food records, while others required 1,724 days, or nearly 5 years! As these data demonstrate, a relatively large number of days of food intake records are required to achieve a certain level of statistical significance *for an individual*. If the individual data are combined into groups, however, fewer days of food intake records are required, as shown in Table 6-8.

The day-to-day variation in an individual's nutrient intake (called within-person variation) has important implications for nutritional assessment. If only one day's intake is determined, then the true long-term nutrient intake may be misrepresented, and, for example, an individual whose vitamin C intake over a period of several days is in fact adequate may be classified as having a low vitamin C intake. The effects of within-person variation on dietary intake must be considered when designing and evaluating studies of the relationship of diet to disease.

TABLE 6-7 *Ranges and Average Number of Days Required to Estimate the True Average Intake* for an Individual*

	Range and Average Number of Days Required					
	Males (*n* = 13)			**Females (*n* = 16)**		
Component	**Minimum**	**Average**	**Maximum**	**Minimum**	**Average**	**Maximum**
Food energy	14	27	84	14	35	60
Protein	23	36	72	23	48	70
Fat	34	57	131	32	71	114
Carbohydrate	10	37	177	16	41	77
Iron	18	68	130	28	66	142
Calcium	30	74	140	35	88	168
Sodium	27	58	140	36	73	116
Vitamin A	115	390	1724	152	474	1372
Vitamin C	90	249	900	83	222	328
Niacin	27	53	89	48	78	126

*The "true" average intake was defined as the 365-day average for individuals.

Source: P. P. Basiotis and coauthors, Number of days of food intake records required to estimate individual and group nutrient intakes with defined confidence. © *Journal of Nutrition* (Vol. 117, p. 1640), American Institute of Nutrition.

TABLE 6-8 *Number of Days Required to Estimate the True Average Intake* for Groups of Individuals*

| | Estimated Number of Days Required for Each group | |
Component	Males (n =13)	Females (n = 16)
Food energy	3	3
Protein	4	4
Fat	6	6
Carbohydrate	5	4
Iron	7	6
Calcium	10	7
Sodium	6	6
Vitamin A	39	44
Vitamin C	33	19
Niacin	5	6

*The "true" average intake was defined as the 365-day average for groups of individuals.

Source: P. P. Basiotis and coauthors, Number of days of food intake records required to estimate individual and group nutrient intakes with defined confidence. © *Journal of Nutrition* (Vol. 117, p. 1641), American Institute of Nutrition.

Cultural Issues

A cultural assessment of the target population is needed before data collection begins. The cultural assessment is undertaken to identify appropriate and inappropriate behaviors within the target population's culture. This is especially important if the assessment involves one-on-one interviews with members of the target population conducted in their home. Issues to consider during the cultural assessment are listed in Table 6-9. The manner in which strangers greet each other, the types of questions that are appropriate to ask, body language during interviews, and other customs differ among cultures.[59] In Japan, for example, a visitor who crosses her legs and shows the bottom of her shoe is being disrespectful. On the Canadian prairie, it is impolite for a guest to wear street shoes in someone's home. The people of North Africa value hospitality and time-honored rituals of eating, including hand washing, clapping, and eating with one's fingers. Visitors to the home of a North African are prepared to eat, whether they are hungry or not, for to refuse food would be insulting.[60]

Survey questions must also be culturally appropriate. For instance, a questionnaire designed to obtain information about the health attitudes and beliefs of Canadian aboriginal adults of the Mi'kmaq First Nation might have included a question that asked, "Do you worry about your diabetes?" If most aboriginal people responded no to this question, a researcher who was unfamiliar with the belief system and culture of the Mi'kmaq people might have concluded that aboriginal people were apathetic about their health. In fact, the question is culturally inappropriate, for in the traditional Mi'kmaq culture, the word *worry* has no meaning. A native person who gave a negative response to the question was likely saying, "I pay attention to my diabetes, but it is a part of my life. I take it day by day." This meaning is far different from what might have been concluded from the original question.[61] Cultural context is important when designing survey questionnaires.

Religion

- **Belief system**—Does this population's religious beliefs affect their food choices, use of alcohol, or other lifestyle decisions?

- **Food rituals**—Are particular foods consumed only during religious events/periods? Are certain foods taboo?

Etiquette and Social Customs

- **Typical greeting**—What is a proper form of address? Is a handshake appropriate? Are shoes worn in the home?

- **Social customs**—Should certain social exchanges occur before the interview or other assessment is undertaken? Are refreshments offered?

- **Direct and indirect communications**—Should a senior household member be expected to answer a question before a junior member does? Are some questions considered taboo by this ethnic population?

Nonverbal Communications

- **Eye contact**—Is eye contact considered polite or rude?

- **Tone of voice**—Does a soft voice have a particular meaning in this culture? A hard voice?

- **Facial expressions**—What do smiles and nods mean?

- **Gestures**—Are certain hand gestures considered rude and offensive? Is it acceptable to cross your legs.

- **Personal space**—Is the realm of personal space wider or narrower than in the North American culture?

- **Touch**—Is touch appropriate? If so, when, where, and by whom?

TABLE 6-9 *Factors to Consider During a Cultural Assessment*

Source: Adapted from M. C. Narayan, Cultural assessment in home healthcare, *Home Healthcare Nurse* 15 (1997): 665.

Case Study 2: Nutritional Status of Independent Elderly Persons

Case Study 2 focused on assessing the nutritional status of an elderly population in the fictional city of Jeffers. The target population was elderly people older than 75 years who live at home. (Review Table 5-1.) The community phase of the assessment obtained information about the types of existing community services aimed at addressing the needs of this group and how many members of the target population used these services. As with the first case study, the community nutritionist is now positioned to collect data about the target population. She first reviews the purpose, goals, and objectives of the assessment and then develops a set of questions aimed at measuring the nutritional status of this target population. These questions are listed in Table 6-10.

This needs assessment is more complex, requires a larger data set, and involves more people than the first case study. The nutritional status assessment methods in this case study include measurements of laboratory, clinical, anthropometric, and dietary outcomes, plus measures of functional status and other risk factors for poor nutritional status. Data are also needed about the population's use of community ser-

Objective (refer to Table 5–1)	Question(s) Asked*	Types of Data	Method of Obtaining the Answer
• Assess the nutritional status of independent elderly persons (> 75 years).	1. Has this population experienced a significant weight loss over time?	Anthropometric data (height, weight)	Scale
	2. Is their weight low or high for height?	Anthropometric data (height, weight)	Scale
	3. Have they experienced a significant change in skinfold measurement?	Anthopometric data (skinfolds)	Calipers
	4. Has this population experienced a significant reduction in serum albumin?	Laboratory data	Phlebotomy
	5. Does this population have nutrition-related disorders (e.g., osteoporosis, arthritis)?	Clinical data	Medical chart
	6. Does this population have an inappropriate food intake?	Dietary data	24-hour recall
	7. Does this population use any dietary supplements?	Dietary data	24-hour recall and survey
	8. Does this population use nutrition support?†	Dietary data	Survey

TABLE 6-10 *Case Study 2: Nutritional Status of Independent Elderly Persons*

vices. Remember, the assessment's purpose is to (1) obtain data about the nutritional status of these independent elderly persons and (2) link these outcomes with the services described in the community phase of the assessment to determine where services can be improved.

As in the first case study, each group of questions is tied to a specific objective, and the material in the table helps the community nutritionist plan the data collection. How does the community nutritionist make a decision about the assessment method, given the important factors described previously in this chapter? Several strategies can be used in decision making. First, he might turn to the medical or nutrition literature for information about accepted, standard assessment tests such as serum retinol for vitamin A status or serum ferritin for iron deficiency. Or, he might conduct a search of MEDLINE or other databases for articles that describe studies of the method's validity, reliability, sensitivity, and variation. (Consult the boxed insert for Internet Resources.) Next, he might speak with colleagues who are working with similar populations to determine their approach to the assessment of the target group.

He might consider a standard measure of health or nutritional status. These measures, called **nutrition status indicators,** are quantitative measures that serve as guides "to screen, diagnose, and evaluate interventions in individuals."[62] Nutrition status indicators are often used to estimate the magnitude of a nutrition problem, its distribution within the population, its cause, and the effects of programs and policies designed to alleviate the problem. They are used by researchers, program planners, health professionals, and policy makers for analyzing health and nutrition problems.

Nutrition status indicator
A quantitative measure used as a guide to screen, diagnose, and evaluate interventions in individuals.

Objective (refer to Table 5–1)	Question(s) Asked*	Types of Data	Method of Obtaining the Answer
	9. What is the functional status‡ of this population?	Clinical data	Survey
	10. What kinds of medications does this population take?	Clinical data	Survey
• Determine which community services are used by this population.	1. How many members of this population use: social services? mental health services? nutrition support services? home health care? home-delivered meals? federal/state assistance programs? housing and home heating assistance programs? other?	Community conditions	Survey
	2. How many members use services provided by dietitians and other health professionals?	Community conditions	Survey

*The population about which questions are being asked consists of elderly persons over 75 years old who live at home in the fictional city of Jeffers.

†Nutrition support refers to enteral and total parenteral nutritional support.

‡Functional status refers to the individual's ability to bathe, dress, groom, use the toilet, eat, walk or move about, travel (outside the home), prepare food, and shop for food or other necessities.

Internet Resources

To locate published studies about the target population or diet assessment methods:

Grateful Med www.nlm.nih.gov

MEDLINE www.nlm.nih.gov

To locate information about dietary validation studies:

Dietary Assessment Calibration/Validation Register www-dacv.ims.nci.nih.gov

To locate information about food composition databases:

Food Composition Resource List for Professionals www.nal.usda.gov/fnic/pubs/bibs/gen/97fdcomp.htm

International Network of Food Data Systems (INFOODS) www.crop.cri.nz/foodinfo/infoods/infoods.htm

Because there is no single "best" indicator, several may be used in the nutritional needs assessment. For example, major indicators of poor nutrition status in older U.S. adults include a weight loss of 5 percent or more of body weight in 1 month, being underweight or overweight, or having a serum albumin below 3.5 grams per deciliter, an inappropriate food intake, a midarm muscle circumference below the 10th percentile, and folate deficiency, among others.[63] Several of these nutrition status indicators were used in this case study. Nutrition status indicators exist for vitamin, mineral, and protein status and energy intake. Efforts are under way to develop accurate and reliable indicators of hunger and overall nutrition status. Table 6-11 lists the core nutrition status indicators recommended for the assessment of populations that are difficult to survey, such as homeless people, migrants, and institutionalized persons.[64]

Finally, he considers any factors particular to the target population that should influence his decision. Consider questions 6, 7, and 8 under the first objective in Table 6-10, which address the target population's nutritional intake. The community nutritionist reviewed the literature related to dietary intake methods and spoke with colleagues who routinely deal with this age group in their practices. He decided on an interviewer-administered 24-hour recall. His main concern was that

TABLE 6-11 *Indicators for the Assessment of Nutritional Status Among Migrant Workers, the Homeless, and Other Difficult-to-Sample Populations*

Nutrition Concern	Indicator
Obesity	Weight for length Weight for height
Hypercholesterolemia	Serum total cholesterol
Hypertension	Brachial artery pressure
Iron status	Complete blood count
Food insecurity	Food consumption: • Patterns over time • Types of foods • Frequency of consumption • Variety of foods Self-reports of: • Food sufficiency • Constraints to obtaining food Sources of food
Drug-nutrient interactions	Assessment of specific nutrient(s) affected by drugs and alcohol
Protein-energy malnutrition	Height Weight for height Weight change Skinfolds Body circumference Edema
Folate status as a marker for quality of diets limited in variety and quantity of foods	Serum and red blood cell folate concentrations
Vitamin A status	Serum retinol

Source: S. A. Anderson, Core indicators of nutritional state for difficult-to-sample populations. © *Journal of Nutrition*: Vol. 120, p. 1585, American Institute of Nutrition.

some individuals in the sample might not remember to record all the food and beverages they consumed during the day (as is required for completing food diaries) or be able to estimate their food intake on a food frequency questionnaire. All things considered, the interviewer-administered 24-hour recall was considered the dietary method most likely to yield good estimates of usual dietary intake in this target population in this setting. The community nutritionist also elected to develop a diet questionnaire to be administered by the trained interviewer at the same time as the 24-hour recall. This questionnaire would ask specific questions about the individual's intake of dietary supplements (for example, vitamins, minerals, shark cartilage, bee pollen, and so forth) and use of nutrition support (for instance, parenteral, enteral solutions). Thus, decisions about which method to use are arrived at by searching the scientific literature, talking with colleagues, and considering any challenges posed by the target population (for example, they do not speak English).

Putting It All Together

This phase of the community needs assessment focuses on obtaining data about the target population. It is designed to find answers to questions about how extensive the nutritional problem is and how it developed. The data collected during this phase are coded, entered into the computer, checked for errors, and analyzed using accepted statistical methods. Attention to quality during the data collection process is important and ensures that the assessment's findings are valid.[65]

Once the data are analyzed, a new issue arises: the choice of reference data against which the assessment's outcomes can be compared. For example, if the nutritional problem being evaluated is the nutritional status of children whose mothers participate in the WIC program and the nutritional status indicators are height, weight, and hemoglobin, the community nutritionist might choose reference growth data for children published by the U.S. National Center for Health Statistics (NCHS) to evaluate the children's growth and development. The NCHS curves for evaluating the physical growth of children are based on a large, nationally representative sample of U.S. children. For evaluating the iron status of children, the community nutritionist might use reference data for hemoglobin from the NHANES II survey for children of the same age and sex.[66] The values obtained for the target population during the community needs assessment can be compared against the reference data. In Case Study 2, reference data for evaluating skinfold measurements among the independent elderly sample could be compared with skinfold standards derived from NHANES.[67]

Nutrient intake data are usually compared with the Dietary Reference Intakes (DRIs), Recommended Daily Allowances (RDAs),[68] or, in Canada, the Recommended Nutrient Intakes (RNIs).[69] Reference data for comparison purposes can also be obtained from countries where national surveys of dietary intakes have been carried out. In situations in which no reference data for a country exist, the FAO and/or WHO requirements for nutritional intake can be used.[70]

Statements drawn from the data collected about the target population and from comparisons with reference data are then organized and added to the final report described in Chapter 5. In this manner, the final report contains information about the target population and about the community in which it lives and works.

Community Learning Activity

Activity 1

Return to Activity 1 in Chapter 5's Community Learning Activity. For the nutritional problem that was your choice and for which you have already developed a purpose, goals, and objectives, develop a set of questions to be asked about the target population. Then, specify a method for obtaining answers to those questions. Why did you choose those particular methods?

Activity 2

Return to Case Study 1. Review the purpose, goals, and objectives for this case study, shown in Table 5-1 on pages 156 and 157, and the questions asked about the target population in Table 6-1 on page 187. Next, do the following activities:

1. Develop a survey questionnaire designed to collect answers to the questions. Use other surveys in the literature or on the Internet to give you ideas about how your survey questions should be worded. For example, your survey should not simply ask, "Which community services related to CHD do you use?" It should ask a specific question:

   Which of the following community services are you participating in now?
   Check all that apply:
   _____ Weight management program
   _____ Smoking cessation program
   _____ Personal trainer
   _____ Fitness program
   _____ Other

2. Administer the questionnaire to family members and friends.
3. Answer these questions:
   a. How long did the questionnaire take to complete? Was this too much time? If yes, how would you shorten the questionnaire?
   b. Did your questions yield the answers you were looking for? If not, how should the questions be modified?
   c. How you would you modify the questionnaire for use with these populations: 30-year-old white women, black teenagers, female workers at a furniture refinishing plant, new Asian immigrants?

Activity 3

Use a search engine—for example, AltaVista (altavista.digital.com), Infoseek (www.infoseek.com), or Lycos (www.lycos.com)—to conduct a search of Internet resources for focus groups. Enter the keywords "focus groups" (use the parentheses as part of the entry to narrow the search). You should find dozens of sites offering tips on holding focus group sessions. What kinds of companies, organizations, and other groups use focus groups to obtain information? What kinds of information did they gather? How would you use the information found on the Internet to design a focus group session to find out what consumers think about snack foods?

Professional Focus | The Well-Read Community Nutritionist

Here is a sobering statistic: More than 20,000 biomedical journals are published each month. If each journal published just 20 research articles, the conscientious community nutritionist would need to browse through 400,000 articles per month, or more than 13,000 articles per day, to stay abreast of current scientific findings! This figure doesn't even include editorials, review articles, and letters to the editor—all valuable reading. How can you, as a busy community nutritionist, handle this volume of information, keep informed, and maintain your sanity? While we don't have all the answers, we offer the following suggestions to help you cope with the onslaught of medical, health, and nutrition information. Let's begin with a few good reasons for reading the literature.

Ten Good Arguments for Reading Journals

Consider the following 10 good reasons for reading journals regularly:[1]

1. To impress others
2. To keep abreast of professional news
3. To understand pathobiology
4. To find out how a seasoned health practitioner handles a particular problem
5. To find out whether to use a new or existing diagnostic test, survey instrument, or educational tool with your patients or clients
6. To learn the clinical features and course of a disorder
7. To determine etiology or causation
8. To distinguish useful from useless or even harmful therapy
9. To sort out claims concerning the need for and the use, quality, and cost-effectiveness of clinical and other health care
10. To be titillated by the letters to the editor

Regular reading of the literature, especially in your area of specialization, is a must. There is no other way to learn about the latest scientific findings, the merits of a particular intervention or assessment instrument, or current legislation and its potential impact on your programs and clients. In short, to be an effective community nutritionist, you must constantly increase and update your knowledge base through regular perusal of journals, which brings us to our next question.

Which Journals Should You Read?

There is no hard and fast rule about the "best" journals to read, since so much depends on the type of work you are doing and the needs of your clients. Some journals will appear on your "must read regularly" list; others can be spot-checked every month or two. While nutrition journals will take priority, other specialty journals, particularly in the disciplines of epidemiology, health education, and medicine, are important. Presented here is a list of journals, newsletters, and other publications that will help you stay abreast of current developments that may be useful to you in delivering community programs:

Nutrition Journals

American Journal of Clinical Nutrition
Canadian Journal of Dietetic Practice and Research
Human Nutrition: Applied Nutrition
Journal of the American College of Nutrition
Journal of the American Dietetic Association
Journal of Nutrition
Journal of Nutrition Education
Nutrition Reviews
Nutrition Today

Nutrition Newsletters

Community Nutrition Institute's *Nutrition Week*
Dairy Council Digest
FDA Consumer
Harvard Health Letter
Nutrition Action
Nutrition and the M.D.
Ross Timesaver
Tufts University Diet & Nutrition Letter
University of California at Berkeley Wellness Letter

Specialty Journals

American Journal of Epidemiology
American Journal of Health Promotion
American Journal of Public Health
Health Education
Health Education Quarterly
Journal of the American Medical Association
Lancet
New England Journal of Medicine
Preventive Medicine
Public Health Reports
Science

Other Publications

American Council on Science and Health Reports
Science News

Continued

Read consumer magazines, newspapers, and journals in other disciplines. Read anything you can get your hands on!

How to Get the Most Out of a Journal

There is no best way to "read" a journal. Although you will want to develop your own reading style, consider the following points. A glance at the table of contents will point you to pertinent research articles and briefs for in-depth reading. In journals that you subscribe to personally, highlight special interest articles in the table of contents with a colored marking pen. (This simple act may make it easier for you to remember, for example, where you spotted that article on monounsaturated fat and type 2 diabetes mellitus.)

After scanning the table of contents, check the professional updates and news features to keep informed about key players, committees, conferences, and events in your discipline. Consult review articles for extensive coverage of current issues. The journal's editorials will expose you to the controversies surrounding a study's findings or the implications of the findings for practitioners. Regular reading of the letters to the editor will help you appreciate the flaws of study designs and will expose you to the questions raised by scientists and practitioners in interpreting study results.

In choosing articles for in-depth reading, be selective and discriminating. You have only so much time. Select those articles that appear to be directly relevant to your needs, but allow time for other articles of interest. Refrain from beating yourself about the head because you don't have time to read as much as you think you should. Be organized and disciplined in your reading, but accept that no one can read everything.

How to Tease Apart an Article

Most research articles have the same basic format with specific sections to help you assimilate the material:[2]

- **Abstract or summary.** Provides an overview of the study, highlights the results, and indicates the study's significance. It should contain a precise statement of the study's goal or purpose.
- **Introduction.** Presents background information such as the history of the problem or relevant clinical features. It reviews the work of other scientists in the area and describes the rationale for the study.
- **Methods.** Describes the study design, selection of subjects, methods of measurement (for example, the diet assessment instrument used), specific hypotheses to be tested, and analytical techniques (for instance, the method used to measure blood cholesterol, method of statistical analysis).
- **Results.** Details the study's outcomes. The results are typically presented in tables, graphs, charts, and figures that help summarize the study's findings.
- **Discussion.** Provides an analysis of the meaning of the findings and compares the study's findings with those of other researchers. The discussion includes a critique of the work: What were the limitations of the study design? What problems occurred that may have affected the study outcome? What were the study's strengths?
- **Conclusions/implications.** Some, but not all, articles include a short section that summarizes the findings or considers how the study results can be applied to practice. This section may also comment on directions for future research.
- **References/bibliography.** Cites the relevant work of other scientists that was considered in conducting the study or interpreting the results.

Reading an article involves more than simply scanning the abstract and flipping to the last paragraph of the discussion section for the author's summary statements. If you have decided that the article is important to you, take time to read the methods sections carefully, for the substance of the work is outlined there. Any new information presented in the article is only as good as the method by which it was obtained. Was the hypothesis clearly stated? Was the study design clearly described? Were the methods used appropriate for testing the hypothesis? How were the data collected and analyzed? Once you learn to critically review the methods section, you will find that in some cases you do not need to read further. The study was so poorly designed or seriously flawed that the results lack validity.

One other important precept remains. Learn to form your own opinions about the study findings presented in the articles you read. Do not automatically assume that the findings are valid merely because the study was published by a leading researcher or research team. Reading the letters to the editors and talking about the study results with your colleagues are good ways to help you assess the validity of a study.

What Else Should You Read?

What else should you read to be an informed, effective community nutritionist? Everything! Well, everything you can get your hands on: newsletters, books, consumer magazines, food labels, newspapers, advertising, menus, junk mail, Internet Web sites.

Are we serious? Absolutely. We said at the outset (in Chapter 1) that one of your roles as a community nutritionist is to improve the nutrition and health of individuals living in communities. To do this well, you must be able to draw on many diverse elements within your community and culture to shape a program that meets a nutritional or health need. Let's say that you have been asked to design, implement, and evaluate a program to reduce the prevalence of obesity among schoolchildren in your community. To develop a nutrition and fitness program that appeals to children, you must be able to speak their language and get inside *their* culture. Which reading material will help you find the right approach, the right action figures, the right

tone? Everything from the *Journal of the American Dietetic Association* to children's books, advertising inserts in your local newspaper, the newspaper comics, fast-food menus, T-shirt logos, a *Newsweek* article on latchkey children, a government publication on quick snack ideas—the list is endless. You never know when something you read in a totally unrelated area is the exact thing you need to help convey a nutrition message to your clients.

References

1. D. L. Sackett, How to read clinical journals. I. Why to read them and how to start reading them critically, *CMA Journal* 124 (1981): 555–58.
2. S. H. Gehlbach, *Interpreting the Medical Literature—A Clinician's Guide* (New York: Macmillan, 1982), pp. 1–15.

References

1. A. R. Folsom and coauthors, Physical activity and incidence of coronary heart disease in middle-aged women and men, *Medicine & Science in Sports and Exercise* 29 (1997): 901–9.
2. T. D. Dye and coauthors, Unintended pregnancy and breast-feeding behavior, *American Journal of Public Health* 87 (1997): 1709–11.
3. J. Brug and coauthors, The relationship between self-efficacy, attitudes, intake compared to others, consumption, and stages of change related to fruit and vegetables, *American Journal of Health Promotion* 12 (1997): 25–30.
4. J. Lykkesfeldt and coauthors, Ascorbic acid and dehydro-ascorbic acid as biomarkers of oxidative stress caused by smoking, *American Journal of Clinical Nutrition* 65 (1997): 959–63.
5. Federation of American Societies for Experimental Biology, Life Sciences Research Office, prepared for the Interagency Board for Nutrition Monitoring and Related Research, *Third Report on Nutrition Monitoring in the United States*, Vol. 1 (Washington, D.C.: U.S. Government Printing Office, 1995), pp. 1–17.
6. E. Velempini and K. D. Travers, Accessibility of nutritious African foods for an adequate diet in Bulawayo, Zimbabwe, *Journal of Nutrition Education* 29 (1997): 120–27.
7. J. F. Pope and coauthors, Adolescents' self-reported motivations for dietary changes during pregnancy, *Journal of Nutrition Education* 29 (1997): 137–44.
8. D. Albanes and coauthors, Effects of supplemental β-carotene, cigarette smoking, and alcohol consumption on serum carotenoids in the Alpha-Tocopherol, Beta-Carotene Cancer Prevention Study, *American Journal of Clinical Nutrition* 66 (1997): 366–72.
9. M. Whitehead, Tackling inequalities: A review of policy initiatives, in *Tackling Inequalities in Health: An Agenda for Action*, eds. M. Benzeval, K. Judge, and M. Whitehead (London: King's Fund, 1995), pp. 22–52.
10. R. S. Kington and J. P. Smith, Socioeconomic status and racial and ethnic differences in functional status associated with chronic diseases, *American Journal of Public Health* 87 (1997): 805–10.
11. L. E. Montgomery and coauthors, The effects of poverty, race, and family structure on US children's health: Data from the NHIS, 1978 through 1980 and 1989 through 1991, *American Journal of Public Health* 86 (1996): 1401–5.
12. B. D. Weiss and coauthors, Health status of illiterate adults: Relation between literacy and health status among persons with low literacy skills, *Journal of the American Board of Family Practitioners* 5 (1992): 257–64.
13. R. Perez-Escamilla and coauthors, *Community Nutritional Problems Among Latino Children in Hartford, Connecticut* (Hartford: University of Connecticut and the Hispanic Health Council, 1997).
14. D. Cohen, *Consumer Behavior* (New York: Random House, 1991), pp. 76–127.
15. A. A. Davies-Adetugbo, Sociocultural factors and the promotion of exclusive breastfeeding in rural Yoruba communities of Osun State, Nigeria, *Social Science in Medicine* 45 (1997): 113–25.
16. W. Jackson, *Research Methods: Rules for Survey Design and*

Analysis (Scarborough: Prentice Hall Canada, 1988), pp. 28–33.

17. C. E. Woteki and coauthors, Selection of nutrition status indicators for field surveys: The NHANES III design, *Journal of Nutrition* 120 (1990): 1440–45.

18. J. H. Sabry, Purposes of food consumption studies, in M. E. Cameron and W. A. Van Staveren, *Manual on Methodology for Food Consumption Studies* (Oxford: Oxford University Press, 1988), pp. 25–31.

19. I. H. E. Rutishauser, Practical implementation, in Cameron and Van Staveren, *Manual on Methodology for Food Consumption Studies*, pp. 223–45.

20. V. J. Schoenbach, Appraising health risk appraisal [editorial], *American Journal of Public Health* 77 (1987): 409–11.

21. D. H. Gemson and R. P. Sloan, Efficacy of computerized health risk appraisal as part of a periodic health examination at the worksite, *American Journal of Health Promotion* 9 (1995): 462–66.

22. D. R. Anderson and M. J. Staufacker, The impact of worksite-based health risk appraisal on health-related outcomes: A review of the literature, *American Journal of Health Promotion* 10 (1996): 499–508.

23. L. Breslow and coauthors, Development of a health risk appraisal for the elderly (HRA-E), *American Journal of Health Promotion* 11 (1997): 337–43.

24. K. W. Smith and coauthors, The validity of health risk appraisal instruments for assessing coronary heart disease risk, *American Journal of Public Health* 77 (1987): 419–24.

25. J. J. Korelitz and coauthors, Health habits and risk factors among truck drivers visiting a health booth during a trucker trade show, *American Journal of Health Promotion* 8 (1993): 117–23.

26. Information about screening was adapted from J. M. Last, *Public Health and Human Ecology* (East Norwalk, CT: Appleton & Lange, 1987), pp. 12–15; and U.S. Department of Health and Human Services, National Institutes of Health, *Report of the Expert Panel on Population Strategies for Blood Cholesterol Reduction*, NIH Pub. No. 90-3046 (Bethesda, MD: National Institutes of Health, 1990), pp. 101–2.

27. A. Barrocas and coauthors, Appropriate and effective use of the NSI Checklist and Screens, *Journal of the American Dietetic Association* 95 (1995): 647–48; Nutrition Screening Initiative, *Implementing Nutrition Screening and Intervention Strategies* (Washington, D.C.: Nutrition Screening Initiative, 1993); and Nutrition Screening Initiative, *Incorporating Nutrition Screening and Interventions into Medical Practice*, Pub. No. A7241/6-94 (Washington, D.C.: Nutrition Screening Initiative, 1994).

28. E. R. Monsen, *Research: Successful Approaches* (Chicago: American Dietetic Association, 1992).

29. I. Seidman, *Interviewing as Qualitative Research: A Guide for Researchers in Education and the Social Sciences*, 2nd ed. (New York: Teachers College Press, 1998), pp. 63–78.

30. R. A. Krueger, *Focus Groups: A Practical Guide for Applied Research* (Thousand Oaks, CA: Sage, 1994).

31. C. E. Basch, Focus group interview: An underutilized research technique for improving theory and practice in health education, *Health Education Quarterly* 14 (1987): 411–48.

32. R. B. Masters and coauthors, The use of focus groups in the design of cholesterol education intervention programs, *American Journal of Health Promotion* 8 (1993): 95–97.

33. J. L. Kristeller and R. A. Hoerr, Physician attitudes toward managing obesity: Differences among six specialty groups, *Preventive Medicine* 26 (1997): 542–49.

34. C. S. Haignere and coauthors, One method for assessing HIV/AIDS peer-education programs, *Journal of Adolescent Health* 21 (1997): 76–79.

35. E. Helsing, On the clear need to meet and learn to speak clearly: Statement from the World Health Organization, *American Journal of Clinical Nutrition* 65 (suppl.) (1997): 1098S–99S.

36. R. S. Gibson, *Principles of Nutritional Assessment* (New York: Oxford University Press, 1990), pp. 3–20.

37. J. M. Karkeck, Improving the use of dietary survey methodology, *Journal of the American Dietetic Association* 87 (1987): 869–71.

38. The discussion of food consumption was adapted from Gibson, *Principles of Nutritional Assessment*, pp. 21–54.

39. H. R. H. Rockett and G. A. Colditz, Assessing diets of children and adolescents, *American Journal of Clinical Nutrition* 65 (suppl.) (1997): 1116S–22S.

40. R. S. Burke and coauthors, Nutrition studies during pregnancy, *American Journal of Obstetrics and Gynecology* 46 (1943): 38–52; and B. S. Burke and H. C. Stuart, A method of diet analysis, *Journal of Pediatrics* 12 (1938): 493–503.

41. The discussion of the validity and reliability of diet assessment methods was adapted from G. Block, A review of validations of dietary assessment methods, *American Journal of Epidemiology* 115 (1982): 492–505.

42. G. H. Beaton and coauthors, Sources of variance in 24-hour dietary recall data: Implications for nutrition study design and interpretation, *American Journal of Clinical Nutrition* 32 (1979): 2456–2559.

43. The discussion of the problems with the 24-hour dietary recall method was adapted from R.-L. Karvetti and L.-R. Knuts, Validity of the 24-hour dietary recall, *Journal of the American Dietetic Association* 85 (1985): 1437–42; S. M. Garn and coauthors, The problem with one-day dietary intakes, *Ecology of Food and Nutrition* 5 (1976): 245–47; and J. L. Forster and coauthors, Hypertension prevention trial: Do 24-h food records capture usual eating behavior in a dietary change study? *American Journal of Clinical Nutrition* 51 (1990): 253–557.

44. W. C. Willett and coauthors, Reproducibility and validity of a semiquantitative food frequency questionnaire, *American Journal of Epidemiology* 122 (1985): 51–65; and W. C. Willett and coauthors, Validation of a semi-quantitative food frequency questionnaire: Comparison with a 1-year diet record, *Journal of the American Dietetic Association* 87 (1987): 43–47.

45. R. J. Coates and C. P. Monteilh, Assessments of food-frequency questionnaires in minority populations, *American Journal of Clinical Nutrition* 65 (suppl.) (1997): 1108S–15S.

46. L. Sampson, Food frequency questionnaires as a research instrument, *Clinical Nutrition* 4 (1985): 171–78; S. N. Zulkifli and S. M. Yu, The food frequency method for dietary assessment, *Journal of the American Dietetic Association* 92 (1992); 681–85; R. R. Briefel and coauthors, Assessing the nation's diet: Limitations of the food frequency questionnaire, *Journal of the American Dietetic Association* 92 (1992): 959–62; and G. Block and A. F. Subar, Estimates of nutrient intake from a food frequency questionnaire: The 1987 National Health Interview Survey, *Journal of the American Dietetic Association* 92 (1992): 969–77.

47. G. P. Sevenhuysen and L. A. Wadsworth, Food image processing: A potential method for epidemiological surveys, *Nutrition Reports International* 39 (1989): 439–50; and T. A. Fox and coauthors, Telephone surveys as a method for obtaining dietary information: A review, *Journal of the American Dietetic Association* 92 (1992): 729–32.

48. S. K. Kumanyika and coauthors, Dietary assessment using a picture-sort approach, *American Journal of Clinical Nutrition* 65 (suppl.) (1997): 1123S–29S.

49. L. Kohlmeier and coauthors, Computer-assisted self-interviewing: A multimedia approach to dietary assessment, *American Journal of Clinical Nutrition* 65 (suppl.) (1997): 1275S–81S.

50. Gibson, *Principles of Nutritional Assessment*, pp. 292–97.

51. Ibid., pp. 178–79.

52. Ibid., pp. 155–62.

53. M. Nestle, *Nutrition in Clinical Practice* (Greenbrae, CA: Jones Medical Publications, 1987), pp. 64–65.

54. The discussion of sensitivity and specificity was taken from R. H. Fletcher and coauthors, *Clinical Epidemiology—The Essentials* (Baltimore, MD: Williams & Wilkins, 1982), pp. 46–48.

55. G. Block and A. M. Hartman, Issues in reproducibility and validity of dietary studies, *American Journal of Clinical Nutrition* 50 (1989): 1133–38.

56. C. A. Woodward and L. W. Chambers, *Guide to Questionnaire Construction and Question Writing* (Ontario: Canadian Public Health Association, 1983), pp. 1–33.

57. W. Willett, *Nutritional Epidemiology* (New York: Oxford University Press, 1990), pp. 3–51.

58. P. P. Basiotis and coauthors, Number of days of food intake records required to estimate individual and group nutrient intakes with defined confidence, *Journal of Nutrition* 117 (1987): 1638–41.

59. M. C. Narayan, Cultural assessment in home healthcare, *Home Healthcare Nurse* 15 (1997): 663–70.

60. T. Barer-Stein, *You Eat What You Are: A Study of Ethnic Food Traditions* (Toronto: McClelland & Stewart, 1979), pp. 13–23.

61. K. D. Travers, Using qualitative research to understand the sociocultural origins of diabetes among Cape Breton Mi'kmaq, in *Chronic Diseases in Canada* (Ottawa: Health Canada, 1995), pp. 140–43.

62. J.-P. Habicht and D. L. Pelletier, The importance of context in choosing nutritional indicators, *Journal of Nutrition* 120 (1990): 1519–24.

63. Nutrition Screening Initiative, *Keeping Older Americans Healthy at Home* (Washington, D.C.: Greer, Margolis, Mitchell, Burns & Associates, 1996).

64. S. A. Anderson, Core indicators of nutritional state for difficult-to-sample populations, *Journal of Nutrition* 120 (1990): 1559–600.

65. M. Stouthamer-Loeber and W. B. van Kammen, *Data Collection and Management: A Practical Guide* (Thousand Oaks, CA: Sage, 1995), pp. 114–18.

66. The discussion of reference data was adapted from Gibson, *Principles of Nutritional Assessment*, pp. 209–46 and 349–76; and F. E. Johnston and Z. Ouyang, Choosing appropriate reference data for the anthropometric assessment of nutritional status, in *Anthropometric Assessment of Nutritional Status*, ed. J. H. Himes (New York: Wiley, 1991), pp. 337–46.

67. Nutrition Screening Initiative, *Keeping Older Americans Healthy at Home*.

68. Food and Nutrition Board, Dietary Reference Intakes (DRIs) for calcium, phosphorus, magnesium, vitamin D, and fluoride, *Nutrition Today* 32 (1997): 182–88; Food and Nutrition Board, Institute of Medicine, *Dietary Reference Intakes for Thiamin, Riboflavin, Niacin, Vitamin B6, Folate, Vitamin B12, Pantothenic Acid, Biotin, and Choline* (prepublication copy) (Washington, D.C.: National Academy Press, 1998); and Food and Nutrition Board, National Research Council, *Recommended Dietary Allowances*, 10th ed. (Washington, D.C.: National Academy Press, 1989).

69. T. K. Murray and J. L. Beare-Rogers, Nutrition recommendations, 1990, *Journal of the Canadian Dietetic Association* 51 (1990): 391–93.

70. Gibson, *Principles of Nutritional Assessment*, pp. 137–53.

Program Planning for Success

Outline

Learning Objectives

After you have read and studied this chapter, you will be able to:

- Describe six factors that can trigger program planning.
- Describe seven steps in designing, implementing, and evaluating nutrition programs.
- Discuss three reasons for conducting evaluations of programs.
- Discuss three major principles to consider when preparing an evaluation report.

Something To Think About...

One should get as many new ideas as possible. – *Isabel Archer in Henry James's*
The Portrait of a Lady

Introduction

In the early 1990s, the state of Arizona foresaw a gap in services. Data obtained from the 1988 National Health Interview Survey on Child Health showed that at least two-thirds of children between the ages of 1 week and 5 years were placed in child care. Some of these children had special health care needs. Arizona's own data predicted that by the year 2000, about 24,000 children would have developmental delays and other special needs in Arizona, and many of these children would participate in child care programs. The problem was that available staff training for child care workers on providing quality nutrition services was limited.

The Arizona Department of Health Services was determined to find a solution to the problem. It organized a team to develop a course that would help child care workers be more aware of the challenges involved in feeding children with special health care needs.[1] The team was composed of health professionals with expertise in pediatric medicine, occupational therapy, speech therapy, child care programs, and community nutrition. A parent of a child with special needs was included on the team. The Arizona Department of Health Services, Office of Nutrition Services, obtained funding for the project through a Maternal and Child Health Improvement Project Field Grant and linked with the Arizona Self-Study Project and the Office of Women's and Children's Health to develop the program.

The community nutritionists assigned to the project had a leadership role in designing and evaluating the course. They reviewed research studies and other technical material related to feeding children with special needs, determined who would use the course materials, assessed the knowledge and skill levels of child care workers, and determined the types of nutritional problems child care workers were likely to encounter in day care and other settings. They evaluated the course and made adjustments in course materials. The end product was Project CHANCE, a course designed and tested by health professionals to meet the training needs of child care workers.[2]

This chapter presents an overview of program planning, the process of designing a program to meet a nutritional need or fill a gap in services. It describes the process of program evaluation, explaining why evaluation is important, who conducts the evaluation, how evaluation findings are used, and how to prepare an evaluation report. Some concepts such as the types of evaluation are introduced in this chapter and elaborated on in the chapters that follow. Topics such as developing the intervention and marketing plan, choosing nutrition messages, managing personnel and data, and identifying funding sources are discussed in the remaining chapters of Section II.

Factors That Trigger Program Planning

The decision to develop a nutrition program or modify an existing program is usually made in response to some background event. Perhaps the community needs assessment revealed that some elderly people were not taking advantage of the community's Meals-on-Wheels program or that a large number of children living in an inner-city neighborhood had low intakes of vitamins A and C. In other cases, as shown in Figure 7-1, the stimulus for program planning may be a mandate handed down from an organization's national office. For example, when the national office of the American Heart Association (AHA) chose nutrition and physical fitness as the organization's national health promotion initiatives for 1 year, state AHA offices then determined whether their existing programs met the organization's mandate in these areas.

Research findings sometimes trigger the planning process. The report of the National Cholesterol Education Program on the detection, evaluation, and treatment of high blood cholesterol in adults, published in 1988, described the findings of major studies that linked high blood cholesterol levels to coronary heart disease (CHD) risk. A subsequent 1990 report described population strategies for reducing blood total cholesterol. These reports led some hospitals and municipal health departments to review their programs for reducing CHD risk within their communities and, in some cases, to develop new programs to promote cardiovascular health.[3] Findings from the Bogalusa Heart Study, an epidemiological study begun in Louisiana in 1974, led to the development of the Heart Smart Family Health Promotion program, a school-based program that targeted high-risk elementary schoolchildren and their families. The program was designed to involve parents in improving the eating and activity patterns of schoolchildren.[4]

FIGURE 7-1 *Factors That Trigger Program Planning*

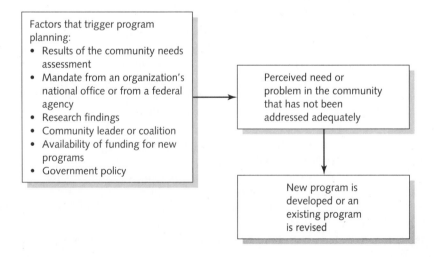

The concerns of a well-known community leader or coalition may stimulate program planning, as when a community activist helps state agencies plan alcohol and drug treatment programs under the new welfare reform law. Government policy and the availability of new funding can also stimulate program planning. When the U.S. Department of Agriculture increased its funding for the WIC Farmers Market Program, the National Commission on Small Farms called for expansion of the program to every state—an action that will likely motivate states where the program has not been implemented to offer the program to eligible mothers and children.[5] Regardless of the impetus, the community nutritionist considers developing a program when there is a nutritional or health problem in the community that has not been resolved adequately.

Steps in Program Planning

The community nutritionist reviews her organization's mission statement before developing or modifying a program. A **mission statement** is a broad declaration of the organization's purpose and a guideline for future decisions; it provides an identity that says, "This is what this organization is all about."[6] The community nutritionist ensures that all programs fulfill her organization's mandate. If the match between the mission statement and the program concept is good, then she has a reasonable level of confidence about gaining senior management's support for it. If the match is not good, it will be difficult to justify the resources, time, and expense of a new program and to secure funding for it. Consider, for example, the FirstRate Spa and Health Resort, mentioned in Chapter 1 (see Table 1-5 on page 21). Its mission statement reads, "The FirstRate Spa and Health Resort works to enhance the health and fitness of its members by promoting physical activity, healthy eating, and self care." The director of health promotion will have difficulty obtaining internal support for a new program whose participants are not spa clients.

> **Mission statement** A broad statement or declaration of an organization's purpose or reason for being.

The program planning process consists of several steps, as shown in Figure 7-2. The first step is a review of the results of the community needs assessment. In step 2, goals and objectives that specify the expected outcomes of the program are defined. In step 3, a program plan that describes the intervention, the appropriate nutrition messages for the target population, and how the program will be marketed is developed. In steps 4 and 5, decisions are made about the management system, budgeting, and potential funding sources for program activities. The program is implemented in step 6 and evaluated in step 7. Evaluation focuses on program elements such as the nutrition education and marketing materials and the program's effectiveness—in other words, evaluation activities are designed to determine whether the program accomplished what it was designed to accomplish. Finally, after the program's effectiveness has been verified, colleagues, community leaders, and the community at large are notified of its success.

Step 1: Review the Results of the Community Needs Assessment

The community needs assessment provides information about the target population's nutritional problem or need. It is a major impetus for program planning.

FIGURE 7-2 *Steps in Program Planning*

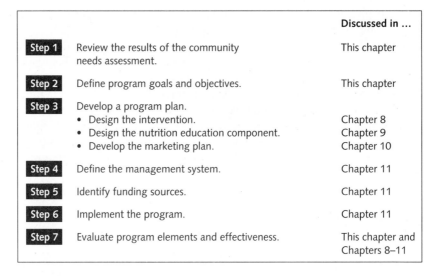

		Discussed in ...
Step 1	Review the results of the community needs assessment.	This chapter
Step 2	Define program goals and objectives.	This chapter
Step 3	Develop a program plan. • Design the intervention. • Design the nutrition education component. • Develop the marketing plan.	Chapter 8 Chapter 9 Chapter 10
Step 4	Define the management system.	Chapter 11
Step 5	Identify funding sources.	Chapter 11
Step 6	Implement the program.	Chapter 11
Step 7	Evaluate program elements and effectiveness.	This chapter and Chapters 8–11

Consider the needs assessment undertaken by Project MANA, a program that provides emergency food relief to Hispanics, Caucasians, and migrant workers in the Incline Village area of Nevada. Project MANA staff were alerted to a problem with the Thanksgiving basket's food items, which traditionally included turkey, stuffing, and cranberry sauce. Some Spanish-speaking families were not familiar with cranberry sauce and did not know how to cook with it. The outcome of the needs assessment was to change an existing program so that families were offered a choice, the traditional Thanksgiving basket or a new basket containing chicken, salsa, and tortillas.[7]

When the community assessment identifies a gap in services, a new program may be developed to fill the gap. A needs assessment of the food and nutrition situation of low-income Hispanic children living in inner-city Hartford, Connecticut, found that about 53 percent of caregivers did not breastfeed their infants. The assessment report called for the development of culturally sensitive campaigns that promote breastfeeding and inform caregivers about the appropriate times for introducing weaning foods in this population. "Lactancia, Herencia y Orgullo" ("Breastfeeding, Heritage and Pride"), an education program based at the Hispanic Health Council, was designed to increase breastfeeding among low-income Hispanic women.[8]

Some results of the community needs assessments for Case Studies 1 and 2, first described in Chapter 5, are shown in Tables 7-1 and 7-2. The tables summarize key findings of the assessments and indicate areas where interventions are needed. (The tables are not meant to be comprehensive, and other information could have been presented.) The results of the needs assessment for Case Study 1 indicate a low level of awareness of CHD risk factors among women and a typical pattern of low activity levels and high dietary fat intakes—factors that contribute to CHD risk and could serve as points for intervention. The results for Case Study 2 reveal several nutritional problems among the city's elderly population. Note the language and organization of the summaries. They were written for a broad audience,

Results of Needs Assessment

The city of Jeffers and its four adjoining municipalities have a population of 612,000. The city's economic base is light manufacturing and service industries. Many residents have moved from traditional ethnic neighborhoods near downtown to the independent "bedroom" communities outside the city proper.

It is an ethnically diverse city. A majority (46 percent) of the population is non-Hispanic white, but there are several major ethnic groups: blacks (22 percent), Hispanics (17 percent), Portuguese (7 percent), and Asian (6 percent).

A survey of women aged 18–72 years, conducted at five medical clinics in the major metropolitan area, found that about two-thirds (64 percent) do not know that CHD is the leading cause of death among women. Three out of four cannot identify two major CHD risk factors. About one-third (32 percent) of women smoke cigarettes—a figure higher than the state average of 27 percent.

Although women claim to be eating a low-fat diet and exercising regularly, 38 percent of all women surveyed are overweight. This figure is similar to the state average of 36 percent, but higher than the national average of 34 percent. About 50 percent of black women and 45 percent of Hispanic women are overweight.

Women's mean intake of fat is 35 percent of total energy. This intake is above the recommended intake level of 30 percent. (Women's mean intake of fat reported in the Continuing Survey of Food Intakes of Individuals [CSFII], 1996, was 31.2–33.7 percent of total energy.)

Weight management programs and classes are available through private health/fitness clubs and dietitians in private practice. Nearly two-thirds (64 percent) of women report that they are trying to lose weight, but only 8 percent of women participate in organized classes and programs.

The local affiliate of the heart association offers some programming for the general public. Brochures describing the leading risk factors for CHD are available at a majority of hospitals and medical clinics. Only 14 percent of women were aware that some local restaurants feature "heart smart" meals on their menus.

Women indicate that they want to do more to reduce their CHD risk and improve their health. No current programs are designed specifically to help women reduce their CHD risk.

TABLE 7-1 *Results of the Community Needs Assessment for Case Study 1: Women and Coronary Heart Disease (CHD)*

which will likely include policy makers, community leaders, and the general public.

Step 2: Define Program Goals and Objectives

The next step in the program planning process is to define the program goals and objectives. Goals are broad statements of desired changes or outcomes. They provide a general direction for the program. Objectives are specific, measurable actions to be completed within a specified time frame. An objective has four components: (1) the action or activity to be undertaken, (2) the target population, (3) an indication of how success will be measured or evaluated, and (4) the time frame in which the objective will be met. There are three types of objectives, as listed here:[9]

- **Outcome objectives**—These are measurable changes in a health or nutritional outcome, such as an increase in knowledge of folate-rich food sources, a decrease in blood total cholesterol concentration, an increase in serum ferritin concentration, or a change in functional status. An outcome objective might state,

TABLE 7-2 *Results of the Community Needs Assessment for Case Study 2: Nutritional Status of Independent Elderly Persons (> 75 years)*

Results of Needs Assessment

The city of Jeffers has a population of 434,000. This is 71 percent of the total metropolitan area population (that is Jeffers plus its four adjoining municipalities). The number of elderly persons > 75 years living in the city of Jeffers is 60,760 (14 percent), an increase of 12 percent since 1995. More than half (56 percent) of these elderly people are living independently, typically in their own homes or apartments.

The overall nutritional status of the 34,025 independent elderly is fairly good, but several key nutritional problems emerged:

- About 2 percent or 680 elderly persons, mostly black and Hispanic women, had serum vitamin A concentrations < 20 μg/dL. At these low levels, impairment of immune function and dark adaptation and development of ocular lesions is likely.

- Twenty percent of white women and 38 percent of white men had low hemoglobin concentrations. Among non-Hispanic black men, 62 percent had low hemoglobin levels. (The Centers for Disease Control and Prevention criteria for low hemoglobin is < 13.5 g/dL in men and < 12.0 g/dL in women.)

- One in four elderly men and one in five elderly women had serum LDL-cholesterol concentrations ≥ 4.14 mmol/L.

- About two-thirds (64 percent) of independent elderly persons have hypertension.

- More than half (56 percent) reported using laxatives for the relief of constipation.

- About 12 percent of independent elderly persons reported having difficulty performing two or more personal care activities (for example, bathing, dressing, using the toilet, getting in and out of bed or chair, eating).

Fewer than one in five (18 percent) use community-based services such as personal care services, homemaker services, and adult day care.

The number of independent elderly persons living in Jeffers who participate in federal assistance programs is shown here:

Program	Number (%) of Participants
Food Stamps	3,240 (32%)
Social Security	8,912 (88%)
Medicare/Medicaid	7,190 (71%)
Supplemental Security Income	1,215 (12%)
Veterans Benefits	810 (8%)

Delivery of health care and nutritional services has traditionally occurred through the state Department of Social Services.

"By the year 2000, decrease salt and sodium intake so at least 65 percent of home meal preparers prepare foods without adding salt" or "By 2000, reduce iron deficiency to less than 10 percent among low-income children under 2 years."[10]

- **Process objectives**—These are measurable activities carried out by the community nutritionist and other team members in implementing the program. They specify the manner in which the outcome objectives will be achieved. A process objective might state, "Each community nutritionist will conduct two nutrition lectures per week over the course of the 3-month program."

- **Structure objectives**—These are measurable activities surrounding the budget, staffing patterns, management systems, use of the organization's resources, and coordination of program activities. A structure objective might read, "Each community nutritionist will submit an itemized statement of expenses related to conducting the program on the last day of each month for the next 12 months."

In this chapter, we are mainly interested in outcome objectives. (Process and structure objectives will be described in more detail in Chapter 11.) The community nutritionists who participated in Case Study 1 developed two broad goals and several outcome objectives, listed here, for a program designed to help women in the city of Jeffers reduce their CHD risk:

Case Study 1: Women and CHD Risk

Goals	Outcome Objectives
Reduce death from CHD among women	1. Decrease the number of CHD deaths among women to a rate of no more than 115 per 100,000 within 5 years. 2. Decrease the number of women who smoke by 33 percent within 2 years. 3. Decrease the mean fat intake of women by 3 percent (from their current intake of 35 percent of total energy) within 1 year.
Increase women's awareness of CHD risk factors.	1. Increase the number of women who can specify CHD as the leading cause of death in the city of Jeffers by 50 percent within 1 year. 2. Increase the number of women who can specify two major CHD risk factors by 25 percent within 1 year.

These goals and objectives are the framework around which the program elements such as the type of intervention, the nutrition messages, and the marketing campaign are built. The goals and outcome objectives are the basis for determining whether the program was effective—that is, whether the program was successful in raising women's awareness of CHD risk factors and in reducing death from CHD among women.[11]

Step 3: Develop a Program Plan

Using the goals and objectives as a guide, the community nutritionist develops a program plan, which consists of a description of the proposed intervention, the nutrition education component, and the marketing plan. Other factors may also be described in the program plan: the number of clients expected to use the program; the staff, equipment, and material resources required to administer the program; the facilities available for staff offices and teaching rooms; and the level of staff training required prior to implementing the program. The community nutritionist plans how the program will work after asking many questions: What criteria determine a client's eligibility for the program? What federal, state, or local regulations must be considered in administering the program? What educational materials are needed to convey important nutrition messages? What provisions should be made for follow-up? Who will be responsible for implementing the program and determining its effectiveness? How much money is required to administer the program? Well-designed programs are scrutinized before being made available to the target population.

The program plan is usually developed after reviewing existing programs and talking to colleagues and other professionals who have worked with similar programs or with the target population. One easy way to network with colleagues and stay informed about new programs and services is to join one or more listservs. Listservs are types of Internet mailing lists to which people subscribe, much like a magazine or newspaper.[12] They are convenient, electronic bulletin or message boards for exchanging ideas and information. Some popular listservs for community nutritionists are given in the boxed insert.

Step 4: Develop a Management System

In this context, management refers to two types of structures needed to implement the program: personnel and data systems. The personnel structure refers to the employees responsible for overseeing the program and determining whether it meets its objectives. The structure of the data management system refers to the manner in which data about clients, their use of the program, and the outcome measures are recorded and analyzed.

An important part of the program planning is calculating the "management" costs of the program. Both direct costs—such as the salaries and wages of program personnel, materials needed, travel expenses, and equipment—and indirect costs—such as office space rental, utilities, and janitorial services—must be determined to identify the true cost of a program.

Step 5: Identify Funding Sources

Community nutritionists in nonprofit organizations and government agencies face many challenges in securing funding for all aspects of a program. Money may be available in the current year's budget for staff time to develop the program's format, choose nutrition education messages, and plan a marketing campaign, but there may not be sufficient money to print educational materials or for personnel to pretest a survey instrument. At this point in the program planning process, the community nutritionist reviews the program elements (for example, educational materials, marketing campaign) and considers whether outside funding in the form of cash grants or in-kind contributions from partners can be found. He identifies the area where financial support is needed, reviews possible funding sources, and prepares and submits a grant application for funding. The grant-writing process requires him to be clear about the purpose of his funding request, have a specific need, and be able to demonstrate how the funds will be used to enhance the program's effectiveness.

Step 6: Implement the Program

Implementation The set of activities directed toward putting a program into effect.

This is the action phase of the program planning process. **Implementation** refers to "the set of activities directed toward putting a program into effect."[13] The format of the program has been finalized, educational materials printed, the marketing plan prepared, and staff trained. Now it is time to put the program into operation, to link the program goals with the plan of action.

Listservs for Community Nutritionists

For General Information About Listservs

SparkLIST   **sparklist.com**

Tile.Net   **tile.net/lists**

Listservs for Community Nutritionists

Dietetics-L (the listserv available to members of the American Dietetic Association [ADA])

To subscribe:
1. Send an E-mail message to ncndlib@eatright.org
2. Use this subject line: Subscribe Dietetics-L
3. In the message area, type:

> subscribe Dietetics-L
> Your member number (this is your six-digit member number for the ADA)
> Your E-mail address (for example, dmorris@pangea.ca)

Dietetics Online Listserv ©

To subscribe:
1. Send an E-mail message to majordomo@empnet.com
2. In the message area, type:
   > subscribe dietetics-online

Public Health Nutrition Listserv

To subscribe:
1. Send an e-mail message to: listproc@u.washington.edu
2. In the message area, type: subscribe phnutr-1<your name>
   > NOTE: Omit the "<>" marks on the preceding line. The <your name> feature is
   > the name by which people know you.

Purdue Food and Nutrition Listserv

To subscribe:
1. Send an E-mail message to almanac@ecn.purdue.edu
2. In the message area, type:
   > set address <your_E-mail_address> (for instance, set address dmorris@pangea.ca)
   > Subscribe fnspec_mg

Implementing the program as conceived is challenging, and glitches in program delivery are inevitable. Perhaps no one on the team thought about modifying handouts for Portuguese-speaking clients who participate in the program, or no one was aware of cultural barriers to teenage girls participating in an inner city's after-school fitness program. The key to successful implementation is to observe all aspects of program delivery and consider ways in which delivery can be improved.

EDUCATED GUESS

Cartoon by John Chase.

Step 7: Evaluate Program Elements and Effectiveness

Evaluation The measurable determination of the value or degree of success in achieving specific objectives.

Evaluation refers to the use of scientific methods to judge and improve the planning, monitoring, effectiveness, and efficiency of health, nutrition, and other human service programs. The purpose of program evaluation is to gather information for making decisions about redistributing resources, changing program delivery, or continuing a program.[14] It takes the guesswork out of planning and implementing programs and occurs throughout the program planning process. The next sections describe the purposes and uses of evaluations.

However beautiful the strategy, you should occasionally look at the results.
– Winston Churchill

Why Evaluation Is Necessary. Although the immediate purpose of program evaluation is to help managers make decisions about the short- and long-term operation of their programs, evaluations also serve to inform the community at large about a program's success or failure. When a community-based nutrition program succeeds, nutritionists in other locations across the country want to know how the lessons can be used in their communities. Likewise, when a program fails, community nutritionists want to examine its flaws and figure out how to avoid them in the future.

Program evaluations force community nutritionists to determine whether they are progressing toward their initial goals and whether these goals are still appropriate. Evaluations may be used for administrative purposes: to determine whether some elements of a program should be changed, to identify ways in which interventions can be improved, to pinpoint weaknesses in program content, to meet certain accountability requirements of the funding agency or senior management, to ensure that program resources (such as supplies, equipment, personnel, facilities) are being used properly, or to assess cost-benefit factors. They may be undertaken to test innovative approaches to a nutrition or public health problem, to meet policy or planning purposes, or to support the advocacy of one program over another. Finally, evaluations may be undertaken to determine whether objectives have been met or whether priorities need to be changed.[15] The finding that a program is not accomplishing its objectives signals a need to consider whether the program is worthwhile or whether its goals can be accomplished in some other fashion.[16] Consult Table 7-3 for a list of reasons for undertaking evaluations.[17]

TABLE 7-3 *Reasons for*
Undertaking Evaluations

Evaluation to Improve Your Program

- To improve methods of placing clients in various activity programs
- To measure the effect of your program or the extent of client progress in your program
- To assess the adequacy of program goals
- To identify weaknesses in the program content
- To measure staff effectiveness
- To identify effective instructional, leadership, or facilitation techniques
- To measure the effectiveness of resources (such as materials, supplies, equipment, or facilities)

Evaluation to Justify Your Program or to Show Accountability

- To justify the budget or expenditures
- To show the need for increased funds
- To justify staff, resources, facilities, etc.
- To justify program goals and procedures
- To account for program practices
- To compare program outcomes against program standards

Evaluation to Document Your Program in General

- To record client attendance and progress
- To document the nature of client involvement and interaction
- To record data on clients who drop out and those who complete the program
- To document the major program accomplishments
- To list program weaknesses expressed by staff or others
- To list leader/therapist functions and activities
- To describe the context or atmosphere of the treatment setting
- To file supportive statements and testimonies about the program

Source: A. D. Grotelueschen and coauthors, *An Evaluation Planner* (Urbana: Office for the Study of Continuing Professional Education, University of Illinois at Urbana-Champaign, 1974). Permission to reprint granted by Arden D. Grotelueschen.

How Evaluation Findings Are Used. Evaluation findings have many uses. Sometimes they are used to influence an executive or politician who has the authority to distribute resources and shape public policy. For example, in preparing its position on Medical Nutrition Therapy, the American Dietetic Association (ADA) reviewed the literature on the economic benefits of nutrition services in acute, outpatient, home, long-term, and preventive care and in the care of pregnant women, infants, children, and older adults. Its evaluation of the impact of nutrition services led to the development of a platform on the benefits of nutrition services in health care. The platform was used in the ADA's grassroots lobbying campaign of members of Congress and state legislatures for Medical Nutrition Therapy.[18]

Evaluation findings alert managers and policy makers to the need for expanding or refining programs. For instance, in the United States, the Food Research and Action Center conducts an annual evaluation of the efforts of each state and the District of Columbia to provide nutritious summer meals to children of low-income families. Evidence of a decline in a state's participation is a signal for state action to develop innovative ways of increasing children's participation in the program.

Generally, evaluation findings are used at two levels. Not only are they applied to an immediate problem and hence used by managers and program staff who are focused on that problem, but they are also used to shape policies and services beyond the scope of the original problem. In other words, evaluations have many different audiences, some of whom may be directly involved in the program, while others are not involved at all or may be concerned with the program at some future date.

Who Conducts the Evaluation. Evaluations may be carried out by program staff, other agency staff, or outside consultants. The evaluator may be intimately familiar with all aspects of a program, because she manages or is involved with it, or she may have a limited knowledge of the program. Regardless, the evaluator is responsible for all aspects of the evaluation from negotiating the evaluation focus to collecting data and preparing the final report. Because evaluations often occur in politically charged atmospheres, where program stakeholders fret about the evaluation's outcome and its ramifications, the evaluator must be sensitive to this environment. In the final analysis, the evaluator must be able to recognize what has to be done and remain objective about the evaluation and its findings.[19]

The Program Evaluation Process. The purpose and scope of an evaluation depend on the questions being asked about the program. An evaluation may focus on one particular program element—for example, determining whether screening every client for high blood cholesterol concentration is cost-effective or whether a significant portion of the target population is aware of the nutrition messages appearing in posters placed in city buses. Or the evaluation may be comprehensive and examine the design of the program, how it is delivered, and whether it is being used properly. Evaluation occurs across all areas of program planning, from design to implementation, and there is no one method for carrying out an evaluation, as each must be tailored to the organization or department in which it is conducted.

When evaluating their programs, community nutritionists begin by asking the following questions:[20]

- Did the intervention reach the target population?
- For which participants was the program most effective?
- For which participants was the program least effective?
- Was the intervention implemented according to the original program plan?
- Was the program effective—that is, did it accomplish what it was supposed to accomplish?
- How much does the program cost?
- What are the program's costs relative to its effectiveness and benefits?

The answers to these and other questions help the community nutritionist design a better, more effective program and formulate recommendations for colleagues and community leaders. Recommendations can be made about the suitability of program goals and objectives, whether the objectives should be changed, whether the program can be applied successfully in another setting or among a different group of participants, and how the program should be changed to improve delivery.[21]

Communicating Evaluation Findings. Because the primary purpose of evaluations is to provide information for decision making, you do not want your findings to go unused. If they are stored away in a filing cabinet or dumped into the "circular file," a program may continue missing the mark or a success story may go unnoticed. With careful planning and work, you can ensure that the main findings of your evaluation get the attention they deserve.

Even as you begin the actual evaluation, you should be thinking about the final report! As the evaluation progresses, make notes of how problems were handled and which documents or materials were used. Retain copies of survey instruments and computer printouts as reference items or for use in the report's appendix. When you begin preparing your report, keep these three rules in mind:[22]

- Communicate the information to the appropriate potential users.
- Ensure that the report addresses the issues that the users perceive to be important.
- Be sure that the report is delivered in time to be useful and in a form that the intended users can easily understand.

Program planning is best done in teams composed of people with different skill sets, ideas, opinions, and perspectives. The more extensive the program planning, the greater the chance the program will succeed.

With these issues in mind, prepare your report, which may be either informal (for example, a short memorandum) or formal (such as a full report). Even if the report is only a three-page memorandum, it should be concise and understandable and give the user what he needs to make a decision. Our focus in this discussion is on the formal report, which tends to have a particular organization, as outlined here:[23]

1. **Front cover.** The front cover should provide (a) the title of the program and its location, (b) the name of the evaluator(s), (c) the period covered by the report, and (d) the date the report was submitted. The front cover should be neat and formatted attractively.
2. **Summary.** Sometimes called the executive summary, this section of the report is a brief overview of the evaluation, explaining why and how it was done and listing its major conclusions and recommendations. Typically, this section is prepared for the individual who does not have time to read the full report. Therefore, it should not be longer than one or two pages. Even though the summary appears first in the report, it should be written *last*.
3. **Background information.** This section places the program in context, describing what the program was designed to do and how it began. The amount of detail provided in this section will depend on the needs and knowledge of the users. If most readers are unfamiliar with the program, it should be described in some detail; if most readers are involved with the program, this section can be kept short. A typical outline for this section might include the following:

   - Origin of the program
   - Goals of the program
   - The program's target population
   - Characteristics of the program materials, activities, and administrative procedures
   - Staff involved with the program

4. **Description of the evaluation.** This section states the purpose of the evaluation, including why it was conducted and what it was intended to accomplish. Here you define the scope of the evaluation and describe how it was carried out. This section establishes the credibility of the evaluator and the evaluation findings (much like the Methods section of a research paper). Technical information about the evaluation design and analysis is presented here. Technical language should be kept to a minimum, however. Refer readers to appendices for specific technical information and copies of any instruments used in the evaluation study. A general outline for this section might include the following headings:

   - Purposes of the evaluation
   - Evaluation design
   - Outcome measures
       Instruments used
       Data collection and analysis procedures
   - Process measures
       Instruments used
       Data collection and analysis procedures

5. **Results.** This section presents the results of the outcome or process evaluation. It is appropriate to present data or summarize findings in tables, figures, graphs, or charts. Before you begin writing this section, you should have already analyzed the data, tested for statistical significance (if appropriate), and prepared the tables, figures, and other illustrations.

6. **Discussion of results.** The results of the evaluation study are interpreted in this section, which should address two key issues: How certain is it that the program caused the results? How good were the results? The Results section explores some of the reasons why a certain outcome was reached and how the program compares to similar programs. Any strengths and weaknesses of the program are described here.

7. **Conclusions, recommendations, and options.** This section is an influential part of the report, as it outlines the major conclusions that can be drawn from the evaluation and lists a suggested course of action for enhancing the program's strengths and dealing with its flaws. The recommendations should address specific aspects of the program and follow logically from your interpretation of the evaluation findings. Preparing a list of recommendations about the program's delivery or impact is especially important when the actual results differ from the predetermined objectives.

Once your report is written, you must decide how best to distribute it. Several options are available. You may send the full report to your immediate supervisor, division director, and board of directors and a copy of the executive summary to interested community groups. You may inform the media and general public by distributing a press release and posting key findings on the organization's Web site. In some cases, the strategies for publishing the evaluation findings will have been specified upfront by the primary client; if not, you may suggest the formats that best communicate the findings to various audiences. Figure 7-3 shows how the findings of the evaluation study might be distributed.

Audience/Users	Technical Report	Executive Summary	Technical Professional Paper	Popular Article	News Release, Press Conference	Public Meeting	Media Appearance	Staff Workshop	Brochure	Memorandum	Personal Presentation
Possible Communication Form											
Program Administrators	✔	✔	✔	✔	✔			✔		✔	✔
Board Members, Trustees, Other Management Staff		✔		✔							
Advisory Committees	✔	✔	✔								
Funding Agencies	✔	✔									✔
Community Groups		✔		✔		✔					
Current Clients				✔		✔	✔				
Potential Clients							✔				
Political Bodies (e.g. City Councils, Legislatures)		✔		✔							
Program Service Providers (e.g. Nutritionists, Health Educators)		✔		✔				✔	✔	✔	✔
Organizations Interested in the Program Content				✔	✔						
Media					✔	✔					

The Challenge of Multicultural Evaluation. Multiculturalism poses some unique and difficult problems for program evaluation. Conducting a fair and democratic evaluation in a multicultural environment requires striking a balance between the rights of minority culture groups and the rights of the larger culture—a complex policy issue.

What does multiculturalism mean for the evaluator? First, the evaluator must strive to remain neutral in the face of competing minority interests. This is especially true when stakeholders have strong views about the evaluation outcome, try to influence the outcome, or downplay the possible contribution of the evaluation process.[24] Second, the evaluator must search out and define the views and interests of the minority groups to ensure that their needs are being met. When the minority group is defined as "poor" or "powerless," the evaluator has an obligation to recognize the views and interests of this group.

Finally, the evaluator must be sensitive to the cultural differences that make implementing the evaluation difficult. The manner in which questions are asked, or

FIGURE 7-3 *Forms of Communicating Evaluation Findings to Various Audiences*

Source: L. L. Morris and coauthors, *How to Communicate Evaluation Findings*, pp. 9–10, copyright © 1987 by Sage Publications. Reprinted by permission of Sage Publications, Inc.

the questions themselves, may be barriers to obtaining reliable data about the program's impact. Muslims, for example, are not comfortable answering personal questions about health and diet from a member of the opposite sex.[25] The Inuit of Nunavik, Quebec, are reluctant to answer questions about diet because they believe that thinking or talking too much about beluga whales and geese—traditional foods harvested by the Inuit—may result in their disappearance.[26] Perhaps the best message about multiculturalism was given by E. R. House at the University of Colorado at Boulder: "Treat minority cultures as you would be treated. Sooner or later, everyone may be a minority."[27]

Spreading the Word About the Program's Success

In her book *The Popcorn Report*, Faith Popcorn discusses the leading consumer trends of the decade and describes the importance of developing a vision of the future. One of her book's chapters bears the title "You Have to See the Future to Deal with the Present."[28] These are apt words for community nutritionists who are responsible for identifying a community's nutritional needs and developing programs to meet those needs. A good, effective nutrition program is not achieved by accident but by planning today to meet the needs of tomorrow.

The state of Arizona was following Popcorn's advice when it determined that existing training programs could not help child care workers acquire the skills needed to provide proper nutrition for the state's children with special health care needs. Its response to this needs assessment was to develop a program—Project CHANCE—to provide the necessary training. But the department's work didn't end when the course manual was published. The final step in the program planning process was to let stakeholders, community leaders, community nutritionists, child care program directors, and other interested parties know that the course was available. To do this, the department listed information about the course in national newsletters. It helped community colleges incorporate the course into their curricula. It had the course materials translated into Spanish to increase its dissemination. The course was listed as a resource for schools in USDA's Healthy School Meals Program on the Internet and described in a presentation at ADA's 1995 annual meeting.[29] All of these activities served to get the word out to community nutritionists and other experts across the country that the Project CHANCE program had been designed, implemented, tested, and was ready for use in other states.

Entrepreneurship in Program Planning

Program planning is one of the most exciting aspects of the community nutritionist's job. It requires a steady mix of creative juices and offers opportunities to learn new skills and work with people in public relations, marketing, design, and communications. Go back to Chapter 1 and review Table 1-6 on page 25. Three-quarters of the entrepreneurial activities listed there are essential aspects of the planning process. The excitement of planning stems from the constantly changing environment. *You* might be the first community nutritionist in your area to link clients in your weight management program with support groups and sound nutrition information on the Internet. Or maybe you convince a popular local television personality to help raise awareness of the risk factors for osteoporosis. Maybe you forge a partnership with a local company that never supported health promotion activities in the past. Perhaps your team introduces inner-city schoolchildren to the new "Apple Jane and the Cucumber King" board game—a program designed to encourage children to eat several fruits and vegetables daily. The possibilities for being an entrepreneur in this area are boundless.

Community Learning Activity

Activity 1

Study the results of the needs assessment for Case Study 2, shown in Table 7-2. Imagine that you have been charged with developing a program to improve the nutritional status of this elderly population. Define at least one goal and two or more outcome objectives for each of the following areas:

1. Nutritional status
2. Community-based programs and services
3. Partnerships (Partnerships apply to the linkages that your agency should develop or strengthen to implement this program.)

Refer to Table 5-1 on page 156 for additional information about the parameters of the community nutritional needs assessment for this case study.

Activity 2

Use the Internet to obtain information about the elderly. For example, access the Web sites for the Food and Nutrition Service (www.usda.gov/fcs), the National Center for Health Statistics (www.cdc.gov/nchswww), and Fedstats (www.fedstats.gov) to obtain data on participation rates of the elderly in food assistance programs such as the Food Stamp Program. For Canadian statistics, access Statistics Canada at www.statcan.ca. (You must have downloaded the Adobe Acrobat® Reader before you can view and print tables from these Web sites.) How would you use these data to design a program to improve the nutritional status of this population?

Professional Focus   *Time Management*

Time is a nonrenewable resource, a precious commodity. "Time is life," writes Alan Lakein in his bestselling book, *How to Get Control of Your Time and Your Life*.[1] "It is irreversible and irreplaceable. To waste your time is to waste your life, but to master your time is to master your life and make the most of it."

Many of us have difficulty deciding how to spend our time. Every day we must choose among dozens of activities and duties, all competing for our time and attention: family, friends, work, school, shopping, sports, television, movies, hobbies. The list is endless. Given that we all have a limited amount of time available each day, how can we choose from the multitude of jobs around us? The key to effective time management, argues Lakein, is *control*.

Control Is Essential

Control means recognizing that it is easy to become overwhelmed by the number of decisions we face about how we spend our time. The secret to controlling time effectively is making conscious decisions about which activities deserve our attention and effort—and which don't. Taking control of your time means *planning* how you will spend your time. Planning starts with deciding what your priorities are.

Quadrant II Is Where the Action Should Be

Stephen Covey, author of *The 7 Habits of Highly Effective People*, writes that the essence of good time management can be summarized in the following phrase: Organize and execute (in other words, *plan*) around priorities. Covey developed a time management matrix, shown in Figure 7-4, to demonstrate how most people spend their time.[2]

Covey defines activities as either "urgent" or "important." Urgent activities demand immediate attention. Important activities have to do with results; they require a more proactive approach than urgent activities. Look at the time management matrix. Quadrant I activities are both urgent and important. These activities are called "crises" or "problems." All of us have Quadrant I activities in our lives. The problem with focusing on Quadrant I activities, though, is that they tend to lead to stress, burnout, and a sense that we are always putting out brushfires.

Some people spend quite a bit of their time in Quadrant III, thinking that they are in Quadrant I. They are reacting to things that are urgent and appear important but that actually yield short-term results. People who spend their time in Quadrant III are likely to feel out of control. Covey believes that people who devote most of their time to Quadrant III and IV activities lead irresponsible lives, because they do not manage themselves and their actions.

Highly effective people spend their time in Quadrant II activities. It is through these activities that we accomplish the truly important things in life.

It's as Easy as ABC

The first step in taking control of your time is setting priorities—determining those things that are most important to

	Urgent	Not Urgent

Important	**Quadrant I**
	Activities:
	  Crises
	  Pressing problems
	  Deadline-driven projects
		Quadrant II
		Activities:
		  Prevention
		  Relationship building
		  Recognizing new opportunities
		  Planning, recreation

Not Important	**Quadrant III**
	Activities:
	  Interruptions, some calls
	  Some mail, some reports
	  Some meetings
	  Proximate, pressing matters
	  Popular activities
		Quadrant IV
		Activities:
		  Trivia, busy work
		  Some mail
		  Some phone calls
		  Time wasters
		  Pleasant activities

FIGURE 7-4 *Time Management Matrix*™

Source: Excerpted from *The 7 Habits of Highly Effective People*, p. 151, by Stephen R. Covey. Copyright © 1989 Stephen R. Covey. Reprinted with permission by Franklin Covey Co., www.FranklinCovey.com. All rights reserved.

you. Lakein calls this process the ABC Priority System. "A" activities have a high value; "B" items, a medium value; and "C" items, a low value.

Take 3 minutes to write down all of the things you should or would like to accomplish in the next 3 months. These activities probably fall into several categories: personal, family, community, social, school, career, financial, and spiritual goals. Don't hesitate to include some things that are off-the-wall. Now, for each item, assign a value of A, B, or C. Remember, the A activities are those with the highest value in your life. In addition, rank each activity within each category: A-1, A-2, and so forth.

Your A activities are top priorities, the items you want to find time for. The key to finding time for them lies not in prioritizing the activities on your schedule but in scheduling your priorities. In other words, if your A-1 activity is learning Japanese, then schedule time for it, even if you can only spend 10 or 15 minutes a day.

Keep three things in mind as you carry out this task. First, only you can decide what your priorities are. No one else can do this for you. Second, your A list will probably change over time and should be reviewed periodically. What is important to you now may not be important to you in 6 months. Finally, because planning in advance is the best thing you can do for yourself, write down your priorities. "A daily plan, in writing, is the single most effective time management strategy, yet not one person in ten does it. The other nine will always go home muttering to themselves, 'Where did the day go?' "[3]

The Top 10 Time Wasters

People from different jobs and disciplines have similar problems with managing time. A survey of 40 sales representatives and 50 engineering managers from 14 countries identified the following activities as the top time wasters:[4]

- Telephone interruptions
- Drop-in visitors
- Meetings (scheduled and unscheduled)
- Crises
- Lack of objectives, priorities, and deadlines
- Cluttered desk and personal disorganization
- Ineffective delegation
- Attempting too much at once
- Indecision and procrastination
- Lack of self-discipline

These are not the only time bandits by any means. Paper work, inadequate staffing, too much socializing, and travel all undermine our ability to manage time well. The key to managing these time wasters is to take control of them. Schedule your telephone calls for certain periods of the day and do not accept interruptions during "quiet" times, unless an emergency arises. Minimize interruptions from colleagues and visitors. Attend only meetings directly related to your work.

Learn to Say No

When you set your priorities—that is, when you say yes to one thing—you must say no to something else. You will always be saying no to something. You can never do everything that you want to do or everything everyone else wants you to do! The key to managing your time effectively is to learn to say no, firmly and courteously. Sometimes this can be difficult, for a task that is a B or C priority for you may be an A priority for someone important to you. When this happens, take a moment to consider the consequences to you, the other person, and your mutual relationship if you say no. In many cases, you can reach a compromise and maintain goodwill by deciding together how to rank an activity.

Work Smarter, Not Harder

It is a myth that the harder you work the more you accomplish. Many workaholics fall into this trap. They are more attuned to time spent working than quality time spent working. Avoid the trap of thinking that time spent working is automatically time spent valuably. "If it is not quality time and quality work, you are wasting time."[5]

Mastering the principles of good time management can take several months or even years. If you experience difficulty wresting control of your time, don't give up. Keep working at it. Although your progress may seem slow, you are further along than if you had never done it at all.

References

1. The description of the ABC Priority System was adapted from A. Lakein, *How to Get Control of Your Time and Your Life* (New York: Penguin, 1973).
2. Covey, Stephen R., *The 7 Habits of Highly Effective People*. New York: Simon & Schuster. © 1989 Stephen R. Covey. Used with permission. All rights reserved.
3. The quotation is from A. Mackenzie, *The Time Trap* (New York: AMACOM, 1990), p. 41.
4. M. LeBoeuf, Managing time means managing yourself, in *The Management of Time*, ed. A. D. Timpe (New York: Facts on File Publications, 1987), pp. 31–36.
5. The quotation is from J. C. Levinson, *The Ninety-Minute Hour* (New York: Dutton, 1990), p. 168.

References

1. Arizona Department of Health Services, Office of Nutrition Services, *Project CHANCE—A Guide to Feeding Young Children with Special Needs* (Phoenix: Arizona Department of Health Services, 1995).
2. L. Rider, MS, RD, and L. C. Patty, RD, CPM, personal communication, May 1998.
3. National Cholesterol Education Program, Report of the National Cholesterol Education Program Expert Panel on detection, evaluation, and treatment of high blood cholesterol in adults, *Archives of Internal Medicine* 148 (1988): 36–69; and National Cholesterol Education Program, *Report of the Expert Panel on Population Strategies for Blood Cholesterol Reduction*, NIH Publication No. 90-3046 (Bethesda, MD: National Cholesterol Education Program, National Institutes of Health, U.S. Department of Health and Human Services, 1990).
4. A. S. Pickoff, G. S. Berenson, and R. C. Schlant, Introduction to the symposium celebrating the Bogalusa Heart Study, *The American Journal of the Medical Sciences* 310 (suppl. 1) (1995): S1–S2; C. C. Johnson and T. A. Nicklas, Health ahead—The Heart Smart Family approach to the prevention of cardiovascular disease, *The American Journal of the Medical Sciences* 310 (suppl. 1) (1995): S127–S132.
5. Community Nutrition Institute, New states will participate in WIC Farmers Market, *Nutrition Week* 28 (February 27, 1998): 6.
6. K. M. Bartol and D. C. Martin, *Management* (New York: McGraw-Hill, 1991), pp. 156–66.
7. A. L. Martinez, *Hunger in Latino Communities* (Washington, D.C.: Congressional Hunger Center and the Congressional Hispanic Caucus Institute, 1995).
8. R. Perez-Escamilla, D. A. Himmelgreen, and A. Ferris, *Community Nutritional Problems Among Latino Children in Hartford, Connecticut* (Storrs and Hartford: University of Connecticut and the Hispanic Health Council, 1997).
9. Information about developing goals and objectives was adapted from K. L. Probert, ed., *Moving to the Future: Developing Community-Based Nutrition Services* (Washington, D.C.: Association of State and Territorial Public Health Nutrition Directors, 1996), pp. 21–24.
10. Public Health Service, *Healthy People 2000: Midcourse Review and 1995 Revisions* (Washington, D.C.: U.S. Department of Health and Human Services, 1995), pp. 166–67.

11. A. Fink, *Evaluation Fundamentals: Guiding Health Programs, Research, and Policy* (Newbury Park, CA: Sage, 1993), pp. 1–17.

12. N. Estabrook, *Teach Yourself the Internet in 24 Hours* (Indianapolis: Sams, 1997), pp. 91–99.

13. G. J. Gordon and M. E. Milakovich, *Public Administration in America*, 5th ed. (New York: St. Martin's, 1995), p. 369.

14. P. H. Rossi and H. E. Freeman, *Evaluation—A Systematic Approach* (Beverly Hills, CA: Sage, 1985), pp. 19–27.

15. A. D. Spiegel and H. H. Hyman, *Strategic Health Planning: Methods and Techniques Applied to Marketing and Management* (Norwood, NJ: Ablex, 1991), pp. 324–25.

16. H. J. Rubin and I. S. Rubin, *Community Organizing and Development*, 2nd ed. (New York: Macmillan, 1992), pp. 410–13.

17. A. D. Grotelueschen and coauthors, *An Evaluation Planner* (Urbana: Office for the Study of Continuing Professional Education, University of Illinois at Urbana-Champaign, 1974).

18. Health care reform legislative platform: Economic benefits of nutrition services, *Journal of the American Dietetic Association* 93 (1993): 686–90.

19. D. J. Caron, Knowledge required to perform the duties of an evaluator, *Canadian Journal of Program Evaluation* 8 (1993): 59–78.

20. Rossi and Freeman, *Evaluation–A Systematic Approach*, p. 18.

21. Fink, *Evaluation Fundamentals*, p. 167.

22. L. L. Morris and coauthors, *How to Communicate Evaluation Findings* (Newbury Park, CA: Sage, 1987), pp. 9–10.

23. Ibid., pp. 20–22.

24. M. O'Brecht, Stakeholder pressures and organizational structure for program evaluation, *Canadian Journal of Program Evaluation* 7 (1992): 139–47.

25. C. Kemp, Islamic cultures: Health-care beliefs and practices, *American Journal of Health Behavior* 20 (1996): 83–89.

26. J. D. O'Neil, B. Elias and A. Yassi, Poisoned food: Cultural resistance to the contaminants discourse in Nunavik, *Arctic Anthropology* 34 (1997): 29–40.

27. E. R. House, Multicultural evaluation in Canada and the United States, *Canadian Journal of Program Evaluation* 7 (1992): 153.

28. F. Popcorn, *The Popcorn Report* (New York: HarperCollins, 1991), p. 12.

29. L. Rider and L. C. Patty, personal communication, May 1998.

Chapter 8

Designing Community Nutrition Interventions

Outline

Learning Objectives

After you have read and studied this chapter, you will be able to:

- Describe five factors to consider when designing a community nutrition intervention.
- Describe three levels of intervention.
- Explain how evaluation research is used to improve the intervention design.
- Discuss five theories and models of consumer health behavior.

Something To Think About...

We need to have visions, visions of healthy and safe lives in healthy and safe communities—even visions that seem impossible. Robert Kennedy used to say often, "Some people see things as they are and ask, 'Why?' I dream things that never were and ask, 'Why not?'" We need to dream things that never were and ask, "Why not?" – *Barry S. Levy*, American Journal of Public Health, *February 1998, p. 191*

Introduction

In 1937 an inventor introduced a new product to the grocery store: the shopping cart. Until that time, people had shopped for their groceries using a small bag or basket. The inventor perceived the convenience and ease of using a cart on wheels for this activity and advertised his product with the question "Can you imagine winding your way through a spacious food market without having to carry a cumbersome shopping basket on your arm?"[1] Unfortunately, most people answered, no, they could not imagine doing this! They refused to accept the innovation. When queried about their behavior, customers claimed that the shopping cart looked like a baby carriage, and it made them feel weak and dependent. To get around this perception, the inventor hired women and men of various ages to come into his supermarkets and use the shopping carts to buy their groceries. This simple approach had the desired effect. Other customers saw the carts being used and elected to use them, too.[2]

This story illustrates two aspects of designing interventions. First, you must have information about your target population and why they do what they do. Understanding the behavior of your target population is an important step in developing strategies to influence—and eventually change—their **behavior.** Second, you must have an arsenal of tools for influencing behavior. Your toolbox might contain posters and table tents, cooking demonstrations, newspaper articles, and a health fair, or, as in the case of the shopping cart inventor, you might choose people from the target population to "model" the desirable behavior.

Recall from Chapter 1 that an intervention is a health promotion activity aimed at changing the behavior of a target population. This chapter describes the factors to consider when designing an intervention: program elements such as goals and objectives; the levels and types of interventions; the dietary habits, values, attitudes, and beliefs of the target population; the theories of consumer behavior; and the results of evaluation research. Study each of these topics before reading how community nutritionists working on Case Study 1 put all of the elements together.

Choose an Intervention Strategy

The first step in designing an intervention is to review the program's goals and objectives, which specify the program outcomes. At this point in the design process,

Behavior The response of an individual to his or her environment.

Intervention strategy An approach for achieving the program's goals and objectives.

your overall goal is to have a rough outline of what the intervention might look like. Sketching the details will come later.

Remember, the intervention or **intervention strategy** is the approach for achieving the program's goals and objectives. It addresses the question of *how* the program will be implemented to meet the target population's nutritional need. The intervention strategy can be directed toward one or more target groups: individuals, communities, and systems. Systems are the large, integrated environments in which all of us live and work. Targeting a system for intervention usually involves changing a public, corporate, or school policy, although it can include reorganizing a department to improve the manner in which a program is delivered.

The intervention strategy can also encompass one or more levels of intervention: (1) building awareness, (2) changing lifestyles, and (3) creating supportive environments for change.[3] Level I interventions focus on increasing awareness of a health or nutritional topic or problem. Awareness programs can be very successful in helping change attitudes and beliefs and increasing knowledge of risk factors, but they seldom result in actual behavior changes. Level II interventions are designed to help participants make lifestyle changes such as quitting smoking, being physically active, eating more fruits and vegetables and less fat, and managing stress. These interventions can be successful when they call for small changes over time and when they use a combination of behavior modification and education. Level III interventions work toward creating environments that support the behavior changes made by individuals. Thus, a company's policy to promote low-fat and high-fiber foods in the company cafeteria makes it easier for employees who are trying to lose weight or lower their blood cholesterol concentration to make healthy food choices at work.

Examples of intervention strategies are shown graphically in Table 8-1. Intervention activities that increase awareness among individuals include health fairs, screenings, flyers, posters, table tents, newsletters, and Internet Web sites. Special events, Web sites, radio advertising, and television public service announcements promote awareness across the entire community. An example of a special event that increased awareness about heart disease risk among women was the Mother/Daughter Walk sponsored by the Heart and Stroke Foundation of Manitoba. Food labeling is an example of a system intervention that can increase awareness. When the Food and Drug Administration authorized a health claim for folate on food product and dietary supplement labels, it recognized the ability of product labels to inform women of childbearing age about the relationship between adequate folate intakes and reduced risk of neural tube defects.[4]

Level II interventions reach individuals through one-on-one counseling and small-group meetings. These interventions usually involve a formal program of assessing the individual's current attitudes, beliefs, and behaviors; setting goals for behavior change; developing the skills needed to change behavior; providing support for change; and evaluating progress. Examples of Level II interventions for communities are fitness programs in primary and secondary schools and health promotion programs for all city employees—activities that cut across broad sectors of the community. Systems interventions at this level include company incentives for employees to join local fitness clubs and the formation of a wellness committee composed of community and business leaders.

	Level of Intervention		
Target group	**Level I: Build Awareness**	**Level II: Change Lifestyles**	**Level III: Create a Supportive Environment**
Individuals	• Health fairs • Health screenings • Flyers, posters, table tents • Internet Web sites • Special events	• One-on-one counseling • Small-group sessions	• Worksite cafeteria programs • Peer leadership
Communities	• Media announcements • Internet Web sites • Special events	• Fitness programs in schools • Health promotion programs for city employees	• Municipal policy that supports food gleaning • Point-of-purchase labeling • Tax incentives for companies with health promotion programs
Systems	• Health claims on food labels • Legislation	• Company incentives for employees to join local fitness clubs • Formation of a community-based wellness committee	• Medicare coverage of medical nutrition therapy • School policy that restricts access to candy and soft drink machines • Legislation

TABLE 8-1 *Examples of Intervention Strategies*

Examples of Level III interventions that target individuals include worksite health promotion and cafeteria programs. Identifying peer leaders who can model behavior change and talk about how they changed their lifestyles is another way of creating a supportive environment. In the community at large, supportive environments are created through policies that support gleaning (a food recovery program), "point-of-purchase" labeling, and tax incentives for companies with health promotion programs. At the system level, supportive environments occur when Medicare coverage for medical nutrition therapy is approved, when school policy restricts access to candy and soft drink machines to lunch hours only, and when eligibility requirements for food assistance programs are broadened.

Study the Target Population

When designing an intervention, study the target population's eating patterns and beliefs, values, and attitudes about foods and health. Some information about the target population was collected during the community needs assessment. Additional information can be obtained by conducting a library search of the literature, reviewing existing programs that deal with the target population, networking with

⊚ Internet Resources

Arbor Nutrition Guide	**arborcom.com**
Food and Nutrition Information Center	**www.nal.usda.gov/fnic**
Educational Materials Database	
HACCP* Training Programs and	
Resources Database	
NET Program and Products Database	
WIC Development Materials	
Food and Nutrition Service	**www.usda.gov/fcs**
Guide to Community Preventive Services	**web.health.gov/communityguide**
MEDLINE	**www.nlm.nih.gov**
National Center for Chronic Disease Prevention	
and Health Promotion	**www.cdc.gov/nccdphp**
Tufts Nutrition Navigator	**navigator.tufts.edu**

*HACCP = Hazard Analysis and Critical Control Point.

colleagues who work with the group, and posting queries about the target population on Internet listservs. Consult the boxed insert for Internet addresses related to MEDLINE, the Food and Nutrition Information Center databases, and other useful sites for obtaining food and nutrition information.

The purpose in studying the target population is to understand their values, why they believe what they believe, and why they eat what they eat. For example, if your target population is overweight, sedentary, middle-class adults aged 30 to 50 years living in an urban area, you might examine the typical meal shown in Figure 8-1 and ask these questions: Why do they choose pizza, making it the top lunch or dinner entree in the United States in 1996, over steak, which was the most popular entree in 1987?[5] Why do they prefer soft drinks rather than milk at meals? Why do they buy organic foods, which are often more expensive than comparable foods not labeled organic? Why does one-third of this group use some form of alternative medicine, including dietary supplements such as guarana and shark cartilage, even when there is no scientific evidence that some of these products are effective? Why do more than two-thirds of this group never tell their physician about their use of these products?[6] Answering these questions provides clues about the target population's beliefs, values, and food patterns and about the interventions that may succeed in changing their behavior.

The target population's food-related behavior is important. For humans, foods are more than simply a source of nutrients and nourishment. They are also used to express friendliness and hospitality, maintain and strengthen personal relationships, enhance social status, relieve stress, and express religious and cultural beliefs.[7] Foods have symbolic meanings for humans, and the symbolic values we give foods, together with other factors, influence the decisions we make about

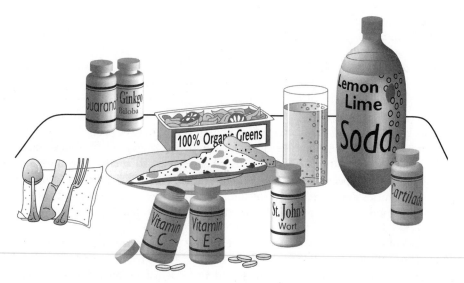

FIGURE 8-1 A *Typical North American Meal*

them. Many environmental, sociocultural, and personal factors, some depicted in Figure 6-1 on page 184 and others described later, affect the food intake and nutritional status of the target population.

Food Supply and Food Availability. Food choices are influenced by the types and amounts of foods available in the food supply. Food availability is affected by the food distribution system, types of imported foods, facilities for food processing and production, and the regulatory environment. The target population's penchant for pizza is likely due in part to pizza's widespread availability in supermarkets, convenience stores, and fast-food restaurants. Their beliefs that "natural" is good and freshness is important steer them toward organic foods, producing a boom in the sales of organic products to the tune of $3.5 billion in 1996.[8]

Income and Food Prices. Income and food prices are two economic factors that affect food consumption. The relationship between income and food consumption is expressed by the Engel function, which is named for Ernst Engel, a Prussian mining engineer who was interested in sociological issues. Engel published a study in 1857 showing that the poorer a family is, the greater the proportion of income it must spend on food purchases. The modern equivalent of this function states that as a consumer's income increases, the proportion of income spent on food decreases. Data from the 1992 Consumer Expenditure Survey reveal that households with incomes over $50,000 spent 10 percent of their income on food, whereas the poorest households with incomes of $5,000 to $9,999 spent about one-third of their disposable income on food.[9]

Food prices affect consumption patterns. Households with higher incomes have more money to spend on food and choose whatever foods they want, regardless of price. They tend to buy convenience cuts of meat, poultry, and fish, which are more expensive than large roasts and whole chickens and fish.[10] Hence, the typical meal shown in Figure 8-1 reflects the buying power of a target population with a comfortable income. Low-income households are more likely to have limited food budgets and to be concerned with price and value.[11]

Social group A group of people who are interdependent and share a set of norms, beliefs, values, or behaviors.

Sociocultural Factors. Food choices are strongly influenced by **social groups.** Primary social groups such as families, friends, and work groups are more likely to affect behavior directly, and within this category, the family exerts the most influence.[12] This is not surprising, as the family is the first social group to which an individual belongs, and under most circumstances, it is the group to which an individual belongs for the longest time. The family is a paramount source of values for its members, and its values, attitudes, and traditions can have lasting effects on their food choices. This is especially true for children and teenagers. The calcium intakes of teenagers, for example, are higher in families in which teenagers perceive their parents' attention, care, support, and understanding than in families with low family connectedness.[13]

The culture in which we live affects our food behavior, and many of our food habits arise from its traditions, customs, belief systems, technologies, values, and norms. Culture dictates how foods are stored, processed, consumed, disposed of, and even which foods are considered edible. North of the U.S.-Mexican border, for example, insects are seldom eaten, but in Mexico, the appetizer *los gusanos fritos* (fried caterpillars) may grace the menu in the finest restaurants. In the arctic region of Nunavik, near the Hudson Strait, the Inuit eat *niqituinnaq*, or real, natural, "country" food. Eating niqituinnaq is important, especially for older people, who prefer *ignuak* (fermented meat), seal, and other country food because it protects against disease and restores health. The Inuit rely on information handed down from generation to generation to know which animals are healthy and can be hunted for food.[14] The typical meal shown in Figure 8-1 reflects many cultural influences in North America, including the desire for good taste and convenience and the presence of a sophisticated food distribution and storage system. It does not, however, give a complete picture of the typical North American diet, for it gives no hint of the popularity of ethnic foods, which are consumed by 90 percent of U.S. consumers.[15]

Religious beliefs also affect the food choices of millions of people worldwide. Many religions, including Islam, Hinduism, Buddhism, Judaism, and Seventh-Day Adventism, specify the foods that may be eaten and how they should be prepared. Hinduism, for example, is the principal religion of India, and its food laws are steeped in ritual and meaning. Because the caste system is an integral part of Indian society, there are strict guidelines on how and with whom foods should be consumed. For Hindus, the focus of mealtime is eating, not conversation or socializing.[16] In North America, the daily food choices of Christians are not generally dictated by the basic doctrines of the Roman Catholic and Protestant churches, although some Christian church sacraments such as Holy Communion use food (bread and wine) symbolically. The principal dietary practices of several major religions are summarized in Table 8-2.

Food Preferences, Cognitions, and Attitudes. The environmental and sociocultural factors just described shape many of our personal attributes such as food preferences, cognitions, and attitudes. These attributes in turn affect our food choices. Preferences or likings for certain tastes and foods appear to develop quite

Religion	Food Laws and Dietary Practices	
Buddhism	The central tenet of Buddhism is vegetarianism, which stems from the dual concepts of Karuna (compassion) and Karma (action, conduct). In the eyes of a Buddhist, to eat meat is to destroy the seeds of compassion. Foods of plant origin are considered appropriate for consumption except for the "five pungent foods": garlic, leek, scallion, chives, and onion. These foods are considered unclean and are believed to generate lust when eaten cooked and rage when eaten raw.	**TABLE 8-2** *Food Laws and Dietary Practices of Several Major Religions*
Hinduism	The Hindus believe that food was created by the Supreme Being for the benefit of humans. Many Hindus are vegetarians, but some, particularly in cold, northern areas of India, eat meat except for beef, which is prohibited.	
Islam	Islamic food laws are derived from the Koran, the divine book given by Allah (the Creator) to Muhammad (the Prophet). Islamic food laws prohibit the consumption of unclean foods, such as carrion or dead animals, flowing or congealed blood, swine, animals slaughtered without pronouncing the name of Allah on them, animals killed in a manner that prevents their blood from being fully drained from their bodies, intoxicants of all types, carnivorous animals with fangs (e.g., lions, dogs, wolves, and tigers), birds with sharp claws (birds of prey, such as falcons, eagles, owls, and vultures), and land animals without ears (e.g., frogs and snakes).	
Judaism	The traditional dietary laws of Judaism prohibit the consumption of swine, carrion eaters, and shellfish and specify other dietary practices, such as the ritual slaughtering of animals and the ritual breaking of bread at each Sabbath meal. The term *kosher* indicates that the food so labeled was not derived from any prohibited animal, bird, or fish; the animal or bird was slaughtered by the appropriate ritual method; the meat was salted to remove the blood; and milk and meat were prepared in separate utensils and containers and not cooked together.	
Seventh-Day Adventism	Seventh-Day Adventists believe the body is the temple of the Holy Spirit. Thus, their dietary practices focus on health. Vegetarianism is the foundation of their dietary standard, although not all adherents are strict vegans. Other dietary standards call for abstaining totally from alcoholic beverages and avoiding certain "hot" spices and condiments such as pepper and chili, aged cheeses, and caffeine-containing beverages. The church recommends that its members eat a wholesome diet, consisting of whole grains, fruits, nuts, vegetables, a little milk, and occasionally eggs.	

Sources: For Buddhism, Y. Huang and C. Y. W. Ang, Vegetarian foods for Chinese Buddhists, *Food Technology* 46 (1992): 105–8; for Hinduism, A. Kilara and K. K. Iya, Food and dietary habits of the Hindu, *Food Technology* 46 (1992): 94–104; for Islam, M. M. Chaudry, Islamic food laws: Philosophical basis and practical implications, *Food Technology* 46 (1992): 92–93, 104; for Judaism, *The New Encyclopaedia Britannica* (Chicago: Encyclopaedia Britannica, 1985), pp. 444–51, and the Macropaedia, pp. 968–79; for Seventh-Day Adventism, G. C. Bosley and M. G. Hardinge, Seventh-Day Adventists: Dietary standards and concerns, *Food Technology* 46 (1992): 112–13.

early in humans. Not surprisingly, parents and their children tend to have similar food preferences.[17] Food preferences are shaped not only by family eating patterns but also by regional tastes. Mexican food is popular in western regions of the United States, whereas the Northeast prefers Italian foods.[18]

Food choices are affected by our **cognitions** or what we think. It seems logical that consumers who have learned about food composition and healthful eating practices have the knowledge base needed to select foods for good health, but consumers do not always practice what they have learned. The American Dietetic Association's 1997 Nutrition Trends Survey found that 79 percent of consumers

Cognitions The knowledge and awareness we have of our environment and the judgments we make related to it.

Simply knowing which food choices consumers make does not tell us how they make their decisions.

believe that nutrition is important to health, but only 4 in 10 say they are doing all they can to eat healthfully. And although 81 percent say exercise is important to good health, only 43 percent indicate that they are physically active on a regular basis.[19] So, although knowledge can influence food-related behavior, the effect does not appear to be large. Having access to information about food, nutrition, and health does not guarantee that consumers will adopt healthy behaviors.[20]

One of the complex areas of consumer behavior is the relationship of **attitude** to behavior. Early attitude research, conducted at the turn of the century, suggested that an individual's behavior was determined to a great extent by his attitude toward that behavior. Beginning in the 1930s, however, some researchers began to suspect that there was not a predictable relationship between attitude and any given behavior. By the 1970s, some investigators concluded that attitudes could not be used to predict behavior; they maintained that the inconsistencies between attitude and behavior could be explained by any number of other variables, such as competing motives and individual differences.[21] Today, attitudes are believed to influence behavior indirectly.

Attitude The positive or negative evaluation of performing a behavior.

Health Beliefs and Practices. Beliefs about foods, diet, and health influence our food choices. Two examples of the power of health beliefs can be drawn from different cultures. In Senegal, pregnant women avoid spicy condiments and foods, but they indulge their cravings for curdled milk, palm oil, meat, butter, and the traditional millet porridge. The Senegalese believe that if pregnant women are not allowed to satisfy their food cravings, their babies may be born with birthmarks.[22]

According to traditional Chinese beliefs, illnesses are caused by an excess of either yin (the dark, cold, feminine aspect) or yang (the bright, hot, masculine aspect) energy. Foods and herbs, which themselves may be either yin or yang, are prescribed to treat certain symptoms of disease. Yang foods such as hot soups made from chicken, pork liver, or oxtail are prescribed to treat the clinical manifestations of excess yin such as dry cough, muscle cramps, and dizziness. Yin foods include fish, most vegetables, and some fruits including bananas; they are used to treat the symptoms of hives and dry throat, which are believed to be caused by excess yang energy.[23] Thus, health beliefs influence many consumer health practices such as choosing leisure-time activities, eating comfort foods during illness, and using dietary supplements.

Draw from Current Research on Consumer Behavior

Simply knowing the factors that affect consumers' food choices does not tell us *how* consumers make their decisions. Many theories have been proposed to explain the decision-making process as it relates to health. Such theories are important because they suggest the questions that community nutritionists should ask to understand why consumers do what they do. Theories are sometimes presented in the form of *models*, which are simple images of the decision-making process. It is not possible to describe every health behavior theory in this chapter, but five deserve mention: Stages of Change Model, Health Belief Model, Theory of Reasoned Action, Social Cognitive Theory, and the Diffusion of Innovation Model. Each theory is described briefly in the following sections, and an example is given of how the theory can be applied to real-life situations.

The Stages of Change Model

The Theory. The most widely used Stages of Change Model is the Prochaska and DiClemente transtheoretical model, originally developed to understand smoking cessation. The model is founded on three assumptions: (1) behavior change involves a series of different steps or stages, (2) there are common stages and processes of change across a variety of health behaviors, and (3) tailoring an intervention to the stage of change in which people are at the moment is more effective than not considering the stage people are in. The transtheoretical model identifies five stages through which people move, although they don't always pass through them in a linear fashion and relapses are common:[24]

- **Precontemplation**—The individual is either unaware of or not interested in making a change.
- **Contemplation**—The individual is thinking about making a change, usually within the next 6 months.
- **Preparation**—The individual actively decides to change and plans a change, usually within 1 month. The individual may have already tried changing in the recent past.

- **Action**—The individual is trying to make the desired change and has been working at making the change for less than six months.
- **Maintenance**—The individual sustains the change for six months or longer.

The model resembles a spiral in some respects, with people moving around the spiral until they eventually achieve maintenance and termination. People in the contemplation stage are typically seeking information about a behavior change, whereas people in the maintenance stage are less likely to be looking for information and more likely to be searching for methods of strengthening the behavior. The key here is to develop an intervention strategy that will meet people's needs according to which stage they are in. Both individuals and communities can be assessed for their state of readiness to change.[25]

The Application—Individuals. Fitness centers are interested in developing and implementing programs that promote fitness and health. A community nutritionist working at the Smithfield Fitness Club had read about a study that found an association between the presence of chronic disease such as hypertension, diabetes, heart disease, and dyslipidemia and readiness to change behavior in the areas of physical activity, fat intake, fruit and vegetable intake, and smoking. Study subjects were HMO members over the age of 40 years, a population much like the community nutritionist's own club members. The study found that members who were at highest risk of adverse health outcomes had the greatest readiness to change. A surprising finding was that members with heart disease were more ready to change their behavior to reduce disease risk than were members with diabetes.[26] Findings of another study found that people in the precontemplation, contemplation, and preparation stages consumed fewer fruits and vegetables than people in the action or maintenance stages.[27]

The community nutritionist used these research findings and the Stages of Change Model to alter certain aspects of her health promotion programming. To increase awareness among clients in the precontemplation stage, she added special lectures on heart disease and hypertension to the roster of special events and developed brochures describing the risk factors for hypertension, diabetes, and heart disease. For clients in the contemplation and preparation stages, she offered four cooking demonstrations featuring low-fat recipes prepared with local fresh fruits and vegetables; the cooking demos were given in the fitness center's food center on Friday nights, when club attendance was high. These activities were designed to show clients who were mainly thinking about making changes that cooking the low-fat way was fun and easy. For clients in the action stage—that is, those who had participated in one of the special events or attended a cooking demonstration—she offered a 20-minute individual counseling session to answer questions about reading food labels, identifying food sources of fat, and calculating fat intake. These actions were all designed to facilitate and support change among club members.

The Application—Communities. A recent university-sponsored study found that the prevalence of eating disorders among high school students in the city of Scottsville was nearly double the state average. A group of university researchers, community nutritionists, the city's chief medical officer, and a school nurse met to discuss what could be done to address the problem. The group determined that the city was in the precontemplation stage of readiness to change: no discussion of the problem among stakeholders (for example, parents, teachers, students, administrators, health authorities) had taken place, no plan for addressing the problem had been developed, and no activities had been undertaken to reduce the prevalence of eating disorders. The group believed that the problem was urgent and some action should be taken. As a first step in moving the community to the contemplation stage, the group decided to hold a citywide meeting of stakeholders and key community leaders to assess their perception of the problem, identify resources for addressing the problem, and discuss desirable actions.

The Health Belief Model

The Theory. The Health Belief Model was developed in the 1950s by social psychologists with the U.S. Public Health Service as a means of explaining why people, especially people in high-risk groups, failed to participate in programs designed to detect or prevent disease.[28] The study of a tuberculosis screening program led G. M. Hochbaum to propose that participation in the program stemmed from an individual's perception of both his susceptibility to tuberculosis and the benefits of screening. Furthermore, an individual's "readiness" to participate in the program could be triggered by any number of environmental events, such as media advertising. Since Hochbaum's analysis, the Health Belief Model has been expanded to include all preventive and health behaviors, from smoking cessation to complying with diet and drug regimens.

The Health Belief Model has three components.[29] The first is the perception of a threat to health, which has two dimensions. An individual perceives that he or she is at risk of contracting a disease and is concerned that having the disease carries serious consequences, some of which may be physical or clinical (for example, death or pain), while others may be social (such as infecting family members or missing time at work). The second component is the expectation of certain outcomes related to a behavior. In other words, the individual perceives that a certain behavior (for example, choosing low-fat foods to facilitate weight loss) will have benefits. Bound up in the perception of benefits is the recognition that there are barriers to adopting a behavior (for instance, choosing low-fat foods requires skill in label reading and knowledge of food composition). The third component is **self-efficacy** or "the conviction that one can successfully execute the behavior required to produce the outcomes."[30] A key tenet of losing weight, for example, is the belief that one *can* lose weight. Other variables such as education, income level, sex, age, and ethnic background influence health behaviors in this model, but they are believed to act indirectly.

Self-efficacy The belief that one *can* make a behavior change.

The Application. The National Cancer Institute (NCI) recommends increasing fruit and vegetable intake to five or more servings a day, reducing dietary fat intake to less than 30 percent of energy intake, and increasing fiber intake to between 20 and 30 grams daily as a means of lowering cancer risk.[31] Since these recommendations were published in 1988, Americans have increased their fruit and vegetable intake slightly, but their intake of dietary fat remains above the recommended level, and obesity is a widespread public health problem. The question for community nutritionists and other practitioners in health promotion is, Which factors promote dietary change in the general population?

Researchers in Washington state sought an answer to this question. They surveyed adults aged 18 years and older about their beliefs and health practices using the Cancer Risk Behavior Survey, which consists of questions on risk factors for cancer, including dietary habits, alcohol consumption, sun exposure, smoking behavior, and preventive cancer screening. The researchers found that adults who believed strongly in a connection between diet and cancer and who were knowledgeable about the NCI recommendations made a greater number of positive dietary changes than those who had little belief. Having knowledge of the fat and fiber content of foods and perceiving social pressure to eat a healthful diet did not predict who made behavioral changes in dietary patterns or weight.[32]

Knowing that beliefs play a key role in motivating people to make lifestyle changes, the community nutritionists with the Mississippi affiliate of the American Cancer Society increased the funding for their public awareness campaign. They designed three posters and a public service announcement for television that reinforced the message that smoking, sun exposure, and dietary patterns are linked with increased cancer risk. This strategy was designed to influence the beliefs of people in high-risk groups.

The Theory of Reasoned Action

The Theory. The Theory of Reasoned Action, sometimes called the Theory of Planned Behavior, was developed by Icek Ajzen and Martin Fishbein. It "predicts a person's intention to perform a behavior in a well-defined setting."[33] The theory is a fundamental model for explaining social action and can be used to explain virtually any health behavior over which the individual has control. According to the model, behavior is determined directly by a person's intention to perform the behavior. **Intentions** are the "instructions people give to themselves to behave in certain ways."[34] They are the scripts that people use for their future behavior. In forming intentions, people tend to consider the outcome of their behavior and the opinion of significant others before committing themselves to a particular action. In other words, intentions are influenced by attitudes and **subjective norms.** Attitudes are determined by the individual's belief that a certain behavior will have a given outcome and by an evaluation of the actual outcome of the behavior. Subjective norms are determined by the individual's normative beliefs. In forming a subjective norm, the individual considers the expectations of various other people.

A modification of the theory, called the Theory of Trying, was proposed by Richard P. Bagozzi, who argued that more is needed to produce a behavior change than an expression of intention.[35] In the new model, shown in Figure 8-2, such factors as past experience (success or failure) with the behavior, the existence of mechanisms for coping with the behavior outcome (for instance, having a strategy for not meeting a weight loss goal), and emotional responses to the process all influence the intention to try a behavior. Bagozzi and his colleagues hypothesized that when intentions are well formed, they are strong mediators of behavior; when intentions are poorly formed, however, their influence on behavior is diminished, while that of attitudes grows stronger.[36]

Intentions A determination to act in a certain way.

Subjective norms The perceived social pressure to perform or not perform a behavior.

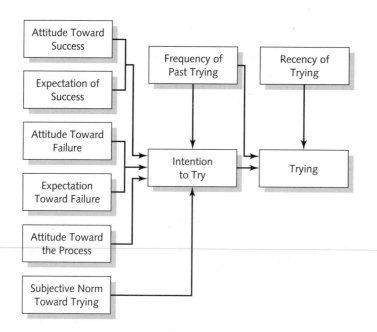

FIGURE 8-2 *The Theory of Trying*

Source: R. P. Bagozzi, The self-regulation of attitudes, intentions, and behavior, *Social Psychology Quarterly* 55 (1992): 179. Used with permission of R. P. Bagozzi and the American Sociological Association.

The Application. Dieting is a common method of trying to achieve an acceptably slim body shape. Young boys and girls often express the same dissatisfaction with their body shapes as do older adolescents and adults. Girls, in particular, express intentions to diet more frequently than boys do.[37] Research has shown, however, that even when the intention to diet is high, people have difficulty sticking with a weight loss program.[38]

The community nutritionist at the Fairlawn Weight Management Center was experiencing a high dropout rate among adolescents participating in the center's "Get Fit Now" program, which included sessions on the principles of balanced eating, controlling eating impulses, and physical activity. Drawing on the principles of the theory of reasoned action, he decided to survey the program participants about their intentions to lose weight, attitudes about their body shapes, level of self-esteem, expectations related to success, support from family and friends, and their perception of their ability to control eating and lose weight. He used the survey results to add certain components to his "Get Fit Now" program. For example, he added two sessions, one to help participants clarify whether the time was right to lose weight and one to boost their coping skills so they could handle lapses. And he paired some participants, who had not been successful losing weight in the past, with others who had reached previous weight loss goals. These actions were meant to improve the participants' intentions to master their eating habits and lose weight.

Social Cognitive Theory

The Theory. Social Cognitive Theory (SCT) explains behavior in terms of a model in which behavior, personal factors such as cognitions, and the environment interact constantly, such that a change in one area has implications for the others. For example, a change in the environment (the loss of a spouse's support for a weight loss effort) produces a change in the individual (a decrease in the incentive to lose weight) and consequently a change in behavior (a low-fat eating pattern is abandoned). The theory was developed to explain how people acquire and maintain certain behaviors.

The major concepts in SCT, many of which were formulated by Albert Bandura, and their implications for interventions are given in Table 8-3. In this context, the environment includes both the social realm (family, friends, peers, coworkers) and the physical realm (the workplace, the layout of a kitchen). A strength of SCT is that it focuses on certain target behaviors rather than on knowledge and attitudes.[39]

The Application. The prevalence of pica—a craving for and ingestion of nonnutritive substances such as clay, dirt, ice, baby powder, or laundry starch—among pregnant African-American women has been reported to range from 8 percent to 65 percent. Some reports have linked pica with low maternal hemoglobin levels and anemia.[40] A community nutritionist with the South Carolina Special Supplemental Nutrition Program for Women, Infants, and Children (WIC) had evidence that the prevalence of pica was quite high among the state's African-American WIC clients. She also recalled reading the results of a study in which peer counseling and motivational tapes were used to enhance breastfeeding among WIC clients.[41] Drawing on these research findings and the principles of SCT, the community nutritionist designed a course that used peer counseling to model healthy eating patterns and provide support for women who were trying not to practice pica. Although her budget did not allow for the cost of producing a videotape on the topic, she developed simple educational materials that explained what pica was and how to substitute other behaviors for it. Her idea drew on the SCT concepts of environment, situation, expectations, and observation.

The Diffusion of Innovation Model

The Theory. People often cannot or will not change their behavior, and many do not adopt innovations easily (recall the story about shopping carts at the beginning of this chapter). Even so, some people are more daring than others. Such people are the vanguard in the diffusion of innovation, the process by which an innovation spreads and involves an ever-increasing number of individuals within a population.[42] The Diffusion of Innovation Model was developed by E. M. Rogers and F. F.

Concept	Definition	Implications
Environment	Factors that are physically external to the person	Provide opportunities and social support
Situation	Person's perception of the environment	Correct misperceptions and promote healthful norms
Behavioral capability	Knowledge and skill to perform a given behavior	Promote mastery learning through skills training
Expectations	Anticipatory outcomes of a behavior	Model positive outcomes of health behavior
Expectancies	The values that the person places on a given outcome and incentives	Present outcomes of change that have a functional meaning
Self-control	Personal regulation of goal-directed behavior or performance	Provide opportunities for self-monitoring and contracting
Observation	Behavioral acquisition that occurs by watching the actions and outcomes of others' behavior	Include credible role models of the targeted behavior
Reinforcements	Responses to a person's behavior that increase or decrease the likelihood of recurrence	Promote self-initiated rewards and incentives
Self-efficacy	The person's confidence in performing a particular behavior	Approach behavior change in small steps; seek specificity about the change sought
Emotional coping responses	Strategies or tactics that are used by a person to deal with emotional stimuli	Provide training in problem solving and stress management

Source: C. L. Perry, T. Baranowski, and G. S. Parcel, How individuals, environments, and health behavior interact: Social Learning Theory, in *Health Behavior and Health Education—Theory, Research, and Practice,* ed. K. Glanz, F. M. Lewis, and B. K. Rimer (San Francisco: Jossey-Bass, 1990), p. 166. Copyright 1990 by Jossey-Bass, Inc., Publishers. Used with permission.

TABLE 8-3 *Key Concepts in Social Cognitive Theory and Their Implications for Behavioral Intervention*

Shoemaker in the 1970s to explain how a product or idea becomes accepted by a majority of consumers. The model consists of four stages:[43]

- **Knowledge**—The individual is aware of the innovation and has acquired some information about it.
- **Persuasion**—The individual forms an attitude either in favor of or against the innovation.
- **Decision**—The individual performs activities that lead to either adopting or rejecting the innovation.
- **Confirmation**—The individual looks for reinforcement for his or her decision and may change it if exposed to counter-reinforcing messages.

Innovations spread throughout a population largely by word of mouth. The speed of diffusion is a function, in part, of the number of people who adopt the innovation. Consumers can be classified according to how readily they adopt new ideas or products. *Innovators* adopt the innovation quite readily, usually without input from significant others. Innovators perceive themselves as popular and are financially privileged. This group is small. Like innovators, *early adopters* are integrated into the community and are well respected by their families and peers. Opinion leaders are often found in this group. Members of the *early majority* tend to be cautious in adopting a new idea or product, and persons in the *late majority* are skeptical and usually adopt an innovation only through peer pressure. Finally, the *laggards* are the last to adopt an idea or product, although they will adopt it eventually. Members of this group tend to come from small families, to be single and older, and to be traditional.[44]

The Application. A community nutritionist with Nutrition in Action, a company owned and operated by three registered dietitians, was concerned about several participants in her "Heart Healthy Living" program. She perceived their disinterest in making the kinds of dietary changes that would help lower their risk of having another heart attack. To boost their interest and enthusiasm for low-fat eating and cooking, she hit on the idea of contacting a popular local chef who had recently been interviewed on local television about the challenges he faced after surviving a heart attack. During the interview the chef had indicated that he was just learning how to prepare low-fat foods and expected his new skills to make their way to his restaurant's menu. His comments agreed with those made by chefs who were surveyed about their food science knowledge and practice; that is, many chefs want to provide good nutrition to their customers but often lack the necessary knowledge and skills.[45] Believing the chef was a good early adopter, the community nutritionist convinced him to join the group and expand his low-fat cooking repertoire. She believed his enthusiasm would be catching and would influence participants who resisted adopting innovations related to cooking.

Conduct Evaluation Research

Formative evaluation The process of testing and assessing certain elements of a program before it is implemented fully.

Evaluation occurs right from the beginning of the design phase. It is often necessary and prudent to pilot-test certain design elements during the development phase. This process, called **formative evaluation,** helps pinpoint and eliminate any kinks in the proposed delivery system or intervention before the program is implemented fully. Formative evaluation can be used to assess educational materials in terms of the appropriateness of language used, accuracy and completeness of the contents, and readability.[46]

For example, program planners in an inner-city district of Philadelphia found that most pregnant women don't consider breastfeeding their newborns and that those who do breastfeed for only a few weeks. They decided to conduct a formative evaluation before they agreed on the final intervention design. They surveyed potential program clients about their family's support for breastfeeding. The planners learned that few new mothers had a close, female relative living nearby who could help explain how to breastfeed an infant, and many new mothers in the younger age groups had boyfriends or husbands who had negative attitudes about breastfeeding. The results of this formative evaluation led the planners to add two pieces to the program intervention: (1) the provision of an experienced mother to befriend the new mother prior to the infant's birth and (2) sessions with husbands and boyfriends to help change their attitudes about breastfeeding.

Put It All Together: Case Study 1

Putting together all of the elements described in this chapter begins with reviewing the results of the needs assessment and the program goals and objectives. At this point in the program planning process, we use broad brush strokes to paint a picture of what an intervention strategy for Case Study 1 might look like. We chose health promotion activities and completed the grid shown in Table 8-4. It is not necessary

Target Group	Level of Intervention		
	Level I: Build Awareness	Level II: Change Lifestyles	Level III: Create a Supportive Environment

Individuals	• Internet Web site
	• Posters, brochures
		• "Heart Smart and
		  Satisfying"
		  cooking course
		• Smoking
		  cessation course
		• Smoking
		  reduction course	

Communities	• Special media
	  events
	• Television and
	  radio public
	  service
	  announcements
	• Anti-smoking
	  campaigns in
	  schools and
	  worksites
			• Work to secure
			  legislation that
			  prohibits smoking
			  in city restaurants

Systems			• Work to secure
			  legislation that
			  prohibits
			  tobacco-related
			  advertising at all
			  sporting events

TABLE 8-4 *Case Study 1: Intervention Strategies for a Program Designed to Reduce Coronary Heart Disease Risk Among Women*

or even practical to expect to fill in every box in the grid, because budgets, the number of staff available, and other factors will limit the scope of the intervention. Note that we chose to increase awareness at the individual and community level (to meet program goal 2) and build skills at the individual level (to meet program goal 1). We elected to promote supportive environments at the community and system levels.

We decided to conduct formative evaluation research to obtain information about the target population's skills. In focus group sessions, we asked women whether they had participated in cooking and smoking cessation courses, whether these programs worked, and what they found valuable about the course materials. We asked about their expectations related to reducing the risk of coronary heart disease (CHD). The results of these focus group discussions led us to consider adding an innovative element to our intervention strategy: a smoking reduction course, in which the goal would not be to get people to quit smoking but rather to help them cut down on the number of cigarettes smoked daily and to try adopting positive lifestyle behaviors (for example, low-fat eating, moderate exercise). Before finalizing the intervention strategy, we would plan to conduct a review of the literature related to smoking and health behaviors, develop a course outline, and then evaluate this element of the intervention strategy.

As the overall intervention strategy began to take shape, we could see how our health promotion activities had been influenced by the theories of consumer behavior.

The Internet Web site, posters, and brochures were aimed at reaching people who were in the contemplation and preparation stages of change. Our proposed activities in the policy arena—namely, working to secure legislation calling for smoke-free restaurants and a ban on advertising sponsored by tobacco companies at athletic events—were aimed at people in the maintenance stage. Some aspects of the "Heart Smart and Satisfying" cooking course, particularly those related to helping participants make simple dietary changes, reinforce healthful eating habits, and cope with setbacks, drew on the principles of the Health Belief Model and Social Cognitive Theory. And we could envision forming partnerships with other organizations such as the local affiliate of the lung and cancer associations to achieve some of our goals. The next step would be to develop the nutrition education component (described in Chapter 9).

Use Entrepreneurship to Steer in a New Direction

We already have a wealth of information about risk factors, epidemiology, healthy lifestyles, motivations for behavior change, and appropriate educational messages. The challenge comes in thinking of new ways of delivering health messages and services to vulnerable populations. One example of innovative thinking is The Well, a community-based drop-in wellness center for black women in a low-income housing complex in Los Angeles. The vision for this project emerged from a planning retreat attended by black women who were community leaders and activists. Their discussions focused on how to improve the poor health status of black women living in the district and led to the founding of The Well in 1994. The Well operates in partnership with the James Irvine Foundation, UCLA Psychology Department, John Wesley Community Health Institute, Inc., Fitness Funatics, and other groups to deliver services and programming in nutrition, fitness, HIV/AIDS education and intervention, substance abuse, family planning, pregnancy, and parenting. Support groups called sister circles work to empower black women to take charge of their health. The Well's affiliations with UCLA, the landlord of the housing project, and the National Black Women's Health Project ensures its continued success in meeting the health and nutritional needs of its clients.[47]

Successful community interventions such as The Well have been made, but many people in the communities are not getting the message. What is needed to ensure the success of community interventions? Manning Feinleib, an associate editor of the *American Journal of Public Health*, writes that we need a better understanding of the community factors that influence change and the reasons why consumers resist change.[48] When you plan community interventions, think of new ways to reach your target audience. Plan strategies for finding out why your clients are resisting a behavior change. Apply your creativity to influencing people to achieve behavior change.

Community Learning Activity

Activity 1

Conduct an Internet search for information and materials related to the elderly. Use the following keywords: program planning, community nutrition, consumer behavior. (You may use Boolean operators such as "AND," "OR," and so forth, to help narrow your search.) First, launch your searches from the following sites: Tufts Nutrition Navigator (navigator.tufts.edu) and Arbor Nutrition Guide (arborcom.com). Then, conduct a search using a popular search engine such as AltaVista or Excite. Describe the findings of your search. How would you rate the usefulness of the materials and information you found?

Activity 2

Review the information in Chapters 5 through 7 regarding Case Study 2. Review the goals and objectives you developed for this case study (see Chapter 7's Community Learning Activity). Complete the following activities:

1. Develop a grid similar to the one shown in Table 8-4.
2. Prepare a summary of your overall intervention strategy.
3. Describe any partnerships you believe should be formed to develop and implement the program. Why are these partnerships essential?

Professional Focus   *Being an Effective Speaker*

Public speaking ranks number one on most people's list of most dreaded activities. It causes churning stomachs, sweaty palms, dry mouths, and outright fear among many competent professional women and men. Many people would sooner have a root canal than give a 30-minute speech at a convention! If you feel this way about public speaking, take heart. You are not alone. More important, you *can* master the art of public speaking.

Tips for making an effective presentation are presented here. These basic principles apply to many different situations, from a formal presentation at a scientific meeting to an informal update for your colleagues at a staff meeting. They also apply to many teaching situations.

Things to Do Before Your Presentation

Public speaking is a skill just like any other skill. Even so, you may find that you expect more of yourself when it comes to public speaking than you would in other settings. If you are learning to downhill ski, you don't expect to be skiing the double–black diamond trails and moguls at the end of your first season. If you are learning to speak Spanish, you don't expect to converse fluently after only a few lessons. So, why should you expect to be a first-rate public speaker after a handful of presentations? In the same way that you acquire any skill, public speaking requires practice, evaluation, and more practice and evaluation. It is an ongoing process. Even when you become skilled at public speaking, there will be room for improvement. In the beginning, try to remove a little pressure by remembering that you are working toward acquiring a skill that can only be gained by doing. You will improve over time. To accelerate your competency curve, follow these rules:

1. Organize your presentation around this basic principle of effective speaking: First, tell your audience what you are going to tell them; then, tell them what you want to tell them; and finally, tell them what you told them! This strategy lets your audience know precisely what your presentation will cover and helps them remember the main points.

2. Prepare your visual aids so that they present your ideas effectively. Here are a few suggestions on preparing slides from the Federation of American Societies for Experimental Biology (FASEB):[1]

- **Clear purpose.** An effective slide should have a main point or central theme.

- **Readily understood.** The slide's main point should be readily understood by the audience. If it is not, the audience will be trying to figure out what the slide has to say and will not be listening to the speaker.

- **Simple format.** The slide should be simple and uncluttered. Avoid slides that present large amounts of data such as columns of numbers or many lines of text. Limit lines of text to 10.

- **Free of nonessential information.** Information not directly related to the slide's main point should be omitted.

- **Digestible.** The audience is capable of assimilating only so much information from a slide. It is better to have only a small amount of information (even just one sentence) than to cram numerous points onto a single slide.

- **Graphic format.** Some information is best presented graphically. In addition, the use of graphs and charts provides a visual change from slides containing only text.[2]

- **Visible.** Because most meeting rooms were not designed for projection, some people sitting in the back of the room may not be able to view the slides over the heads of those in front. For this reason, horizontal slides are more appropriate than vertical slides.

- **Legible.** Studies of projected image size and legibility show that the best slide template is about 42 spaces wide (9 centimeters) by 14 single-space lines (6 centimeters). The best type for slides is at least 5 millimeters or 14 points.

- **Thumbspot slides.** Hold the slide as it reads correctly to you. Then, write a sequence or other identifying number on the lower left corner of the mount. This spot tells the projectionist the correct corner to grasp when placing the slide in a carousel tray.

- **Integrated with verbal text.** Slides should support and reinforce the verbal text. Conversely, the verbal text should lay a proper foundation for the slide.

Similar principles apply to preparing overheads. Keep the information simple, and limit the amount of information provided. To improve legibility, type the master copy in large type. Like slides, overheads should be integrated with your text.

3. Rehearse your presentation several times—you would do no less for a piano recital! If the presentation is formal, use a table or desk as a podium. Time the presentation from start to finish, including your opening and closing remarks,

and adjust your presentation as needed. A general rule of thumb is that you should plan to spend about one minute per slide or overhead. If the presentation is informal, write down the key points you want to make and practice saying them. Rehearsing your presentation ensures that you know your material and how you want to present it.

4. Use mental imaging to boost your self-confidence. What is mental imaging? It's a technique used by many successful businesspeople, politicians, actors, and athletes to develop and strengthen a positive mental picture of their performance. It is a way to relieve stress and reduce anxiety about speaking in public. It sounds hokey, but it works.

Picture Nikki Stone, a freestyle ski aerialist. She is preparing to do her routine at the 1998 Nagano Olympics, a performance watched by millions of spectators worldwide. She stands ready to begin. What is going through her head during those few seconds before she launches downhill into her routine? Is she thinking, "Oh, no, there's one spot where I always mess up. I'll never do it right. My coach, family, friends, and country will be humiliated." It isn't remotely possible that Stone's thought processes took this tack. You can be sure that when she stood poised to begin her routine in the Olympics, she pictured every move from beginning to end, all done flawlessly. It is an image she worked to cultivate both mentally and physically. The result was a gold medal performance.

You can use mental imaging to boost your performance and quell those butterflies in your stomach. On several occasions before your presentation, walk yourself mentally through the speech from beginning to end. Picture being introduced, standing up and walking to the podium, adjusting the microphone, smiling, giving your opening remarks, asking for the first slide, and so on, right down to the very end of your presentation. Picture giving your presentation and handling questions at the end with complete confidence. The key to using mental imaging successfully is to use it *whenever* a negative thought about your presentation intrudes. If a mental picture of you passing out behind the podium surfaces suddenly, use mental imaging to squash it. Instead, force yourself to picture a confident, in-control YOU. Allow no negative thoughts about your presentation to take form. Encourage only positive thoughts. Mental imaging takes a little practice, but it is worth the effort. You will find that because you *think* you are more confident, you *are* more confident.

Things to Do During Your Presentation

Use the following techniques to ensure that you give a first-rate presentation:

- **Smile.** A smile will go a long way toward helping you relax and making you appear user-friendly to your audience. This is especially important when dealing with the general public.
- **Use eye contact.** Regardless of the size of the audience, select one person with whom to establish eye contact. Let your eyes dwell on this one individual a few moments, and then move on to another person. This gives the appearance of a one-on-one interactive discussion, which engages the audience in your presentation and helps ensure that they are listening to you.
- **Use gestures.** Gestures give energy to your presentation and provide additional emphasis for key points. Practice making them during your rehearsals. Exercise a little common sense here—wild arm movements and pirouettes will detract from your presentation.
- **Control the pace.** While a steady pace will ensure that you complete your speech on time, it may make your audience sleepy. Vary the pace to keep the audience interested in what you are saying.
- **Use pauses.** Pauses, like gestures, can be used for emphasis. A well-timed pause keeps your audience engaged and allows them a moment to process what you've just said.
- **Vary the volume and pitch.** Changing the volume and pitch of your voice has more auditory appeal for the audience than listening to a monotone voice.

Finally, two other points deserve mention. First, remember that the purpose of your presentation is to share information with your audience. Your listeners will generally be much less critical of your performance than you are. Being an effective speaker means that your audience is listening to your messages and absorbing the material you present. Second, despite all the tips and techniques listed here, you will want to develop your own style. Learn to be yourself *and* a relaxed, confident speaker.

References

1. The suggestions for preparing good slides were taken from the FASEB Call for Papers for the 1993 Annual Meeting, pp. 30–31.
2. J. W. King and J. Rupnow, A primer on using visuals in technical presentations, *Food Technology* 46 (1992): 157–70.

References

1. The man who put your groceries on wheels, *New York Post,* April 28, 1977, p. 37.

2. D. Cohen, *Consumer Behavior* (New York: Random House, 1981), p. 427.

3. M. P. O'Donnell, Definition of health promotion: Part II: Levels of programs, *American Journal of Health Promotion* 1 (1986): 6–9.

4. Food and Drug Administration, Food labeling: Health claims and label statements; folate and neural tube defects; final rule, *Federal Register* 61 (March 5, 1996): 8752–81.

5. A. E. Sloan, Food industry forecast: Consumer trends to 2020 and beyond, *Food Technology* 52 (1998): 37–38, 40, 42, 44.

6. The statistic on consumer use of alternative therapies was taken from Alternative Medicine: Time for a second opinion, *Harvard Health Letter* 23 (1997): 1–3.

7. M. L. Axelson, The impact of culture on food-related behavior, *Annual Review of Nutrition* 6 (1986): 345–63.

8. Sloan, *Food Technology.*

9. B. Senauer, E. Asp, and J. Kinsey, *Food Trends and the Changing Consumer* (St. Paul, MN: Eagan, 1991), pp. 133–53; and Interagency Board for Nutrition Monitoring and Related Research, *Nutrition Monitoring in the United States,* Vol. 1 (Washington, D.C.: U.S. Government Printing Office, 1995), pp. 59–60.

10. Interagency Board for Nutrition Monitoring and Related Research, *Nutrition Monitoring in the United States, Chartbook 1,* DHHS Pub. No. 93-1255-2 (Hyattsville, MD: Department of Health and Human Services, 1993), p. 55.

11. M. Krondl and D. Lau, Social determinants in human food selection, in *The Psychobiology of Human Food Selection* (Westport, CT: AVI, 1982), pp. 139–51.

12. Cohen, *Consumer Behavior,* pp. 76–127.

13. D. Neumark-Sztainer and coauthors, Correlates of inadequate consumption of dairy products among adolescents, *Journal of Nutrition Education* 29 (1997): 12–20.

14. J. D. O'Neil, B. Elias, and A. Yassi, Poisoned food: Cultural resistance to the contaminants discourse in Nunavik, *Arctic Anthropology* 34 (1997): 29–40.

15. A. E. Sloan, Way beyond burritos, *Food Technology* 52 (1998): 24.

16. A. Kilara and K. K. Iya, Food and dietary habits of the Hindu, *Food Technology* 46 (1992): 94–104.

17. J. Borah-Giddens and G. A. Falciglia, A meta-analysis of the relationship in food preferences between parents and children, *Journal of Nutrition Education* 25 (1993): 102–7.

18. Regional food preferences outlined in study of consumer eating trends, *Journal of the American Dietetic Association* 90 (1990): 1727.

19. Eating patterns and behaviors, *Food & Nutrition News* 69 (1997): 1.

20. I. M. Parraga, Determinants of food consumption, *Journal of the American Dietetic Association* 90 (1990): 661–63.

21. I. Ajzen and M. Fishbein, *Understanding Attitudes and Predicting Social Behavior* (Upper Saddle River, NJ: Prentice Hall, 1980), pp. 13–27.

22. C. S. Wilson, Nutritionally beneficial cultural practices, *World Review of Nutrition and Dietetics* 45 (1985): 68–96.

23. American Dietetic Association and American Diabetes Association, *Ethnic and Regional Food Practices: Chinese American Food Practices, Customs, and Holidays* (Chicago: American Dietetic Association, 1990), pp. 2–3.

24. M. K. Campbell, Stages of change: The transtheoretical model, in *Charting the Course for Evaluation: How Do We Measure the Success of Nutrition Education and Promotion in Food Assistance Programs? Summary of Proceedings* (Alexandria, VA: U.S. Department of Agriculture, 1997), pp. 19–21.

25. The description of the Stages of Change Model was adapted from R. J. Budd and S. Rollnick, The structure of the readiness to change questionnaire: A test of Prochaska & DiClemente's transtheoretical model, *British Journal of Health Psychology* 1 (1996): 365–76; M. H. Read, Age, dietary behaviors and the stages of change model, *American Journal of Health Behavior* 20 (1996): 417–24; and D. J. Bowen, S. Kinne, and N. Urban, Analyzing communities for readiness to change, *American Journal of Health Behavior* 21 (1997): 289–98.

26. R. G. Boyle and coauthors, Stages of change for physical activity, diet, and smoking among HMO members with chronic conditions, *American Journal of Health Promotion* 12 (1998): 170–75.

27. M. K. Campbell and coauthors, Stages of change and psychosocial correlates of fruit and vegetable consumption among rural African-American church members, *American Journal of Health Promotion* 12 (1998): 185–91.

28. I. M. Rosenstock, The Health Belief Model: Explaining health behavior through expectancies, in *Health Behavior and Health Education—Theory, Research and Practice,* ed. K. Glanz, F. M. Lewis, and B. K. Rimer (San Francisco: Jossey-Bass, 1990), pp. 39–62.

29. The discussion of the Health Belief Model was adapted from A. Caggiula, Health Belief Model, in *Charting the Course for Evaluation,* pp. 15–16; and J. H. Storer, C. M. Cychosz, and D. F. Anderson, Wellness behaviors, social identities, and health promotion, *American Journal of Health Behavior* 21 (1997): 260–68.

30. A. Bandura, *Social Learning Theory* (Upper Saddle River, NJ: Prentice Hall, 1977), p. 79.

31. R. R. Butrum, C. K. Clifford, and E. Lanza, NCI dietary guidelines: Rationale, *American Journal of Clinical Nutrition* 48 (1988): 888–95.

32. R. E. Patterson, A. R. Kristal, and E. White, Do beliefs, knowledge, and perceived norms about diet and cancer predict dietary change? *American Journal of Public Health* 86 (1996): 1394–1400.

33. W. B. Carter, Health behavior as a rational process: Theory

of Reasoned Action and Multiattribute Utility Theory, in *Health Behavior and Health Education*, pp. 63–91.

34. R. P. Bagozzi and Y. Yi, The degree of intention formation as a moderator of the attitude-behavior relationship, *Social Psychology Quarterly* 52 (1989): 266–79.

35. R. P. Bagozzi, The self-regulation of attitudes, intentions, and behavior, *Social Psychology Quarterly* 55 (1992): 178.

36. Ibid., pp. 178–204.

37. M. Conner, E. Martin, and N. Silverdale, Dieting in adolescence: An application of the theory of planned behaviour, *British Journal of Health Psychology* 1 (1996): 315–25.

38. C. A. Pratt and coauthors, A multivariate analysis of the attitudinal and perceptual determinants of completion of a weight-reduction program, *Journal of Nutrition Education* 24 (1992): 14–20.

39. D. B. Abrams and coauthors, Social learning principles for organizational health promotion: An integrated approach, in *Health and Industry—A Behavioral Medicine Perspective*, ed. M. F. Cataldo and T. J. Coates (New York: Wiley, 1986), pp. 28–51.

40. A. J. Rainville, Pica practices of pregnant women are associated with lower maternal hemoglobin level at delivery, *Journal of the American Dietetic Association* 98 (1998): 293–96.

41. S. M. Gross and coauthors, Counseling and motivational videotapes increase duration of breast-feeding in African-American WIC participants who initiate breast-feeding, *Journal of the American Dietetic Association* 98 (1998): 143–48.

42. E. M. Rogers and F. F. Shoemaker, *Communication of Innovations—A Cross-Cultural Approach* (New York: Free Press, 1971).

43. As cited in Cohen, *Consumer Behavior*, pp. 429–30.

44. S. Ram and J. N. Sheth, Consumer resistance to innovations: The marketing problem and its solutions, *Journal of Consumer Marketing* 6 (1989): 5–14.

45. G. Reichler and S. Dalton, Chefs' attitudes toward healthful food preparation are more positive than their food science knowledge and practices, *Journal of the American Dietetic Association* 98 (1998): 165–69.

46. J. L. Breault and R. Gould, Formative evaluation of a video and training manual on feeding children with special needs, *Journal of Nutrition Education* 30 (1998): 58–61.

47. K. A. E. Brown and coauthors, The Well: A neighborhood-based health promotion model for black women, *Health & Social Work* 23 (1998): 146–52.

48. M. Feinleib, Editorial: New directions for community intervention studies, *American Journal of Public Health* 86 (1996): 1696–98.

Principles of Nutrition Education

Outline

Learning Objectives

After you have read and studied this chapter, you will be able to:

- Develop a nutrition education plan for a program intervention.
- Design nutrition messages.
- Describe how to evaluate Internet resources.
- Describe four strategies for increasing program participation.

Something To Think About...

To cease smoking is the easiest thing I ever did. I ought to know because I've done it a thousand times. – *Mark Twain*

Introduction

Consumers are bombarded daily with dozens of health messages. Some are short and sweet: "Just Do It!" "Just Say No!" Others are complex and require time to process and understand. For example, an article in a consumer health magazine titled "The B Vitamin Breakthrough" describes new research related to homocysteine and the risk of coronary heart disease (CHD) and takes 20 minutes to read and study.[1] Regardless of the format in which the health message is delivered—television and radio commercials, Internet and billboard advertising, print (magazines, newspapers, brochures, books) and electronic media (television, CD-ROM, the Internet), even sports clothing and T-shirts—the "successful" health or nutrition message has a favorable impact on the target audience. It gets them to examine their belief system, evaluate the consequences of a certain behavior, or *change* their behavior.

This chapter describes how the community nutritionist develops the nutrition education component of an intervention. **Nutrition education** is an instructional method that promotes healthy behaviors by imparting information that individuals can use to make informed decisions about food, dietary habits, and health.[2] This chapter discusses the principles of nutrition education. Two topics are as important in this as in previous chapters: the target population—what they think, feel, believe, want, and do—and entrepreneurship.

Developing a Nutrition Education Plan

The nutrition education plan outlines the strategy for disseminating the intervention's key messages to the target population. The key nutrition messages may be designed to change consumer behavior, as in the "5 a Day for Better Health" message to "Eat five to stay alive." (The "5 a Day for Better Health" program is a national media program designed to get Americans to eat five or more servings of fruits and vegetables daily. The program is a joint venture of the National Cancer Institute [NCI] and the Produce for Better Health Foundation, which has more than 800 members.[3]) Nutrition messages may also inform or educate. When community nutritionists lobby legislators for more funding to support a statewide network of food banks and soup kitchens, they use nutrition messages to educate legislators about food insecurity and inform them of the consequences to young children of being malnourished.

Nutrition education An instructional method for promoting healthy food-related behavior.

The nutrition education plan is a written document that describes the needs of the target population, the goals and objectives for intervention activities, the program format, the lesson plans (including instructional materials such as handouts, videos, and so forth), the nutrition messages to be imparted to the target population, the marketing plan, any partnerships that will support program development or delivery, and the evaluation instruments.[4] These aspects of the plan are often organized into a manual that can be used by staff who work with the program. Having a detailed nutrition education plan can prevent confusion, especially when new staff members join the team or a substitute instructor must be found on short notice.

A nutrition education plan is developed for each intervention target group. Table 9-1 illustrates the similarities among nutrition education plans developed for individuals, communities, and systems. Nutrition education plans for individuals and communities are identical, the same activities being appropriate to both types of interventions. At the systems level, the nutrition education plan might properly be called a strategy. System level strategies do not require formal lesson plans or program identifiers such as logos or action figures per se, but they may draw on these program elements to reinforce key messages. A marketing plan is as essential for system level activities as for those aimed at individuals and communities.

The next section describes the development of a nutrition education plan for the "Heartworks for Women" program, a health promotion activity designed to help individuals (women) reduce their CHD risk. Consult Table 9-1 as you read this section.

Nutrition Education to Reduce CHD Risk: Case Study 1

The senior manager responsible for developing, implementing, and evaluating the intervention for reducing CHD risk among women (Case Study 1) reviews the proposed intervention levels and activities outlined in Table 8-4 on page 253. The senior manager's goal is to develop a coordinated plan for carrying out the intervention and evaluating its effectiveness. She begins by reviewing the proposed intervention activities and the expertise and time commitments of her staff. She

TABLE 9-1 *Activities Related to Developing a Nutrition Education Plan*

	Target Group		
Activity	**Individuals**	**Communities**	**Systems**
Assess needs	✔	✔	✔
Set goals and objectives	✔	✔	✔
Specify the format	✔	✔	✔
Develop a lesson plan	✔	✔	
Specify nutrition messages	✔	✔	✔
Choose program identifiers	✔	✔	
Develop a marketing plan	✔	✔	✔
Specify partnerships	✔	✔	✔
Conduct evaluation research	✔	✔	✔

decides to organize intervention activities into two areas: smoking, which she assigns to team 1, and nutrition, which she assigns to team 2. Team 1 members have expertise in health promotion, medicine, and epidemiology; team 2 consists mainly of community nutritionists and a health educator. Table 9-2 shows how the manager divided the intervention activities. The manager expects the teams to collaborate on some activities, such as developing content for the Web site, and to take leadership on others. Some activities, such as creating education materials like flyers and brochures, will be developed by each team for its respective programs. The senior manager designates a leader for each team.

The community nutritionist assigned to lead team 2 is given responsibility for developing the nutrition component of three intervention activities: the "Heartworks for Women" program, the nutrition content of the Web site, and the "smoking reduction" program. Her team will work with team 1 to develop the smoking reduction program, which will likely include messages and activities in the areas of nutrition and physical activity. The teams will work together to secure antismoking legislation, because this is a departmental activity. The team leader maps a strategy for the activities she's been assigned, breaking each major activity into smaller ones. For the "Heartworks for Women" program, for example, she indicates that goals and objectives must be developed, a format chosen, nutrition messages specified, and evaluation research conducted. Next, she reviews the interests, skills, and current assignments of her staff and assigns a staff member to take responsibility for each of the major activities. The next sections describe how the "Heartworks for Women" program was developed.

Refer to Appendix J for tips on designing a Web site.

Assess the Needs of the Participants

The community nutritionist responsible for developing the "Heartworks for Women" program first identifies the target population's educational needs. She asks these questions: What learning style is best suited for the potential program participants? What kinds of instructional tools (for example, videos, printed handouts) will have the greatest impact with this group? Can Internet activities be incorporated into lesson plans? Will participants be comfortable in group settings? Should some individual nutrition counseling be provided? Where should group sessions be taught? The answer to these questions can be obtained by reviewing the data obtained during the community needs assessment and by conducting formative evaluation research. For instance, focus groups can be organized to gather

Target Group	Team 1 (Smoking)	Team 2 (Nutrition)
Individuals	Smoking cessation course Smoking reduction course?	"Heartworks for Women" course Smoking reduction course?
Community	Special media events TV and radio public service announcements Antismoking campaigns in schools and worksites	Special media events
Systems	Legislative activities	Legislative activities

TABLE 9-2 *Case Study 1: Assignment of Responsibility for Carrying Out Intervention Activities*

information not obtained previously from members of the target population. Focus group participants might be asked about suitable locations for group activities, whether they have access to the Internet, and what they would like to learn about diet and CHD risk.

The community nutritionist also draws on her knowledge of the principles of adult education. **Adult education** is a process whereby adults undertake "systematic and sustained learning activities for the purpose of bringing about changes in knowledge, attitudes, values or skills."[5] The term *adult education* is a generic term that refers to formal, informal, vocational, and continuing education for the purpose of learning. For adults, learning is an intentional, purposeful activity. Adult learners approach learning differently than children do, and they require different motivations for learning. Adult learners learn best when the subject matter is tied directly to their own realm of experience, and their learning is facilitated when they can make connections between their past experiences (a parent died of cancer) and their current concerns (will eating a low-fat, high-fiber diet protect me from cancer?).[6] Adults are motivated to learn by the relevance of the topic to their lives, and they stick with a program or activity because they believe it will help them meet their learning goals.[7] In short, an effective program takes into account the learning style and motivations of the target population.

> **Adult education** The process whereby adults learn and achieve changes in knowledge, attitudes, values, or skills.

Set Goals and Objectives

The next step is to develop goals and objectives for the program. Recall that the "Heartworks for Women" program is a Level II intervention, meaning that it is a skills-building program. It will be designed to address three of the outcome measures specified in Chapter 7: (1) Decrease women's mean fat intake to 32 percent from their current intake of 35 percent of total energy within 1 year, (2) increase the number of women who can specify CHD as the leading cause of death in the city of Jeffers by 50 percent within 1 year, and (3) increase the number of women who can specify two major CHD risk factors by 25 percent within 1 year. The "Heartworks for Women" program is not the only means of achieving these objectives; some Level I activities are designed to address these objectives at the individual and community levels.

After reviewing the larger goals and objectives (described in Chapter 7), the community nutritionist determines that the "Heartworks for Women" program has two goals: (1) to educate individuals about the contributions of diet to CHD risk and (2) to build skills related to low-fat cooking and eating. Specific objectives are as follows:

- Increase awareness of the relationship of diet to CHD risk so that by the end of the course the number of participants who can name two dietary factors that raise blood total cholesterol will increase to 75 percent from 25 percent.
- Increase knowledge of dietary sources of fat so that by the end of the course the number of participants who can name three major sources of dietary fat will increase to 75 percent from 30 percent.
- Increase knowledge of low-fat cooking methods so that by the end of the course the number of participants who can describe and use five low-fat cooking methods will increase to 75 percent from 60 percent.

- Increase label-reading skills so that by the end of the course the number of participants who can specify accurately the fat content of foods using the nutrition information provided on food labels will increase to 75 percent from 20 percent.

Using these objectives as a guide, the community nutritionist sketches a rough outline of the program sessions, as shown in Table 9-3. The outline shows the link between the program objectives and the individual sessions. Session 1, for example, will provide general information about the major CHD risk factors and the contribution of diet to CHD risk. In this manner, the community nutritionist can be certain that any information that must be imparted to participants to meet the program objectives has been included in the program outline.

Specify the Program Format

Program formats vary, depending on what the program is intended to accomplish and the resources available to implement it. The process for choosing a format is much like that for choosing an intervention strategy: Begin with the big brush strokes. The format might consist of only three didactic lectures, or it might require six lectures and two cooking demonstrations, or it might involve three individual counseling sessions and 10 group sessions. The community nutritionist chooses a format that suits the topic and the amount of information that must be presented. She anticipates making some changes to the program format after analyzing the results of evaluation research and estimating projected program costs.

Results of the community needs assessment and focus group sessions showed that most potential participants for the "Heartworks for Women" program have at least a high school education. Many have participated in group classes (for example, weight management). About 15 percent of those surveyed in focus groups had access to the Internet. The community nutritionist considers these and other factors

Program Objective	Proposed Session to Meet Objective
Increase awareness of the relationship of diet to CHD risk so that by the end of the course 75% of participants can name two dietary factors that increase blood total cholesterol.	• Introduction to course
Increase knowledge of dietary sources of fat so that by the end of the course 75% of participants can name three major sources of dietary fat.	• Major food sources of visible and hidden fats • Low-fat meats • Low-fat dairy products • Shopping for low-fat foods
Increase knowledge of low-fat cooking methods so that by the end of the course 75% of participants can describe and use five low-fat cooking methods.	• Low-fat meats • Low-fat dairy products • Fruits and vegetables • Reading restaurant menus
Increase label reading skills so that by the end of the course 75% of participants can specify accurately the fat content of foods using the nutrition information provided on food labels.	• How to read food labels • Shopping for low-fat foods

TABLE 9-3 *Case Study 1: Rough Outline Showing the Link Between the Objectives and Proposed Sessions for the "Heartworks for Women" Program*

when choosing a format. She decides upon an 8-week program designed to fulfill the goals and objectives outlined previously. The program will consist of 90-minute sessions in which participants will have an opportunity to set target dietary goals, try new behaviors, and assess their success. The key strategy will be to seek small behavioral changes. The participants' skill level at entry and readiness to learn will be evaluated at the beginning of the program. The sessions will be organized as follows:

Session 1: Getting Started
Session 2: Looking for Fat in All the Right Places
Session 3: Cooking Meat the Low-Fat Way
Session 4: Dairy Goes Low-Fat
Session 5: Focus on Fruits and Vegetables
Session 6: Reading Food Labels
Session 7: Grocery Shopping Made Easy
Session 8: Reading Restaurant Menus

When choosing a format, the community nutritionist considers many details related to implementing the program. If the format calls for individual counseling, the facility must have private rooms for this activity. Likewise, if the format calls for small group sessions, there should be conference rooms or classrooms for teaching groups. If cooking demonstrations are included in the format, the facility should have counters, sinks, electrical outlets, and other equipment. Her decision about the final program format is influenced by the availability of facilities, equipment, and staff.

Develop Lesson Plans

The community nutritionist considers the instructional method (for instance, group sessions, one-on-one counseling) best suited for teaching low-fat cooking skills in the "Heartworks for Women" program. She chooses to present the material in group sessions, knowing that the participants will learn from one another.[8] Moreover, a program consisting mainly of group sessions is more likely to fit within the budget, because group sessions tend to be less costly than individual counseling. She develops objectives, selects instructional materials, and specifies other materials such as goal sheets required for each lesson. Table 9-4 shows this information for the first two sessions.

She must decide whether to develop nutrition education materials herself, use existing materials, or do both. To save time, she reviews existing programs and their nutrition education materials to determine whether they can be used with or adapted for this population. For example, in Session 1, which describes the major risk factors for CHD, the community nutritionist elects to develop her own handout on homocysteine because she cannot locate one, but she plans to use an existing brochure that describes other leading CHD risk factors. In Session 2, she plans to use a dietary fats chart developed by a leading food company and used widely in nutrition counseling. For Session 3, which will demonstrate low-fat cooking methods for meat, she decides to use reproducible masters and other materials from the *Lean 'N Easy* educational kit developed by the National Cattlemen's Beef Association.[9] (See Figure 9-1.) When selecting program materials, factors such as the cost of purchasing existing educational programs and materials must be weighed

Session	Title	Session Objectives	Instructional Materials	Learning Activities	Nutrition Messages
1	Getting Started	At the end of the session, participants will be able to: • Describe the program's two goals and four objectives. • Describe five major risk factors for CHD.	• Participant information form • Description of course goals and objectives • Handout: "Is Homocysteine a Risk Factor for Heart Disease?" • Handout: "Recipe for Summer Salsa and Baked Pita Chips" • Brochure: "Get the Facts About Heart Health"	• CHD and nutrition knowledge pretest • Handout: "Am I Ready for Change?" • Taste test = Summer Salsa and Baked Pita chips (a low-fat recipe)	• Diets high in fat and saturated fat raise blood total cholesterol. • High blood cholesterol levels are a risk factor for CHD. • Low-fat cooking is easy.
2	Looking for Fat in all the Right Places	• Define four types of dietary fats. • Describe the major food sources of dietary fat. • Describe the major sources of fat in the typical U.S. diet.	• Handout: "Definitions of Fats" • Handout: "Dietary Fats Chart" • Handout: Goal sheet	• Dietary Fats Quiz • Completion of goal sheet: Reducing fat intake	• Choose low-fat foods more often than high-fat foods.

against the time required to produce educational materials in-house and the cost of duplicating materials for participants.

TABLE 9-4 *Case Study* 1: *Lesson Plans for the First Two Sessions of the "Heartworks for Women" Program*

Specify the Nutrition Messages

Nutrition messages should be specified for each lesson plan. The messages should convey a simple, easy-to-understand concept related to low-fat eating and cooking: "Choose low-fat dairy products," "Choose lean cuts of meat," and so forth. One or two of these messages may be used in the nutrition education plan developed for the community- or systems-level interventions. The nutrition message for Session 5—"Eat five or more servings of fruits and vegetables every day"—serves double-duty: it is used in the individual session for the "Heartworks for Women" program, and it is a key nutrition message used in the media campaign to build awareness among women. More detailed information about nutrition messages is given later in this chapter.

Choose Program Identifiers

The community nutritionist and her teammates choose program identifiers such as the program name, a logo, an action figure, or a **tag line.** Tag lines are short, simple messages that convey a key theme of the program. They are typically used in

Tag line A simple, short message that conveys a key intervention message and is used on promotional materials.

FIGURE 9-1 *An Example of a Reproducible Master*

Source: *Lean 'N Easy* Educational Kit (Chicago: National Cattlemen's Beef Association, 1994). Copyright © 1994 National Cattlemen's Beef Association. Used with permission.

Handout 2

PORTION SIZE GUIDELINES

Lean 'N Easy

HOW MANY SERVINGS A DAY?

The *Food Guide Pyramid* recommends that servings from the Meat, Poultry, Fish, Dry Beans, Eggs and Nuts group be limited to about two to three servings per day, or the equivalent of five to seven ounces.

WHAT COUNTS AS A SERVING?

A serving is defined as two to three ounces of cooked lean meat, poultry or fish. Estimate about four ounces of boneless, trimmed and skinless raw to get three ounces of cooked meat, fish or poultry. One-half cup of cooked dry beans, one egg or two tablespoons of peanut butter count as one ounce of lean meat.

THE THREE-OUNCE RULES OF THUMB

○ A three-ounce cooked serving is about the same size as a deck of cards.

○ A three-ounce cooked serving is about the size of an average woman's palm. It may be helpful for you to weigh a three-ounce serving, and do a side-by-side comparison to judge its size as a future reference.

○ Imagine your dinner plate divided into three sections. A three-ounce serving of meat, poultry, or fish should account for just 1/3 of the plate. Another 1/3 should contain starches, like potatoes or rice, and the other 1/3 should contain vegetables.

○ One half of a cooked, boneless, skinless whole chicken breast is about three ounces. A three-ounce serving of chicken* is also equal to:

2 thighs = 2 drumsticks = 1 drumstick and 1 thigh

*All chicken parts assume bone and skin have been removed.

Copyright © 1994, NATIONAL LIVE STOCK AND MEAT BOARD. May be duplicated for instructional purposes.   17-808C2

promotional materials such as flyers and brochures. The tag line "Good Food for Good Health" might appear on departmental stationery, along with the program name and logo. These elements give the program its own identity and promote a sense of ownership among participants. The program name is important. It is usually selected after consultation with colleagues and members of the target population.

Develop a Marketing Plan

"If you don't exist in the media, for all practical purposes you don't exist," remarked National Public Radio's Daniel Schorr.[10] The community nutritionist develops a

Focus group sessions can be used to obtain the target population's opinions and impressions about program elements. Here, high school students offer their views on a program's name, logo, and tag line.

marketing plan to promote the "Heartworks for Women" program to the target population. Details about this plan are presented in Chapter 10.

Specify Partnerships

Forming partnerships with grocery stores, retail establishments like Target Stores, government agencies, nonprofit organizations, and other groups is one way of controlling the cost and increasing the reach of programs.[11] The community nutritionist with the "Heartworks for Women" program established a partnership with a local grocery store chain to use one of its stores as the setting for one session on shopping for low-fat foods and reading labels. She also networked with a national food company to obtain complimentary nutrition education materials for the course.

Conduct Formative Evaluation

Formative evaluation should be conducted throughout the program design process. Examples of formative evaluation research include the focus group sessions held early in the design phase and additional focus group testing of dietary messages and program instructional materials such as the handout on homocysteine for Session 1 and the dietary fats chart for Session 2. Here, the community nutritionist invites members of the target population to review these materials for reading level and ease of understanding. In other words, the target population helps determine whether the materials are appropriate and useful. Educational materials such as brochures, fact sheets, and handouts should be checked for reading grade level using a readability formula such as the SMOG grading test, shown in Appendix K.

At least one trial run of the program should be completed to ensure that the lectures fit within the designated time frame, the dietary messages are understandable and appropriate for the target population, and the nutrition education plan is sufficiently detailed to curtail glitches. The results of formative evaluation are used to change and improve program delivery.

Designing Nutrition and Health Messages

Consumers process nutrition and health messages at different levels, depending on their interest in the topic and their past experience with similar messages. Their responses to nutrition and health messages can be viewed as a continuum. On one end is a state of mindless or passive attention, where consumers pay little attention to the message; at the other end is the state of active attention, where consumers respond to the message by thinking about it and considering what it might mean for their own health and well-being.[12] The important question is, How can nutrition messages be formulated to influence consumer behavior? In this section, several general ideas for designing nutrition and health messages are outlined and the specific messages developed by the Dietary Guidelines Alliance for the "It's All About You" campaign are described.

General Ideas for Designing Messages

Studies of consumer behavior suggest several ways of designing nutrition and health messages to grab consumers' attention.[12] For example, present information in a novel or unusual fashion. The "5 a Day for Better Health" campaign logo, shown in Figure 9-2, appears on aprons, bookmarks, buttons, clothing (caps, T-shirts, golf shirts, jackets), pens, magnets, coloring books, key chains, watches, patches, mugs, tote bags, and magnetic memo boards.[13] No matter what the medium, the message to eat more fruits and vegetables reaches consumers in unexpected ways.

Use language that says to the consumer, "Listen to this. It's important." For example, if you are giving a radio interview to promote the "Heartworks for Women" program, you might say something such as "How many listeners know the number of women who die from heart attacks every year?" This approach is a cue for listeners to pay attention to the message; it engages their thought processes.

FIGURE 9-2 *Logo for the "5 a Day for Better Health" Campaign*

Source: This logo appears on the National Cancer Institute's Web site at dccps.nci.nih.gov/5aday and on the Produce for Better Health Foundation's Web site at www.5aday.com. Used with permission of the Produce for Better Health Foundation.

Use language that is immediate. A good example is available from the antidrug program: "This is your brain. . . . This is your brain on drugs. . . . Any questions?" Messages that use verbs in the present tense and demonstratives such as *this*, *these*, and *here* make consumers pay attention. Whenever possible, avoid using qualifiers such as *perhaps*, *may*, and *maybe* that express uncertainty. Consumers prefer straightforward statements such as "High-fat diets raise blood cholesterol levels" rather than tentative statements such as "A diet high in fat may increase blood cholesterol levels." Of course, the challenge for community nutritionists is deciding when strong, clear statements about research findings can be made and when more conservative statements are appropriate.

Think about the target population and consider the style and format of messages that will get their attention. Children, for example, are not small adults, and they process information differently than adults do. A common mistake when presenting information to children has been to assume that children will reject a behavior portrayed as unhealthy or bad (for instance, skipping breakfast, snacking on junk foods all day).[14] Sometimes a behavior is appealing to children and teenagers precisely because it is portrayed as bad or unhealthy! When designing messages for children and teenagers, go directly to the source: Ask them which messages they respond to and what type of messages they prefer. Also, start early. Messages directed to children and adolescents have the potential for a lasting effect. The "Gimme 5" program, for example, is a "5 a Day" program that targets high school students. Nutrition messages are delivered through "Fresh Choices" (a school meal and snack program), "Raisin' Teens" (a program for parents), and other program activities.[15]

The "It's All About You" Campaign

The Dietary Guidelines Alliance is a consortium of professional organizations such as the American Dietetic Association; trade organizations such as the National Cattlemen's Beef Association, National Dairy Council, and Food Marketing Institute; and federal government agencies such as the U.S. Department of Agriculture and the U.S. Department of Health and Human Services. These varied groups formed a partnership to promote positive, simple, consistent messages to help consumers achieve healthy, active lifestyles.[16]

The messages developed by the alliance were derived from focus group discussions with consumers who were asked, "Why aren't Americans doing more to eat right and get fit?" The answer was perhaps not too surprising: Consumers cite many reasons for not eating better and being more active—lack of time and energy, no "will power," bad habits, lack of motivation, the perception that "healthful" foods cost more and are hard to find, and the belief that being fit requires expensive equipment or club membership. Moreover, the focus group participants had clear ideas about how nutritionists and other health professionals should communicate with them. A summary of their opinions on effective communications is listed here:

- **"Give it to me straight."** Use simple, straightforward language. Don't use technical, scientific jargon or complicated instructions.
- **"Make it simple and fun."** Make it clear that eating healthful diets and being physically active are not time-consuming, complicated chores. Place the emphasis on improving habits, not on trying to achieve perfection.

- **"Explain what's in it for me."** Make the benefits of healthy lifestyles clear. Use outcomes such as being happier and having more energy to help motivate consumers.
- **"Stop changing your minds."** Be consistent in making recommendations. The wealth of dietary recommendations and scientific findings are overwhelming and confusing for most consumers.

The ideas and opinions expressed by focus group participants led to the creation of the "It's All About You" campaign. The campaign's nutrition messages, shown along with the campaign logo in Figure 9-3, were designed to be simple, appealing, and "action-able." In tests of the nutrition messages, consumers reported that the messages were positive, motivating statements that got their attention. The messages can be paired with supporting tips that provide ideas for specific behavioral changes. For example, the message "Be Realistic" can be paired with supporting tips such as "Park your car in the farthest spot" or "To save calories and fat, use a cooking spray

FIGURE 9-3 *Nutrition Messages and Logo from the "It's All About You" Campaign of the Dietary Guidelines Alliance*

Copyright © 1996 The Dietary Guidelines Alliance. Used with permission.

Make healthy choices that fit your lifestyle so you can do the things you want to do.

Be Realistic

Make small changes over time in what you eat and the level of activity you do. After all, small steps work better than giant leaps.

Be Adventurous

Expand your tastes to enjoy a variety of foods.

Be Flexible

Go ahead and balance what you eat and the physical activity you do over several days. No need to worry about just one meal or one day.

Be Sensible

Enjoy all foods, just don't overdo it.

Be Active

Walk the dog, don't just watch the dog walk.

instead of oil to sauté foods." These messages can be incorporated into a variety of programs as a way of helping consumers do more to achieve good health.

Implementing the Program

After the program has been designed and tested, it is ready for implementation. The goal of this phase of the planning process is to deliver as faithfully as possible the program laid out in the nutrition education plan. There will be glitches, of course. Perhaps an ingredient or cooking utensil was omitted from the list of materials needed for a cooking demonstration, or no overhead projector was available at the facility, or a program flyer featured the wrong time for the session. Anything can happen. A record should be made of any unexpected problems so that a strategy for preventing them can be developed for future programs. Once the program is under way, the community nutritionist and her team work to keep the program running smoothly. Two questions arise during this phase: How can program participation be increased? And how can the program be improved?

Enhancing Program Participation

Let's state the obvious: The higher the level of participation in a program the better. High participation increases the likelihood that a program will be effective and that people will be involved with it for a sufficient amount of time to make a behavior change. But what is participation? And how do program planners maximize participation in their programs? **Participation** refers to the number of people who take part in a health promotion activity. If the health promotion activity is a group educational program, participation refers to the number of people at the end of each educational session. If the activity is a newsletter, participation is the number of people who receive the newsletter times the number of newsletter editions or mailings per year. Participation rates vary, depending on how new the activity is and whether an incentive for participating is offered. For group education sessions offered on a voluntary basis and without incentives, the participation rate may range from 5 percent to 35 percent. A newsletter may have a much higher participation rate: 85 to 95 percent.[17]

> **Participation** The number of people who take part in a health promotion activity.

What can be done to improve participation rates? First, understand the target population and their needs and interests. Second, use evaluation research to improve the program design. Make the activity enjoyable and relevant to the target population's needs. Remove barriers to participation. Remember, people participate in health promotion activities for different reasons: to have fun, be with friends or family, learn something new, be challenged, fulfill a goal, or seek support. Find ways to help them see the immediate benefits of participating. Make it easy for people to sign up for or attend the activity. Schedule the activity at a convenient time. Third, use incentives for participating. Incentives range from formal recognition for achieving goals to prizes and treats such as T-shirts, refrigerator magnets, and cookbooks. Fourth, build ownership of the program among participants by using slogans, action figures, and logos to enhance the program's identity. Finally, promote, promote, promote—in other words, make the program highly visible for the target population.

Conducting Summative Evaluation

Summative evaluation
Research conducted at the end of a program that helps determine whether the program was effective and how it might be improved.

Summative evaluation is conducted at the end of the program and provides information about the effectiveness of the program.[18] Summative evaluation is designed to obtain data about the participants' reaction to all aspects of the program, including the topics covered, the instructors or presenters, any instructional materials, the program activities (for example, cooking demonstrations, taste tests), the physical arrangements for the program (including the location, room temperature, availability of parking), registration procedures, advertising and promotion, and any other aspect of the program. Participants are asked to rate these program elements, perhaps scoring their assessment on a five-point rating scale. They may be asked to express in their own words the aspects of the program they liked most or least and to suggest ways in which the program can be improved.[19] As with formative evaluation, the data obtained from summative evaluation is used to improve the program's delivery and effectiveness and to make the program an inviting place for learning.[20]

Evaluating Internet Resources

As the Internet continues to grow and becomes integrated into community nutrition practice, it is likely that community nutritionists will be expected to evaluate information obtained from the Internet. Perhaps your clients will have questions about information they found on the World Wide Web, or your organization asks you to choose 5 to 10 sites to link to its own Web page, or you wish to include Web sites on a handout for program participants. For whatever reason, community nutritionists must have skills for evaluating Internet information.

Information is rampant on the Internet. In a sense, the Internet *is* information, and the information is continually being revised and created. Internet information exists in many forms (for instance, facts, statistics, stories, opinions), is created for many purposes (for example, to entertain, to inform, to persuade, to sell, to influence), and varies in quality from good to bad. One method for determining whether the information found on the Internet is reliable and of good quality is the CARS Checklist. The acronym CARS stands for credibility, accuracy, reasonableness, and support. Each of these is summarized here:[21]

- **Credibility.** Check the credentials of the author (if there is one!) or sponsoring organization. Is the author or organization respected and well known as a source of sound, scientific information? Evidence of a lack of credibility includes no posted author and even the presence of misspelled words or bad grammar. A credible sponsor will use a professional approach to designing the Web site.
- **Accuracy.** Check to ensure that the information is current, factual, and comprehensive. If important facts, consequences, or other information are missing, the Web site may not be presenting a complete story. Evidence of a lack of accuracy include no date on the document, the use of sweeping generalizations, and the presence of outdated information. Watch for testimonials masquerading as scientific evidence. This is a common method for promoting questionable products on the Internet.

- **Reasonableness.** Evaluate the information for fairness, balance, and consistency. Does the author present a fair, balanced argument supporting his ideas? Are his arguments rational? Has he maintained objectivity in discussing the topic? Does the author have an obvious—or hidden—conflict of interest? Evidence of a lack of reasonableness includes gross generalizations ("Foods not grown organically are all toxic and shouldn't be eaten") and outlandish claims ("Kombucha tea will cure cancer and diabetes").
- **Support.** Check to see whether supporting documentation is cited for scientific statements. Are there references to legitimate scientific journals and publications? Is it clear where the information came from? An Internet document that fails to show the sources of its information is suspect.

Check the boxed insert for a list of Internet addresses that provide help on evaluating Internet information. These principles can be applied to evaluating other forms of information.

 ## A *Resource* List *for Evaluating Internet Information*

Bibliography on Evaluating Internet Resources	**refserver.lib.vt.edu/libinst/critTHINK.HTM**
Milton's Web Site for Evaluating Information Found on the Internet	**milton.mse.jhu.edu:8001/research/education/net.html**
Evaluating Web Resources	**www.science.widener.edu/~withers/webeval.htm**
Thinking Critically About World Wide Web Resources	**www.library.ucla.edu/libraries/college/instruct/critical.htm**
Evaluating Internet Research Sources	**www.sccu.edu/faculty/R_Harris/evalu8it.htm**
Evaluating World Wide Web Information	**thorplus.lib.purdue.edu/research/classes/gs175/3gs175/evaluation.html**
How to Critically Analyze Information Sources	**www.library.cornell.edu/okuref/research/skill26.htm**
Evaluating Quality on the Net	**www.tiac.net/users/hope/findqual.html**

Entrepreneurship in Nutrition Education

Creativity and innovation—the twin elements of entrepreneurship—can be applied to many aspects of nutrition education, from the development of action figures and logos to the use of new communications media such as the Internet and CD-ROM. Program planning is definitely a good venue for practicing entrepreneurship. Consider the smoking reduction program mentioned earlier in this chapter. This program idea represents a new approach to reducing CHD risk. It focuses on smokers—a target population that contributes the lion's share toward the costs of treating heart attacks and stroke—but in a new way. Rather than trying to get smokers to quit smoking altogether, the program is designed to build positive health behaviors in other areas of their lives (for example, eating lower-fat foods, taking a walk three days a week, learning to manage stress). The program's goals are to help smokers cut down on the number of cigarettes smoked and to reduce their CHD risk by acquiring some of the positive health behaviors seen typically in nonsmokers. Before this program idea can be implemented, however, many questions must be asked: Has this approach been tried before? Does evidence from the literature indicate that it will work? Are there sufficient data to make dietary and physical activity recommendations to smokers? What do the experts in this area think about this approach? What is the theoretical framework that supports this approach? If the approach has never been tried, should it be undertaken as a formal, scientific study? Some of these questions fall into the policy arena, but all must be addressed before program development can move forward.

Examine Mark Twain's comment at the beginning of this chapter. It highlights a perpetual problem in the field of health promotion: How can we help consumers change their behavior? One approach to motivating consumers is to design effective nutrition messages and programs.

Community Learning Activity

Activity 1

Review the elements of the "Heartworks for Women" program described in this chapter. If you were the team 2 leader, which elements would you place on the Web site? Why?

Activity 2

Review at least three of the Internet sites shown in the boxed insert in this chapter. Using the information you acquire there, review and evaluate one Web site whose address is shown anywhere in this textbook. Next, conduct a Web search using any one of the following keywords: shark cartilage, St. John's wort, guarana, chromium picolinate, the "Water of Life cure," or melatonin. Choose one Web site found in your search and evaluate it, using the principles you have learned about evaluating Internet information. How would you rate the site? What would you say to a client about the site? What can be done to improve the site's quality and reliability?

Activity 3

For this activity, you will work in teams of about five to eight members. Your assignment is to develop a foods and nutrition screen saver. Imagine that you are planning to meet next week with two software specialists who will develop the computer code for the screen saver. Outline the basic elements of the project (for example, goals, objectives, target audience, nutrition messages, graphics, and so forth) so that you can communicate your "vision" to the software specialists. Now, present your ideas to the rest of the class.

Activity 4

Continue with the program development process for Case Study 2. Develop a nutrition education plan for your program. You should be able to specify goals and objectives, the program format, lesson plans, nutrition messages, program identifiers, and partnerships.

Professional Focus   *Being an Effective Writer*

The written word, Rudyard Kipling once observed, is "the most powerful drug used by mankind." It inspires, educates, and engages us. Because it is so powerful, we want to be sure that we can express ourselves well, regardless of the audience for whom we are writing or the topic being discussed.

But how do you become a good writer? Any number of strategies, some of which are described here, can help you improve your writing skills, but one of the most important steps you can take is to practice. To learn to express yourself clearly and concisely, you must practice, practice, practice. Fortunately, most of us have ample opportunity to practice our craft, for our jobs or schooling require us to write many types of documents: business letters, informal memos, reports, study proposals, scientific articles, fact sheets for the general public, and project updates, to name just a few. Mastering the basic principles of grammar and syntax is also important. Knowing how to use these tools properly will serve you well in all writing situations.

Three Basic Rules of Writing

Although there are different types of writing—for example, professional and business writing versus writing copy for an advertisement—several basic principles apply:

1. **Know what you want to say.** A well-known fiction writer once remarked, "I don't find writing particularly difficult; it's figuring out what I want to say that's so hard!" If you don't know what you want to tell the reader, your writing will meander around and leave the reader bewildered and dissatisfied. Your first step then is to decide what point or points you want to make. Jot them down before you begin to write your article or report. Once you clarify in your own mind precisely what it is that you want to say, the writing itself will flow more smoothly, and the reader will follow your thinking more easily.
2. **Eliminate clutter.** "The secret of good writing is to strip every sentence to its clearest components."[1] Every word in every sentence must serve a useful purpose. If it doesn't, mark it out. Consider the approach used by Franklin D. Roosevelt to convert a federal government memo into plain English. The original blackout memo read as follows:[2]

   Such preparations shall be made as will completely obscure all Federal buildings and non-Federal buildings

occupied by the Federal government during an air raid for any period of time from visibility by reason of internal or external illumination.

"Tell them," Roosevelt said, "that in buildings where they have to keep the work going to put something across the windows." By changing gobbledygook into plain English, Roosevelt made the memo simple and direct—and comprehensible. This strategy is important in all types of writing. Consider this statement from one computer company's "customer bulletin": "Management is given enhanced decision participation in key areas of information system resources." What on earth does that mean? It might mean "The more you know about your system, the better it will work," or it could mean something else.[3] The wording is so jumbled that the customer can't be sure what the company is trying to tell her and will probably take her business elsewhere as a result.

Whenever possible, eliminate clutter and jargon. To free your writing from clutter, clear your head of clutter. Clear writing comes from clear thinking.

3. **Edit, edit, edit.** A well-written piece does not occur by accident. It is crafted through diligent editing. Writers often edit their manuscripts eight, nine, ten, or more times. Editing pares the piece to the bare bones. It makes the writing stronger, tighter, and more precise. The paragraph shown in Figure 9-4 is an example of an edited manuscript. This particular paragraph was taken from William Zinsser's book *On Writing Well*. What you see in the figure is a draft that had already been edited four or five times. Zinsser says that he is "always amazed at how much clutter can still be profitably cut."[4]

Different Strokes for Different Folks

Not all writing assignments are the same. Some require the formal language of the scientific method, whereas others are meant to entertain (*and* inform). Choose the style, format, and tone of voice appropriate for the piece you are writing.

Writing for the General Public

Writing for the general public is an important part of the community nutritionist's job. Newspaper and magazine articles, fact sheets, brochures, pamphlets, posters, and Web sites can all be used to teach and inform consumers. In addition to the basic writing principles outlined in the previous section, two things are important to bear in mind when writing for the general public:

1. The most important sentence in any piece of writing is the first one. If it doesn't engage the reader and induce him to read further, your article or brochure is dead. Therefore, the lead sentence must capture the reader immediately. Consider this lead to an article about designer tomatoes: "Strap on your goggles, consumers, this one's getting messy."[5] The reader wonders instantly why tomatoes should be stirring up trouble. Or this lead from an article about dieting: "Dieters are a diverse group, but they share one common goal: to make their current diet their last one. Unfortunately, lasting results aren't what most get. In fact, the odds are overwhelming—9 to 1—that people who've lost weight will gain it back."[6] Anyone who has tried to lose weight will find this lead enticing. In addition to capturing the reader's attention, the lead must do some work. It must provide a few details that tell the reader why the article was written and why she should read it.

2. Know when to close. Choosing an endpoint is as important as choosing a lead. A closing sentence works well when it surprises the reader or makes him think about the article's topic. An article about the challenge of change concluded with this comment, made by a man who participated in Dean Ornish's lifestyle program for high-risk heart disease patients: "What Ornish has given me is that opening and an understanding of what members of the group have said many times: A longer life may be important, but a better life is of the essence."[7] This closing remark works because it is personal and thought-provoking.

Writing for Professional Audiences

Materials written for professional groups must conform to a more rigorous, traditional format and style than those aimed at consumers. Scientific articles, for example, have specific subheadings and a formal tone (refer to the Professional Focus for Chapter 6). The best way to learn how to write these types of documents is to study published articles. Once

---

This carelessness can take any number of ~~different~~ forms. Perhaps a sentence is so excessively ~~long and~~ cluttered that the reader, hacking his way through ~~all~~ the verbiage, simply doesn't know what _it_ ~~the writer~~ means. Perhaps a sentence has been so shoddily constructed that the reader could read it in any of _several_ ~~two or three different~~ ways. ~~He thinks he knows what the writer is trying to say, but he's not sure.~~ Perhaps the writer has switched pronouns in mid-sentence, or ~~perhaps he~~ switched tenses, so the reader loses track of who is talking ~~to whom,~~ or ~~exactly~~ when the action took place. Perhaps Sentence B is not a logical sequel to Sentence A -- the writer, in whose head the connection is ~~perfectly~~ clear, has not _bothered to provide_ ~~given enough thought to providing~~ the missing link. Perhaps the writer has used an important word incorrectly by not taking the trouble to look it up ~~and make sure~~. He may think that "sanguine" and "sanguinary" mean the same thing, but ~~I can assure you that~~ the difference is a bloody big one ~~to the reader.~~ _The reader_ He can only ~~try to~~ infer ~~as to~~ (speaking of big differences) what the writer is trying to imply.

FIGURE 9-4 *An Example of an Edited Manuscript*

Source: Copyright © 1976, 1980, 1985, 1988, 1990, 1994, 1998 by William K. Zinsser. From *On Writing Well*, 4th ed., published by HarperCollins. Reprinted by permission of the author.

Continued

again, editing is important. When possible, ask your colleagues to review your manuscripts. Their comments will help you identify places where your meaning is unclear.

Reading and Writing

Russell Baker, who won the Pulitzer Prize for his book *Growing Up*, was asked to write a piece about punctuation as part of a series on writing published by the International Paper Company. Baker began his piece by saying, "When you write, you make a sound in the reader's head. It can be a dull mumble—that's why so much government prose makes you sleepy—or it can be a joyful noise, a sly whisper, a throb of passion." He went on to speak of the importance of punctuation in letting your voice speak to the reader. "Punctuation," he wrote, "plays the role of body language. It helps readers hear you the way you want to be heard."[8]

How can you learn to master the rules of punctuation and the principles of writing? Read. The more you read, the better you write. The better you write, the better you can communicate. The better you communicate, the better you inform, inspire, and educate. To improve your writing skills, consider adding one or more of the following resources to your professional library:

- *On Writing Well* by W. Zinsser
- *The Elements of Style* by W. Strunk, Jr., and E. B. White
- *The Chicago Manual of Style*, by the University of Chicago Press
- *The Elements of Grammar* by M. Shertzer
- *Fowler's Modern English Usage* by H. W. Fowler

References

1. W. Zinsser, *On Writing Well* (New York: Harper & Row, 1988), p. 7.
2. Ibid., p. 8.
3. Ibid., p. 153.
4. Ibid., p. 10.
5. B. Carey, Tasty tomatoes: Now there's a concept, *Health* 7 (1993): 24.
6. C. Simon, The triumphant dieter, *Psychology Today*, June 1989, p. 48.
7. G. Leonard, A change of heart, *In Health* 5 (1992): 51.
8. The quotation by Russell Baker was taken from an advertisement that appeared in *Discover*, April 1987, pp. 30–31.

References

1. M. Mason, The B vitamin breakthrough, *Health* 9 (1995): 69–73.
2. L. K. Guyer, Outcome-based nutrition education, in *Nutrition and Food Services for Integrated Health Care: A Handbook for Leaders* (Gaithersburg, MD: Aspen, 1997), pp. 206–41.
3. The "Eat five to stay alive" message was taken from P. Jaret, Only 5 a day, *Health* 12 (1998): 78; information about the "5 a Day for Better Health" program was adapted from J. Heimendinger, 5 a Day, in *Charting the Course for Evaluation: How Do We Measure the Success of Nutrition Education and Promotion in Food Assistance Programs?* (Summary of Proceedings) (Alexandria, VA: U.S. Department of Agriculture, 1997), pp. 56–57.
4. K. L. Probert, ed., *Moving to the Future: Developing Community-Based Nutrition Services* (Washington, D.C.: Association of State and Territorial Public Health Nutrition Directors, 1996), pp. 25–34.
5. A. C. Tuijnman, Concepts, theories, and methods, in *International Encyclopedia of Adult Education and Training*, 2nd ed. (New York: Pergamon, 1996), pp. 3–8; and C. J. Titmus, Adult education: Concepts and principles, in *International Encyclopedia of Adult Education and Training*, pp. 9–17.
6. K. Wagschal, I became clueless teaching the genXers, *Adult Learning* 8 (1997): 21–25.
7. D. R. Garrison, Self-directed learning: Toward a comprehensive model, *Adult Education Quarterly* 48 (1997): 18–33; and R. J. Wlodkowski, Motivation with a mission: Understanding motivation and culture in workshop design, *New Directions for Adult and Continuing Education* 76 (1997): 19–31.
8. P. Cranton, Types of group learning, *New Directions for Adult and Continuing Education* 71 (1996): 25–32.
9. *Lean 'N Easy* education kit (Chicago: National Cattlemen's Beef Association, 1996).
10. As cited in T. MacLaren, Messages for the masses: Food and nutrition issues on television, *Journal of the American Dietetic Association* 97 (1997): 733–34.
11. A. L. Eldridge and coauthors, Development and evaluation of a labeling program for low-fat foods in a discount department store foodservice area, *Journal of Nutrition*

Education 29 (1997): 159–61; and K. J. Harris and coauthors, Community partnerships: Review of selected models and evaluation of two case studies, *Journal of Nutrition Education* 29 (1997): 189–95.

12. R. L. Parrott, Motivation to attend to health messages, in E. Maibach and R. L. Parrott, eds., *Designing Health Messages: Approaches from Communication Theory and Public Health Practice* (Thousand Oaks, CA: Sage, 1995), pp. 7–23.

13. These items are all available for purchase on the National Cancer Institute's 5 a Day for Better Health Web site: dccps.nci.nih.gov/5aday.

14. E. W. Austin, Reaching young audiences, in *Designing Health Messages: Approaches from Communication Theory and Public Health Practice*, pp. 114–41.

15. T. A. Nicklas and coauthors, Development of a school-based nutrition intervention for high school students: Gimme 5, *American Journal of Health Promotion* 11 (1997): 315–22.

16. *Reaching Consumers with Meaningful Health Messages: A Handbook for Nutrition and Food Communicators* (Chicago: Dietary Guidelines Alliance, 1996).

17. L. S. Chapman, Maximizing program participation, *The Art of Health Promotion* 2 (1998): 1–8.

18. J. J. Moran, *Assessing Adult Learning: A Guide for Practitioners* (Malabar, FL: Krieger, 1997), pp. 1–20.

19. A. E. Goody and C. E. Kozoll, *Program Development in Continuing Education* (Malabar, FL: Krieger, 1995), pp. 62–68.

20. P. A. Sissel, Participation and learning in Head Start: A sociopolitical analysis, *Adult Education Quarterly* 47 (1997): 123–37.

21. R. Harris, Evaluating Internet research sources [on-line], available at www.sccu.edu/faculty/R_Harris/evalu8it.htm; and Milton's Web [on-line], available at milton.mse.jhu.edu:8001/research/education/net.html.

Chapter 10

Marketing Nutrition for Health Promotion and Disease Prevention

Outline

Learning Objectives

After you have read and studied this chapter, you will be able to:

- Develop a marketing plan.
- Conduct a situational analysis.
- Describe how to apply the four P's of marketing to the development of a marketing strategy.
- Explain how social marketing is used to promote community interventions.

Something To Think About...

I can give you a six-word formula for success: "Think things through—then follow through." – *Edward Rickenbacker*

Introduction

On her way home from work, a New York woman hears a familiar jingle on the car radio reminding her which soap to use for beautiful skin. Halfway around the world, in a remote Sri Lankan village, another woman hears a message on the community radio that teaches her about oral rehydration therapy for her child's diarrhea. Both of these women are part of a target audience for a well-planned marketing campaign. But while the New York woman is listening to traditional Madison Avenue marketing, the woman in Sri Lanka is a "consumer" receiving a message grounded in social marketing.[1] The same basic principles lie behind both the commercial and the social approaches to marketing. The aim of both approaches is to strengthen the fit between the products, services, and programs offered and the needs of the population. As you will see, marketing is for everyone, regardless of their job description. Whether you are a dietitian in private practice seeking referrals from physicians for new clients, a public health nutritionist developing nutrition education materials for pregnant teens, or a community nutritionist coordinating citywide screenings for hypertension, you can use marketing strategies. This chapter provides an overview of basic marketing principles and strategies.

What Is Marketing?

Buying and selling have a long history, but comprehensive and systematic marketing research developed only in recent decades.[2] Peter Drucker, a management consultant, is credited with demonstrating the benefits of marketing to business. In the 1950s, he suggested that the primary focus of any business should be the consumer, not the product.[3]

Most people think that marketing means selling and promotion. Perhaps that is not too surprising, given that everyday someone is trying to sell us something. Selling and promotion are only one part of marketing. So, what is marketing? **Marketing** is the process by which individuals and groups get what they need and want by creating and exchanging products and values with others.[4] Informally, marketing is the process of finding and keeping customers.[5] Many companies are very successful at commercial marketing, which employs powerful techniques for selecting, producing, distributing, promoting, and selling an enormous array of goods and services to a wide variety of people in every possible political, social, and economic context.[6]

Marketing The process by which individuals and groups get what they need and want by creating and exchanging products and values with others.

Social marketing The design, implementation, and management of programs that seek to increase the acceptability of a social idea or practice among a target group.

Social marketing draws on many of the techniques and technologies of commercial marketing, except that it seeks to increase the acceptability of a social idea or practice among a certain group of people—the target population.[7] Social marketing, described in more detail later in this chapter, is a strategy for changing consumer behavior. It combines the best elements of consumer behavior theory with marketing tools and skills to help consumers change their beliefs, attitudes, values, actions, or behaviors.[8] Social marketing promotes ideas and behaviors as "products." In the nutrition arena, a social practice to be marketed, for example, is summed up in the slogan "Got Milk?" A social idea to be marketed is the theme "Low-fat foods taste good."

Marketing, whether commercial or social, is a tool for managing change. It helps companies, government agencies, nonprofit organizations, dietitians in private practice, and others recognize, define, interpret, and cope with changing consumer values, interests, lifestyles, and purchasing behaviors.[9] The purpose of marketing is to find a problem, need, or want (through marketing research) and to fashion a solution to it. The solution to the problem, need, or want is outlined in the marketing plan, described in the next section.

Develop a Marketing Plan

When completed, the marketing plan will outline the steps for achieving the goals and objectives of the overall intervention strategy and the program plan. It describes precisely how and in what form the nutrition and health messages will be delivered to the target population. Figure 10-1 shows the steps the community nutritionist takes in developing a marketing plan.

First, determine the needs and wants of the target population, because marketing starts with the customer.[10] Some ideas about their needs and wants can be gleaned from the community needs assessment and from focus group sessions held earlier in the program planning process. Additional information can be collected by asking questions of the target population: What are your perceptions about this nutritional problem or need? About this product or service? About this agency or organization? What products or services are you buying? The goal of this step is to build a knowledge base from which to develop a marketing strategy.[11]

Next, specify the benefits of the product or service to the target population. Remember, people generally want intangible things when they buy a product or service: safety, security, happiness, attractiveness, fun. Women who sign up for the "Heartworks for Women" program (Case Study 1) may seek benefits such as reducing their risk of having a heart attack or stroke, learning new cooking skills, enjoying new recipes, and trying something new with a friend. It's the benefit that sells the product or service.[12]

Market Potential customers for a product or service; a group of unique customers who share some common life characteristics.

Target market One particular market segment pinpointed as a primary customer group.

Third, conduct a situational analysis. Analyze your potential **market**—those customers who share some common life characteristics—the environment in which your product or service will be positioned, and the competition.[13] Select a **target market,** which will be the primary, distinct customer group for your product, program, or service. In this step, you may be required to split your target population into smaller groups, each of which will respond differently to a given marketing

FIGURE 10-1 A *Marketing Plan for Health Promotion*

Source: Adapted from S. C. Parks and D. L. Moody, A marketing model: Applications for dietetic professionals, *Journal of the American Dietetic Association* 86 (1986): 40; and J. C. Levinson and S. Godin, *The Guerrilla Marketing Handbook* (Boston: Houghton Mifflin, 1994), pp. 5–7.

strategy. For example, the target population for Case Study 2, which consists of independent elderly people over the age of 75 years, may be very diverse, with some having high incomes and personal computers with Internet access and others living near the poverty line. A marketing strategy that includes the Internet may reach the former but will probably not reach the latter.

Next, develop a marketing strategy for ensuring a good fit between the goals and resources of the organization and the needs and wants of the target population.[14] The marketing strategy specifies a target market and four distinct elements traditionally known as the four P's: product, place, price, and promotion.[15] For example, every time a consumer buys a box of Shredded Wheat rather than a box of Cheerios, she made the purchase at a competitive price in a grocery store, possibly because of a television advertisement. Such a sale was the result of a marketing strategy made months before by Nabisco Brands regarding the issues of product, place, price, and promotion. (These elements are described in a later section.) This phase requires setting goals and objectives to indicate what the marketing strategy is expected to accomplish.

Before the marketing plan can move forward, a budget and timetable must be developed. The budget accounts for all expenditures related to implementing the marketing strategy, such as the cost of designing logos, action figures, brochures, videos, and so forth, and printing all educational and promotional materials. The timetable specifies the marketing activities to be done each month both before the launch of the product, program, or service and after the launch, when the goal is to keep awareness high among the target market.

After all aspects of the marketing plan have been decided, implement the plan according to the original design and then evaluate its effectiveness. Did the marketing strategy reach the right audience at the right time with the right messages? Did the target market's beliefs, attitudes, actions, or behavior change? Use the results of the evaluation to make corrections to the marketing strategy and improve the positioning of the product or service in the marketplace.

Conduct a Situational Analysis

SWOT An acronym that stands for strengths, weaknesses, opportunities, and threats. A situational analysis technique often used in market research.

A situational analysis is a detailed assessment of the environment including an evaluation of the consumer, the competition, and any other factors that may affect the program or business.[16] This step, which is critical to the ultimate success of the entire marketing plan, is sometimes referred to as a **SWOT analysis.** SWOT, an acronym for Strengths, Weaknesses, Opportunities, and Threats, requires a description of the present state of the business or agency, including its strengths and weaknesses, programs, and services, as well as the threats and opportunities present in the external environment (competition, pending legislation, and the like).[17]

Getting to Know Your Market

The first step in the marketing process is the identification and analysis of all "consumers" of your product or service—one's current and potential markets. Consumers can be categorized as one of three types: users of services, referral sources, or other decision makers.[18] The users of services are the clients themselves or potential consumers. For example, the Special Supplemental Nutrition Program for Women, Infants, and Children (WIC) program identifies its users as low-income pregnant or breastfeeding women and mothers of children under 5 years of age. The National Dairy Council targets a different market—"leader groups," such as educators, dietitians, health teachers, and dental hygienists.[19] Users of the "Heartworks for Women" program are women aged 18 years and older.

Referral sources include anyone who refers clients or customers to you. They may include physicians, social workers, teachers, and former clients. Other decision makers are those who influence the client's decision to use a service or join a program. Such people would include family members (spouses, parents) and third-party payers, among others. It is important to identify these three types of users for your particular setting. Table 10-1 illustrates how users or consumers of the "Heartworks for Women" program are classified.

Type	Consumers	Characteristics	Benefits
Users	Women living in in the city of Jeffers	Age 18+ years	Improved quality of life; decreased risk of CHD; better health
Referral sources	Cardiologists Other physicians Social workers Former clients	Health care providers, coworkers	Delayed CHD development
Decision makers	Spouse/significant other, coworkers	Age 18+ years	Want spouse, significant other, or coworker to be "healthy"

TABLE 10-1 *Worksheet to Identify Consumers: Case Study 1*

Source: Adapted from C. B. Matthews, Marketing your services: Strategies that work, *ASHA Magazine* 30 (1988): 23.

Market Research: Target Markets

An inherent part of the situational analysis is to determine and target your clients or audiences. Each target market should be viewed as a separate and different audience. For example, one survey of dietitians in private practice found they listed the following as their target populations.[20]

- Overweight women between 20 and 40 years
- Athletic men
- Middle-class women for weight reduction
- People interested in sports nutrition
- People in need of modification of diet and lifestyle
- Individuals wanting basic nutrition information and modified diets
- Physicians for referrals
- Corporations to provide employee workshops

Ideally, if resources are available, you would develop a specific marketing strategy for each target audience. Once a population group has been identified for targeting, it is important to determine its prevailing patterns of lifestyle, eating, drinking, working conditions, attitudes toward nutrition and health, and current and past state of health.[21]

Most programs or organizations find it unrealistic to service the total target market effectively. For this reason, actual and potential markets should be divided further into distinct and homogeneous subgroups, a process called **market segmentation.** Market segmentation offers the following benefits:[22]

- A more precise definition of consumer needs and behavior patterns
- Improved identification of ways to provide services to population groups
- More efficient utilization of nutrition and health education resources through a better fit between products, programs, and services and consumers

As an example of market segmentation, consider as your potential market, the population of adults 45 years or older—sometimes referred to as 45+ consumers. This total market can be divided into several more homogeneous parts, some of which are shown in Figure 10-2. Each of these segments can then be reached with a distinct marketing strategy (to be discussed shortly).

Market segmentation The separation of large groups of potential clients into smaller groups with similar characteristics. Advantages include simpler, more accurate analysis of each group's needs and more customized delivery of service.

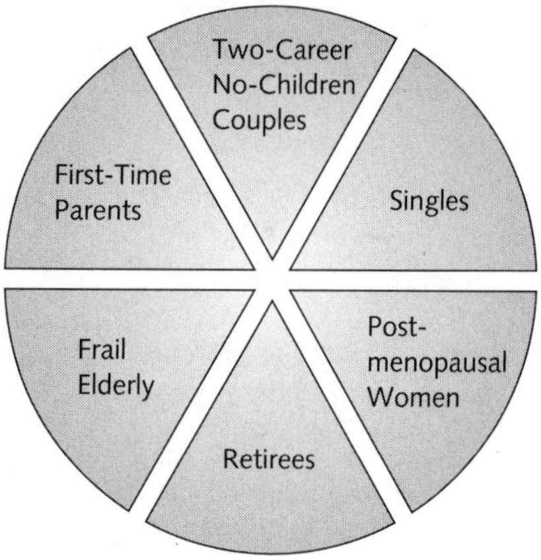

FIGURE 10-2 *Market Segmentation: The 45+ Consumers*

Market Research: Market Segmentation

Market research enables community nutritionists to target specific groups for health promotion and disease prevention in terms of their geography, demography, and psychography. This type of analysis is helpful in many ways: (1) targeting those at risk, (2) carrying out strategic marketing planning, (3) developing marketing media strategies, (4) examining the feasibility of various promotional tools (for example, direct mailing of nutrition education materials), or (5) determining the appropriate mix of nutrition programs and services to offer based on demographics (such as, concentration of women, infants, children, and the elderly).

Four classes of variables are typically used for market segmentation:[23]

1. Geographical segmentation refers to the grouping of people according to the location of their residence or work (region, county, census tract). This can be done on a simple geographical basis or according to other variables, such as population density or climate.
2. Demographic segmentation is the grouping of individuals on the basis of such variables as age, sex, income, occupation, education, family size, religion, race, marital status, and life cycle stage.
3. Psychographic segmentation is based on such criteria as personal values, attitudes, opinions, personality, behavior, lifestyle, and level of readiness for change.
4. Behavioristic segmentation is based on such criteria as purchase frequency and occasion, benefits sought, and attitude toward product.

An understanding of demographics is essential to the development and targeting of nutrition and public health programs. Consider the public health agency or wellness center with programs that promote such products as smoking cessation, low-fat diets, fitness programs, maternal-infant care, cancer prevention, hypertension screening, diabetes education, infant mortality reduction, and prevention of AIDS.[24] The community agency or wellness center must first analyze the demographics of the areas served by these programs in order to identify its clients. The information to categorize and examine includes the following:

- Total population of the area that the program is intended to serve
- Rate of change of the population
- Age and sex distribution
- Racial, ethnic, and religious composition
- Socioeconomic status
- Housing information
- Fertility patterns

Of these, age distribution of the population, trends over time, and fertility rates are particularly important to public health nutrition programs. Major segmentation variables for consumer markets are given in Table 10-2. Figure 10-3 summarizes the steps required for market segmentation and target marketing.

Obviously, the situational analysis demands a significant amount of market research. This research includes the use of both primary (direct) and secondary (indirect) data. Primary data are new data collected for the first time through random sampling surveys, questionnaires and qualitative methods such as personal interviews and focus groups. Table 10-3 describes six methods frequently used to collect primary data about a market.

Secondary data are those gathered by government agencies, private market research companies, and nonprofit organizations. Federal government sources of secondary data include the Bureau of Economic Analysis, Bureau of Justice Statistics, Bureau of Labor Statistics, Census Bureau, National Center for Health Statistics, National Technical Information Service, Social Security Administration, and U.S. Department of Agriculture. (Many of these were described in Chapter 5.) The *American Statistics Index*, *Statistical Abstract of the United States*, and *Survey of Current Business* are excellent sources of business and general economic statistics.

Private market research companies include A. C. Nielsen and America's Research Group, both of which conduct consumer behavior surveys. Some nonprofit groups such as the American Dietetic Association and other professional associations are sources of secondary data. Directories such as *Dun and Bradstreet*, *Moody's Manuals*, and *Thomas Register* provide data on businesses in a given market area. The *Sales Management Annual Survey of Buying Power* provides local

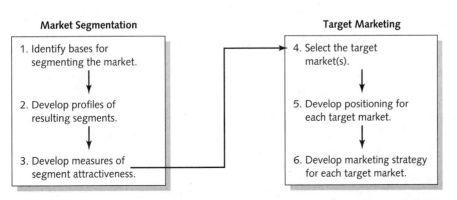

Market Segmentation

1. Identify bases for segmenting the market.

2. Develop profiles of resulting segments.

3. Develop measures of segment attractiveness.

Target Marketing

4. Select the target market(s).

5. Develop positioning for each target market.

6. Develop marketing strategy for each target market.

FIGURE 10-3 *Steps in Market Segmentation and Target Marketing*

Source: Adapted with permission from P. Kotler and R. N. Clarke, *Marketing for Health Care Organizations* (Upper Saddle River, NJ: Prentice Hall, 1987), p. 234.

TABLE 10-2 *Major Segmentation Variables for Consumer Markets*

Variable	Typical Breakdowns
Geographical	
Region	Pacific, Mountain, West North Central, West South Central, East North Central, East South Central, South Atlantic, Middle Atlantic, New England
County size	A, B, C, D
City size	Under 5,000, 5,000–20,000, 20,000–50,000, 50,000–100,000, 100,000–250,000, 250,000–500,000, 500,000–1,000,000, 1,000,000–4,000,000, over 4,000,000
Density	Urban, suburban, rural
Climate	Northern, southern
Demographic	
Age	Under 6, 6–11, 12–19, 20–34, 35–49, 50–64, 65+
Sex	Male, female
Family size	1–2, 3–4, 5+
Family life cycle	Young, single; young, married, no children; young, married, youngest child under 6; young, married, youngest child 6 or over; older, married, with children; older, married, no children under 18; older, single; other
Income	Under $2,500, $2,500–$5,000, $5,000–$7,500, $7,500–$10,000, $10,000–$15,000, $15,000–$20,000, $20,000–$30,000, $30,000–$50,000, over $50,000
Occupation	Professional and technical; managers, officials, and proprietors; clerical, sales; artisans, forepersons; operatives; farmers; retired; students; homemakers; unemployed
Education	Grade school or less, some high school, graduated high school, some college, graduated college
Religion	Catholic, Protestant, Jewish, other
Race	Asian, black, Hispanic, American Indian, white
Nationality	American, British, French, German, Italian, Japanese, Latin American, Middle Eastern, Scandinavian
Social class	Lower-lower, upper-lower, lower-middle, upper-middle, lower-upper, upper-upper
Psychographic	
Lifestyle	Straight, swinger, longhair, yuppie, conservative, liberal
Personality	Compulsive, gregarious, authoritarian, ambitious, leader, follower, independent, dependent
Behavioristic	
Purchase occasion	Regular occasion, special occasion
Benefits sought	Quality, service, economy, convenience, health
User status	Nonuser, ex-user, potential user, first-time user, regular user
Usage rate	Light user, medium user, heavy user
Loyalty status	None, medium, strong, absolute
Readiness stage	Unaware, aware, informed, interested, desirous, intending to buy
Attitude toward product	Enthusiastic, positive, indifferent, negative, hostile

Source: Adapted from P. Kotler and R. N. Clarke, *Marketing for Health Care Organizations* (Upper Saddle River, NJ: Prentice Hall, 1987), p. 237.

Method	Description
Mail survey	The most frequently used technique in market research; often misused. Tips to improve reliability: Use homogeneous sample, keep length reasonable, pretest and rewrite as necessary.
Telephone survey	Can involve errors as in mail surveys. Helpful to ask, Is the true meaning of the question being reflected by the interviewer, or is it distorted? Is the wording of the question likely to elicit a biased response?
Internet survey	Gathers data from people with E-mail accounts or who visit a particular Web site; can obtain highly specific information about people who use the Internet.
Personal interview	A recommended supplement to mail or telephone surveys; helpful in observing subtle feedback that would otherwise be unavailable.
Consumer panel	Often used to test new products using the same persons in several tests; a problem is that panel members do not always represent the buying population and may not always give honest answers.
Focus group	Frequently used to gather information from small homogeneous groups; allows the researcher to see how people view a product, intervention, or other issue; serves as a communications bridge between the researcher and the people the researcher is trying to reach; not to be used to persuade, convince, or reach a consensus.

TABLE 10-3 *Methods Used to Collect Primary Data*

Source: Adapted from S. C. Parks, Research techniques used to support marketing management decisions, in *Research: Successful Approaches,* ed. E. R. Monsen (Chicago: American Dietetic Association, 1992), pp. 308–10. © The American Dietetic Association. Reprinted with permission. Also, J. Sterne, *World Wide Web Marketing* (New York: Wiley, 1995).

information on population, income, and retail establishments. If you can't locate the secondary data you require, consider working with a specialist or information broker who can provide research services for a fee.[25]

A number of vendors also provide information about various segments of a population based on demographics, geography, and psychographics. Each vendor has a clustering system that groups individuals on the basis of like characteristics. This technique uses the statistical method called cluster analysis to classify neighborhoods by their residents' demographics, attitudes, media habits, and buying patterns.

Analyzing the Environment

The next step in the situational analysis is to identify any external environmental factors or social trends that may influence the needs of the program or organization. Such issues include health care reform, legislative and regulatory changes, shifting demographics, and the competition. For example, a declining birth rate and a growing population of older people are affecting the composition of many communities. Changes in the typical family, including older, first-time parents, the high divorce rate, and increasing numbers of working mothers, also affect the needs of a given community.

The general age of the target area is also influential in determining needed programs. In an area largely inhabited by retirement-age people, classes on prenatal nutrition counseling will have far less impact than, for instance, a class on heart-healthy cooking for one or two. Another significant trend is the maturing of the baby boom generation, which must be considered when developing or expanding any public health program.

Analyzing the Competition

Once you have a thorough understanding of your consumers, you must determine how your existing competitors are positioned in the marketplace. What are their strengths and weaknesses? What is the attitude of your target market toward the competition? Table 10-4 shows the results of such an analysis completed by a community nutritionist who specializes in cardiovascular disease.

Your aim is to find a **market niche** for your program or service in which your strengths can be matched with the needs of your particular target market. To satisfy those needs, you must know and understand your target audience so well that your service provides the perfect fit and "sells" itself, setting you apart from other providers in the same market and improving your **competitive edge.** Examples of competitive edges include your area of expertise, professional image, size, location, and customer service. You will want your target market to perceive that it can benefit from using your services as opposed to those of your competition.

Develop a Marketing Strategy

The four elements of the marketing strategy—product, place, price, and promotion—are usually referred to as the **marketing mix.** The development of the appropriate marketing mix should result directly from the previous step of the marketing process—the analysis of the consumer, environment, and competition. Once the needs of the target audience have been identified and analyzed, the set of four Ps can be constructed. The primary focus of the four Ps is the identified target market, as shown in Figure 10-4. Successful marketers get the right product, service, or program to the right place at the right time for the right price.[26]

Product

The term **product** refers to all of the characteristics of the product or service that are to be exchanged with the target market. Characteristics such as style, special

TABLE 10-4 *Worksheet for Analyzing the Competition*

Competitor	Strengths	Weaknesses	My Competitive Advantage
Wellness center	Great location; personable staff	Large and diverse; no known specialty; little follow-up after program completion	I specialize in cardiovascular nutrition counseling; I provide 6 months of follow-up (phone calls, newsletter).
Sports medicine clinic	Personable staff; good programming in sports nutrition; good name recognition	Outdated brochures and videos; nearly impossible to find parking	I maintain up-to-date educational resources; I travel to the client in my "Nutritionist-on-Wheels" mobile.

Source: Adapted from C. B. Matthews, Marketing your services: Strategies that work, *ASHA Magazine* 30 (1988): 23.

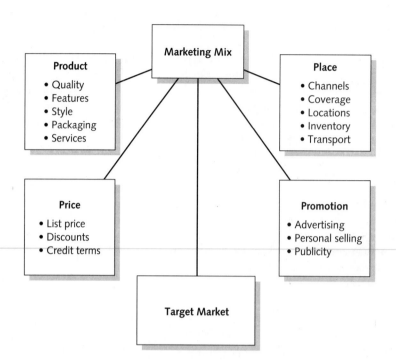

FIGURE 10-4 *The Four Ps of the Marketing Mix*

Source: Adapted with permission from M. Ward, *Marketing Strategies: A Resource for Registered Dietitians* (Binghamton, NY: Niles & Phipps, 1984), p. 111.

features, packaging, quality, brand names, and options must be designed to fit the needs and preferences of the target market. From a marketing standpoint, the product or service—whether it's a new automobile, diet soft drink, or nutrition class for a congregate meal site—is viewed as a collection of tangible and intangible attributes that may be offered to a market to satisfy a want or need.[27]

In community nutrition, the product is often a service to be delivered. These services should be of high quality, tailored to fit the needs of the target market, and adapted to meet the consumers' social characteristics (for example, culture, ethnicity, language skills). A group of dietitians in private practice delineated their services as including individual counseling for modified diets, weight reduction, sports nutrition, normal nutrition, prenatal diets, and eating disorders; group programs and workshops on nutrition topics; consulting services to schools, health care facilities, supermarkets, health spas, and restaurants; and teaching courses at community colleges and universities.[28]

Product Anything offered in the marketplace to be exchanged for money or something of value. It may be either a tangible good or a service.

Place

Place refers to the actual location where the exchange takes place. Accessibility, convenience, and comfort for the client are the criteria to consider. Are your hours of operation convenient and flexible? Is parking available?

Place also includes the channels of distribution required to deliver the service or product to the consumer. The distribution channels are the intermediaries—individuals, facilities, or agencies—that control or influence the consumer's choice of service or product.[29] Health providers, employers, school boards, voluntary health organizations, shopping malls, and commercial retailers are a few of the important intermediaries for community nutritionists. They are viewed as channels for reaching identified target markets.

In other instances, distribution channels are more like "gatekeepers"—you cannot reach your target except by going through them. Physicians, parents, media program directors, corporate executives, members of Congress, and insurance company decision makers can all be gatekeepers, depending on the service offering and target audience. The overall marketing strategy may include one approach for the client and a different approach to reach the intermediary or gatekeeper.

Distribution channels vary depending on the target market and service provided. Because third-party reimbursement by insurance companies for nutrition services usually requires a physician referral, a dietitian in private practice may want to target obstetricians, cardiologists, or other medical practice groups identified as distribution channels. Another approach the dietitian might take is to target insurance companies or state legislators in an effort to change the current third-party reimbursement system. A wellness program dietitian might identify corporate executives or employers as distribution channels, since approval for a wellness program usually rests with a company's management.

Price

Price includes both tangible costs (fee for service) and intangible commodities (time, effort, inconvenience) that the consumer must bear in the marketing exchange. Once you understand what the consumer perceives as the costs involved in adopting a health behavior or participating in a given program, you have a better chance of influencing the exchange. You may do this by persuading the consumer that the benefits to be received outweigh the perceived costs. Alternatively, incentives (money, groceries, gifts, personal recognition) can be offered to increase motivation and facilitate consumer participation. Likewise, costs can be reduced (less waiting time) or prices discounted to certain groups (senior citizens, students). Any of these tactics can considerably reduce the "price" consumers perceive that they pay for the program or service.

Promotion

The last P in the marketing mix—promotion—refers to the agency or organization's informative or persuasive communication with the target market. What do you want to say? To whom do you want to say it? When do you want to say it?[30] The communication messages are designed to have a measurable effect on the knowledge, attitude, and/or behavior of the target market.[31] The medium, message content, and message format are chosen to complement the target market's communication needs.[32]

Promotion has four general objectives:[33]

- To inform and educate consumers about the existence of a product or service and its capabilities (what the community program has to offer)
- To remind present and former users of the product's continuing existence (for example, prenatal nutrition counseling)

- To persuade prospective purchasers that the product is worth buying (improved health status, other benefits)
- To inform consumers about where and how to obtain and use the product (accessibility, location, and time)

Although people frequently think that promotion is limited to advertising, promotion actually includes much more than advertising, as shown in Table 10-5. As one author has noted:

> Marketers generally agree that although mass media approaches are appropriate for developing consumer awareness in the short term, face-to-face programs such as workplace encounters are more effective (though not always cost-effective) in changing behavior in the long term. In designing any communication policy, the marketer commonly considers the effects that may be achievable through the use of a mix of approaches, capitalizing on the strengths of each.[34]

The four most common promotional tools are advertising, sales promotion, personal selling, and publicity.

Advertising is standardized communication in print or electronic media that is purchased.[35] Examples of advertising media include telephone directory yellow pages, billboards, newspapers, radio, trade and professional journals, magazines, and the Internet. The role of advertising is to communicate a concise and targeted message that ultimately stimulates action by the carefully defined audience. Advertising can reach large numbers of people in many locations and can help build image. In advertising, you control the nature and timing of the message because you are buying the time or space. Which media you choose will depend on the characteristics of your target market, the size of your budget, and the goals of your advertising campaign. The advantages and limitations of the major media categories are listed in Table 10-6. The boxed insert offers tips for working with the media.

Advertising Any paid form of nonpersonal presentation and promotion of ideas, goods, or services by an identified sponsor.

Personal Promotion Aids	Media and Public Events	Graphic/Print Materials
One-to-one communication	Public relations	Logos
Networking	Publicity	Brochures
Business cards	News releases	Flyers
Letters	Press kits	Portfolios
Résumés	Media interviews	Proposals
Letters of reference	News conferences	Posters
Use of a name	Press briefings	Banners
Seminars	Special events	Audiovisual aids
Workshops	Celebrities	Giveaways (T-shirts,
Consulting	Advertising	mugs, tote bags,
Gifts	Direct mail	product samples)
	Public speaking	Internet Web sites
	Writing	Statement stuffers
	Contests	Catalogs
		Screen Savers

TABLE 10-5

Promotional Tools

Source: Adapted from K. K. Helm and J. C. Rose, *The Competitive Edge: Marketing Strategies for the Registered Dietitian* (Chicago: American Dietetic Association, 1986), p. 45. © The American Dietetic Association. Reprinted with permission. Also, J. Kremer and J. D. McComas, *High-Impact Marketing on a Low-Impact Budget* (Rocklin, CA: Prima, 1997), pp. 237–92.

TABLE 10-6

Advantages and Limitations of Major Media Categories

Medium	Advantages	Limitations
Newspapers	Flexibility; timeliness; good local market coverage; broad acceptance; high believability	Short life; poor reproduction quality; small "pass-along" audience
Television	Combines sight, sound, and motion; appealing to the senses; high attention; high reach	High absolute cost; high clutter; fleeting exposure; less audience selectivity
Direct mail	Audience selectivity; flexibility; no ad competition within the same medium; personalization	Relatively high cost; "junk mail" image
Radio	Mass use; high geographical and demographic selectivity; low cost	Audio presentation only; lower attention than television; nonstandardized rate structures; fleeting exposure
Magazines	High geographical and demographic selectivity; credibility and prestige; high-quality reproduction; long life; good pass-along readership	Long ad purchase lead time; some waste circulation; no guarantee of position
Outdoor	Flexibility; high repeat exposure; low cost; low competition	No audience selectivity; creative limitations
Internet	Low cost; high selectivity; personal; is interactive one-on-one marketing	Not always statistically representative; can't verify user identity

Source: Adapted from P. Kotler and R. N. Clarke, *Marketing for Health Care Organizations* (Upper Saddle River, NJ: Prentice Hall, 1987), p. 451; and J. Sterne, *World Wide Web Marketing* (New York: Wiley, 1995), pp. 239–68.

Sales promotion Short-term incentives to encourage purchases or sales of a product or service.

Personal selling/ communication Oral presentation in a conversation with one or more prospective purchasers for the purpose of making sales or building goodwill.

The Professional Focus at the end of Chapter 8 offers numerous tips for effective public speaking.

Sales promotion consists of such things as coupons, free samples, point-of-purchase materials, and trade catalogs. The use of these activities as part of the promotion strategy encourages potential consumers to purchase or use a particular product or service.

Personal selling or communication can be done through small group meetings, counseling sessions and nutrition classes, formal presentations to organizations or community groups, displays and booths at health fairs and related conferences, and telephone conversations or personal meetings with the public and other professionals (such as your referral sources). Unlike the standardized message presented in advertising, the message presented in personal selling can be tailored to fit the needs of the particular individual or group. Personal selling also offers other advantages:[36]

- Direct contact provides positive feedback to the listener through both the verbal and the nonverbal gestures of the communicator.
- The interpersonal contact facilitates the transfer of the message better than alternate methods.
- The communicator can ensure comprehension by asking questions and monitoring responses.
- The communicator has an opportunity to probe for resistance to change and then is in a position to address each issue.
- The listener may believe that someone is now in a position to monitor his behavior and thus hold him accountable for it.

Media Tips

Louis Pasteur said "chance favors the prepared mind." When working with the media, always be prepared to provide credible nutrition information. Besides keeping current with the various scientific and trade journals and the popular press, you'll need to consider radio and television news and talk shows as well. (Chapter 6's Professional Focus offered tips for becoming a media monitor.) By doing so, everyone benefits: the media by providing information of value to their audiences, the public by receiving accurate information, and you as a community nutritionist by enhancing your image and visibility as a professional. As you work with the media, consider the following tips:

- Be sure that the information you supply is accurate—check and recheck all names, dates, facts, and figures. This helps establish your credibility with the media.
- Become familiar with the format and types of coverage of the various media—television and radio news and talk shows, newspapers, newsletters and local publications, and trade publications and magazines. Adapt your messages to the format of the media you choose. Identify how a particular audience will benefit from the information you provide.
- Nurture good press relations with the media contact people (reporters, editors, program directors, and producers) in your area. Keep a list of the names, addresses, phone numbers, and deadlines of the media in your community. They are the "gatekeepers" to your target audiences.
- Write a "pitch" letter or "news release" about a topic to the medium of your choice. Consider your letter or release to be your personal sales representative. Convince your readers that the information you possess is of value and interest to their audiences. Writers learn to capture their readers' attention with a news "hook" in the first paragraph, followed by answers to the five Ws: who, what, when, where, and why.
- Occasionally, you may send an "FYI" (for your information) piece about a research report or recently published article to the people on your contact list. This alerts them to potential newsworthy topics or events they may find useful either now or at a later date. Make yourself available to them for follow-up information or assistance.
- Be consumer oriented. Keep your audience in mind as you prepare your media information. Be both an authority on nutrition information and an entertainer, when appropriate. Get the audience interested and involved with your message. Consider using visuals to support your messages.
- When working with television or radio, consider your appearance—dress with professional style. Practice your presentation as much as possible beforehand. You'll want your facial expressions, body language, and voice to show enthusiasm and animation for your topic.

Adapted from G. A. Levey, Communicating with the media, in *Communicating as Professionals*, R. Chernoff, ed. (Chicago: American Dietetic Association, 1986), pp. 57–60; and *The Competitive Edge: Marketing Strategies for the Registered Dietitian* (Chicago: American Dietetic Association, 1986), pp. 54–62.

Publicity Nonpersonal stimulation of demand for a product, service, or business unit by planting commercially significant news about it in a published medium or obtaining favorable presentation of it on radio, television, or stage that is not paid for by the sponsor.

Publicity, or public relations, is used to create a positive image of an individual or organization in the mind of the consumer. The American Dietetic Association receives publicity every March during National Nutrition Month activities.

Publicity tools include articles in newsletters or local newspapers, informational brochures and newsletters, radio and television interviews, other forms of public speaking, displays, posters, audiovisual materials, thank-you notes, Internet Web sites, and other tools that present a favorable image of the organization or professional to the target market.

Public service announcements (PSAs) can also be used as a form of publicity. They offer the advantage of being free of charge and have the potential of reaching a large audience. PSAs are brief messages—often only 30 to 60 seconds in length—used to promote programs, activities, and services of federal, state, or local governments and the activities of nonprofit organizations when they are regarded as serving a community interest.[37] A nutrition message can be a PSA with credit given to the organization submitting it. Here are five tips for creating effective PSAs:[38]

- Produce announcements of professional quality. One professional PSA is better than several low-quality ones.
- Get the audience involved. Sound effects, questions, repetition, and humor are sometimes more effective at grabbing the attention of the audience than the factual approach.
- Market the service offered, not just the PSA topic. Offer a brochure, a toll-free information hotline, or a screening service.
- Simplify the response action needed. Advertise a local phone number or an easy-to-remember PO box number.
- Develop and nurture a good rapport with the television and radio public service directors. They may be able to improve the PSA's production quality and increase scheduled air time.

Two additional promotional tools deserve mention: direct mail and word-of-mouth referrals. Many marketers develop brochures, newsletters, fliers, and other promotional items that are mailed to specific targeted groups or geographical areas. Others claim that word-of-mouth referrals—having associates and clients do some of the promoting for you—are one of the most important promotional strategies. A successful program or service will generate its own word-of-mouth publicity because satisfied consumers will tell others about the program and may encourage their participation.

The promotional strategies used by individuals or organizations will vary depending on the populations they are trying to reach, the goals of the program, and the resources available for promoting it. Check the boxed insert for a list of Internet addresses related to the Food Guide Pyramid—a widely-promoted educational tool— and cultural diversity.

Monitor and Evaluate

Evaluation is the key to the success of any marketing program. Methods include tracking changes in volume or net profit, referral sources, and customer satisfac-

⟲ Internet Resources

Internet Addresses for Food Guide Pyramids

Asian Diet Pyramid	www.news.cornell.edu/science/Dec95/st.asian.pyramid.html
Latin American Diet Pyramid*	www.oldwayspt.org/html/p_latin.htm
Mediterranean Diet Pyramid	www.oldwayspt.org/html/p_med.htm
Miscellaneous Food Guide Pyramids†	www.kde.state.ky.us/bmss/odss/dscn/intpyr.htm
Puerto Rican Food Guide Pyramid	www.hispanichealth.com/pyramid.htm
Vegetarian Food Guide Pyramid‡	www.eatright.org/adap1197.html
USDA Food Guide and other Pyramids§	www.nal.usda.gov/fnic/Fpyr/pyramid.html

Internet Addresses for Cultural Diversity

Cultural Diversity—Eating in America‖	www.ag.ohio-state.edu/~ohioline/lines/food.html
Working with Culturally Diverse Audiences#	www.osu.orst.edu/dept/ehe/diverse.html
The Web of Culture	www.webofculture.com

*Asian and Vegetarian pyramids are also available on this site sponsored by Harvard University and Oldways Preservation and Exchange Trust.
†This Web site is sponsored by the Kentucky Department of Education, School and Community Nutrition. It provides links to food guide pyramids posted by some universities, food companies, and trade organizations on the Internet.
‡This Web site is sponsored by the American Dietetic Association.
§USDA = U.S. Department of Agriculture. Scroll down and click on the line for Ethnic/Cultural and Special Audience pyramids.
‖This site is sponsored by Ohio State University's College of Food, Agricultural, and Environmental Sciences.
#This site is sponsored by Oregon State University Extension Home Economics.

tion.[39] In this step of the marketing process, you need to ask the following questions: Are you accomplishing your goals? Who benefited from the service? What changes in knowledge, attitudes, and practices occurred? What are the actual costs of providing the service? What changes are warranted to make the service more effective in the future? For example, once you understand the profile of clients who were successful with a given program, your promotional efforts can be targeted more effectively.

If your marketing plan included a thorough situational analysis and you carefully translated these results into a marketing mix for your targeted audience, a periodic assessment may be all that is needed. Marketing is an ongoing process, however, and situations change—sometimes affecting your marketing strategy. If so, you may need to reevaluate your objectives and goals. Are they still achievable? Do you need to take an alternative direction? Redirect your strategy?

In *Guerilla Marketing*, J. C. Levinson cautions against abandoning one's marketing plan once it begins generating favorable results. Instead he advises an ongoing commitment to your marketing orientation and continued investment in your marketing strategy. He offers "ten truths you must never forget":[40]

1. The market is constantly changing.
2. People forget fast.
3. Your competition isn't quitting.
4. Marketing strengthens your identity.
5. Marketing is essential to survival and growth.
6. Marketing enables you to hold on to your old customers.
7. Marketing maintains morale.
8. Marketing gives you an advantage over competitors who have ceased marketing.
9. Marketing allows your business to continue operating.
10. You have invested money that you stand to lose.

Social Marketing: Community Campaigns for Change

Social marketing makes a comprehensive effort to influence the acceptability of social ideas in a population, usually for the purpose of changing behavior.[41] Health ideas such as cardiovascular fitness, low-fat eating, and hypertension screening can be sold in the same way as presidential candidates or toothpaste.[42] Examples of social marketing include the public service messages produced by electronic and print media such as messages intended to change behavior related to smoking, hypertension, use of seat belts, teenage pregnancy, driving after drinking, drug use, safe sex, suicide, and similar concerns.

Whereas traditional marketing seeks to satisfy the needs and wants of targeted consumers, social marketing aims to change their attitudes and/or behavior. To do so, the marketing process must be followed in its entirety.[43] Marketers identify four types of behavior change, listed here in order of increasing difficulty:

- **Cognitive change.** A change in knowledge is the easiest to market, but there appears to be little connection between knowledge change and behavior change. Examples include campaigns to explain the nutritional value of different foods or campaigns to expand awareness of government programs such as The Food Stamp Program or WIC.
- **Action change.** This kind of change is more difficult to achieve than cognitive changes, as the individual must first understand the reason for change and then invest something of value (time, money, energy) to make the change. Screening programs for hypertension, hypercholesterolemia, or breast cancer involve this type of change.
- **Behavior change.** This type of change is more difficult to achieve than either cognitive or action changes because it costs the consumer more in terms of personal involvement on a continuing basis—for example, the adoption of a low-fat, high-fiber daily diet or the addition of regular physical exercise to one's lifestyle.
- **Value change.** This type of change is the most difficult to market. An example would be population control strategies to persuade families to have fewer children.

Social marketing goes beyond advertising. It seeks to bring about changes in the behavior as well as in the attitudes and knowledge of the target audience.[44] Social marketing can be applied to a wide variety of social problems but is particularly appropriate in three situations:[45]

1. When new research data and information on practices need to be disseminated to improve people's lives. Examples include campaigns for cancer prevention, breastfeeding promotion, and childhood immunization.
2. When countermarketing is needed to offset the negative effects of a practice or promotional efforts of companies for products that are potentially harmful—for example, cigarettes, alcoholic beverages, or the promotion of infant formula in developing countries.
3. When activation is needed to move people from intention to action—for example, motivating people to lose weight, exercise, or floss their teeth. Without movement, there is no marketing.[46]

Social Marketing at the Community Level

The Pawtucket Heart Health Program (PHHP) is presented here as an example of how marketing principles can be used in the planning, implementation, and evaluation of a community-wide social marketing campaign. The Pawtucket "Know Your Cholesterol" campaign was one of the earliest cholesterol awareness and screening efforts.[47] The following objectives for the campaign were formulated based on national random sampling data of both the general population and physicians:

- Physician education
- Increased awareness among the general population of blood cholesterol as a risk factor for coronary heart disease
- Increased numbers of people knowing their cholesterol level as a result of attending PHHP-sponsored screening, counseling, and referral events (SCOREs)
- Large numbers of people showing reductions in their blood cholesterol level at two-month follow-up measurements

In segmenting the community of Pawtucket, Rhode Island, adults were a primary focus because demographics (gender and age) showed that awareness levels were equivalent for men and women in the national sample. Cardiologists, family practice physicians, and physicians in general practice were targeted for direct mail educational packages and grand rounds presentations on blood cholesterol and heart disease at the community hospital. Middle-aged men who had previous contact with the PHHP were also the focus of a direct mail and telemarketing campaign to attend SCOREs during the campaign period.

Early program development steps included a pilot test of a self-help "nutrition kit" on lowering cholesterol levels and pretesting of the SCORE protocol at the community hospital. Promotional tools selected to reach the general and targeted audiences included newspapers; print media distributed through worksites, churches, and schools; direct mail and telemarketing; and SCORE delivery at worksites, churches, and various other community locations. Television and radio were not used to avoid spillover into the control community.

In the marketing mix, SCOREs were initially priced at $5 per person for both an initial and follow-up measurement. The researchers reasoned that people who had already paid for a second measurement would be more likely to have a follow-up test than if they had to pay for it separately. Price reductions and specials were also offered. Promotional publicity strategies included the "kick-off" SCORE at a St. Patrick's Day parade and six weekly advice columns in the local newspaper.

Results showed that 1,439 adults attended 39 SCOREs. Sixty percent were identified as having elevated serum cholesterol levels. Two months after the campaign, 72.3 percent of these persons had returned for a second measurement. Nearly 60 percent of this group had reduced their serum cholesterol levels.

The essential components of the campaign's marketing strategy were integrated into the ongoing activities of the PHHP. During the first 2 years following the campaign, over 10,000 persons had their blood cholesterol level measured, were given information on dietary management of high serum cholesterol, and were referred to physicians when necessary. A later survey of local physicians showed they were more aggressive in initiating either diet or drug therapy than physicians in the neighboring community or those who had participated in the national survey. The local physicians cited increased patient requests for blood cholesterol measurements and/or nutrition information as the major reason for changing their practice. The researchers concluded that informed consumers had influenced changes in their physicians' treatment of high serum cholesterol levels. Their overall conclusion was that "a well-functioning marketing operation can lead to more effective and efficient use of resources and improved consumer satisfaction. . . . **Health marketing** has the potential of reaching the largest possible group of people at the least cost with the most effective, consumer-satisfying program."[48]

> **Health marketing** Health promotion programs that are developed to satisfy consumer needs, strategized to meet as broad an audience as is in need of the program, and thereby enhance the organization's ability to effect population-wide changes in targeted risk behaviors.

A Marketing Plan for "Heartworks for Women": Case Study 1

Team 2 reviewed the results of the community needs assessment (see Table 7-1), the goals and objectives for the intervention strategy (see Table 8-4), and the program goals (see Chapter 9). The following needs and wants of the target population were identified in the community needs assessment and in focus group discussions: want to stop smoking (one participant's view: "I want to get this monkey off my back."); want to feel better; want to look better; don't want to have a heart attack (one participant remarked, "My mother died of a heart attack at the age of 44. She was shopping at the mall and just went like that. I don't want that to happen to me."). Some women wanted better health and more information about low-fat cooking (one participant said, "I just buy whatever is convenient to cook. Junk food is quick and easy.").

The benefits of the "Heartworks for Women" program, shown in Table 10-1, addressed these needs and wants. The results of the situational analysis revealed one major target segment: working women (64 percent of the total market), who were mostly aged 25 to 58 years. Two smaller segments were university/college students (20 percent), aged 18 to 24 years; and an "other" category (16 percent), which included stay-at-home mothers, retirees, and other women who did not work outside the home or go to school. The latter group had two main age segments: 18 to

Think about novel ways to reach your target population. Here, an eye-catching city bus promotes the "Got Milk?" message.

30 years and 55+ years. Drawing on an analysis of the broad environment in which these women live and work, team 2 found both positive and negative elements in the environment. On the positive side, smoking was not allowed in government buildings, and many women were aware of low-fat eating messages in the media. On the negative side, the media feature smoking as attractive and seldom present women, especially middle-aged women, as fit and active. Competition for the "Heartworks for Women" program included weight management programs offered by all private health/fitness clubs and counseling provided by 11 dietitians in private practice in the areas of weight management and cholesterol reduction.

The objectives for the marketing strategy were designed to meet the broad goals outlined for the "Heartworks for Women" program: (1) increase women's awareness of the relationship of diet to CHD risk, and (2) build skills related to low-fat eating and cooking. Specific objectives of the marketing strategy for the program are as follows:

- The "Heartworks for Women" program will be offered in 60 companies within the city by the end of 1 year.
- The "Heartworks for Women" program will be offered in all five universities and colleges within the city by the end of 1 year.
- At least 100,000 women living in the city of Jeffers will be exposed to "Heartworks for Women" messages through the following channels: advertising (city bus), flyers, promotional brochures, posters, radio interviews, newspaper articles, and the Internet Web site.

The marketing mix for the "Heartworks for Women" program is shown in Figure 10-5. It focuses on worksites and universities/colleges as the primary gates for delivering program messages and promoting the program. Partnerships with local private health/fitness clubs provide another opportunity for boosting the program's visibility in the community. Promotion of the program will occur through flyers, brochures, posters in company cafeterias, press releases related to special events, the painted city bus, the Internet Web site, and the Mother/Daughter Walk. The program will be priced competitively at $60 per participant.

FIGURE 10-5

Marketing Mix for the "Heartworks for Women" Program

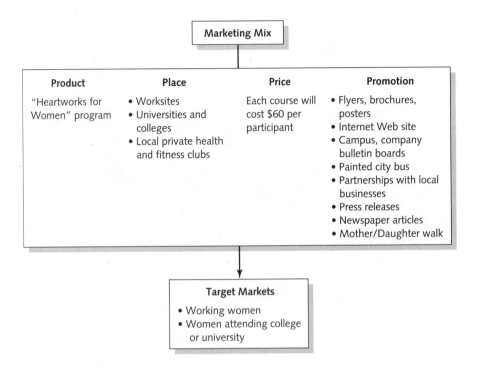

Marketing Mix			
Product	**Place**	**Price**	**Promotion**
"Heartworks for Women" program	• Worksites • Universities and colleges • Local private health and fitness clubs	Each course will cost $60 per participant	• Flyers, brochures, posters • Internet Web site • Campus, company bulletin boards • Painted city bus • Partnerships with local businesses • Press releases • Newspaper articles • Mother/Daughter walk

Target Markets
• Working women
• Women attending college or university

The projected budget for the marketing strategy must take into account both staff time for designing materials and the cost of printing and duplicating promotional materials such as flyers, posters, and brochures. (Additional information about the projected budget for the marketing strategy is presented in the next chapter.) The timeline allows team 2 to schedule all marketing activities prior to the program launch date. An example of a timetable is shown in Figure 10-6. Of course, in real life, the timetable is considerably larger than this, because it shows all activities for the program, including those designed to sustain awareness and increase participation over the life of the program.

As with all other aspects of program planning, the marketing strategy must be evaluated. In this example, the summative evaluation conducted 6 months after the launch determined that only 35 companies had signed on for the program, and most of these were large companies with a human resources department and more than 200 employees. Team 2 realized that two changes should be made in the marketing strategy. First, an important segment of the target group—women who worked in light manufacturing companies with no human resources department and fewer than 100 employees—were not aware of the program. Thus, the marketing strategy and the program format had to be redesigned to be more attractive to this group. For example, rather than offering a full eight-session course, small companies were offered only two or three sessions, each given in 45-minute sessions held over the lunch hour. Each individual session cost only $5 per participant. The marketing mix was changed accordingly. Second, the decision to paint a city bus with the program logo and one nutrition message had been made early in the planning process, after

Launch ↓

Marketing Tool	Jan.	Feb.	Mar.	Apr.	May	Jun.	Jul.	Aug.	Sept.	Oct.	Nov.	Dec.
Bulletin boards												
Campus			✔	✔	✔			✔	✔	✔		
Worksite			✔	✔	—————————————→							
Flyers			✔	✔				✔	✔			
Brochures			✔	——————————————————→								
City bus				✔	———————————————→							
Internet Web site			✔	——————————————————→								
Mother/Daughter walk									✔			
Press releases				✔					✔			
Radio announcement				✔					✔			
Newspaper articles				✔					✔			

team 2 learned of its success in another city. However, the chosen bus route was one that included a small mall and two large residential sections in an affluent part of town. Team 2 decided to choose another bus route, one that circulated through two industrial parks which featured light manufacturing businesses. This bus route would provide more program visibility for women in the target market and for small companies, which were more difficult to recruit to the program than large companies. Adjustments in the marketing strategy would be made again in another 6 months when another evaluation was planned.

FIGURE 10-6

Marketing Timetable for the "Heartworks for Women" Program

Entrepreneurship Leads the Way

The challenge for the remainder of the decade and into the next century will be to use the marketing strategies described in this chapter to remind consumers of the benefits of good health and to motivate them to make behavior changes. The important thing is to stay focused on the needs and wants of the "consumer" of health promotion activities. Remember, too, that consumers are a diverse lot, and broad target groups such as men aged 18 to 49 years or teenagers or postmenopausal women no longer describe today's consumer market. Consumers can and should be grouped into smaller, better-defined categories. In today's marketplace, the Internet makes one-to-one marketing a reality. Even mass marketing, a successful marketing tool of the past, is giving way to selective, target marketing. Tomorrow's consumer promises to be an independent thinker, highly educated and sophisticated, demanding, a seeker of innovation, and a pursuer of wellness.[49] Marketing helps capture this changing profile, and entrepreneurship helps plan for it.

Community Learning Activity

Activity 1

1. For Case Study 1, described in this and previous chapters, describe how some promotional activities for the "Heartworks for Women" program can be tied to the larger marketing plan developed for the entire intervention strategy (that is, one that includes both the antismoking and nutrition components).
2. Describe how you would use the organization's Internet Web site to promote the program.
3. Figure 10-6 shows a marketing timetable for the "Heartworks for Women" program. In April and again in September, marketing activities are designated for press releases, radio announcements, and newspaper articles, all coinciding with either the program launch or the Mother/Daughter walk. Describe the groundwork that must be done to ensure that the media are aware of these events.

Activity 2

Imagine that you have developed a foods and nutrition screen saver. (Refer to Community Learning Activity 3 in Chapter 9). Develop a marketing plan for the screen saver, providing a detailed description of the four P's.

Activity 3

For Case Study 2, the focus of previous Community Learning Activities in this section:

1. Identify your consumers: the users, referral sources, and key decision makers within this group. Are there any environmental factors that you need to consider?
2. What sources of data might you use to better understand your target group?
3. Describe each element of your marketing mix: product, place, price, and promotion.

Professional Focus | *Communicating Nutrition and Health Fraud*

At the same time that science has shown that to some extent we really are what we eat, many consumers are more confused than ever about how to translate the steady stream of new findings about nutrition into healthful eating.[1] Each additional nugget of nutrition news that comes along raises new concerns: Is caffeine bad for me? Does feverfew prevent headache? Should I take vitamin supplements? Will diet pills work? Can a sports drink improve my performance? Do pesticides pose a hazard? As a community nutritionist, you will find that many consumers turn to you as the resident expert on these questions and more.

Some manufacturers of food and nutrition-related products, as well as many members of the media, feed into the confusion by offering a myriad of unreliable products and misleading dietary advice targeted to health-conscious consumers. Unfortunately, many consumers fall prey to this vast array of information. Americans spend some $30 billion annually on medical and nutritional health fraud and quackery, up from $1 billion to $2 billion in the early 1960s.[2] Consider that college athletes alone may spend as much as $400 a month on nutritional supplements, even though most of the products pitched to serious exercisers are useless and, in some cases, potentially harmful.[3] According to the Federal Trade Commission, Americans also spend $6 billion each year on diet products. Because deceptive marketing of such items is so widespread, in 1997 the agency launched Operation Waistline—a long-term education and law enforcement program designed to alert consumers to misleading weight-loss claims and bring legal action to companies that violate the law.[4] This Professional Focus provides a sieve with which you can help consumers separate valid from bogus nutrition information.

Money down the drain is just one of the problems stemming from misleading dietary information. While some fraudulent claims about nutrition are harmless and make for a good laugh, others can have tragic consequences. Swallowing false claims about nutritional products has been known to bring about malnutrition, birth defects, mental retardation, and even death in extreme cases. Negative effects can happen in two ways. One is that the product in question causes direct harm. Even a seemingly innocuous substance such as vitamin A, for instance, can cause severe liver damage over time if taken in large enough doses. The other problem is that spurious nutritional remedies build

false hope and may keep a consumer from obtaining sound, scientifically tested medical treatment. A person who relies on a so-called anticancer diet as a cure, for example, might forgo possible lifesaving interventions such as surgery or chemotherapy.

Part of the confusion consumers experience stems from the way the media choose to interpret the findings of scientific research. A good case in point is the controversy over whether people should take supplement pills containing large doses of the antioxidant, beta-carotene. Since the early 1990s, the media have delivered a steady stream of news stories hailing the benefits of beta-carotene in, among other things, preventing cancer. Supplement manufacturers were quick to get on the bandwagon, offering consumers an ever-growing assortment of beta-carotene pills.

In 1996, a flurry of headlines threatened to pull the pedestal out from under beta-carotene. The *New York Times,* for example, said, "Studies Find Beta-Carotene, Taken By Millions, Can't Forestall Cancer or Heart Disease."[5] A 12-year trial called the Physicians' Health Study found that beta-carotene supplements had no effect on rates of cancer or heart disease in some 22,000 doctors who participated.[6] At the same time, researchers decided to end another large study ahead of schedule. Known as the Beta-Carotene and Retinol Efficacy Trial, the investigation indicated that beta-carotene supplements raised the risk of lung cancer among participants, most of whom were smokers and former smokers.[7] In other words, the new results and some of the news reports that covered them implied that scientists had deceived the public by overstating the value of beta-carotene.

The beta-carotene story illustrates how news reports based on just one study can leave the public with the impression that scientists can't make up their minds. It seems as if one week scientists are saying that beta-carotene is good, and the next week the word is that it is harmful. The truth is that few experts and no major health organizations have recommended that the general public take beta-carotene supplements because, as the studies that created such a stir underscored, the answers on beta-carotene are not all in. Most scientists have been emphasizing all along that eating fruits and vegetables is the key to reaping any potential benefits from beta-carotene and other antioxidants.

Contrary to what some headlines imply, reputable scientists do not base dietary recommendations for the public on

Continued

the findings of just one or two studies. Scientists are still conducting research to determine whether vitamin supplements do in fact help to prevent disease and, if so, in what dosages. Scientists design their research to test theories, such as the notion that taking a vitamin tablet is associated with a lower risk of cancer.

Even if an experiment is carefully designed and carried out perfectly, however, its findings cannot be considered definitive until they have been confirmed by other research. Testing and retesting reduce the possibility that the outcome was simply the result of chance or an error or oversight on the part of the experimenter. When making dietary recommendations for the public, experts pool the results of different types of studies and consider all of them before coming to any conclusions. The dietary guidelines spelled out in the 1,300-page *Diet and Health* report are based on the results of hundreds of studies. Thus, if a story in the paper or on television advises making a dramatic change in your diet based on just one study, exercise a little caution and a lot of common sense. The study may make a good story, but it may not be strong enough evidence for a radical diet or lifestyle change.

Consumers sometimes ask why the government doesn't prevent the media from delivering misleading nutrition information, but the government lacks the power to do so. The First Amendment guarantees freedom of the press, which means that people may express whatever views they like in the media, whether sound, unsound, or even dangerous. By law writers cannot be punished for publishing misinformation unless it can be proved in court that the information has caused a reader bodily harm.

Fortunately, most professional health groups maintain committees to combat the spread of health and nutrition misinformation. In addition, many professional organizations have banded together to form the National Council Against Health Fraud (NCAHF), which has branches in many states. The NCAHF monitors radio, television, and other advertising, and investigates complaints. The list of Internet Web sites on the next page can serve as sources for your own inquiries about the authenticity of information in nutrition and health-related areas. Remember, too, that although the Internet is an excellent source of an incredible amount of health information, it has also become one of the fastest growing outlets for health fraud. The quality of the information found on the Web is only as good as its

source. If you suspect that a Web site is relaying fraudulent information or selling bogus products on the Internet, e-mail the FDA at otcfraud@cder.fda.gov.

Unlike journalists, purveyors of products are bound by law to make only true statements about their wares. The FDA has the authority to prosecute companies that display false nutrition information on product labels or in advertising materials. Combating health fraud, however, is an overwhelming task requiring enormous amounts of time and money. As one FDA official put it, "Quack promoters have learned to stay one step ahead of the laws either by moving from state to state or by changing their corporate names."[8]

It's not always easy to separate the nutrition wheat from the chaff, given that many misleading claims are supposedly backed by scientific-sounding statements, but you can offer your clients some tips to help them tell whether a product is bogus. The following red flags can help you spot a quack:

- *The promoter claims that the medical establishment is against him or her and that the government won't accept this new "alternative" treatment.*

If the government or medical community cannot accept a treatment, it is because the treatment has not been proved to work. Reputable professionals do not suppress knowledge about fighting disease. On the contrary, they welcome new remedies for illness, provided the treatments have been carefully tested.

- *The promoter uses testimonials and anecdotes from satisfied customers to support claims.*

Valid nutrition information comes from careful experimental research, not from random tales. A few persons' reports that the product in question "works every time" are never acceptable as sound scientific evidence.

- *The promoter uses a computer-scored questionnaire for diagnosing "nutrient deficiencies."*

Those programs are designed to show that just about everyone has a deficiency that can be reversed with the supplements the promoter just happens to be selling, regardless of the consumer's symptoms or health.

- *The promoter claims that the product will make weight loss easy.*

Unfortunately, there is no simple way to lose weight. In other words, if a claim sounds too good to be true, it probably is.

- *The promoter promises that the product is made with a "secret formula" available only from this one company.*

Legitimate health professionals share their knowledge of proven treatments so that others can benefit from it.

- *The treatment is provided only in the back pages of magazines, over the phone, or by mail-order ads in the form of news stories or 30-minute commercials (known as infomercials) in talk-show format.*

Results of studies on credible treatments are reported first in medical journals and administered through a physician or other health professional. If information about a treatment only appears elsewhere, it probably cannot withstand scientific scrutiny.

Internet Addresses

Center for Food Safety and Applied Nutrition
vm.cfsan.fda.gov/list.html
Federal Trade Commission's Consumer Response Center
www.ftc.gov/bcp/conline/fraud.htm
National Council Against Health Fraud
www.ncahf.org
National Fraud Information Center
www.fraud.org
Quackwatch
www.quackwatch.com

References

1. This discussion was adapted from M. A. Boyle and G. Zyla, *Personal Nutrition*, 4th ed. (St. Paul, MN: West, 1999), Chapter 1.
2. Position of the American Dietetic Association, Identifying food and nutrition misinformation, *Journal of the American Dietetic Association* 88 (1988): 1589–91; S. H. Short, Health quackery: Our role as professionals, *Journal of the American Dietetic Association* 94 (1994): 607–608.
3. S. H. Short, 1994; R. M. Philen and coauthors, Survey of advertising for nutritional supplements in health and bodybuilding magazines, *Journal of the American Medical Association* 268 (1992): 1008–1011.
4. Federal Trade Commission News, FTC Announces 'Operation Waistline'—a Law Enforcement and Consumer Education Effort Designed to Stop Misleading Weight Loss Claims, March 25, 1997.
5. G. Kolata, Studies Find Beta-Carotene, Taken By Millions, Can't Forestall Cancer or Heart Disease, *New York Times*, January 19, 1996, p. A16.
6. Brigham and Women's Hospital News, Physicians' Health Study Shows Beta-Carotene Supplements to Be Without Benefit or Harm, January 18, 1996.
7. National Cancer Institute Office of Cancer Communications, Beta-carotene and vitamin A halted in lung cancer prevention trial, January 18, 1996.
8. Top 10 Health Frauds, *FDA Consumer* (October 1989): 29–31.

References

1. The opening vignette was adapted from E. Clift, Social marketing and communication: Changing health behavior in the Third World, *American Journal of Health Promotion* 3 (1989): 17–24.
2. P. F. Basch, *International Health* (New York: Oxford University Press, 1990), p. 285.
3. S. C. Parks and D. L. Moody, A marketing model: Applications for dietetic professionals, *Journal of the American Dietetic Association* 86 (1986): 37–43.
4. G. H. G. McDougall, P. Kotler, and G. Armstrong, *Marketing*, 2nd ed. (Scarborough: Prentice Hall Canada, 1992), pp. 4–10.
5. J. Trivers, *One Stop Marketing* (New York: Wiley, 1996), p. 31.
6. Clift, Social marketing, p. 17. The margin definition was adapted from P. Kotler, *Marketing in Non-Profit Organizations*, 2nd ed. (Upper Saddle River, NJ: Prentice Hall, 1982), p. 6.
7. McDougall, Kotler, and Armstrong, *Marketing*, p. 478.
8. P. Kotler and E. L. Roberto, *Social Marketing: Strategies for Changing Public Behavior* (New York: Free Press, 1989), pp. 24–61.
9. E. S. McKay, *The Marketing Mystique* (New York: AMACOM, 1994), pp. 3–16.
10. Trivers, *One Stop Marketing*, p. 44.
11. P. Francese, *Marketing Know-How: Your Guide to the Best Marketing Tools and Sources* (Ithaca, NY: American Demographics Books, 1996), pp. 15–29.
12. J. C. Levinson and S. Godin, *The Guerrilla Marketing*

Handbook (Boston: Houghton Mifflin, 1994), p. 7.

13. Trivers, *One Stop Marketing*, p. 135.

14. G. E. A. Dever, *Community Health Analysis*, 2nd ed. (Gaithersburg, MD: Aspen, 1991), pp. 252–53.

15. E. J. McCarthy, S. J. Shapiro, and W. D. Perreault, *Basic Marketing*, 6th ed. (Homewood, IL: Irwin, 1992), pp. 26–52.

16. The discussion of the marketing planning process was adapted from Matthews, Marketing your services: Strategies that work, *ASHA Magazine* 30 (1988): 21–25.

17. K. K. Helm and J. C. Rose, *The Competitive Edge: Marketing Strategies for the Registered Dietitian* (Chicago: American Dietetic Association, 1986), pp. 17–18.

18. Matthews, Marketing your services, p. 23.

19. M. Ward, *Marketing Strategies: A Resource for Registered Dietitians* (Binghamton, NY: Niles & Phipps, 1984), p. 63.

20. Ibid.

21. W. Lancaster, T. McIllwain, and J. Lancaster, Health marketing: Implications for health promotion, *Family and Community Health* (February 1983): 47.

22. Ibid., p. 45.

23. Dever, *Community Health Analysis*, pp. 253–64.

24. The section on demographics and psychographics was adapted from Dever, *Community Health Analysis*, pp. 255–77.

25. S. C. Parks, Research techniques used to support marketing management decisions, in *Research: Successful Approaches*, ed. E. R. Monsen (Chicago: American Dietetic Association, 1992), p. 306.

26. S. W. Brown and A. P. Morley, Jr., *Marketing Strategies for Physicians: A Guide to Practice Growth* (Oradell, NJ: Medical Economics Books, 1986).

27. W. D. Novelli, Health care, politicians, toothpaste: All can be marketed the same way, *Marketing News* 14 (1981): 1, 7.

28. Ward, *Marketing Strategies*, p. 81.

29. Novelli, Health care, p. 7.

30. Matthews, Marketing your services, p. 25.

31. P. Kotler, *Marketing Management* (Upper Saddle River, NJ: Prentice Hall, 1980), Chapter 3.

32. Kotler, *Marketing in Non-profit Organizations*, as cited by Parks and Moody, A marketing model, p. 40.

33. Dever, *Community Health Analysis*, p. 252.

34. J. A. Quelch, Marketing principles and the future of preventive health care, *Millbank Memorial Fund Quarterly/Health and Society* 58 (1980): 317, as cited by Dever, *Community Health Analysis*, p. 283.

35. The section on promotional tools was adapted from Ward, *Marketing Strategies*, pp. 17–22.

36. The list of advantages is from T. Golaszewski and P. Prabhaker, Applying marketing strategies to worksite health promotion efforts, *Occupational Health Nursing* 32 (1984): 188–92.

37. Ward, *Marketing Strategies*, p. 19.

38. The list of five tips was adapted from ibid., pp. 72–73.

39. Matthews, Marketing your services, p. 25.

40. J. C. Levinson, *Guerilla Marketing* (Boston: Houghton Mifflin, 1984), pp. 26–27.

41. The margin definition is from Kotler and Clark, *Marketing for Health Care Organizations*, pp. 29–31.

42. Novelli, Health care, p. 7.

43. Dever, *Community Health Analysis*, p. 280.

44. This synopsis of the social marketing process was adapted from Clift, Social marketing, p. 18.

45. The three situations described here are from K. F. Fox and P. Kotler, The marketing of social causes: The first 10 years, *Journal of Marketing* 44 (1980): 24–33.

46. P. E. Smith, Cost-benefit analysis and the marketing of nutrition services, in *Benefits of Nutrition Services: A Costing and Marketing Approach* (Columbus, OH: Ross Laboratory, 1987), p. 30.

47. The discussion of the PHHP was adapted from Lefebvre and Flora, Social marketing, pp. 309–11.

48. Ibid., p. 302.

49. S. Roberts, The new consumer, *Marketing Magazine* 14 (1998): 12–13.

Program Management and Grant Writing

Outline

Learning Objectives

After you have read and studied this chapter, you will be able to:

- Describe the four functions of management.
- Discuss the roles of process, outcome, structure, and fiscal evaluation in managing programs.
- Outline the general process for grantsmanship, and name three reasons why grant proposals are commonly rejected.

Something To Think About...

Do you want to be a positive influence in the world? First, get your own life in order. Ground yourself in the single principle so that your behavior is wholesome and effective. If you do that, you will earn respect and be a powerful influence. Your behavior influences others through a ripple effect. A ripple effect works because everyone influences everyone else. Powerful people are powerful influences.

If your life works, you influence your family.

If your family works, your family influences the community.

If your community works, your community influences the nation.

If your nation works, your nation influences the world.

If your world works, the ripple effect spreads throughout the cosmos.

— Lao Tzu, Tao Te Ching (The Tao of Leadership)

Introduction

Community nutritionists must be good planners and managers. One of the past presidents of the American Dietetic Association (ADA), Judith L. Dodd, said this about dietitians' need for management expertise: "It's not enough to have technical knowledge. Other skills are necessary, whether you want to call them leadership skills or communication skills or simply survival skills. We must know how to communicate, how to negotiate, how to persuade, and how to work with various groups within any of our market environments."[1] In other words, community nutritionists must have good management skills. **Management** refers to the process of achieving organizational goals through planning, organizing, leading and controlling.[2] In this chapter we examine the functions of management and consider how they are used by community nutritionists. We also review the principles of grant writing, because community nutritionists often seek extramural funding for program activities.

The Four Functions of Management

The main activities and responsibilities of managers can be grouped into four areas or functions, as shown in Figure 11-1. *Planning* is the forward-looking aspect of a manager's job; it involves setting goals and objectives and deciding how best to achieve them. *Organizing* focuses on distributing and arranging human and nonhuman resources so that plans can be carried out successfully. *Leading* involves influencing others to carry out the work required to reach the organization's goals. *Controlling* is the function that regulates certain organizational activities to ensure that they meet established standards and goals. This section describes each management function.

Planning

A colossal marketing blunder made headlines around the world in the spring of 1993. Hoover, a subsidiary of the Maytag Corporation, launched a promotional campaign in Ireland and Britain in which consumers who purchased any of its

Management The process of achieving organizational goals through engaging in the four major functions of planning, organizing, leading, and controlling.

household appliances (worth at least $150 U.S.) were offered two free return plane tickets to Europe or the United States (valued at about $500 U.S.). Hoover's aim was to stimulate interest in its products, a feat it readily achieved. For reasons not entirely clear, Hoover's management failed to anticipate the number of customers who would accept the promotion's restrictions and demand the promised air tickets. The result was a $30 million loss, the firing of three top executives, and a place in management textbooks as "one of the great marketing gaffes of all time."[3]

What is the lesson in this story for community nutritionists? In two words: *plan ahead*. To those of us standing outside the Hoover snafu, it seems incredible that no one foresaw the shortfall between anticipated revenues and actual expenses. What happened on the inside may never be fully known. Regardless, the magnitude of the event suggests a glitch in the planning process, a failure of managers to calculate accurately the direct and indirect costs associated with the promotional campaign.

Planning A process that helps find solutions to problems; the basis for good management and the key to success.

For the individual manager or managerial team, **planning** involves deciding what to do and when, where, and how to do it. It focuses on future events and finding solutions to problems. Planning is ongoing. It involves performing a number of activities in a generally logical, predetermined sequence and considering a variety of solutions before a plan of action is chosen.[4]

Strategic planning Long-term planning that addresses an organization's overall goals.

Types of Planning. There are different types of planning. **Strategic planning** is broad in scope and addresses the organization's overall goals. Strategic planning occurs over a period of several years and is usually undertaken by the organization's senior managers. The focus of strategic planning is on formulating objectives; assessing past, current, and future conditions and events; evaluating the organization's strengths and weaknesses; and making decisions about the appropriate course of action. The American Dietetic Association's 1996–1999 strategic plan for "Creating the Future" has three main initiatives: policy, member development, and public education. The plan provides direction for the organization in its mission to improve the public's health and the viability of the profession.[5]

FIGURE 11-1 *The Four Functions of Management*

Source: K. M. Bartol and D. C. Martin, *Management*, p. 7. Copyright 1991 by McGraw-Hill, Inc. Used with permission of McGraw-Hill, Inc.

The strategic plan guides the development of operational plans. **Operational planning** is short term and typically is done by midlevel managers. It deals with specific actions, expenditures, and controls and with the timing of these activities in a formal, structured process.[6] After setting policy as a major strategic initiative, for example, ADA set into motion an operational plan to obtain data it could use in meeting its policy goal. It commissioned the Lewin Group to determine the cost savings associated with Medical Nutrition Therapy services and then launched its first public relations campaign to educate policy makers about the study findings.[7]

Another type of planning is **project management.** Community nutritionists who implement diabetes education programs, conduct citywide hypertension screenings, and assess the iron status of inner-city school children are all involved in project management. *You* have participated in project management through the Community Learning Activities in this section. Project management involves coordinating a set of activities that are typically limited to one program or intervention. It requires setting goals and objectives and outlining the project's **critical path,** the series of tasks and activities that will take the longest time to complete. Any delay in the activities on this path will delay the project's completion.[8] Managers work to spot and then remove bottlenecks, thus improving work flow along the critical path.[9] Refer to the case study later in this chapter for a discussion of the critical path for the "Heartworks for Women" program.

Evaluation as a Planning Tool. Evaluation is fundamental to every step of community assessment and program planning. Recall from Chapter 5 that the community needs assessment itself is an evaluation of a population's health or nutritional status. It is designed to find answers to questions about who in the community has a nutritional problem, how the problem developed, what programs and services exist to address the problem, and what can be done to alleviate the problem. The evaluation tools of the community needs assessment are the health risk appraisal, focus group discussions, screenings, surveys, interviews, and direct assessments of nutritional status, as we saw in Chapter 6. During program planning, evaluation occurs at every step. In the design stage, managers develop goals and objectives for the program to determine its impact and effectiveness (Chapter 7). They conduct formative evaluation to achieve a good fit between the program and the target population's needs (Chapter 8), to develop appropriate nutrition messages for the target population (Chapter 9), and to design a marketing plan for the program's target market (Chapter 10). In this section, we describe evaluations that occur during the implementation of the program and when the program is completed. Managers use evaluation findings to plan changes in programs, interventions, and staff activities.

Process Evaluation. Monitoring how a program operates helps managers answer questions and make decisions about what services to provide, how to provide them, and for whom. This type of evaluation is called **process evaluation** and involves examining program activities in terms of (1) the age, sex, race, occupation, or other demographic variables of the target population; (2) the program's organization, funding, and staffing; and (3) its location and timing.[10] Process evaluation focuses on program *activities* rather than *outcomes.*

Operational planning Short-term planning that focuses on the activities and actions required to meet the organization's goals.

Project management A plan that coordinates a set of limited-scope activities around a single program or intervention.

Critical path The series of tasks that take the longest amount of time to complete.

Process evaluation A measure of program activities or efforts—that is, how a program is implemented.

Process evaluation is gaining recognition as a tool to help managers make good decisions. Through process evaluation, managers can systematically exclude the various explanations that may arise for a given outcome. If the program appears to have had no effect, process evaluation can determine which, if any, of the following problems was the reason:[11]

- The program was not properly implemented (meaning that program staff were not fully effective in implementing the program).
- The program could not be implemented properly in some participants (suggesting that compliance with the program protocol was a problem for some participants).
- Some participants had difficulty accessing the program.

Alternatively, if the program has had a beneficial effect on the defined outcomes, process evaluation can determine whether the effect was due, in fact, to the program or to one of the following:

- The greater receptivity of some participants or target groups compared with others
- Competing interventions

Process evaluation focuses on how a program is delivered. In the course of conducting a process evaluation, the evaluator examines the target population to determine how they were attracted to the program and to what extent they participated. In addition, the program is evaluated for bias in terms of how participants were served—that is, whether the target group received too much or too little coverage by the program. Such information can be obtained by examining the records kept on program participants and by conducting surveys of the target population and the program participants. The participant records should reveal which services were used and how often. Surveys help define the characteristics of clients who used the program, compared with those who dropped out or refused to use the program.

Process evaluation deals with activities that are planned to occur. In Case Study 1, an intervention strategy was designed to increase women's awareness of coronary heart disease (CHD) risk factors and reduce their death from CHD. Several process objectives were developed, including these three:

- Provide the eight sessions of the "Heartworks for Women" program over a 12-week period at each participating worksite.
- Obtain a CHD risk factor knowledge test from each participant at the beginning and end of the program.
- Obtain an estimate of fat intake using a 24-hour dietary recall completed by each participant at the beginning and end of the program.

We might examine the third process objective listed above—to obtain an estimate of usual fat intake using a 24-hour recall of all participants at the beginning and end of the program—and consider how the planned activity compared with the actual results. If 66 employees entered the program and 24-hour recalls were obtained on 57 of them as planned, we can calculate the percentage of activities attained, using the following formula:

$$\frac{\text{Actual activities}}{\text{Planned activities}} \times 100 = \text{Percentage of activities attained}$$

$$\frac{57}{66} \times 100 = 86.4 \text{ percent}$$

Thus, 86.4 percent of the planned 24-hour recalls were actually obtained during the specified time period. A number of questions arise immediately: Why weren't recalls obtained from all participants? Was there a problem scheduling employees to see the dietitian and, if so, why? Were the instructions to participants unclear? The information obtained through process evaluation signals the program manager that additional planning is required to improve the program's delivery.

Another example of process evaluation comes from the Child and Adolescent Trial for Cardiovascular Health (CATCH), an intervention that targeted the school environment, staff, students, and the students' families in four centers located in California, Louisiana, Minnesota, and Texas. CATCH was designed to assess the effectiveness of school foodservice programs, physical education, classroom instruction, and family activities to reduce CHD risk in elementary schoolchildren. Process evaluation was used to assess the level of standardization of teacher instruction, the number of CATCH classroom activities completed out of the total number expected for each school session, the number of promotional activities sponsored by the foodservice operation, and the number of support visits made by CATCH staff to schools over the course of the year. Process evaluation determined that the fourth-grade classes had the lowest rate of completion of classroom activities, whereas third-grade classes had the highest rate of completion. The number of cafeteria promotional events ranged from about 6 to 14 and the number of support visits by CATCH staff ranged from 2.5 to 19.7. The overall findings of the process evaluation provided information that helped intervention planners improve the intervention.[12]

Impact Evaluation. **Impact evaluation** is used to determine whether and to what extent a program or an intervention contributed to accomplishing its stated goals. It describes the specific effect of program activities on the target population. In the "Heartworks for Women" program, the skills-building intervention in Case Study 1, the impact of the program would be the knowledge about CHD risk and major sources of dietary fat acquired by women who participated in the program. The impact evaluation would assess whether women had learned the key risk factors for heart disease and whether they could describe the types of fat in the diet, among other things. In the CATCH intervention, described previously, one unexpected finding of impact evaluation was that students in classes where teachers modified the CATCH sessions to suit their needs learned more about diet, health, and heart disease than students in classes headed by teachers who did not modify their sessions. Perhaps these teachers were more motivated, confident, and creative and had better communication skills than teachers who made no changes in the lesson plans. The finding led the intervention planners to consider changing the model on which the intervention was based.[13]

Impact evaluation focuses on immediate indicators of a program's success. Depending on the program's goals and objectives, it might examine variables such as beliefs, attitudes, decision-making skills, self-esteem, self-efficacy, and knowledge.[14]

Impact evaluation The process of determining whether the program's methods and activities resulted in the desired immediate changes in the client.

Outcome Evaluation. The purpose of **outcome evaluation** (also referred to as summative evaluation) is to determine whether the program or intervention had an effect on the target population's health status, food intake, morbidity, mortality, or other outcomes. Outcome evaluations are a challenging managerial activity, for they require technical skills in survey design and analysis. The problems associated with outcome

Outcome evaluation The process of measuring a program's effectiveness in changing one or more aspects of nutritional or health status.

evaluations arise from the difficulty of determining whether a particular effect was "caused" by the intervention and was not due to some extraneous factor. It is possible that factors beyond the control of the program staff influenced the outcome significantly. Such confounding factors might include secular trends within the community, the occurrence of unexpected events such as a natural catastrophe, or certain characteristics of clients (for example, they tended to "self-select" for the program).[15]

Case Study 1 included a skills-building program, the "Heartworks for Women" program, and nutrition messages directed at women in the community at large. Did the intervention strategy accomplish its goals and objectives? The intervention manager and team 2 leader plan outcome evaluations that target the three time frames (1, 2, and 5 years) specified in the objectives. The first evaluation will be undertaken 12 months after the startup of the intervention and program. Data will be collected from women who enrolled in the "Heartworks for Women" program and from women in the broader community. The purpose of the evaluation will be to determine whether the nutrition and smoking messages delivered through the intervention resulted in a *behavior change* in the target population. In other words, did women who participated in the "Heartworks for Women" program reduce their intake of dietary fat by 3 percent? Did the number of women who smoke decrease by 33 percent? (Refer to the program objectives described in Chapter 7 on page 221.) The findings of the outcome evaluation will be used to modify the intervention strategy, nutrition messages, marketing plan, and other program elements.

Outcome evaluation, like impact evaluation, is tied to the program's goals and objectives. It is designed to account for a program's accomplishments and long-term effectiveness in terms of a health change in the target population. Outcome evaluation measures can include serum ferritin levels, percent body fat, calcium intake, stroke prevalence, blood pressure, and use of home food services, depending on the nature of the program.

Structure Evaluation. Structure refers to personnel and environmental factors related to program delivery such as the training of personnel, the adequacy of the facility, the use of equipment such as overhead projectors and skinfold calipers, the storage of participant records, and the like.[16] In preparing a structure evaluation of the "Heartworks for Women" program in Case Study 1, the senior manager specifies several structure objectives, three of which are listed here:

- The 12-month operating budget for the program is $274,500.
- Monthly operating budgets for the next 12 months will not show a variance exceeding 0.02 percent of the total operating budget.
- In-house educational materials will be used at a monthly rate that does not exceed 10 percent of the total stored supply.

These and other structure objectives provide targets that regulate the resources needed to deliver the program. The third objective, for example, helps staff members plan for the use of educational materials. It guides session instructors in monitoring their use of the materials, and it helps the staff member who coordinates the supply and storage area maintain an adequate supply of materials for the program. The manager uses the findings from structure evaluation to make changes in the internal processes that support the staff and program activities.

Fiscal or Efficiency Evaluation. The purpose of **fiscal evaluation** is to determine how program outcomes compare with their costs. There are two types of efficiency evaluations: cost-benefit analysis and cost-effectiveness analysis.

A cost-benefit analysis requires that managers estimate both the tangible and intangible benefits of a program and the direct and indirect costs of implementing that program, as summarized in Table 11-1 for the "Heartworks for Women" program. Once these have been specified, they must be translated into a common measure, usually a monetary unit. In other words, the cost-benefit analysis examines the program outcomes in terms of money saved or reduced costs. A prenatal program that costs $200 to produce per participant and results in reduced medical costs of $800 can be expressed as a cost-benefit ratio of 1:4. For every one dollar required to produce the program, there is a four dollars savings in medical costs.[17]

The second type of fiscal evaluation is cost-effectiveness analysis. Unlike cost-benefit analysis, which reduces a program's benefits and costs to a common monetary unit, cost-effectiveness analysis relates the effectiveness of reaching the program's goals to the monetary value of the resources going into the program. With this type of evaluation, similar programs can be compared to one another or ranked in order of their cost per program goal. A cost-effectiveness analysis would be done, for example, to determine which of two methods of intervention—individual dietary counseling or group nutrition education classes—produces a desired outcome for less cost.

> **Fiscal** or **efficiency evaluation** The process of determining a program's benefits relative to its cost.
>
> Refer to Chapter 13 for a discussion of a cost-benefit analysis of the Special Supplemental Nutrition Program for Women, Infants, and Children (WIC).

Organizing

Organizing is the process by which carefully formulated plans are carried out. Managers must arrange and group human and nonhuman resources into workable units to achieve organizational goals. To do this, certain structures must be in place to guide employees in their activities and decision making. Imagine the confusion that would result if you had a problem and didn't know who your supervisor was! In

Type	Costs	Benefits
Direct	Personnel—salaries Utilities (telephone, Internet) Travel (to/from worksites) Office supplies Postage Equipment Instructional materials Design Printing Promotion and advertising	Revenue from program
Indirect	Personnel—benefits Rental of office space Utilities Equipment depreciation Maintenance and repairs Janitorial services	Increased exposure for the program and the organization within the community Reduced use of health care services by employees of participating worksites

TABLE 11-1 *Costs and Benefits Considered in a Cost-Benefit Analysis of the "Heartworks for Women" Program*

this section, we describe organization structures and how people are managed as part of the process of organizing.

Organization Structures. Just as a building's layout—the arrangement of offices, windows, hallways, restrooms, and stairwells—affects the way people work, so does an organization's structure.[18] The **organization structure** is the formal pattern of interactions and activities designed by management to link the tasks of employees to achieve the organization's goals. The word *formal* is used intentionally here to denote management's official operating structure in contrast to the informal patterns of interaction that exist in all organizations.[19] In developing the organization structure, managers consider how to assign tasks and responsibilities, how to define jobs, how to group individual employees to carry out certain tasks, and how to institute mechanisms for reporting on progress.

In organizing their employees, managers may find an **organization chart** helpful. Organization charts give employees information about the major functions of departments, relationships among departments, channels of supervision, lines of authority, and certain position titles within units.[20] According to the organization chart for a typical public health department (shown in Appendix B-3), the community nutritionist who coordinates the maternal, infant, and child care programs reports to the department manager, the director of nutrition, who in turn reports to the medical health officer. Nontraditional organization charts, in which employees have roughly the same rank, also exist. A "web" format, such as the one shown in Figure 11-2, has no strict hierarchy of authority. This format has the advantage of promoting teamwork and consensual decision making, but it can create confusion

Organization structure The formal pattern of interactions designed by management to link the tasks of individuals and groups to achieve organizational goals.

Organization chart A line diagram that depicts the broad outlines of an organization's structure.

FIGURE 11-2 *Nontraditional Organization Chart for Nutrition in Action*

In this nontraditional organization structure, employees have roughly the same rank, and there is no strict hierarchy of authority.

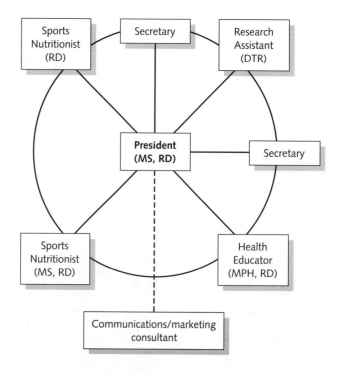

over authority and responsibility. This design seems to work well in small companies and organizations.

Organization charts help establish lines of communication and procedures, but they do not depict rigid systems. An organization's informal structure, depicted humorously in Figure 11-3, is often quite powerful and sometimes represents a more realistic picture of how the organization actually works. The informal structure arises spontaneously from our interactions, brief alliances, and friendships with coworkers throughout the organization.[21]

Another essential dimension of organization structure is departmentalization or the manner in which employees are clustered into units, units into departments, and departments into divisions or other larger categories. Departmentalization directly influences how managers carry out their duties, supervise their employees, and monitor group dynamics and perspectives. An important aspect of departmentalization is the **span of management** or **span of control,** which refers to the number

Span of management or **span of control** The number of subordinates who report directly to a specific manager.

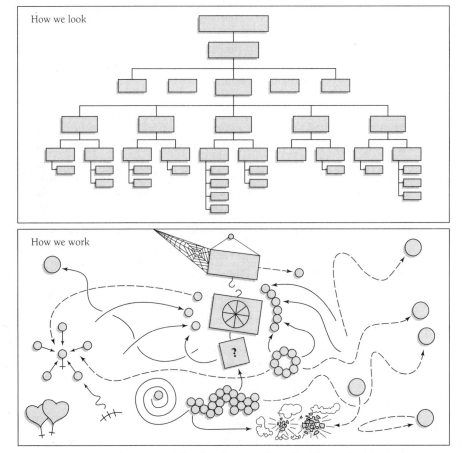

FIGURE 11-3 *The Typical Organization Chart on Paper and in Practice*

Source: ORGANIZATIONAL BEHAVIOR: Concepts and Applications, 4th ed. by Gray/Starke, © 1988. Reprinted by permission of Prentice-Hall, Inc., Upper Saddle River, NJ.

How an organization looks on paper differs from how it really works. Whereas the formal organization structure shows the lines of authority and how communications should travel, the informal organization reveals the actual patterns of social interaction between employees in different units of the organization.

of subordinates who report directly to a specific manager.[22] Deciding how many employees should report to a single manager is not a trivial decision. When a manager must supervise a large number of employees directly, she may feel overwhelmed and have difficulty coordinating tasks and keeping on schedule; with too few employees to supervise, she may feel underutilized and disaffected. The ideal span of management has not been identified precisely. Some researchers argue that the range is about 5 to 25 employees, depending on the level of organization; theoretically, lower-level managers can supervise more employees directly than managers higher in the hierarchy. Napoleon once remarked: "No man can command more than five distinct bodies in the same theater of war." Most management experts now recommend that only three to seven subordinates be supervised directly.[23]

Another method of coordinating an organization's activities is through **delegation** or the assignment of part of a manager's work to others. "Everyone knows about delegating, but not too many people do it, and fewer still do it well."[24] The inability to delegate work to others has felled many a fast-track manager.[25] Some people do the work themselves because they don't realize they can delegate it, don't know how to get the most out of a computer, or don't appreciate how delegating work makes *them* work more efficiently. Table 11-2 gives tips on how managers can delegate some activities to their employees.

Line and staff relationships also help clarify an organization's structure. A person in a **line position** has direct responsibility for achieving the organization's goals and objectives. The term *staff* is commonly used to refer to the groups of employees who work in a particular unit or department—for example, the director of nutrition in a public health department is said to supervise the nutrition staff. In management practice, however, *staff* has a particular meaning. An employee in a **staff position** assists those in line positions. To some extent, the concept of line and staff positions is a holdover from eighteenth- and nineteenth-century management theory, but the terms are still used today in many organizations to help employees identify individuals with the authority and responsibility for fulfilling the organization's mandate.

Delegation The assignment of part of a manager's work, along with both the responsibility and the authority necessary to achieve the expected results, to others.

Line position A position with the authority and responsibility for achieving the organization's main goals and objectives.

Staff position A position whose primary purpose is to assist those in line positions.

TABLE 11-2 *Tips for Delegating Your Work to Others*

- The first step to take when delegating your work to subordinates is to evaluate their skill levels and the difficulty of the task. The trick is to select the employee with the appropriate skills who will find the task challenging but not frustrating.

- Once you have chosen the best employee, give him or her all the information needed to do the job well. Be honest about the work; if it's drudgery, say so.

- Do these three things to ensure success:

  1. Give the employee *responsibility* for completing the project or task.
  2. Hold the employee *accountable* for the results.
  3. Provide the employee with the *authority* to make needed decisions, direct others to help with the project, and carry out actions as required.

- Make sure the employee has the necessary materials, equipment, time, and funds to complete the project.

- Evaluate the employee's progress periodically and be prepared to accept a less-than-perfect result.

Source: Adapted from L. Baum, Delegating your way to job survival, *Business Week,* November 2, 1987, p. 206.

Job Design and Analysis. Managers in community nutrition are responsible for **job design,** that is, determining the various duties associated with each job in their area. They sometimes need to conduct a **job analysis** to determine the purpose of a job, the skill set and educational background required to carry it out, and the manner in which the employee holding that job interacts with others. The formal outcome of a job analysis is the preparation of a job description.[26]

A job description helps employees understand what is expected of them and to whom they report. Although there is no standard format, most job descriptions include the items shown in Table 11-3.[27] The job description serves as a basis for rating and classifying jobs, setting wages and salaries, and conducting a performance appraisal (described later). It helps organizations comply with the accrediting, contractual, legal, and regulatory directives of the government or other institutions, and it provides a basis for decisions about promotions and training. In addition, it can be adapted for use in recruiting and hiring prospective employees. For example, the text of a position announcement to be published in a newspaper or journal can be taken from the job description. A sample job description for a nutritionist with the FirstRate Spa and Health Resort is given in Figure 11-4.

Job design The specification of tasks and activities associated with a particular job.

Job analysis The systematic collection and recording of information about a job's purpose, its major duties, the conditions under which it is performed, and the knowledge, skills, and abilities needed to perform the job effectively.

Human Resource Management. Paying attention to the people who produce a product or service is what human resource management is all about. Community nutritionists report being involved in several aspects of managing people, including recruiting people for the organization and evaluating the performance of their employees.[28] Both these activities are described in the next paragraphs.

- **Staffing.** Staffing is the set of human resource activities designed to recruit individuals to help meet the organization's goals and objectives. Recruitment has two distinct phases: (1) attracting applicants and (2) hiring candidates.[29] Both direct and indirect recruiting strategies are often used to attract applicants. Direct methods include media-based advertising in newspapers and journals, mailing personalized letters to potential applicants, and participating in job recruitment

---

- **Job title.** The job title clarifies the position and its skill level (for example, *director, supervisor, assistant,* or *secretary*).

- **Immediate supervisor.** The job description states the position and title of the immediate supervisor.

- **Job summary.** The job summary is a short statement of the purpose of the job and its major tasks and activities.

- **Job duties.** This section of the job description is a detailed statement of the specific duties of the job and how these duties are carried out.

- **Job specifications.** This is a list of minimum hiring requirements, usually derived from the job analysis; it includes the skill requirements for the job, such as educational level, licensure requirements, and experience (usually expressed in years), and any specific knowledge, advanced training, manual skills, and communication skills (both oral and written) required to do the job. The physical demands of the job, if any, including working conditions and job hazards, are included here.

TABLE 11-3 *Components of a Job Description*

Source: Adapted from J. G. Liebler and coauthors, *Management Principles for Health Professionals,* 2nd ed. (Gaithersburg, MD: Aspen Publishers, 1992), pp. 181–82.

FIGURE 11-4 *Sample Job Description*

**FirstRate Spa and Health Resort
Department of Human Resources**

Position Title: Nutritionist

Responsible to: Director of health promotion

Responsibilities: Under the general direction of the director of health promotion, the nutritionist is responsible for developing, implementing, and evaluating the spa's risk reduction programs.

Job Duties: The nutritionist has the following duties and responsibilities:
• Develops, implements, and evaluates the spa's risk reduction programs in the areas of smoking cessation, blood cholesterol reduction, stress management, and diabetes control.
• Assists the director in developing marketing strategies for the risk reduction program.
• Assists the director in developing budgets.
• Tracks program revenue and expenses.
• Works with other spa nutritionists and staff to develop marketing tools and teaching/instructional aids.
• Participates as required in coordinating research projects.

Qualifications and Experience: The nutritionist should have the following qualifications and experience:
• A university degree in nutrition
• Certification as a registered dietitian
• Two years experience teaching community-based courses or programs

Skills: The nutritionist should have the following skills:
• Demonstrated ability to write and give presentations
• Experience using word processing and spreadsheet software is desirable
• Experience using the Internet is desirable
• Ability to work with people of all ages and backgrounds

Date: June 1999

---

fairs. Indirect methods are activities that keep the organization's name in the public's eye such as collaborative projects with other institutions and training sessions for professionals.

Affirmative action Any activity undertaken by employers to increase equal employment opportunities for groups protected by federal equal employment opportunity (EEO) laws and regulations.

An important aspect of recruiting is **affirmative action,** which includes all activities designed to ensure and increase equal employment opportunities for groups protected by federal laws and regulations. A significant law in this area is Title VII of the Civil Rights Act of 1964, which forbids employment discrimination on the basis of sex, race, color, religion, or national origin. Most organizations have some type of affirmative action plan that details its goals and policies related to hiring, training, and promoting protected groups. In fact, organizations with federal contracts exceeding $50,000 and with 50 or more employees *must* file an affirmative action plan with the Department of Labor's Office of Federal Contract Compliance Programs; others may develop such plans on a voluntary basis.[30]

The hiring process involves conducting job interviews with potential candidates, screening applicants, selecting the best candidate, checking references, and, finally, offering the position. In some cases, two or three suitable candidates will be invited back for interviews with key managers in other departments or divisions before a final candidate is chosen.

• **Evaluating job performance.** Providing feedback to employees about their performance is essential to maintaining good working relationships and can occur informally at any time. In the daily course of solving unexpected problems and reviewing the progress on a project, managers in community nutrition have ample opportunity to update their employees about their performance. In fact, the best managers spend up to 40 percent of their time in face-to-face interactions with their employees that allow time for performance feedback.[31]

The **performance appraisal** is a formal method of providing feedback to an employee. It is designed to influence the employee in a positive, constructive manner and involves defining the organization's expectations for employee performance and measuring, evaluating, and recording the performance compared with those expectations. The performance appraisal is used to help define future performance goals, determine training needs and merit pay increases, and assess the employee's potential for advancement. Because employees are often uneasy about the review process, managers must take care to be objective in their rating of performance. One way to reduce employees' discomfort is to let them know in advance how the process works and what to expect during the performance appraisal interview. Employees are most likely to have a favorable impression of the process when they are aware of the evaluation criteria used, have the opportunity to participate in the process, and are able to discuss their career development.[32]

> **Performance appraisal** A manager's formal review of an employee's performance on the job.

The performance appraisal interview is a good time for reviewing and changing both organizational and personal goals. Managers who spend time helping subordinates develop their personal goals are more likely to achieve their organization's goals, since setting challenging personal goals is linked strongly with performance.[33]

The key to conducting a good performance appraisal interview is to start with clear objectives, focus on observable behavior, and avoid vague, subjective statements of a personal nature. Consult Table 11-4 for a list of questions that can be used in talking about performance and giving constructive criticism.[34]

> *Employees want to hear from management, and they want management to listen to them.*
>
> – D. Keith Denton

Leading

Leading is the management function that involves influencing others to achieve the organization's goals and objectives. It focuses on the horizon, on asking what and why. Managing focuses on the bottom line, on asking how and when.[35] (Reread Chapter 3's Professional Focus on leadership.) Organizations need strong leaders *and* good managers.

Motivating Employees. An important function of managers is to motivate their employees to "get the job done." Barcy Fox, a vice president with Maritz Performance Improvement Company, comments, "Your employees aren't you. What works for you won't work for them. What motivates you to pursue a goal won't motivate them. Until you grasp that, you won't set attainable goals for your workers."[36] There is no single strategy for motivating employees to perform, and

TABLE 11-4 *How to Provide Constructive Criticism*

Use the process outlined here to offer constructive criticism to your employees without making them defensive. While these questions can be used in a performance appraisal interview, they lend themselves to many other job-related discussions. Most people questioned in this manner make specific suggestions to improve their performance.

- **Initiate.** "How would you rate your performance during the last six months?" or "How do you feel about your performance during the last quarter?"

- **Listen.** If you get a general response like "fine," or "pretty good," or "8 on a 10-point scale," follow up with a more focused question: "What in particular comes to mind?" or "What have you been particularly pleased with?" Your objective is to discuss a positive topic raised by the employee.

- **Focus.** "You mentioned that you were particularly satisfied with . . . . Let's talk further about that aspect of your job."

- **"How" probe.** "How did you approach . . . ?" or "What method did you use?"

- **"Why" probe.** "How did you happen to choose that approach?" or "What was your rationale for that method?"

- **"Results" probe.** "How has it worked out for you?" or "What results have you achieved?"

- **Plan.** "Knowing what you know now, what would you have done differently?" or "What changes would you make if you worked on this again?"

Source: J. G. Goodale, Seven ways to improve performance appraisals, *HRMagazine* 38 (1993): 79. Reprinted with the permission of *HRMagazine* published by the Society for Human Resource Management, Alexandria, VA.

opinions differ on exactly what managers need to do to stimulate motivation. Peter Drucker, a management consultant, argues that managers can do four things to obtain peak performance from their employees:[37]

- Set high standards and stick to them. Employees become demoralized when managers do not hold themselves to the same high standards set for other workers in the organization. Employees are stimulated to work hard with a manager who expects a lot of herself, believes in excellence, and sets a good example.

- Put the right person in the right job. Employees take pride in meeting a challenge and doing a job well. Their sense of accomplishment enhances their performance. Underutilizing employees' skills destroys their motivation.

- Keep employees informed about their performance. Employees should be able to control, measure, and evaluate their own performance. To do this, they need to know and understand the standard against which their performance is being compared. The job description and performance appraisal interview help keep employees informed about what managers expect of them. Keep in mind the unwritten rule of managing people: Put praise in writing. When you must reprimand an employee, do so verbally and in private.

- Allow employees to be a part of the process. One means of motivating employees is to give them a managerial vision, a sense of how their work contributes to the success of the project, program, or organization. In a sense, a manager cannot "give" his own vision to his employees, but he can allow them to help mold and shape the vision, thereby making it theirs as well as his.

Communicating with Employees. Communicating is a critical managerial activity that can take many forms. Verbal communication is the written or oral use of words to communicate messages. Written communications can take the form of reports, résumés, telephone messages, memorandums, procedure manuals, policy manuals, and letters. Oral communications, or the spoken word, usually take place in forums such as telephone conversations, committee meetings, and formal presentations. Good managers are skilled in both of these areas (review the Professional Focuses for Chapters 8 and 9). Nonverbal communication in the form of gestures, facial expressions, and so-called body language is also important. Tone of voice—the *way* something is said as opposed to *what* is said—sometimes relays information more effectively than the actual words. Even the placement of office furniture communicates certain elements of attitude and style. Reportedly, nonverbal communication accounts for between 65 and 93 percent of what actually gets communicated.[38]

Becoming a good communicator means paying attention to people and events, observing the nuances of nonverbal and verbal communication, and becoming a good listener. Open communication in an organization does not occur by accident. It results from the daily use of certain techniques and skills that promote communication. Merck, the pharmaceutical company, uses an approach called Face-to-Face communication. The four rules on which this communication is built are listed in Table 11-5.[39] Communicating effectively with coworkers is a skill that, like other skills described in this book, can be acquired with practice and determination.

The team 2 leader in Case Study 1, for example, learns of a conflict between two staff members—both believe they are the main liaison with the design company hired to prepare the marketing materials. The liaison has a key role in keeping the project on schedule and also has the opportunity to become involved in the organization's other marketing campaigns. The leader determines first that the design company will

Computers are essential to community nutrition practice. They are used to prepare letters, memos, grant proposals, reports, manuals, fact sheets, and brochures; track revenue and expenses; exchange e-mail with colleagues; and access the Internet for timely information.

TABLE 11-5 *Four Rules of Face-to-Face Communication*

- **Be candid.** Any questions or comments that arise during face-to-face interactions (be they performance appraisal interviews or committee meetings) should be addressed truthfully. If appropriate, indicate that you cannot answer the question because of its confidential nature. If you do not know the answer to a question, say so.

- **Be prepared.** Ask participants to submit questions or topics for discussion several weeks or days before the meeting is scheduled to occur. Organize an agenda, preferably a written one. This strategy has the advantage of ensuring that topics of interest to you and your employees are covered. If you develop a written agenda, distribute it to the participants before the meeting so that they can arrive prepared.

- **Encourage participation.** Conduct the face-to-face meeting in a relaxed and friendly atmosphere where people feel comfortable expressing their ideas.

- **Focus on listening, not judging.** When sensitive topics arise, try not to appear upset or judgmental. Listen carefully to questions and the discussion they prompt. Pay attention to nonverbal cues and group dynamics. The bottom line for good communications is simple: Treat others with respect.

Source: Adapted from Merck's Face-to-Face Communication, in D. K. Denton, *Recruitment, Retention, and Employee Relations* (Westport, CT: Quorum Books, 1992), pp. 158–59.

want to deal with only one staff member and, second, that the rivalry between these two staff members has been friendly but has the potential to be disruptive. She schedules a meeting with the two of them, asking them to think about the problem and how it might be resolved. During the meeting, the three of them talk about the liaison's role, the responsibilities and activities of each staff member, and certain projects "in the works" within the department. The leader aims to keep the discussion open and amiable. She succeeds in getting the more senior staff person to agree to allow the junior staff person, who has worked previously with the design company, to serve as liaison; she gives the senior staff person the responsibility for developing the Web site content and some design elements. The leader's role in this communication process is to help the staff members appreciate each other's skills and contribution to the organization.

Controlling

Controlling is the management function concerned with regulating organizational activities so that actual performance meets accepted organizational standards and goals.[40] The controlling function involves determining which activities need control, establishing standards, measuring performance, and correcting deviations. Managers set up control systems or mechanisms for ensuring that resources, quality of products and services, client satisfaction, and other activities are regulated properly. Such controls help managers cope with the uncertainties that arise when plans go awry. An unforeseen reduction in the public health department's budget, for example, forces the manager of the nutrition division to reallocate personnel and institute tighter spending controls.

Community nutritionists often have responsibility for program and departmental budgets and managing information. The ADA's role delineation study determined that managers in community or public health programs were responsible for preparing financial analyses and reports, controlling program costs, monitoring a program's financial performance, documenting a program's operations, and making decisions about capital expenditures.[41] Even entry-level community nutritionists are expected to be able to define basic financial terms, read and understand financial reports, maintain cost control of materials and projects, and, in some settings, develop plans to generate revenue.[42] Strong management skills enable community nutritionists to detect irregularities in client service or cost overruns and handle complex projects and programs. Two aspects of the control process are described in the next paragraphs.

Financial and Budgetary Control. The primary tools of financial control are the two types of financial statements: balance sheets and income statements. The **balance sheet** lists the organization's assets and liabilities. Assets include cash, accounts receivable (sales of a product or service for which payment has not yet been received), inventories, and items such as buildings, machinery, and equipment. Liabilities include accounts payable (bills that must be paid to outside suppliers and the like), short- and long-term loans, and shareholders' equity such as common stock. (Government agencies and most small, independent companies do not have shareholders' equity.) The **income statement** summarizes the organization's operations over a specific time period, such as a quarter or a year. It generally lists revenues (the assets derived from selling goods and services) and expenses (the costs incurred in producing the revenue). The difference between revenues and expenses is the organization's profit or loss—sometimes called the "bottom line."

Financial control is typically managed through a budget, which is "a plan for the accomplishment of programs related to objectives and goals within a definite time period, including an estimate of resources required, together with an estimate of the resources available."[43] Budgets can be planned for the organization as a whole or for subunits such as divisions, departments, and programs. **Budgeting** is closely linked to planning. It is the process of stating in quantitative terms (usually dollars) the planned organizational activities for a given period of time. Although the terms *budgeting* and *accounting* are sometimes used interchangeably, they are not the same thing. Accounting is a purely financial activity, whereas budgeting is both a financial and a program planning activity.

Because all expenses of an organization—everything from pencils and paper clips to salaries and benefits—must be accounted for, the budgeting process requires considerable planning and managing. It usually begins with a review of the previous budget

> **Balance sheet** A financial statement that depicts an organization's assets and claims against those assets (liabilities) at a given time.

> **Income statement** A financial statement that summarizes the organization's financial position over a specified period, such as a quarter or a year.

> **Budgeting** The process of stating in quantitative terms, usually dollars, the planned organizational activities for a given period of time.

DILBERT reprinted by permission of United Feature Syndicate, Inc.

period and then proceeds to the development of a new budget. Most programs and projects receive a monthly budgetary review. The primary areas to be examined include total project funding, expenditures to date, current estimated cost to complete the project, anticipated profit or loss, and an explanation of any deviations from the planned budget expenditure. Periodic status summaries help managers control resources and keep to the planning schedule.[44] The internal approval process for a new budget can involve bargaining, compromise, and outright competition for scarce resources. Managers who can justify their budget requests are more likely to be successful in appropriating funds for their program's projects and activities.

Information Control. In the course of your community work, you will need to collect, organize, retrieve, and analyze many types of data and information. You need to know about the health and nutritional status of the population in your district, the demographic profile of your client base, how people in your community use your organization's programs and services, how cost-effective your programs are, and the expenses incurred by your staff in carrying out their program activities, to name only a few. At this point, a distinction must be made between "data" and "information." **Data** are unanalyzed facts and figures. For example, a survey of infants in your community shows that over the past 12 months 36 of 478 infants had birthweights less than 2,500 grams. These data don't mean much on their own; to be meaningful, they must be transformed into **information.** By processing the data further (for example, converting the data into a percentage or calculating the mean weights), the manager can compare these figures with those from previous reporting years. A change in the figures may signal the need to reassess program goals and reallocate resources to prevent low birthweight.

The primary purpose of information is to help managers make decisions. Information is a decision support tool in that it helps managers answer these questions: What do we want to do? How do we do it? Did we do what we said we were going to do? The first question addresses the planning function of management; the second, the organizing function; and the third, the controlling function. Managers begin the decision-making process by organizing the data into usable information. They then identify the means by which the data are to be analyzed. This process involves sorting, formatting, extracting, and transforming the data, usually following certain statistical methods. Finally, managers interpret the output and prepare reports that summarize the information. Decisions are then based on the figures in the reports.[45] Of course, managers' decisions are only as good as their information. Constant vigilance is required to prevent the GIGO (garbage in/garbage out) syndrome.

With the advent of computers, processing information has become both simpler and more complex. It is simpler because many aspects of data analysis are carried out on computers and not by hand. It is more complex because one can choose from hundreds of computer systems and software programs. Because computers can produce reams of data very quickly, managers are apt to become overwhelmed by the magnitude and diversity of the data available to them. To be useful, information must be:[46]

1. **Relevant.** Although it is sometimes difficult to know precisely what information is needed, managers should try to identify the information most likely to help in making decisions.

Data Unanalyzed facts and figures.

Information Data that have been analyzed or processed into a form that is meaningful for decision makers.

2. **Accurate.** If information is inaccurate or incorrect, the quality of the decision will certainly be questionable.
3. **Timely.** Information must be available when it is needed.
4. **Complete.** Information must cover all of the areas important to the decision-making process.
5. **Concise.** Information should be concise, providing a summary of the items central to the decision.

Principles of Grant Writing

Community nutritionists increasingly find themselves seeking extramural funding for program activities and interventions. Perhaps extra funds are needed for printing banners and T-shirts, publishing a newsletter, underwriting the advertising costs for a program launch, buying food supplies for a cooking demonstration or taste test, or marketing nutrition messages on a local television station. Each year both the public and private sectors award billions of dollars in grants to individuals, agencies, organizations, and schools for a variety of activities and projects. According to some, "If you can find an effective solution to a pressing community problem, you can find funders to help you implement it."[47]

Regardless of the need, there is likely a government or community agency, industry trade group, food or drug company, or local business willing to support your program or intervention activities. The challenge is to find that one person who is willing to say, "Yes, my organization is willing to send you a check." How do community nutritionists find out about available funding? They network with colleagues to learn about upcoming funding opportunities. They contact granting agencies for information about deadlines for grant proposals. They call local businesses to seek small grants to support local community projects. (Refer to the boxed insert for a glossary of grant-related terminology.)

Once potential granting sources are identified, the community nutritionist works at **grantsmanship.** Successful grantsmanship requires a combination of good writing skills and perseverance. A clearly written grant proposal is based on a good idea, complements the goals of the funding organization, and serves as a bridge between your vision and the grantor's funds and resources.[48] Experts advise dividing the task of writing a grant proposal into several steps, as outlined here:[49]

Grantsmanship The art of seeking and being awarded funds.

1. **Develop the idea for your proposal.** Explore the problem and the available resources for dealing with the problem. Conduct a formal and informal needs assessment for your proposed project in the target or service area. Survey the related literature. Has this idea been considered in the past? How does your idea differ from or complement the existing resources or activities? Why is it likely to succeed? Be able to describe the problem clearly and concisely and explain the uniqueness of the proposed project or activity in two to three minutes. This will help you when contacting funders by phone or in person.
2. **Identify a funding source.** Funding for grants in the United States is usually derived from one of three sources: (1) governmental agencies, (2) foundations and community trusts, and (3) business and industrial organizations (see

◎ Glossary of Grants

Types of grants

Block grant A grant from the federal government to states or local communities for broad purposes as authorized by legislation. Recipients have great flexibility in distributing such funds as long as the basic purposes are fulfilled.

Capitation grant A grant made to an institution to provide a dependable support base, usually for training purposes. The amount of the grant is dependent on the size of enrollment or number of people served.

Categorical grant A grant similar to a block grant, except funds must be expended within specific categories, such as maternal and child care.

Conference grant A grant awarded to support the costs of meetings, symposia, or special seminars.

Demonstration grant A grant, usually of limited duration, made to establish or demonstrate the feasibility of a theory or an approach.

Formula grant A grant in which funds are provided to specified grantees on the basis of a specific formula, prescribed in legislation or regulation, rather than on the basis of an individual project review. The formula is usually based on such factors as population, enrollment, per capita income, morbidity and mortality, or a specific need. Capitation grants are one type of formula grant, and block grants are also usually awarded on the basis of a formula.

Planning grant A grant made to support planning, developing, designing, and establishing the means for performing research or accomplishing other approved objectives.

Project grant The most common form of grant, made to support a discrete, specified project to be performed by the named investigator(s) in an area representing his or her specific interest and competencies.

Research grant A grant made to support investigation or experimentation aimed at the discovery and interpretation of facts, revision of accepted theories in light of new facts, or the application of such new or revised theories.

Service grant A grant that supports costs of organizing, establishing, providing, or expanding the delivery of health or other essential services to a specified community or area. This may also be done through a block grant.

Training grant A grant awarded to an organization to support costs of training students, personnel, or prospective employees in research, or in the techniques or practices pertinent to the delivery of health services in the particular area of concern.

Related terms

Annual reports A summary of all research and training activities from all funding sources.

Budget The itemized list of expenditures required for an activity.

Consultant A person who is asked to participate in a grant because of expertise in a particular area.

Contract The legal agreement between the grantee and grantor establishing the work to be performed, products to be delivered, time schedules, financial arrangements, and other provisions or conditions governing the arrangement.

Direct costs A budget item that reflects the direct expenditure of funds for salaries, fringe benefits, travel, equipment, supplies, communication, publications, and similar items.

Grant A sum of money comprising an award of financial assistance to recipient individuals or organizations with assurances that a particular use will or will not be made of it.

Grantee The institution, public or private nonprofit corporation, organization, agency, individual, or other legally accountable entity that receives a grant and assumes legal and financial responsibility and accountability both for the awarded funds and for the performance of the grant-supported activity.

Grant-in-aid Another name for a project or formula grant.

Grantor The agency (government, foundation, corporation, nonprofit organization, individual) awarding a grant to a recipient.

Indirect costs Costs that are incurred for common or joint objectives and therefore cannot be identified specifically with a particular project or activity. They are costs involving operational support for mixed purposes (for instance, building maintenance, heating, and cooling).

Proposal A formal written document containing a descriptive narrative of an idea for a proposed program or project and a budget to be submitted to a funding agency.

Request for proposal (RFP) If an agency has a particular area it wishes to fund, it will put out an RFP in that area. The RFP invites the submission of proposals and specifies the requirements that the proposal must meet.

Sources: Adapted from R. Lefferts, *Getting a Grant: How to Write Successful Grant Proposals* (Upper Saddle River, NJ: Prentice Hall, 1982), pp. 156–62; and V. P. White, *Grants: How to Find Out About Them and What to Do Next* (New York: Plenum, 1985), pp. 293–94.

Table 11-6). A basic step in grant seeking is to identify funding agencies with similar interests, intentions, and needs regarding the proposed project or activity. Use personal and professional contacts in your funding search. Also, it is helpful to examine an organization's past pattern of giving. The boxed insert on page 335 provides Internet resources on potential funding sources.

3. **Obtain grant application materials from the chosen funding source.** Follow the application instructions exactly. Make special note of proposal deadlines. Contact the funding agency's grant contact person for assistance during proposal development when necessary to clarify details or answer questions about the application procedures. Identify all requested information. Remember that the basic requirements, application forms, information, and procedures will vary from one funding agency to another.

4. **Plan the project.** Use local resources when addressing local problems. What is your purpose in developing the proposal? Who will be the beneficiaries and how will they benefit? Think through how you will execute the project from start to finish. What resources are needed? Seek the advice of colleagues.

5. **Write the proposal.** Keep three principles in mind as you write: (1) be candid, (2) be brief, and (3) be on target (for example, why is this project relevant to the funder's mission?). The components of a proposal will vary from one funding source to another, but they generally include the following:
   • Title page
   • Table of contents

TABLE 11-6 *Types of Funding Agencies*

Type	Examples
Federal	National Institutes of Health, National Cancer Institute, National Institute on Aging, Department of Health and Human Services, U.S. Agency for International Development, National Heart, Lung, & Blood Institute
State and local	Department of education, department of human services, arts council, local school district
Foundations	Ford Foundation, Spencer Foundation, Rockefeller Foundation, International Life Sciences Institute—Nutrition Foundation
Nonprofit organizations	American Diabetes Association, American Heart Association, American Dietetic Association
Industry	IBM, Eli Lilly, Ross Laboratories, Mead Johnson, National Dairy Council, National Cattlemen's Beef Association

Source: Adapted from M. R. Schiller and J. C. Burge, How to write proposals and obtain funding, in *Research: Successful Approaches,* ed. E. R. Monsen (Chicago, IL: American Dietetic Association, 1992), pp. 49–69.

- Abstract or summary. The proposal summary outlines the proposed project and appears at the beginning of the proposal. The summary is usually one page in length and should be written last to ensure that it contains all the key points necessary to present the objectives of the project. The abstract should also include a needs statement (why the project is important and needs to be done), the main goal of the project, the beginning and ending dates, and the amount of money requested. Many consider the abstract to be the most important page of the proposal. Grant reviewers form their initial impression of your project from the abstract, and it can be critical to the success of your venture.

- Introduction of the organization and grant seeker. This section establishes the credibility of the applicant or organization. What training and experience do you have? Why should the grant monies be awarded to your project or organization? What are your organization's goals, philosophy, and track record with other grantors? What are your success stories?

- Project description. This includes a clear, concise, and well-supported problem statement (or needs assessment) with a review of the current literature related to the problem; a statement of the goals and measurable objectives; project methodology showing how you will achieve the stated objectives; a time line showing how the project will proceed, including activities that will occur and the resources and staff needed to operate the project (inputs). This section also identifies the kinds of facilities, transportation, and support services required (throughputs); explains what will be achieved in measurable terms (outputs); justifies your course of action (why you selected these activities); highlights innovative or unique features of your project; and should convey a sense of urgency regarding the problem.

- Itemized project budget. The well-prepared budget is realistic, justifies all expenses of the project, and is consistent with the proposal narrative.

⊚ Internet Resources for U.S. Funding Sources

Catalog of Federal Domestic Assistance Programs	**family.info.gov/cfda/index.htm**
Cooperative State Research, Education, and Extension Service Grant and Funding Opportunities	**www.reeusda.gov/new/funding.htm**
CRISP (Computer Retrieval of Information on Scientific Projects)	**www-commons.dcrt.nih.gov**
Department of Health and Human Services Gateway Partnership Funding Page	**www.odphp.osophs.dhhs.gov/partner/funding.htm**
ERS Research on Food Assistance and Nutrition*	**www.econ.ag.gov/Briefing/foodasst/contracts.htm**
Federal Information Exchange	**nscp.fie.com**
Federal Register Online via GPO Access	**www.access.gpo.gov/su_docs/aces/aces140.html**
Food and Nutrition Information Center†	**www.nal.usda.gov/fnic**
The Foundation Center	**www.fdncenter.org/2index.html**
Grants Database (Oryx Press)	**www.oryxpress.com/grants.htm**
Grantsmanship Center	**www.tgci.com**
GrantsNet	**www.os.dhhs.gov/progorg/grantsnet**
National Institutes of Health	**www.nih.gov/grants**
National Science Foundation	**www.nsf.gov**
Notices of Funding Availability (Government-wide notices from the *Federal Register*)	**ocd.usda.gov/nofa.htm**

*Abbreviations: ERS = Economic Research Service; GPO = Government Printing Office.
†Click on "Internet Resources" icon at left, then scroll down to "Grant Information."

Itemize overall project costs including such things as salaries, fringe benefits, consultant fees, equipment, computer costs, facilities, materials and supplies, travel, postage, telephone, and indirect costs.

- Project evaluation. The evaluation plan shows how you will measure and communicate outcomes or impact. How will you determine that objectives have been met? How do you plan to disseminate the results? Note any specific evaluation methods required by the grantor.

- Future funding needs. This section reflects long-term project needs and describes a plan for continuation of the project beyond the grant period and/or the availability of other resources necessary to implement the grant.
- Appendixes should be used to provide details about the project, supplementary data, sample data collection instruments, references of cited literature, and information requiring in-depth analysis. (Such materials might otherwise detract from the readability of the proposal, and are better placed in an appendix.) Appendixes also include personnel vitaes, time tables, letters of support and endorsements for the project, legal papers, and reprints of relevant articles written by the grant seeker.

6. **Proofread your proposal.** Check grammar and spelling. Have you included all required information? Review your proposal with colleagues and seek their input and constructive criticism (for instance, have you used unsupported assumptions, jargon, or excessive language in the proposal?). Revise the proposal as necessary.

7. **Submit the proposal to the funding source.** Keep in mind that several submissions may be necessary to one or more funding agencies. Be persistent and resilient. Learn from your rejections.

In summary, the grant seeker requires many qualities, but four are critical to successful grantsmanship: (1) diligence in researching funding sources, (2) creativity in matching your project's goals with those of a funding source, (3) attention to detail in proposal preparation, and (4) persistence in revising and resubmitting proposals to potential funding sources. Remember that in grant writing, "failure often precedes success."[50] The most common reasons why proposals are not funded reflect basic problems with planning and time management: The proposal did not follow the guidelines; the deadline for submitting the application was missed; the proposal lacked a well-conceived plan of action or idea or had errors in the budget estimates. In many cases, a proposal is rejected because it doesn't match the grantor's funding priorities.[51]

Your proposal is more likely to be funded if it has been reviewed by your peers, follows the proposal guidelines exactly, provides useful supplemental materials in an appendix, and includes letters of support from individuals who are important to the project's success. Proposals that reflect a multidisciplinary approach to a problem are rated highly.[52]

Management Issues for "Heartworks for Women": Case Study 1

The development and design of the "Heartworks for Women" program has spanned this entire section, beginning with Chapter 5. We have discussed how it fit into a larger intervention strategy, the choice of program format and nutrition messages, and the elements of the marketing plan. In this section, we describe three management issues related to the program: the critical path, the budget, and grantsmanship.

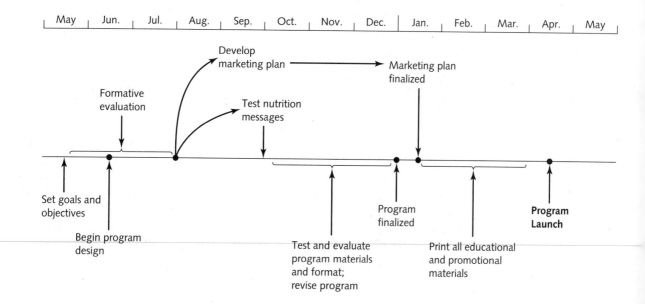

The Critical Path

FIGURE 11-5 *Timeline for Determining the Critical Path for the "Heartworks for Women" Program: Case Study 1*

The community nutritionist who leads team 2 develops a timeline early in the planning stages, when the program plan (described in Chapter 7) is outlined. A variety of formats can be used for this activity. Figure 10-6 on page 305 is one example of a timetable and shows key marketing activities for the "Heartworks for Women" program. (In reality, the marketing timetable would be developed *after* the timeline showing the critical path.) Because the program will be launched the following April, the team 2 leader lays out the paths that must be completed to ensure that all program elements have been finalized before the launch date. She plots a timeline, shown in Figure 11-5. The critical path is program design and evaluation—a process estimated to take 7 months. Two minor paths—developing the marketing plan and testing the appropriateness of the nutrition messages for the target population—are estimated to require several months as well. Any glitches in finalizing the program will produce delays in finalizing the marketing plan. For example, some elements of the marketing plan will be incomplete if the nutrition messages aren't finalized on time.

Several milestones appear on the timetable: early June, when program planning begins; August 1, when the formative evaluation has been completed; December 31, when the program must be finalized; January 15, when the marketing plan is finalized; and April 18, the date of the launch. These are the points at which major tasks must be completed on time, or the program launch date will be delayed.

Once the timeline has been developed, what is the next step for the community nutritionist? First, she shares it with her team members to confirm that it is reasonable and all activities have been accounted for. Next, she breaks down each path into a group of activities that can be monitored. For example, the program launch requires the following tasks be completed: choose and confirm the date, location, and time for the launch; prepare a list of invitees; develop, print, and mail invitations; prepare and mail press releases; prepare and finalize all advertising about the launch; choose a menu; arrange for delivery of food and beverages; outline the launch program (for instance, opening remarks, and the like); contact guest speakers and mail a confirmation letter to each of them; arrange for any decorating of the facility; and arrange interviews with the media. Completion dates for these activities are also determined.

The Operating Budget

The operating budget is a statement of the financial plan for the program. It outlines the revenues and expenses related to its operation. Revenues for the "Heartworks for Women" program will be derived from the enrollment fees for individual participants and the sales of merchandise such as coffee mugs and T-shirts, as shown in Table 11-7. Some expenses such as the design of the program logo and

TABLE 11-7 *Sample Operating Budget for the "Heartworks for Women" Program*

Revenue		
Program income ($60/participant × 75 participants/worksite × 60 worksites)		$270,000
Sales of merchandise		4,500
	Total revenue:	$274,500
Expenses		
Direct expenses		
Personnel—salaries		$163,500
Telephone/Internet		4,250
Equipment		2,000
Travel		1,575
Postage		3,500
Office supplies		8,000
Instructional materials		5,500
Advertising and promotion		10,000
Design		3,750
Printing		3,000
	Total direct expenses:	$205,075
Indirect expenses		
Personnel—benefits (28.7%)		$ 46,925
Rent		7,800
Utilities		1,200
Equipment depreciation		7,200
Maintenance and repairs		2,000
Janitorial services		4,300
	Total indirect expenses:	$69,425
	Total expenses:	$274,500
	Net revenue (expenses):	$0

printing of educational materials are directly related to the program; others such as office rent and electricity contribute indirectly to the program's costs. At the beginning of the fiscal year, the operating budget has a zero balance (net revenue minus total expenses). Expenses will be offset by anticipated revenue. Any discrepancies in either revenue or expenses will be reflected in the operating budget's monthly net balance.

Extramural Funding

Early in the planning process, the team 2 leader identifies several activities for which extramural funding might be obtained. She perceives that some expenses related to the launch, such as the printing of banners and balloons and providing food for invitees, could be underwritten, as might the costs of printing program manuals and training materials. She sets a target figure of about $8,000 and begins networking in her community. She contacts two colleagues whose organizations have sponsored health promotion activities in the past—the manager of a leading supermarket and a registered dietitian who is the district sales representative for a national drug company—who agree to provide financial support for these activities on behalf of their organizations. She expands her network to include the owner of a popular ladies' dress shop, who is interested in "doing some community work." This contact proves profitable—the owner of the dress shop is willing to contribute $1,000 toward the launch. Other commitments are confirmed over a period of several months. The team 2 leader assures all grantors that their contributions will be acknowledged publicly and negotiates details such as the amount of the grant and the timing of the payment.

The Business of Community Nutrition

Whether you work in the public or private sector, you need strong management skills. You must be able to set a direction for your business or program; define goals and objectives; organize the delivery of your product or service; motivate people to help your organization reach its goals; allocate materials, equipment, personnel, and funds to your operations; control data systems; and provide leadership. More than ever before, management and leadership skills are needed to gain a competitive edge in what has become an increasingly competitive health care environment. As you move into the community with your ideas, products, and nutrition services, keep these four strategies for success in mind:[53]

- Continually assess the competitive environment.
- Continually assess your strengths.
- Build organizational skills.
- Build managerial (people and process) skills.

Community Learning Activity

Activity 1

This Community Learning Activity continues your work on Case Study 2:

1. Develop a timeline that shows the critical path and any other paths for activities that must be completed.
2. Specify three process and three structure objectives for your program.
3. Develop an organizational chart that shows the staff needed to implement your program for elderly people. Did you choose a traditional chart over the web format? If so, why?
4. Develop a budget for your program.
5. Briefly outline an idea for obtaining extramural funds for a portion of your program activities. You should be able to write a short (one-paragraph) summary of your idea, list at least one other organization that could partner with you to apply for funding, and estimate the amount of money needed. Use the Internet addresses shown in the boxed insert on page 335 to determine whether a request for funding proposal exists along the lines you propose. While browsing for potential funding sources on the Internet, did you find a source of funds that might be applied to your project in an unexpected way?

Activity 2

Divide the class into 2–4 groups. Each group will outline a management plan for developing an Internet Web site designed to educate consumers about the USDA and other food guide pyramids. Refer to Appendix J, "Tips for Designing Web Sites," and the Internet Resources boxed insert on food guide pyramids (see page 299). Each team sketches a general plan, drawing on the four functions of management. For example, each team sets goals and objectives, specifies the site's content, develops a time line, describes the staff needed to develop, implement, and maintain the Web site, develops a budget, and outlines a plan for evaluating the site, including the formative evaluation needed to get the site up and running. Each team presents its plan to the class. How similar are the plans? If you were a manager and your team's plan differed significantly from your classmates' plans, what steps would you take next?

Activity 3

Describe the contribution of each of the following to the successful community needs assessment and community intervention:

- the target population
- asking questions
- networking
- partnerships
- goals and objectives
- planning
- evaluation
- entrepreneurship

Professional Focus | Lighten Up—Be Willing to Make Mistakes and Risk Failure

"When a man knows he is to be hanged in a fortnight, it concentrates his mind wonderfully."

— Samuel Johnson, 1777

If you knew that you had only a few months to live, would you live differently than you do right now? Would you find time to take dancing lessons, learn to rollerblade, snorkel off the Great Barrier Reef, study the stock market, get a pilot's license, or try your hand at papier-mâché? Would you take more chances and worry less about your image?

Most of us would probably answer that last question in the affirmative. We would *choose* to live differently if we knew that only a few grains of sand were left in the hourglass. Nadine Stair, at the age of 85, said the same thing: "If I had my life to live over again, I'd try to make more mistakes next time. I would relax. I would limber up. I would be sillier than I have been this trip. I know of a very few things I would take seriously. I would take more trips. I would climb more mountains, swim more rivers and watch more sunsets. I would do more walking and looking. I would eat more ice cream and less beans. . . . If I had it to do over again, I would go places, do things and travel lighter than I have. . . . I'd pick more daisies."[1]

Notice that the first thing Nadine Stair said she would do differently "next time" was to try to make more mistakes. Most people work hard to *avoid* making mistakes, and few of us are willing to undertake a venture so risky that mistakes are almost guaranteed and the probability of failure looms large. In our culture, these activities are to be avoided at all costs. If you doubt this, go to your local bookstore or library and search for books on failure, making mistakes, and risk taking. You won't find any of the following books: *The Joy of Failing, The Seven Habits of Complete Nincompoops, The One-Minute Mistake Maker, Fit for Failure* or *100 Risk Takers Who Fell Flat on Their Faces.*

Some People Don't Give Up on Their Goals

Virtually every successful entrepreneur, adventurer, and risk taker *has* made mistakes and *has* failed at some point in his or her struggle to reach a personal or professional goal. In an essay in *Science*, Harold T. Shapiro, president of Princeton University, remarked that "the world too often calls it failure if we do not immediately reach our goals; true failure lies, rather, in giving up on our goals."[2] What would our world be like if the following individuals had given up on their goals?

- In 1842, at a time when most young British women of position were concerned mainly with parties and pending marriages, Florence Nightingale felt a "call" to perform some lifework. Although she sensed that her destiny "lay among the miserable of the world," the precise nature of this vocation eluded her for many years. Not until she was in her early thirties did she begin to pursue a career in nursing despite the persistent objections of her mother and sister, a cultural bias against nursing care, and the resistance of the traditional medical establishment. Over a lifetime of hard work, her determination and vision radically altered the practice and professionalism of nursing. Nightingale was among the first to document and describe hospital conditions, and she became an expert on sanitation. She reformed the health administration of the British army in response to the brutal mortality of the Crimean War and thereby influenced medical practice for years to come. Her reports on proper hospital construction, the training of nurses, and patient care led to the establishment of sanitary commissions and eventually to the public health service.[3]

- Thomas Alva Edison, born in 1847, has been described as "one of the outstanding geniuses in the history of technology."[4] He received very little formal schooling, having been expelled by a schoolmaster as "addled," and was taught history, science, and philosophy primarily by his mother. His fascination with the wireless telegraph as a young boy led to a lifelong enjoyment of experimentation and research. At his death in 1931, he held 1,093 patents on such devices as the incandescent electric lamp, the phonograph, the carbon telephone transmitter, and the motion picture projector.

At one point in his career, Edison struggled to develop a storage battery. "I don't think Nature would be so unkind as to withhold the secret of a *good* storage battery if a real earnest hunt for it is made," he said. "I'm going to hunt."[5] This was no mean feat. He knew what was required of a good battery—it must last for years, should not lose capacity when recharged, and needed to be nearly indestructible. He began by testing one chemical after another in a series of experiments that spanned a decade. When 10,000 experiments failed to give the desired results, Edison remarked, "I have not failed. I've just found 10,000 ways that won't work."[6] He eventually succeeded.

Continued

- When he was a 20-year-old student at Yale University, Fred Smith wrote a term paper that analyzed freight services existing at the time. He concluded that there was a market for a company that moved "high-priority, time-sensitive" goods such as medicines and electronic components. He believed the existing system was cumbersome and failed to respond quickly to consumers' needs. In his paper, he proposed an overnight delivery service based on a "hub-and-spokes" air freight system. His professor was unimpressed with Smith's proposal, citing a restrictive regulatory environment and competition from airlines as major barriers to implementing such a service. The paper earned a grade of C.

  Smith did not give up on his idea, although he could not do anything about it for several years. Eventually, at the age of 29, he founded a company, Federal Express, designed to deliver packages "absolutely, positively overnight." In March 1973, his first planes flew over the eastern United States, carrying a total of six packages. One month later the volume had increased to 186 packages. In its first years, the company nearly folded from a lack of capital, concerns about Smith's leadership, and a formal charge of fraud against him. The company—with Smith at its helm—survived this difficult period. By 1983, its earnings were more than $1 billion. By the late 1980s, Federal Express was handling more than 700,000 packages and parcels daily and had altered American business practices substantially.[7]

The Secret of Success

Risk takers make mistakes and sometimes fail. If they have a common feature, however, it is their willingness and determination to persevere, sometimes against great odds. In our professional lives, it is impossible to avoid risk and the possibility of failure. The trick is to learn how to minimize risk and capitalize on your mistakes. Here are a few points to keep in mind when you are next faced with undertaking a risky venture:

1. Do your homework. There are risks and there are *calculated* risks. The difference between the two is substantial. To prepare for a calculated risk, talk with people who have undertaken similar ventures. Find out about the unexpected problems they experienced and how they handled them.
2. Write down your options and the potential outcome of each. This activity will help you focus on the option that may stand the best chance for success. Then, write down the worst possible thing that could happen if you proceeded with that option. Is it something you can live with? If not, how can the option be changed to protect you or your employees?
3. Learn from your mistakes or failures. We all make them, but we don't all learn from them. In the business world, bankruptcy is often viewed as the ultimate failure. One entrepreneur commented that "if you hadn't been bankrupt at least once, you hadn't really learned much about business."[8] Although it is certainly painful, business failure can be an opportunity for learning new lessons both personally and professionally. The successful entrepreneur and risk taker has the ability to learn from her experiences and regain control of her destiny.
4. Be committed to your goal. Having a high level of commitment to the work at hand is one element that distinguishes the successful entrepreneur from the also-rans.[9]

Words to Work By

In his book of wildlife portraits, the artist Robert Bateman remarked, "A great master teacher once said, 'In order to learn how to draw you have to make two thousand mistakes. Get busy and start making them.'"[10] These are apt words to keep in mind as you begin traveling *your* career path.

References

1. As cited in R. N. Bolles, *The Three Boxes of Life—And How to Get Out of Them* (Berkeley, CA: Ten Speed Press, 1981), p. 377.
2. H. T. Shapiro, The willingness to risk failure, *Science* 250 (1990): 609.
3. C. Woodham-Smith, *Florence Nightingale* (New York: McGraw-Hill, 1951). The quotation in this paragraph was taken from p. 31 of this book.
4. Edison, in *The New Encyclopaedia Britannica: Macropaedia*, vol. 17, 15th ed. (Chicago: Encyclopaedia Britannica, 1985), pp. 1049–51.
5. W. A. Simonds, *Edison—His Life, His Work, His Genius* (London: Kimble & Bradford, 1935), p. 278.
6. As cited by Shapiro, Willingness to risk failure.
7. W. Davis, *The Innovators* (New York: American Management Association, 1987), pp. 361–65.
8. As cited in B. J. Bird, *Entrepreneurial Behavior* (Glenview, IL: Scott, Foresman, 1989), pp. 354–55.
9. Bird, *Entrepreneurial Behavior*, pp. 366–67.
10. R. Bateman, *The Art of Robert Bateman* (Ontario: Penguin Books Canada, 1981), p. 19.

References

1. How the new strategic plan works—for you, *Journal of the American Dietetic Association* 92 (1992): 1070.

2. The definition of management was taken from K. M. Bartol and D. C. Martin, Management (New York: McGraw-Hill, 1991), p. 6. Other definitions appearing in this chapter were also taken or adapted from this textbook.

3. Hoover: It sucks, *The Economist* 327 (1993): 66.

4. J. S. Rakich, B. B. Longest, Jr., and K. Darr, *Managing Health Services Organizations* (Philadelphia: Saunders, 1985), pp. 213–41.

5. President's page: A strategic plan for "Creating the Future," *Journal of the American Dietetic Association* 96 (1996): 78.

6. K. G. Hardy, The three crimes of strategic planning, *Business Quarterly* 57 (1992): 71–74.

7. President's page: Taking stock of ADA's strategic initiatives, *Journal of the American Dietetic Association* 97 (1997): 429, and President's page: Checking the strategic framework road map, *Journal of the American Dietetic Association* 98 (1998): 588.

8. Bartol and Martin, *Management*, pp. 174 and 308.

9. J. Elton and J. Roe, Bringing discipline to project management, *Harvard Business Review* 76 (1998): 153–59.

10. G. E. A. Dever, *Community Health Analysis—Global Awareness at the Local Level*, 2nd ed. (Gaithersburg, MD: Aspen, 1991), pp. 45–73.

11. J. B. McKinlay, The promotion of health through planned sociopolitical change: Challenges for research and policy, *Social Science and Medicine* 36 (1993): 109–17; and J. B. McKinlay, More appropriate evaluation methods for community-level health interventions, *Evaluation Review* 20 (1996): 237–43.

12. S. A. McGraw and coauthors, Using process data to explain outcomes: An illustration from the Child and Adolescent Trial for Cardiovascular Health (CATCH), *Evaluation Review* 20 (1996): 291–312.

13. Ibid.

14. The discussion of impact evaluation was adapted from M. B. Dignan and P. A. Carr, *Program Planning for Health Education and Promotion*, 2nd ed. (Philadelphia, PA: Lea & Febiger, 1992), pp. 143–54; J. P. Elder, E. S. Geller, M. F. Hovell, and J. A. Mayer, *Motivating Health Behavior* (Albany, NY: Delmar, 1994), pp. 183–91; and K. Tones, S. Tilford, and Y. K. Robinson, *Health Education: Effectiveness and Efficiency* (New York: Chapman and Hall, 1990), pp. 124–27.

15. A. Fink, *Evaluation Fundamentals: Guiding Health Programs, Research, and Policy* (Newbury Park, CA: Sage, 1993), pp. 1–4.

16. K. L. Probert, *Moving to the Future: Developing Community-Based Nutrition Services*. (Washington, D.C.: Association of State and Territorial Public Health Nutrition Directors, 1996), pp. 59–66.

17. H. G. Tolpin, Economics of health care, *Journal of the American Dietetic Association* 76 (1980): 217–22; E. Bartlett, Cost-benefit and cost-effectiveness analyses, in *Benefits of Nutrition Services: A Costing and Marketing Approach* (Columbus, OH: Ross Laboratories, 1987), p. 20.

18. D. R. Dalton and coauthors, Organization structure and performance: A critical review, *Academy of Management Review* 5 (1980): 49–64.

19. J. L. Gray and F. A. Starke, *Organizational Behavior—Concepts and Applications*, 4th ed. (Columbus, OH: Merrill/Prentice Hall, 1988), pp. 426–27.

20. J. G. Liebler and coauthors, *Management Principles for Health Professionals*, 2nd ed. (Gaithersburg, MD: Aspen, 1992), pp. 160–97.

21. Gray and Starke, *Organizational Behavior*, pp. 424–57.

22. Bartol and Martin, *Management*, p. 348.

23. D. D. Van Fleet and A. G. Bedeian, A history of the span of management, *Academy of Management Review* 2 (1977): 356–72.

24. J. C. Levinson, *The Ninety-Minute Hour* (New York: Dutton, 1990), p. 51.

25. M. W. McCall, Jr., and M. M. Lombardo, What makes a top executive? *Psychology Today* 17 (1983): pp. 26–31.

26. Bartol and Martin, *Management*, p. 407.

27. Liebler and coauthors, *Management Principles*, pp. 181–83.

28. American Dietetic Association, *Role Delineation for Registered Dietitians and Entry-Level Dietetic Technicians* (Chicago: American Dietetic Association, 1990), pp. 255–90.

29. The discussion of staffing was adapted from Liebler and coauthors, *Management Principles*, pp. 246–66.

30. Bartol and Martin, *Management*, pp. 410–11.

31. F. Rice, Champions of communication, *Fortune* 123 (1991): 111–112, 116, 120.

32. B. R. Nathan and coauthors, Interpersonal relations as a context for the effects of appraisal interviews on performance and satisfaction: A longitudinal study, *Academy of Management Journal* 34 (1991): 352–69.

33. A. A. Chesney and E. A. Locke, Relationships among goal difficulty, business strategies, and performance on a complex management simulation task, *Academy of Management Journal* 34 (1991): 400–24.

34. J. G. Goodale, Seven ways to improve performance appraisals, *HR Magazine* 38 (1993): 77–80.

35. W. Bennis, *On Becoming a Leader* (Reading, MA: Addison-Wesley, 1989), pp. 39–45.

36. R. McGarvey, Goal-getters, *Entrepreneur* 21 (1993): 144.

37. P. F. Drucker, *The Practice of Management* (New York: Harper & Row, 1986), pp. 302–11.

38. Bartol and Martin, *Management*, p. 520.

39. D. K. Denton, *Recruitment, Retention, and Employee Relations* (Westport, CT: Quorum, 1992), pp. 155–71.

40. Bartol and Martin, *Management*, p. 594.

41. American Dietetic Association, *Role Delineation Study*, pp. 273–78.

42. J. Sneed, E. C. Burwell, and M. Anderson, Development of financial management competencies for entry-level and advanced-level dietitians, *Journal of the American Dietetic Association* 92 (1992): 1223–29.

43. T. D. Lynch, *Public Budgeting in America* (Upper Saddle River, NJ: Prentice Hall, 1985), p. 4.

44. D. D. Roman, *Managing Projects—A Systems Approach* (New York: Elsevier Science, 1986), p. 129.

45. H. H. Schmitz, Decision support: A strategic weapon, in *Healthcare Information Management Systems*, ed. M. J. Ball et al. (New York: Springer, 1991), pp. 42–48.

46. Bartol and Martin, *Management*, pp. 704–5.

47. A. Quinn, Writing a grant proposal that works, *Leader in Action*, Spring 1992, pp. 9–16.

48. V. P. White, *Grants: How to Find Out About Them and What to Do Next* (New York: Plenum Press, 1985), p. vii; and Quinn, Writing a grant proposal that works, p. 11.

49. The steps for proposal writing were adapted from D. Richards, Ten steps to successful grant writing, *Journal of Nursing Administration* 20 (1990): 20–23; C. Kemp, A practical approach to writing successful grant proposals, *Nurse Practitioner* 16 (1991): 51–56; and Appendix VI: Developing and Writing Grant Proposals, *1993 Catalog of Federal Domestic Assistance* (Washington, D.C.: U.S. Government Printing Office, 1993), pp. FF1–FF7.

50. Kemp, A practical approach, p. 53.

51. D. Richards, *Journal of Nursing Administration*, pp. 20–23.

52. B. R. Ferrell and coauthors, Applying for Oncology Nursing Society and Oncology Nursing Foundation grants, *Oncology Nursing Forum* 16 (1989): 728–30.

53. A. J. Morrison and K. Roth, Developing global subsidiary mandates, *Business Quarterly* 57 (1993): 104–10.

Community Nutritionists in Action: Delivering Programs

Leon T. can't decide what to do next. "I could go downstairs and finish the pull-toy for young Kevin," he muses. Leon remembers leaving a pile of sawdust and wood chips scattered about the floor down there. "Or I could read the new *Reader's Digest,* just arrived today." Leon pats his shirt pocket for his glasses. Even with the magazine's large print, he can't read it very well without his glasses. Nope, they aren't in his shirt pocket and they aren't in his pants pocket either, or on the end table. "Well, shucks, I can't keep track of anything these days." He stands and starts a search of the easy chair and sofa. No luck.

He moves to the dining table, where he picks among the dirty dishes, magazine flyers, newspapers, and mail. His frustration level mounts quickly when he realizes he can't remember what he's looking for. "Ah, look what I found— my pills!" Momentarily satisfied, he goes to the kitchen to take his medications. He's being treated for high blood pressure, low thyroid, and angina. He's suddenly confused, though, knowing there was something he was aiming to do. What was it?

Things just haven't been the same for Leon since his oldest, dearest friend, Bill, moved to Memphis a year ago to be close to his only daughter. Leon and Bill played golf regularly and met for dinner several nights a week. Now, Leon doesn't seem to have the energy for much of anything, and he no longer enjoys his woodworking shop or reading. At 76, Leon lives alone and likes it that way. But how much longer can he live independently? Does he need assistance with meals, medications, and doctor's visits? Are some of his symptoms due to depression or some other medical condition? Does he have family living nearby, and, if so, what is the nature of his relationship with them?

These are the types of questions the community nutritionist asks when assessing Leon's needs. The community nutritionist aims to determine whether existing nutrition services and programs can improve his quality of life and help him continue to live independently. These programs help "protect" people like Leon from the environmental and social elements that place them at risk for disease and poor health. They offer some security to people who have little money to spend on food, and they help control the problems associated with malnutrition.

This section introduces you to the major food assistance and nutrition programs of the federal government and other noteworthy community-based programs. The section is divided along life cycle lines, with chapters on mothers and infants, children and adolescents, and adults. Chapters on poverty, food insecurity, and international issues in nutrition are included, because these areas are increasingly important in a shrinking global community.

We encourage you to draw on the material you have already learned in studying the chapters that follow. By now, you can appreciate the complexities of policy making, goal setting, program design, and human behavior. You know how challenging the process of program development can be. As you review the programs described in these chapters, consider how *you* would change the delivery of community-based nutrition programs and the policies that influence them.

Chapter 12

Domestic Hunger and the Food Assistance Programs

Outline

Learning Objectives

After you have read and studied this chapter, you will be able to:

- Communicate the current status of food security in the United States.
- Understand the complexity of causes of domestic food insecurity.
- Explain the significance and relevance of food security to dietetics professionals.
- Describe recent domestic hunger policy initiatives.

- Describe the purpose, status, and current issues related to the U.S. food assistance programs.
- Identify the impact of welfare reform in the United States on government food assistance programs.
- Describe actions that individuals might take to eliminate domestic food insecurity.

Something To Think About...

The goal of nutrition is to apply scientific knowledge to feed people, to feed them adequately, and to feed them all. . . . It would be a mistake to think of nutrition simply as a group of empirical prescriptions for personal and social behavior. Rather, nutrition is an agenda, the fulfillment of which requires the knowledge of a number of sciences—biochemistry, physiology, pathology, epidemiology, psychology, economics, and sociology. By its nature, nutrition is a set of scientific disciplines whose end is action. . . . Nutritionists, unlike biochemists or physiologists, but like cardiologists and pediatricians, have to see their science as one whose goal is the benefit of mankind. Action often has to be taken on a reasonable presumption of health benefit without obvious health costs. As researchers, nutritionists have made amazing strides over an extremely short period in the history of science. I have every confidence we will continue to do so. But studying the metabolism of vitamin B-12 in the laboratory rat or the pathology of aging in the gerbil is not enough. While research continues in nutrition and agriculture, we must provide for the people who are here now, in America and the world. It is my hope that the new generation of nutritionists, trained in the sciences that form the basis of nutrition, will also keep this goal clearly in mind. — *Jean Mayer*

Introduction

The problems of overnutrition—obesity, heart disease, cancer, and others—have guided the nutrition and health objectives and recommendations of the economically developed nations, for these are the diseases of our society. Not everyone shares these problems, though. People in developing nations, as well as people in the less-privileged parts of developed nations, suffer the problems of **undernutrition,** which is characterized by chronic debilitating hunger and **malnutrition.** These conditions are most visible in times of **famine,** but they are widespread and persistent even when famine does not occur. They have been with us throughout history, and despite numerous development and assistance programs, they are not disappearing; the number of hungry and malnourished people continues to grow. Hunger and malnutrition can be found among people of both sexes and all ages and nationalities. Even so, these problems hit some groups harder than others.

Everyone has known the uncomfortable feeling of hunger that signals "time to eat" and passes with the eating of the next meal. But many people know hunger more intimately because a meal does not follow to quiet the signal. For them, hunger is a constant companion bringing ceaseless discomfort, weakness, and pain—the continuous lack of food and nutrients. People who live with chronic hunger either have too little food to eat or do not receive an adequate intake of the essential nutrients from the foods available to them—either way, malnutrition ensues. One person in the developing world described hunger as follows:

Undernutrition As used in this chapter, a continuous lack of the food energy and nutrients necessary to achieve and maintain health and protection from disease: both a domestic and a world problem.

Malnutrition The impairment of health resulting from a relative deficiency or excess of food energy and specific nutrients necessary for health.

Famine Widespread lack of access to food due to natural disasters, political factors, or war; character- ized by a large number of deaths due to starvation and malnutrition.

Food security Access by all people at all times to sufficient food for an active and healthy life. Food security includes at a minimum the ready availability of nutritionally adequate and safe foods and an assured ability to acquire personally acceptable foods in socially acceptable ways.

Food insecurity The inability to acquire or consume an adequate quality or sufficient quantity of food in socially acceptable ways, or the uncertainty that one will be able to do so.

For hunger is a curious thing: at first it is with you all the time, waking and sleeping and in your dreams, and your belly cries out insistently, and there is a gnawing and a pain as if your very vitals were being devoured, and you must stop it at any cost, and you buy a moment's respite even while you know and fear the sequel. Then the pain is no longer sharp but dull, and this too is with you always, so that you think of food many times a day and each time a terrible sickness assails you, and because you know this you try to avoid the thought, but you cannot, it is with you. Then that too is gone, all pain, all desire, only a great emptiness is left, like the sky, like a well in drought, and it is now that the strength drains from your limbs, and you try to rise and you cannot, or to swallow and your throat is powerless, and both the swallow and the effort of retaining the liquid tax you to the uttermost.[1]

Today the phenomenon of hunger is being discussed in terms of **food security** or **insecurity.** Food security means that all people at all times have access to enough food for an active, healthy life. The concept of food security includes five components:[2]

- **Quantity:** Is there access to a sufficient quantity of food?
- **Quality:** Is food nutritionally adequate?
- **Suitability:** Is food culturally acceptable and the capacity for storage and prepa- ration appropriate?
- **Psychological:** Does the type and quantity of food alleviate anxiety, lack of choice, and feelings of deprivation?
- **Social:** Are the methods of acquiring food socially acceptable?

A writer in Boston describes food insecurity in more personal terms:[3]

I've had no income and I've paid no rent for many months. My landlord let me stay. He felt sorry for me because I had no money. The Friday before Christmas he gave me ten dollars. For days I have had nothing but water. I knew I needed food; I tried to go out but I was too weak to walk to the store. I felt as if I were dying. I saw the mailman and told him I thought I was starving. He brought me food and then he made some phone calls and that's when they began delivering these lunches. But I had already lost so much weight that five meals a week are not enough to keep me going.

I just pray to God I can survive. I keep praying I have the will to save some of my food so I can divide it up and make it last. It's hard to save because I am so hungry that I want to eat it right away. On Friday, I held over two peas from the lunch. I ate one pea on Saturday morning. Then I got into bed with the taste of food in my mouth and I waited as long as I could. Later on in the day I ate the other pea.

Today I saved the container that the mashed potatoes were in and tonight, before bed, I'll lick the sides of the container. When there are bones I keep them. I know this is going to be hard for you to believe and I am almost ashamed to tell you, but these days I boil the bones until they're soft and then I eat them. Today there were no bones.[4]

This chapter examines the extent of hunger and malnutrition in the United States; Chapter 16 examines the incidence of hunger around the world. Both chapters offer suggestions for personal involvement with the issues presented. As you read, challenge yourself with the following questions: What problems would you attack first in solv- ing the problem of hunger and food insecurity? Should we solve our own hunger problems before tackling problems related to international hunger? These issues are complex and often overwhelming from an individual's standpoint. Remember, however, as you read, that it is better to light one small candle than to curse the darkness.

Who Are the Hungry in the United States?

The United States is one of the wealthiest nations in the world, consumes 40 percent of the world's resources, and is home to only 6 percent of the world's people. Still, it does not meet the food needs of its poor.

Through the late 1960s, hunger was evident among the chronic poor: migrant workers, American Indians, southern blacks, unemployed minorities, and some of the elderly. These groups were hungry during the 1970s and 1980s as well, as were the newly unemployed blue-collar workers. Now, despite the strong economy of the 1990s, hunger is reaching into other segments of the population without regard for age, marital status, previous employment or successes, family ties, or efforts to change the situation. Frequently cited causes include low-paying jobs, food stamp cuts, and the fact that available resources are failing to reach many groups. The millions who experience hunger today in the United States include the following:

- **The young.** One in five U.S. children lives in poverty.[5]
- **The new poor.** Changes in the nation's economy during the 1980s have hurt millions of people who were once members of the middle class—displaced farm families and former blue-collar workers forced out of manufacturing (oil, natural gas, steel, and mining) into the service sector (maintenance and the hotel and restaurant industry). The auto industry, for example, has laid off some 40 percent of its workers since 1980. Only a third of them find jobs that pay as well as their old ones. A job that pays the minimum wage does not lift a family above the federal poverty line, and many such jobs fail to provide fringe benefits to help meet rising health care costs. A minority of the poor in the United States are on welfare; most are working people.
- **The elderly.** Social Security and other programs have pulled many older people out of poverty, but large numbers of older people who cannot work and have no savings or families to turn to are facing rising bills for housing, utilities, food, and health care.
- **The homeless.** As many as 3 million people may now be living on the streets. More than half of the homeless are single mothers with children; many of the homeless are former residents of psychiatric institutions.
- **Low-income women.** Many women live in poverty, struggling to provide child care while working for the minimum wage.
- **Ethnic minorities.** Although the majority of the poor in America are white, the median income of African American and Hispanic families is lower than that of white families.

The most compelling single reason for this hunger is **poverty.**

Poverty The state of having too little money to meet minimum needs for food, clothing, and shelter. As of 1998, the U.S. Department of Health and Human Services defined a poverty-level income as $16,450 annually for a family of four.

Category	National Rate of Poverty Increase
Infants and children under six years old	+63%
The "working poor"*	+56%
Poor married couples (two-parent families)	+47%
Young adults living independently	+13%
Single-parent families (usually headed by women)	+10%
The elderly	−3%

*Families who receive 75% of their income from jobs, as opposed to public assistance benefits.
Source: U.S. Bureau of the Census, *Statistical Updates* (Washington, D.C.: U.S. Government Printing Office, 1989).

Causes of Hunger in the United States

Poverty and hunger are interdependent. Most studies conducted since 1980 attribute the increases in hunger throughout the country to worsening economic conditions among the poor. Table 12-1 shows how poverty has increased among a number of population groups. Why are these increases in poverty occurring? For one thing, federal spending for antipoverty programs was significantly reduced during the 1980s. Between 1982 and 1985, federal expenditures for human services programs were cut by $110 billion in an attempt to reduce the national debt. Programs were eliminated, eligibility requirements were tightened, no adjustments for inflation were allowed, and budgets for the remaining programs were slashed. Some of the most severe reductions came in programs directly affecting those deepest in poverty, including Aid to Families with Dependent Children (AFDC), food stamps, low-income housing assistance, and child nutrition. Funding for the school lunch program was cut by one-third.[6] Cuts in college financial aid effectively blocked a pathway out of poverty taken by many in the past.

Although poverty is the major cause of hunger in the United States, other problems also contribute:

- Alcoholism and chronic substance abuse, which often contribute to increased poverty and malnutrition among not only the afflicted individuals, but their families as well
- Mental illness, loneliness, isolation, depression, and despair, which may result in people ceasing to be concerned for their own physical well-being
- The reluctance of people, particularly the elderly, to accept what they perceive as "welfare" or "charity"
- Delays in receiving requested food stamps and other public assistance benefits
- An increase in the number of single mothers without the means to care for their children
- Poor management of limited family financial resources
- Health problems of old age, precipitating an inability to purchase and prepare food

Feeding the hungry in the United States: Then and now.

- Lack of nutritional adequacy and balance in the food available to hungry people through emergency feeding programs and food assistance organizations, which take what they can get and pass it on to hungry people
- Lack of access to assistance programs because of intimidation, ineligibility, complicated paperwork, and other reasons
- Insufficient community food resources for the hungry
- Insufficient community transportation systems to deliver food to hungry people who have no transportation

Health experts saw similar situations back in the 1930s and 1960s. Let us first look at how programs were developed to handle the problems of hunger and poverty in those times and then at how those programs are working now.

Historical Background of Food Assistance Programs

During the Great Depression of the 1930s, concern about the plight of farmers who were losing their farms and the economic problems facing U.S. families in general led Congress to enact legislation giving the federal government the authority to buy and distribute excess food commodities. A few years later, Congress initiated an experimental Food Stamp Program to enable low-income people to buy food. Then in 1946, it passed the National School Lunch Act in response to testimony from the surgeon general that "70 percent of the boys who had poor nutrition 10 to 12 years ago were rejected by the draft." Despite these programs, in the 1960s, large numbers of people were still going hungry in the United States, and some of them suffered seriously from malnutrition as a result.

As evidence accumulated during the 1960s and 1970s showing that hunger was prevalent in the United States, poverty and hunger became national priorities. Old programs were revised and new programs were developed in an attempt to prevent malnutrition in those people found to be at greatest risk. The Food Stamp Program was expanded to serve more people. School lunch and breakfast programs were enlarged to support children nutritionally while they learned. Feeding programs were started to reach senior citizens. To provide food and nutrition education during the years when nutrition has the most crucial impact on growth, development, and future health, a supplemental food and nutrition program (WIC) was established for low-income pregnant and breastfeeding women, infants, and children who were nutritionally at risk. These and other events in the history of U.S. federal efforts to address domestic hunger are highlighted in the boxed insert.

As a result of these efforts, hunger diminished as a serious problem for the United States. Several studies, including comparative observations made 10 years apart, documented the difference the food assistance programs had made. In a baseline study in the late 1960s, a Field Foundation report stated:

> Wherever we went and wherever we looked we saw children in significant numbers who were hungry and sick, children for whom hunger is a daily fact of life, and sickness in many forms, an inevitability. The children we saw were . . . hungry, weak, apathetic . . . visibly and predictably losing their health, their energy, their spirits . . . suffering from hunger and disease, and . . . dying from them.

Ten years later, in 1977, the same group reported:

> Our first and overwhelming impression is that there are far fewer grossly malnourished people in this country today than there were ten years ago. . . . This change does not appear to be due to an overall improvement in living standards or to a decrease in joblessness in those areas . . . . But in the area of food there is a difference. The Food Stamp program, school lunch and breakfast programs, and the Women-Infant-Children programs have made the difference.[7]

⊙ The History of U.S. Federal Policies Addressing Hunger: An Overview

1930	The USDA and Federal Emergency Relief Administration distribute surplus agricultural commodities as food relief.
1933	Congress creates the Agricultural Adjustment Administration to control farm prices and production and the Federal Surplus Relief Corporation to distribute surplus farm products to needy families.
1935–1942	Congress provides for continued operation of the Federal Surplus Commodities Corporation, which, under the USDA, purchases commodities for distribution to state welfare agencies.
1936–1942	Amendments to the Agricultural Act permit food donations to school lunches.
1939–1943	The Federal Surplus Commodities Corporation initiates an experimental food stamp program.
1946	The National School Lunch Program is established.
1954	The Special Milk Program is established.
1955	The USDA determines that the average low-income family spends one-third of its after-tax income on food.
1961	President John F. Kennedy expands the use of surplus foods for needy people at home and abroad and announces eight pilot food stamp programs.
1964	Congress establishes the national Food Stamp Program. The Social Security Administration establishes the poverty line at three times the cost of the USDA's lowest-cost Economy Food Plan. Since 1969, values have been adjusted according to the Consumer Price Index.
1966	The Child Nutrition Act passes. It initiates the School Breakfast Program, which becomes permanent in 1975. President Lyndon B. Johnson outlines the Food for Freedom program.
1968–1977	The Senate establishes the Select Committee on Nutrition and Human Needs to lead the nation's antihunger efforts.
1968–1970	The Ten-State and Preschool Nutrition Surveys and *Hunger, U.S.A.* report evidence of malnutrition among children in poverty.
1969	President Richard M. Nixon announces a "war on hunger" and holds the White House Conference on Food, Nutrition, and Health. The USDA establishes the Food and Nutrition Service to administer federal food assistance programs.
1971	Results of the Ten-State Survey released to Congress indicate high risk of malnutrition among low-income groups.
1972	Congress authorizes the Special Supplemental Food Program for Women, Infants, and Children (WIC). The Older Americans Act authorizes the Nutrition Program for Older Americans.
1973	Amendments to the Older Americans Act establish congregate and home-delivered meals programs.
1977	The Food and Agricultural Act and Child Nutrition and National School Lunch Amendments are passed.
1981	The USDA establishes a small demonstration project for commodity distribution, the Special Supplemental Dairy Distribution Program, which becomes institutionalized as the Temporary Emergency Food Assistance Program (TEFAP) in 1983.

Continued

⊚ The History of U.S. Federal Policies Addressing Hunger: An Overview—continued

1981–1982	Congress passes the Omnibus Budget Reconciliation Acts, Omnibus Farm Bill, and Tax Equity and Fiscal Responsibility Act, which eliminate, restrict, and reduce food and income benefits.
1984	The President's Task Force on Food Assistance finds little evidence of widespread or increasing undernutrition but concludes that hunger exists and is intolerable in the United States.
1986	The General Accounting Office finds that methodologic flaws discredit findings of the Physician Task Force on Hunger that hunger is prevalent in counties with low food stamp participation rates.
1988	The DHHS publishes the *Surgeon General's Report on Nutrition and Health*, which states that lack of access to an appropriate diet should not be a health problem for any American. Congress passes the Hunger Prevention Act increasing eligibility and benefits for Food Stamps, Child Care, and TEFAP programs.
1989	The House Select Committee on Hunger holds hearings on food security in the United States.
1991	The Mickey Leland Childhood Hunger Relief Act (HR-1202, S-757) is introduced.
1994	Healthy Meals for Healthy Americans Act requires the National School Lunch Program and School Breakfast Program to comply with the Dietary Guidelines for Americans.
1995	School Meals Initiative for Healthy Children ensures that nutrition standards for school meals meet the Dietary Guidelines for Americans by updating nutrition standards for school meals, providing a variety of menu planning alternatives, and streamlining program administration.
1996	Healthy Meals for Children Act grants additional flexibility to school food authorities to meet the nutrition standards established in the Dietary Guidelines for Americans. Schools may thus continue to use the traditional meal pattern or any reasonable approach to menu planning providing the menus meet the U.S. Dietary Guidelines.

Source: M. Nestle and S. Guttmacher, Hunger in the United States: Rationale, methods, and policy implications of state hunger surveys. Reprinted with permission, *Journal of Nutrition Education* 24 (1992): 22S. Society for Nutrition Education, and Position of the American Dietetic Association: Child and adolescent food and nutrition programs, *Journal of the American Dietetic Association* 96 (1996): 915.

Now, however, hunger is on the rise due to growing poverty and cuts in government aid. By the end of the 1980s, as shown in Figure 12-1, 31.5 million people in the United States lived on incomes below the official poverty line ($13,400 for a family of four in 1990); 12.5 million of them lived on incomes of less than half of the poverty line.[8] The number of people living in poverty rose to 36.5 million (13.8 percent of the population) in 1996.[9] The national trends in poverty are shown in Table 12-2. A later section of the chapter takes a closer look at the U.S. food assis-

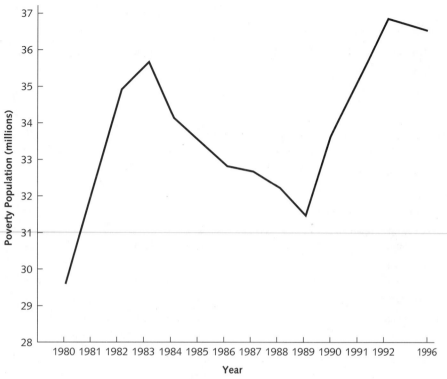

FIGURE 12-1 *Total U.S. Population in Poverty (Millions), 1980–1996*

Source: Adapted from Bread for the World Institute, *Hunger 1997: What Can Governments Do?* (Washington, D.C.: Bread for the World Institute, 1996). Reprinted by permission.

tance programs to explain why federal budget cuts exacerbate the problem of hunger in this country.

Counting the Hungry: The Surveys

We can get a sense of the number of people who are too poor to feed themselves adequately by looking at the group officially counted as poor by the U.S. Bureau of the Census. Data collected by the National Nutrition Monitoring and Related Research Program (NNMRRP) and state and local surveys also provide information about the prevalence of hunger.

The poverty line was developed in 1965 by taking the cost of an emergency short-term diet—called the Thrifty Food Plan—and multiplying it by 3.3. (The factor of 3.3 was used because a 1955 survey had shown that low-income people spent about one-third of their incomes on food.) This figure became the "poverty line," and it is adjusted annually, based on changes in the Consumer Price Index (see Table 12-3). Besides being out of date, the food budget used in the 1965 calculation reflects a diet that is just barely adequate—one designed for short-term use when funds are extremely low. Therefore, people with incomes below the poverty line have less money than is needed to buy even a short-term, emergency diet.

Despite its inadequacies, the official poverty line defines eligibility for most federal assistance programs. Needy individuals with incomes a certain amount above

TABLE 12-2 *National Trends in Poverty, 1970–1995*

	1970	1980	1986	1990	1995
Population in millions	205.1	227.8	241.6	248.7	263.0
Total poverty rate (%)	12.6	13.0	13.6	13.5	13.8
White poverty rate (%)	9.9	10.2	11.0	10.7	11.2
Black poverty rate (%)	33.5	32.5	31.1	31.9	29.3
Hispanic poverty rate	—	25.7	27.3	28.1	30.3
Elderly poverty rate (%)	24.6	15.7	12.4	12.2	10.5
Total child poverty rate (%)	15.1	18.3	20.5	20.6	20.8
White child poverty rate (%)	—	13.9	16.1	15.9	16.2
Black child poverty rate (%)	—	42.3	43.1	44.8	41.9
Hispanic child poverty rate (%)	—	39.7	39.2	36.2	40.0
Poverty rate of people in female-headed households (%)	38.1	36.7	38.3	33.4	32.4
Percentage of federal budget spent on food assistance	0.5	2.4	1.9	1.9	2.4
Total infant mortality rate	20.0	12.6	10.4	9.1	7.6
White infant mortality rate	17.8	11.0	8.9	7.7	6.3
Black infant mortality rate	32.6	21.4	18.0	17.0	15.1
Unemployment rate (%)	4.9	7.1	7.0	5.5	5.4

Source: Adapted from Bread for the World Institute, *Hunger in a Global Economy* (Washington, D.C.: Bread for the World Institute, 1997), p. 108.

the poverty line are automatically ineligible for programs like food stamps or free and reduced-price school meals. Such criteria mean that the programs now in place do not reach all people in need and do not provide enough to allow those they do reach to escape poverty.

Since the early 1980s, numerous studies on hunger have been conducted throughout the United States.* Almost without exception, these studies find hunger to be a serious and rapidly growing problem. Approximately 20 million people in the United States—12 million children and 8 million adults—are suffering from chronic hunger, and the problem is getting worse in all regions of the country. An analysis of 28 state hunger surveys found consistent results across all surveys and evidence for several broad conclusions:[10]

- Food insecurity has become a chronic problem in the United States.
- Food insecurity is not due to food shortages. Hunger results from unequal distribution of economic resources—poverty.

* Studies have been conducted by the U.S. Conference of Mayors, the Center on Hunger, Poverty, and Nutrition Policy at Tufts University, the National Council of Churches, the Citizens' Commission on Hunger in New England, Bread for the World Institute, the President's Task Force on Food Assistance, the Physician Task Force on Hunger, and the Food Research and Action Center, among others.

Household Size	Poverty Guideline* (100% Poverty)[†]
1	$ 8,050
2	10,850
3	13,650
4	16,450
5	19,250
6	22,050
7	24,850
8	27,650

For each additional family member, add $2,800.

TABLE 12-3 *Annual Poverty Guidelines,* 1998

*The poverty guideline for Alaska starts at $10,070 and rises by increments of $3,500, and that for Hawaii starts at $9,260 and rises by $3,220.
[†]This table shows income levels equal to the poverty line (100% of the poverty line). Programs sometimes set program income eligibility at some point above the poverty line. For example, if a program sets income eligibility at 130% of the poverty line, then the cutoff for a family of two living in the 48 contiguous states is $10,850 × 130% = $14,105.
Source: Food and Nutrition Service, USDA Income Eligibility Guidelines, July 1, 1998.

- People who lack access to a variety of resources—not just food—are most at risk of hunger. When income is inadequate to meet the costs of housing, utilities, health care, and other fixed expenses, these items compete with and may take precedence over food.
- The federal poverty level is an inappropriate index of food insecurity, since it is based on a formula that fails to account for changes in the cost of living, regional variations in costs, or unusual expenses that may be required.
- The U.S. social welfare system does not provide an adequate safety net.
- Private charity cannot solve the food insecurity problem. Voluntary activities are limited in expertise, time, and resources and are likely to require government support in order to continue.
- Food insecurity is inextricably linked to poverty, which in turn is inextricably linked to underemployment and the costs of housing and other basic needs.

The NNMRRP, as discussed in Chapter 4, is a valuable tool for assessing the needs of hungry people nationwide and gaining support for public policies to end hunger.[11] The NNMRRP has proposed using indicators of food security or insecurity as a necessary component of the measurement of nutrition status of individuals, communities, and nations. Accordingly, for the third NHANES, questions were developed that focus on the individual experience of food insecurity (see Table 12-4). These questions enable the NNMRRP to relate food security or insecurity to measurements of nutrition and health status.[12] State and local surveys can also use the questions in comparing local population data to the national data collected by NHANES III. Surveys such as NHANES and the Nationwide Food Consumption Survey (NFCS) also provide information about population characteristics that can be useful in designing interventions for target populations.[13]

TABLE 12-4 *Third National Health and Nutrition Examination Survey's Food Insecurity Measures*

Food sufficiency questions in the NHANES III Family Questionnaire administered during the household interview

- Which one of the following statements *best* describes the food eaten by (you/your family)? Do you have *enough* food to eat, *sometimes not enough* to eat, or *often not enough* to eat?
- Thinking about the past month, how many days did (you/your family) have no food or money to buy food?
- Which of the following reasons explain why your family has had this problem?
  a. You did not have transportation?
  b. You did not have working appliances for storing or preparing foods (such as stove, refrigerator)?
  c. You did not have enough money, food stamps, or WIC vouchers to buy food or beverages?
  d. Any other reason?

Food sufficiency questions asked of individuals during the private dietary interview in NHANES III

- Thinking about the past month, how many days did you have no food or money to buy food?
- Is that because there wasn't enough money to buy food or is there another reason?
- During the past month did you skip any meals because there wasn't enough food or money to buy food?
- How many days in the month did you skip any meals because there wasn't enough food or money to buy food?
- Did you skip any meals yesterday because there wasn't enough food or money to buy food?
- During the past month, were there any days when you did not eat at all because there wasn't enough food or money to buy food?
- In the past month, how many days were there when you didn't eat at all?

Source: Adapted from R. R. Briefel and C. E. Woteki, Development of food sufficiency questions for the third National Health and Nutrition Examination Survey. Reprinted with permission, *Journal of Nutrition Education* 24 (1992): 24S–8S, Society for Nutrition Education.

In 1997, the USDA released the government's first national data on the prevalence of food insecurity in the United States. The data were obtained by means of a questionnaire administered as part of the U.S. Census Bureau Current Population Survey in 1995.[14] Based on their answers to 18 key questions, households were categorized in one of four ways:

Hunger is defined here as an uneasy or painful sensation caused by lack of food due to lack of resources to obtain food.

1. *Food secure:* Households with no or minimal evidence of food insecurity
2. *Food insecure without hunger:* Households experiencing uncertain access to sufficient food, concerned about inadequate resources to buy enough food, and who have adjusted by decreasing the quality of food their family eats
3. *Food insecure with moderate hunger:* Households in which the adults have decreased the quantity as well as the quality of food they consume (because of lack of money) to the point where they show clear evidence of a repeated pattern of hunger. This category includes households who have indicated that due to constrained resources their children were not eating enough and that they had, at times, been forced to cut the size of their children's meals in order to make ends meet.
4. *Food insecure with severe hunger:* Households in which children's food intake had been reduced even further, and there were hungry children due to lack of financial resources. In addition, adults' food intake was severely reduced.

Significant findings of the Census Bureau Survey include:[15]

- 11.9 million households (11.9 percent of households) experienced food insecurity during 1995. This represents more than 34 million people. Of these, 7.8 million households (23.5 million people) are food insecure without hunger, 3.3 million households (9.2 million people) experience moderate hunger, and 820,000 households (2 million people) are food insecure with severe hunger.
- food insecurity rates are higher than average in female-headed households, households with children, especially black and Hispanic households, and households in inner city areas (see Table 12-2 on page 356).

This survey will be repeatedly conducted by the Census Bureau to track hunger from year to year. The 1995 estimate of food insecurity can serve as a benchmark against which the impact of welfare reform can be measured in the future.[16] As a follow-up to the 1996 International Conference on Nutrition, the U.S. government has adopted a goal of decreasing by one-half the number of food insecure people in the United States by the year 2015.[17]

Welfare Reform

Welfare in the United States is at a historic crossroads. The challenge to reform the current welfare system resulted in the Personal Responsibility and Work Opportunity Reconciliation Act of 1996.[18] The full impact of the law on the nutrition and social safety nets for low-income people is not yet known. The challenge of the welfare reform legislation is to change the welfare system from an income support-based system to a work-based system with a 5-year time limit on benefits.

As a result of the welfare reform law, the Temporary Assistance for Needy Families (TANF) Program now replaces the former Aid to Families with Dependent Children (AFDC) Program. Under TANF, states determine the eligibility of needy families and the benefits and services those families will receive. However, states face strict work requirements that increase each year. For example, by 2002, a state must have 50 percent of its welfare caseload participating in 30 hours of work activities each week.

The new welfare reform law allows states greater flexibility in creating opportunities for job training and economic security for low-income households. A major target group for job placement and training are single women with children. Two critical issues for states to consider for the long-term successful placement of these women in jobs will be transitional child care assistance and the maintenance of health care benefits.[19]

Federal Domestic Food Assistance Programs Today

Current U.S. federal food assistance and nutrition education programs are summarized in Table 12-5. The Food and Nutrition Service (FNS) administers the 15 U.S. Department of Agriculture (USDA) food assistance programs. The agency's goals are to provide needy people with access to a more nutritious diet, to improve the eating habits of the nation's children, and to stabilize farm prices through the distribution of surplus foods.[20]

TABLE 12-5 *Federal Food Assistance and Nutrition Education Programs**

Program and Year Started	Purpose(s)	Administration and Type of Assistance	Eligibility Requirements
General Food Assistance Programs			
Food Stamp Program, 1964	Improve the diets of low-income households by increasing their food purchasing ability	USDA Direct payments in the form of coupons or electronic benefits transfer (EBT) redeemable at retail food stores	Household eligibility and allotments are based on household size, income, assets, housing costs, work requirements, and other factors
Food Distribution Programs			
Commodity Supplemental Food Program, 1973	Improve the health and nutrition status of low-income pregnant, postpartum and breastfeeding women, infants, and children up to 6 years of age, and elderly persons through the donation of supplemental foods	USDA Formula grants to states; plus, the sale, exchange, or donation of foods	Eligible persons must meet age and income requirements and be determined at nutritional risk by a competent health professional at the local agency
Food Distribution Program on Indian Reservations, 1935	Improve the diets of needy persons living on or near Indian reservations and increase the market for domestically produced foods acquired under surplus removal or price support operations	USDA Project grants; plus, sale, exchange, or donation of foods	Eligible households must be living on or near an Indian reservation that offers the program and must meet certain income and resource guidelines as determined by local authorities
The Emergency Food Assistance Program (TEFAP), 1981	Make food commodities available to states for distribution to needy persons	USDA Formula grants to states	Needy individuals, including those who are low income or unemployed or receive welfare benefits
Commodity Distribution to Charitable Institutions, Summer Camps, Soup Kitchens, and Food Banks, 1967	Make food commodities available to nonprofit, charitable institutions that serve meals to low-income persons on a regular basis	USDA Formula grants to states	Institutions eligible for this program include homes for the elderly, hospitals, soup kitchens, food banks, meals-on-wheels programs, temporary shelters, and summer camps or orphanages not participating in any federal child nutrition program
Food Assistance in Disaster Situations, 1977	Provide commodity foods for shelters and other mass feeding sites; distribute commodity foods to needy households; issue emergency food stamps	USDA Direct distribution of states' commodity food stocks and/or emergency food stamps	Disaster relief organizations and households in need as a result of an emergency (storm, flood, earthquake, civil disturbance, or other disaster)
USDA Food Recovery and Gleaning Initiative, 1997	Collect excess wholesome food for delivery to persons in need	USDA Provision of gleaned and recovered food to supplement existing federal nutrition assistance programs	Nonprofit feeding organizations

Program and Year Started	Purpose(s)	Administration and Type of Assistance	Eligibility Requirements
Child Nutrition Programs			
National School Lunch Program, 1946	Assist states in providing nutritious free or reduced-price lunches to eligible children and encourage the domestic consumption of nutritious agricultural commodities	USDA Formula grants to states	1. All students attending schools where the lunch program is available may participate. 2. Children from families with incomes at or below 130 percent of the poverty level are eligible for free meals. 3. Children from families with incomes between 130 and 185 percent of the poverty level are eligible for reduced-price meals. 4. Children from families with incomes over 185 percent of the poverty level pay full price.
School Breakfast Program, 1966	Assist states in providing nutritious, nonprofit breakfasts for children in public and nonprofit private schools of high school grade and under and in residential child care institutions	USDA Formula grants to states	Eligibility requirements are the same as for the School Lunch Program
Special Milk Program for Children, 1955	Provide cash reimbursement to schools and institutions to encourage the consumption of fluid milk by children	USDA Formula grants to states	1. Eligible schools include public and private nonprofit schools of high school grade and under, and nonprofit residential or nonresidential child care institutions (provided they do not participate in other federal meal service programs). 2. The program is available to all children in participating schools, regardless of their family income (except in those schools operating the program solely for kindergarten children).
Summer Food Service Program for Children, 1968	Assist states in conducting nonprofit food service programs that provide meals and snacks for children in needy areas when school is not in session during the summer and at other times, when area schools are closed for vacation	USDA Formula grants to states	Any child under the age of 18 years or any person over the age of 18 years who is handicapped and who participates in a program established for the mentally or physically handicapped may participate

Continued

Program and Year Started	Purpose(s)	Administration and Type of Assistance	Eligibility Requirements
Child Nutrition Programs—continued			
Child and Adult Care Food Program, 1968	Provide federal funds and USDA-donated foods to nonresidential child care and adult day care facilities and to family day care homes for children New legislation includes a provision to allow after school programs operated by community groups to serve snacks to those aged 12–18 in low-income areas[†]	USDA Formula grants to states; plus the sale, exchange, or donation of foods	Eligible institutions include licensed or approved nonresidential, public or private, nonprofit child care centers; Head Start centers; settlement houses; neighborhood centers; some for-profit child care centers; and licensed or approved private homes providing day care for a small group of children.
Homeless Children Nutrition Program, 1994	Provide free food to homeless children	USDA Cash reimbursement for meals to sponsoring organizations	Public and private nonprofit organizations; homeless children under the age of 6 in emergency shelters
Programs for Pregnant Women, Infants, and Children			
Special Supplemental Nutrition Program for Women, Infants, and Children, 1974	Provide, at no cost, supplemental, nutritious foods, nutrition education, and referrals to health care to low-income pregnant, breastfeeding and postpartum women, infants, and children to 5 years of age who are determined to be at nutritional risk	USDA Formula grants to states	Pregnant, breastfeeding and postpartum women, infants, and children up to 5 years of age are eligible if they are individually determined by a competent professional to be in need of the special supplemental foods provided by the program because they are nutritionally at risk, and they meet an income standard.
WIC Farmers' Market Nutrition Program, 1992	Allow WIC participants to purchase fresh produce at authorized farmers' markets; expand consumers' awareness and use of farmers' markets	USDA Formula grants to states	Same as those of the WIC program
Title V Maternal and Child Health (MCH) Programs, 1935	Promote, improve, and deliver maternal, infant, and child health care services	DHHS Block grants to states	Women of childbearing age, infants and children; eligibility based on state requirements

The FNS administers all its programs in partnership with the states. The individual states determine most details regarding distribution of food benefits and eligibility of participants, and the FNS provides funding to cover most of these administrative costs.

Congress appropriated $38 billion for the FNS to operate the food assistance programs in 1996—up from $33 billion in 1992 and $1.1 billion in 1969, the first year of the agency's operation. As shown in Figure 12-2 on page 364, the Food Stamp Program accounts for the majority of these funds.

Program and Year Started	Purpose(s)	Administration and Type of Assistance	Eligibility Requirements
Nutrition Programs for Older Adults			
Nutrition Program for the Elderly (NPE), 1965	Provide commodity foods and financial support to DHHS's senior nutrition programs	USDA Cash and commodities to state agencies for meals served to seniors	People 60 years of age or over and their spouses
Elderly Nutrition Program (ENP), 1965	Provide congregate and home-delivered meals and related nutrition services to older adults	DHHS Administration on Aging	People 60 years of age or over and their spouses; services targeted to those with greatest economic or social need.
Federal Nutrition Education Programs			
The Extension Food and Nutrition Education Program (EFNEP), 1968	Assist low-income families and youth to acquire the knowledge and skills needed to obtain a healthy diet	USDA Extension service	Low-income families and youth
The Head Start Program, 1965	Provide comprehensive health, education, nutrition and social services to low-income preschool children and their families	DHHS Regional Departments of Health	3 to 5-year-old children from low-income families
Nutrition Education and Training (NET) Program, 1977	Provide nutrition education training to teachers and school food service personnel so that they can teach children about nutrition; provide nutrition education materials for use in classrooms and child care settings	USDA Formula grants to state educational agencies for the development of comprehensive nutrition information and education programs, technical assistance to school districts	Children eligible to participate in the National School Lunch Program or other child nutrition programs
The Food Stamp Nutrition Education Program (FSNEP) or Family Nutrition Program (FNP), 1986	Improve the dietary intake of food stamp recipients through nutrition education activities that increase self-sufficiency	USDA Funding provided by Food Stamp administrative budget with 50 percent matching requirement by participating states	Persons eligible for food stamps

*Most of the USDA programs listed in this table are discussed in this chapter. The maternal-child programs are discussed in Chapter 13; more about the child nutrition programs can be found in Chapter 14, and the nutrition programs for older adults are discussed in Chapter 15.

†Community Nutrition Institute, *Nutrition Week,* June 5, 1998, p. 1.

Source: USDA and DHHS Program Fact Sheets, May 1998.

Other food assistance programs for seniors (Congregate Meals and Home-Delivered Meals) are administered by the Administration on Aging of the Department of Health and Human Services (DHHS). Most of the food assistance programs administered by USDA are discussed in turn in the sections that follow.

FIGURE 12-2 *Food Program Costs, 1980–1995*

Source: USDA

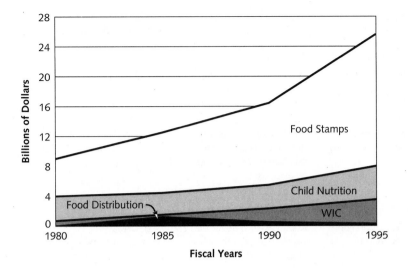

Food Stamp Program

The Food Stamp Program is currently authorized by the Food Stamp Act of 1977 (PL 95–113).

Work requirements: At the time of application and once every 12 months, all able-bodied household members between 18 and 60 years of age and 16- and 17-year-old heads of households who are not in school must register to work. Many adult recipients must participate in employment and training programs.

Net income is all the household's income that counts in figuring food stamps minus the deductions for which the household is eligible. Most households may have up to $2,000 in countable resources (cash, bank accounts, car, stocks/bonds, and so forth). Households may have $3,000 if at least one person is age 60 years or older.

The Food Stamp Program (FSP) is designed to improve the diets of people with low incomes by providing coupons to cover part or all of their household's food budget. The FSP is an entitlement program, meaning that anyone who meets eligibility standards is entitled to receive benefits. Eligibility and allotments are based on income, household size, assets, housing costs, work requirements, and other factors. Most households must have gross monthly incomes below 130 percent of the poverty line and net monthly income at or below 100 percent of the poverty guidelines for their household size (see Table 12-6 for both standards). Households with elderly or persons with disabilities are subject only to the net income test. The majority of food stamp households have gross incomes below the poverty line; over 40 percent have gross incomes below half the poverty line. At least 67 percent of all food stamp households include children (60 percent) or elderly (7 percent).

The USDA issues food stamp coupons through state welfare or human services agencies to households—people who buy and prepare food together. The number of stamps a household receives varies according to its size and income (see Figure 12-3). Food stamp benefits per person average $70.80 a month (76 cents per meal).[21] Recipients may use the coupons like cash to purchase food and seeds at stores authorized to accept them. They cannot buy ready-to-eat hot foods, vitamins or medicines, pet foods, tobacco, cleaning items, alcohol, or nonfood items (except seeds and garden plants) with coupons.

The Food Stamp Program in Puerto Rico was replaced in 1982 by a block grant program (PL 97–35). The territory now provides cash and coupons to eligible participants rather than food stamps or food commodities. The grant can also be used to pay up to 50 percent of administrative costs and to finance special projects related to food production and distribution. For example, a special cattle tick eradication program was funded in 1992.

Gross Monthly Income Eligibility Standards (130% of Poverty Line)			
Household Size	48 States*	Alaska	Hawaii
1	$ 855	$1,070	$ 983
2	1,150	1,438	1,322
3	1,445	1,806	1,661
4	1,739	2,175	2,000
5	2,034	2,543	2,339
6	2,329	2,911	2,678
7	2,623	3,280	3,018
8	2,918	3,648	3,357
Each additional member	+295	+369	+340

Net Monthly Income Eligibility Standards (100% of Poverty Line)			
Household Size	48 States*	Alaska	Hawaii
1	$ 658	$ 823	$ 756
2	885	1,106	1,017
3	1,111	1,390	1,278
4	1,338	1,673	1,539
5	1,565	1,956	1,800
6	1,791	2,240	2,060
7	2,018	2,523	2,321
8	2,245	2,806	2,582
Each additional member	+227	+284	+261

TABLE 12-6 *The Food Stamp Program's Monthly Income Eligibility Guidelines (Gross and Net), 1998*

*Includes District of Columbia, Guam, and Virgin Islands.

Source: United States Department of Agriculture.

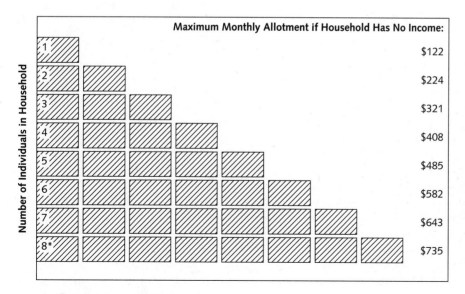

Maximum Monthly Allotment if Household Has No Income:

Number of Individuals in Household	Amount
1	$122
2	$224
3	$321
4	$408
5	$485
6	$582
7	$643
8*	$735

FIGURE 12-3 *Food Stamp Allotments Based on the Thrifty Food Plan (1998)*

Subtract 30% of available monthly income from the maximum food stamp allotment to determine the coupon allotment. For example, a 4-person household with $400 in available income would receive $250 in food stamps. 30% of income = $120. Maximum monthly allotment for a family of 4 = $408 − $120 = $288 food stamp allotment.

*For each additional person, add $92.

Source: United States Department of Agriculture.

Program Spotlight   *The Food Stamp Program*

The Food Stamp Program (FSP) dates back to the food assistance programs of the Great Depression—a time when farmers were burdened with surplus crops they could not sell while thousands stood in breadlines, waiting for something to eat.[1] To help both farmers and consumers, the government began distributing the surplus farm foods to hungry citizens. Today, food stamp benefits enable recipients to buy food in grocery stores. Average characteristics of food stamp households include:[2]

- Just over half of all participants are children (18 or younger).
- 60 percent of food stamp households include children.
- 7 percent of all participants are elderly (age 60 or older).
- 89 percent of all benefits go to households with children or elderly persons.
- 70 percent of households with children are headed by a single parent, most of whom are women.
- The average household size is 2.5 persons.
- The average gross monthly income per food stamp household is $514.
- The average benefit per person is currently $70.80 a month.
- Among adult recipients, women outnumber men by about 2 to 1.
- 41 percent of participants are white; 34 percent are African-American; 18 percent are Hispanic.

Nutrition surveys in the United States have demonstrated consistently that the lower a family's income, the less adequate its nutrition status.[3] Few low-income families obtain an adequate intake of nutrients. The problem is not that people with low incomes don't know how to shop, but that they are unable to buy sufficient amounts of nourishing foods. Some families with low incomes who receive food stamps are probably as skilled as, or even more skilled than, others at food shopping.[4] Thus, income apparently confers the ability to obtain a nutritionally adequate diet, and indeed, the number of households that meet or exceed the RDA rise and fall with income.

The stated objective of the FSP is to "improve the diets of low-income households by increasing their food purchasing ability." Unfortunately, the ability of the FSP to achieve its objective has been questioned.[5] Although the program potentially increases a household's ability to purchase nourishing foods, food stamps may be used to buy most available human foods. The effect of the food stamp purchases on nutrient intakes of participants will vary depending on the nutritional composition of the foods they select.

One of three major problems with the FSP is that benefit allotments are insufficient to meet needs. Many households receiving food stamps still need emergency food by the end of the month because their stamps rarely last the entire month. In a study of 1,922 households, monthly food expenditures of FSP participants, including cash, food stamps, and WIC benefits, averaged just under 80 percent of the value of the Thrifty Food Plan.[6] Results from the Nationwide Food Consumption Surveys have shown that only 12 percent of people purchasing food valued at 100 percent of the Thrifty Food Plan were eating nutritionally adequate diets, indicating that the FSP households spending only 80 percent of the value of the Thrifty Food Plan may be at nutritional risk.[7]

The second major problem is that many households who are eligible for the FSP and in need do not participate. In 1996, an estimated 36.5 million people were living at income levels at or below the poverty line. These people were all potentially eligible to receive food stamps, yet only 25 million people participated in the program.[8] Reasons for nonparticipation include embarrassment about receiving assistance, complex rules and requirements, confusing paperwork, caseworker hostility, and lack of public information about eligibility requirements. Figure 12-4 shows the gap between eligibility and participation in the FSP in selected states.

The USDA is currently trying to improve participation rates by eligibles through a number of outreach efforts targeted at low-participation groups, such as persons who are elderly, working poor, non-English speaking, homeless, or living in rural areas. Outreach efforts include training community workers and volunteers to refer families to food stamp offices, community education and mass media campaigns, and individualized client assistance.[9]

A third major problem with the FSP involves fraud. One of the most promising developments in the efforts to prevent food stamp fraud has been the increasing use of electronic benefit transfer (EBT) to issue food stamp benefits. EBT uses a plastic card similar to a bank card to transfer funds from a food stamp recipient's account to a food retailer's account. With an EBT card, food stamp recipients purchase groceries without the use of paper coupons. EBT keeps an electronic record of all transactions and makes fraud easier to detect. Twenty-three states are now using an EBT system for food stamps. The Welfare Reform Act of 1992 requires all states to convert to an EBT system for food stamp programs by 2002.[10]

To improve the FSP's ability to meet the needs of low-income households, the following steps are recommended:[11]

- Improve and expand outreach about the FSP.
- Lower administrative barriers to participation in the FSP.
- Increase food stamp benefits so that families can afford to eat a nutritionally adequate diet throughout each month.
- Provide nutrition education materials to food stamp households.

References

1. Food and Nutrition Service, *Nutrition Program Facts:* Food Stamp Program (Washington, D.C.: USDA, 1998), p. 3.
2. Nutrition Program Facts, 1998, p. 3.
3. Community Nutrition Institute, *Nutrition Week*, January 22, 1993, p. 6.
4. P. L. Splett, Federal Food Assistance Programs: A step to food security for many, *Nutrition Today*, March/April, 1994, pp. 6–13.
5. B. B. Peterkin and M. Y. Hama, Food shopping skills of the rich and the poor, *Family Economics Review* 3 (1983): 8–12.
6. J. Allen and K. Gadson, Food consumption and nutritional status of low-income households, *National Food Review* 26 (1984): 27–31.
7. Food Research and Action Center, *Community Childhood Hunger Identification Project* (Washington, D.C.: FRAC, 1995), p. 34.
8. B. B. Peterkin, R. L. Kerr, and M. Y. Hama, Nutritional adequacy of diets of low-income households, *Journal of Nutrition Education* 14 (1982): 102–4.
9. Food Research and Action Center, *State of the States: A Profile of Food and Nutrition Programs Across the Nation* (Washington, D.C.: FRAC, December, 1997), p. 3.
10. Community Nutrition Institute, Food assistance: Reducing food stamp benefit overpayments and trafficking, *Nutrition Week*, November 7, 1997, pp. 4–6.
11. Food Research and Action Center, *Community Childhood Hunger*, pp. 33–34.

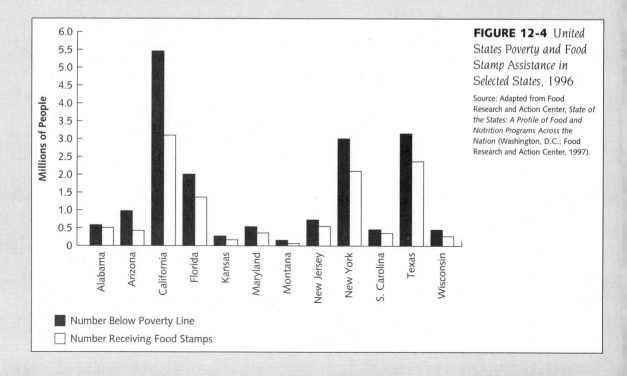

FIGURE 12-4 *United States Poverty and Food Stamp Assistance in Selected States, 1996*

Source: Adapted from Food Research and Action Center, *State of the States: A Profile of Food and Nutrition Programs Across the Nation* (Washington, D.C.: Food Research and Action Center, 1997).

Food Distribution Program on Indian Reservations

Also known as the Needy Family Program, this is the oldest of the FNS programs—dating back to the Great Depression of the 1930s (PL 74–320). This program was the main form of food assistance for all low-income people in the United States until the Food Stamp Program was expanded in the early 1970s. The program provides monthly food packages of commodity foods to low-income American Indian households living on or near reservations. The USDA commodities include canned meats and fish products; vegetables, fruits, and juices; dried beans; peanuts or peanut butter; milk, butter, and cheese; pasta, flour, or grains; adult cereals; corn syrup or honey; and vegetable oil and shortening. A nutrition education plan is currently under way to help participants make more nutritious use of the foods they receive.

Participants may choose from month to month whether they will participate in the Food Stamp Program or the food distribution program. Many prefer the Food Distribution Program on Indian Reservations if they do not have easy access to grocery stores. An average of 123,000 participants were served monthly in 1997; monthly packages were valued at $40.00 per person.

Special Supplemental Nutrition Program for Women, Infants, and Children (WIC)

WIC, authorized in 1974, is a cost-effective program of grants to state and local health agencies. WIC provides supplemental foods to infants, children up to age five, and pregnant, breastfeeding, and non-breastfeeding postpartum women who qualify financially and are at nutritional risk. Financial eligibility is determined by income (between 100 percent and 185 percent of the poverty line or below) or by participation in the Food Stamp or Medicaid programs. Nutritional risks, determined by a health professional, may include one of three types: medically-based risks (anemia, underweight, maternal age, history of high-risk pregnancies); diet-based risks (inadequate dietary pattern); or conditions that make the applicant predisposed to medically-based or diet-based risks, such as alcoholism or drug addiction. Homelessness and migrancy are also considered nutritional risks for purposes of WIC.

The WIC program serves both a remedial and a preventive role. Its services include the following:

- Food packages or vouchers for supplemental food to provide specific nutrients (protein, calcium, iron, and vitamins A and C) known to be lacking in the diets of the target population.
- Nutrition education.
- Referrals to health care services.

Unlike most of the other food assistance programs, WIC is not an entitlement program. Approximately 7.4 million women, infants, and children received monthly WIC benefits in 1997. Due to caps on allocated federal funds, WIC currently reaches only about 60 percent of eligible persons.

Authorization for WIC: Congress created a pilot WIC project in 1972 (PL 92–433) and authorized WIC as a national program as part of the National School Lunch and Child Nutrition Act Amendments of 1975 (PL 94–105).

More about WIC appears in Chapter 13.

WIC Farmers' Market Nutrition Program

The WIC Farmers' Market Nutrition Program (FMNP) was created to accomplish two goals:

- To provide fresh, nutritious fruits and vegetables to WIC participants
- To expand the awareness and use of farmers' markets by consumers

Under the program, now offered in 32 states, coupons are issued to eligible recipients.[22] States may also use their own matching funds to issue FMNP coupons to other groups, such as elderly persons, older children, the homeless, or other low-income people.

Child Nutrition Programs

Many federal programs address the special nutritional needs of children. The following sections describe the USDA's child nutrition programs. Chapter 14 provides more detail about the USDA programs as well as other major federal and private nutrition programs for children and adolescents.

National School Lunch and Breakfast Programs. The National School Lunch Program (NSLP) and the School Breakfast Program provide financial assistance to schools so that every student can receive a nutritious lunch, breakfast, or both. Participating schools get cash payments on the basis of the number of meals served in the free, reduced-price, and full-price categories and also receive food commodities. Program schools must serve meals meeting specified nutritional guidelines and must offer free or reduced-price meals to eligible students. These programs enable students in households with incomes at or below 130 percent of the poverty line to receive meals at no cost and allow students from households with incomes between 130 and 185 percent of the poverty line to receive a reduced-price meal. Children whose families participate in the Food Stamp Program are automatically eligible for free school meals.

Nationally, the USDA runs the programs; on the state level, the programs are run by the state Department of Education. They usually can be implemented with little cost to the school district. The income eligibility guidelines for the NSLP, School Breakfast Program, Child and Adult Care Food Program, Summer Food Service Program, and Special Milk Program are listed in Table 12-7. In 1997, more than 26 million children participated in the NSLP at more than 94,000 schools and residential institutions. Half of these children received their meal free or at a reduced price. Approximately 7.2 million participated in the School Breakfast Program, which is offered in some 70,000 schools and institutions.

Summer Food Service Program for Children. The Summer Food Service Program (SFSP) funds meals and snacks for eligible children during school vacation periods. The program operates in areas where half or more of the children are from households with incomes at or below 185 percent of the poverty level. Sponsors of the program include local schools, government units (e.g., parks and recreation departments), summer camps, community action agencies, and other nonprofit

Congress created a pilot FMNP through the Hunger Prevention Act of 1988 (PL 100–435) and awarded grants to 10 states (Connecticut, Iowa, Maryland, Massachusetts, Michigan, New York, Pennsylvania, Texas, Vermont, and Washington). The FMNP was authorized in 1992 (PL 102–314) as a national program through an amendment to the Child Nutrition Act of 1966.

The NSLP was authorized in 1946 by the National School Lunch Act (PL 79–396). The School Breakfast Program was authorized by the Child Nutrition Act of 1966 (PL 89–642).

The SFSP was authorized in 1975 as an amendment to the National School Lunch Act (PL 94–105).

TABLE 12-7 *Annual Income Eligibility Standards for the Federal Child Nutrition Programs, 1998–1999*

Household Size	Federal Poverty Guidelines 100% of Poverty	Free Meals 130% of Poverty	Reduced-Price Meals 185% of Poverty
1	$ 8,050	$10,465	$14,893
2	10,850	14,105	20,073
3	13,650	17,745	25,253
4	16,450	21,385	30,433
5	19,250	25,025	35,613
6	22,050	28,665	40,793
7	24,850	32,305	45,973
8	27,650	35,945	51,153
Each additional, +	$ 2,800	$ 3,640	$ 5,180

Sources: *Federal Register,* Vol. 62, No. 49, 3/13/97, pp. 11811–13; United States Department of Agriculture.

organizations. The SFSP offers meals that meet the same nutritional standards as those provided by the School Lunch Program at no cost to all children, up to age 18, who attend the program site. Approximately 2.3 million children participated in the SFSP in 1997 at more than 28,000 sites.

Special Milk Program. The Special Milk Program (SMP) provides cash reimbursement for each half-pint of milk served to children in schools and child care institutions that are not participating in the National School Lunch Program. Nearly 140 million half-pints of milk were served in 1997.

> The SMP was incorporated into the Child Nutrition Act of 1966 (PL 89–642).

Child and Adult Care Food Program. The Child and Adult Care Food Program (CACFP) is designed to help public and private nonresidential child and adult day care programs provide nutritious meals for children up to age 12, the elderly, and certain people with disabilities. The program provides cash reimbursements for two meals and one snack per child per day plus one additional meal or snack for children in care for eight or more hours. Sponsors may also receive USDA commodity foods. The CACFP served nearly 2.6 million children and 58,000 adults in March, 1998.

> The CACFP was permanently authorized in 1978 (PL 94–105). The 1987 amendments to the Older Americans Act authorized the Child Care Food Program to change its name and expand its service to include the elderly and persons with disabilities.

Emergency Food Assistance Program

The Emergency Food Assistance Program, which was formerly known as the "Temporary" Emergency Food Assistance Program and retains the acronym TEFAP, has operated now for over a decade. The TEFAP was designed to reduce the level of government-held surplus commodities by distributing them to low-income households to supplement the recipients' purchased food. Most states set eligibility criteria at between 130 and 150 percent of the poverty line. In many states, food stamp participants are automatically eligible for the TEFAP. The types of foods

> The TEFAP was authorized by the Temporary Emergency Food Assistance Act of 1983 (PL 98–8). The 1990 Farm Bill (PL 101–624) reauthorized TEFAP and dropped the word "Temporary" from the program's name.

USDA purchases for TEFAP distribution vary depending on the preferences of states and agricultural market conditions. Foods available in 1998 included more than 40 products, including canned and dried fruits, fruit juice, canned vegetables, meat, poultry, fish, rice, grits, cereal, peanut butter, nonfat dried milk, dried egg mix, and pasta products.

Commodity Supplemental Food Program

The Commodity Supplemental Food Program (CSFP), currently available in approximately 18 states, is a direct food distribution program providing supplemental foods and nutrition education. The CSFP serves a target population similar to WIC and persons 60 years of age and older as well. Recipients may not participate in both WIC and the CSFP. Food packages are designed to suit the nutritional needs of participants and may include canned fruit juice, canned fruits and vegetables, farina, oats, ready-to-eat cereal, nonfat dry milk, evaporated milk, egg mix, dry beans, peanut butter, canned meat, poultry or tuna, dehydrated potatoes, pasta, rice, cheese, butter, honey, and infant cereal and formula. Distribution sites make packages available on a monthly basis. Monthly participation in 1997 averaged more than 370,000 people, including 243,000 elderly persons and more than 127,000 women, infants, and children.

> The CSFP was authorized by the Agriculture and Consumer Protection Act of 1973.

Commodity Distribution to Charitable Institutions

Food commodities are distributed to nonprofit, charitable institutions that serve meals to low-income people on a regular basis. These include homes for the elderly, hospitals, soup kitchens, food banks, meals-on-wheels programs, temporary shelters, and summer camps or orphanages not participating in any federal child nutrition program.

Senior Nutrition Programs

The federal Elderly Nutrition Program (ENP) (Title III) is intended to improve older people's nutrition status and enable them to avoid medical problems, continue living in communities of their own choice, and stay out of institutions. Its specific goals are to provide the following:[23]

- Low-cost, nutritious meals
- Opportunities for social interaction
- Nutrition education and shopping assistance
- Counseling and referral to other social services
- Transportation services

One of the Title III efforts is the Congregate Meals Program. Administrators try to select sites for congregate meals that will attract as many of the eligible elderly as possible. Through the Home-Delivered Meals Program, meals are also delivered to those who are homebound either permanently or temporarily. The home-delivery program ensures nutrition, but its recipients miss out on the social benefit of the congregate meal sites; every effort is made to persuade them to come to the shared meals if they can. The DHHS's Administration on Aging administers these programs.

> The Congregate Meals Program and the Home-Delivered Meals Program were authorized by the Older Americans Act of 1965 (PL 89–73).

All persons 60 years and older (and spouses of any age) are eligible to receive meals from these programs, regardless of their income level. Priority is given to those who are economically and socially needy. In 1995, ENP provided 123 million meals to 2.4 million people at congregate meal sites and 119 million home-delivered meals to 989,000 homebound older persons.

Another program for the elderly is the USDA's Nutrition Program for the Elderly (NPE). The NPE provides cash and commodity foods to local senior citizen centers for use in the Congregate and Home-Delivered Meals Programs. USDA provided reimbursement for an average of more than 20 million meals per month in 1997 at a rate of 56 cents per meal.[24] Chapter 15 provides additional information about these programs.

The Rising Tide of Food Assistance Needs

Emergency food services:
- Soup kitchens
- Church charities
- Surplus food giveaways
- Food banks
- Food pantries
- Prepared and perishable food programs

Despite all of these federal food assistance programs, hunger and the public demand for emergency food assistance have increased in every region of the United States since 1980. The country has become a soup kitchen society to an extent unmatched since the breadlines of the Great Depression. The demand for emergency food assistance has increased steadily over the past 2 decades (see Figure 12-5). Today's worsening economic conditions of the poor are attributed to the federal economic policies of the 1980s. Much of the increased public demand for food assistance is

FIGURE 12-5 *Demand for Emergency Food Assistance*

Donated foods distributed by the Second Harvest network rose from 2.5 million pounds in 1979 to 100 million pounds in 1985 and then to well over 860 million pounds in 1997, reflecting the efforts of private organizations to cope with the rise in hunger. Second Harvest services state food banks.

Source: Second Harvest, *1997 Annual Report;* www.secondharvest.org.

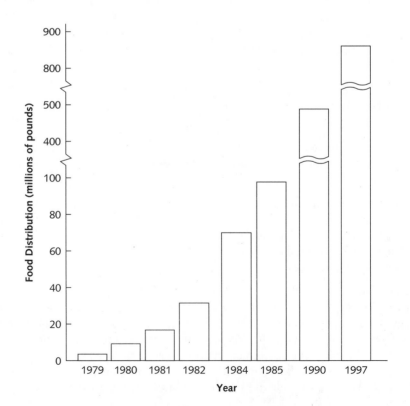

coming from the "new poor"—those who, until recently, had been employed, productive, and financially stable. Many seeking emergency food assistance are families with children, who find that food stamps are unable to meet their food needs.

To help fill the gaps in the federal programs, concerned citizens are working through community programs and churches to provide meals to the hungry. **Second Harvest,** the nation's largest supplier of surplus food, distributed over 860 million pounds of food to **food banks** and other agencies for direct distribution around the nation in 1997.[25] An estimated 26 million people relied on food banks, soup kitchens, and other agencies for emergency food in 1997 (see Figure 12-6). However, even the dramatic increases in the number of food banks, **food pantries, soup kitchens, prepared and perishable food programs,** and other emergency food assistance programs cannot keep pace with the growing number of hungry people seeking food assistance.[26] Each day's supply of meals lasts only for that day, leaving the problem of poverty unsolved; moreover, one out of every five needy people is

Second Harvest A national network to which the majority of food banks belong.

FIGURE 12-6

What is a Community Food Bank?

With extra groceries, hot meals, or day care snacks, hundreds of thousands of individuals benefit from the Food Bank's partnerships with industry and the community.

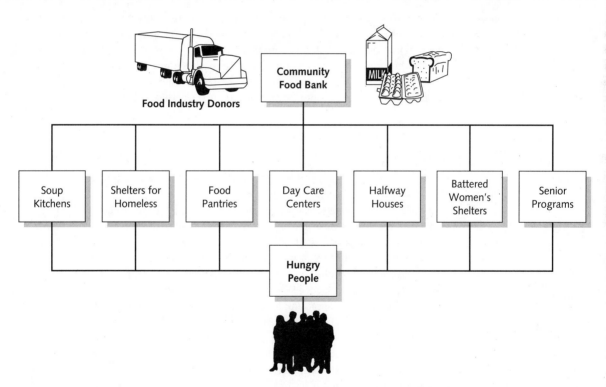

- Industry donors contact the food bank when they have products to contribute: goods that are mislabeled, in damaged packages, underweight, close to expiration, or in oversupply.

- The food bank picks up large-scale donations, cases to trailerloads, from the plant, distribution center, public warehouse, or retailer.

- At the food bank, nonprofit, charitable groups select products to stock their emergency food pantries, soup kitchens, shelters, child care centers, senior citizen programs, rehabilitation centers, and summer camps.

- Donated food reaches people in need: children and the elderly, the unemployed and low-wage earners, the homeless and frail homebound, the disabled and ill.

Food banks Nonprofit community organizations that collect surplus commodities from the government and edible but often unmarketable foods from private industry for use by nonprofit charities, institutions, and feeding programs at nominal cost.

Food pantries Usually attached to existing nonprofit agencies, these pantries distribute bags or boxes of groceries to people experiencing food emergencies. Distributed foods are prepared and consumed elsewhere. Pantries often require referrals or proof of need. There are roughly two food pantries to every soup kitchen.

Soup kitchen Small feeding operations attached to existing organizations, such as churches, civic groups, or nonprofit agencies, that serve prepared meals that are consumed on-site. Soup kitchens generally do not require clients to prove need or show identification.

Prepared and perishable food programs (PPFPs) Nonprofit programs that help to feed people in need by linking sources of unused, unserved cooked and fresh food—like caterers, restaurants, hotel kitchens, and cafeterias—with social service agencies that serve meals to people who would otherwise go hungry. *Foodchain* is a national network of over 131 community-based member programs.

not even receiving meals. These people must scavenge garbage, steal food or money to buy food, or continue to starve.

Food Recovery and Gleaning Initiative

The terms **food recovery** and **gleaning** refer to programs that collect excess wholesome food for delivery to hungry people.* Approximately 96 billion pounds—or over one-quarter of the 356 billion pounds of food produced in this country for human consumption—is lost at the retail and food service levels.[27] In an effort to reduce food wastage, the 1997 National Summit on Food Recovery and Gleaning set a goal of a 33 percent increase in the amount of food recovered nationally by the year 2000. This increase would provide an additional 500 million pounds of food a year to feeding organizations.

The Food Recovery and Gleaning Initiative is intended to supplement existing federal nutrition assistance programs. Currently, USDA is working with the National Restaurant Association to produce a food recovery handbook for its members, enabling schools to donate excess food from the School Lunch Program, encouraging airlines to donate unserved meals, working with the Department of Transportation to develop a comprehensive way to transport recovered foods, facilitating the donation of excess food from the Department of Defense, and providing technical assistance to community-based groups who seek to help.[28]

The Plight of the Homeless

Estimates of the number of homeless range from 600,000 to 3 million. In addition, nearly 3 million people spend more than 70 percent of their income on rent and are at risk of becoming homeless.[29]

The U.S. Conference of Mayors surveyed 29 major cities to assess the status of hunger and homelessness in the urban United States during 1997. Lack of food, inadequate diets, poor nutrition status, and nutrition-related health problems—stunted growth, failure to thrive, low-birthweight babies, infant mortality, anemia, and compromised immune systems—are common among homeless persons.[30] Increasing numbers of people living with the HIV virus are homeless due to the high costs of health care or lack of supportive housing.[31] Tuberculosis is spreading at alarming rates among homeless persons because of the close sleeping arrangements in shelters and on the streets.

The lack of affordable housing leads the list of causes of homelessness identified by the mayors. Without adequate low-income housing, many poor people are forced to choose between shelter and food. Other causes include unemployment, underemployment, poverty, inadequate public assistance benefits, the high cost of health care, substance abuse and lack of needed services, and mental illness and lack of needed services. Figure 12-7 shows the average composition of the survey cities'

*The Good Samaritan Food Donation Act of 1996 encourages the donation of food and grocery products to nonprofit organizations such as homeless shelters, soup kitchens, and churches for distribution to needy individuals. The law provides uniform national protection to citizens, businesses, and nonprofit groups that in good faith donate, recover, and distribute excess food. The law limits the liability of donors to instances of gross negligence or intentional misconduct.

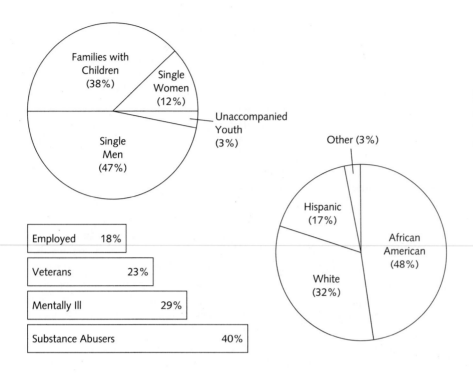

FIGURE 12-7

Demographics of the Homeless Population in 29 Survey Cities

Source: Adapted from The U.S. Conference of Mayors, *A Status Report on Hunger and Homelessness in America's Cities: 1997* (Washington, D.C.: U.S. Conference of Mayors, 1997).

homeless population. Families with children accounted for 38 percent of the homeless in these cities.[32]

Although low-income, homeless people may qualify for any of the USDA's 15 food assistance programs, homeless people who try to receive public assistance face many barriers. For example, transportation to agency offices may not be affordable or accessible. The homeless may also lack the documentation required to apply for the various benefits programs.

Several steps are necessary to end homelessness in the United States. Local communities need to focus on outreach efforts to inform homeless persons of existing services and help them obtain benefits. More 24-hour shelters with programs such as literacy classes, medical services, nutrition education, drug and alcohol rehabilitation, job training and placement, and child care, as well as food and shelter, must be established. Longer-term solutions include improving the economy and creating jobs that pay adequately. Improving the Food Stamp Program and increasing funding for it and for WIC are other necessary measures. Other steps cited by the Conference of Mayors as crucial for ending homelessness include making more housing available at affordable prices and providing comprehensive services to vulnerable populations.

Food recovery Such activities as salvaging perishable produce from grocery stores and wholesale food markets; rescuing surplus prepared food from restaurants, corporate cafeterias, and caterers; and collecting non-perishable, canned or boxed processed food from manufacturers, supermarkets, or people's homes. The items recovered are donated to hungry people.

Gleaning The harvesting of excess food from farms, orchards, and packing houses to feed the hungry.

Poverty Trends Among Vulnerable Groups: A Focus on Children

Poverty and hunger affect certain socioeconomic, geographic, and demographic groups more than others. According to one report, those most at risk of hunger are female-headed households, children, the rural poor, farm workers, the homeless, and older people.[33] African-American, Hispanic, and American Indian peoples in all these groups have consistently been vulnerable to hunger.[34]

The hunger surveys of the past decade reported an increasing demand for emergency food by families with young children. As a result, the Food Research and Action Center (FRAC) designed the Community Childhood Hunger Identification Project (CCHIP) to document the prevalence of hunger in children. CCHIP has implemented a number of surveys across the nation among low-income families (with incomes at or below 185 percent of the federal poverty line) with at least one child under the age of 12. The studies document conditions such as participation in food assistance programs, finances, and children's health.[35]

CCHIP documents hunger using a hunger index—a scale of eight questions (see Table 12-8) that indicate whether a household is affected by food insecurity, food shortages, perceived food insufficiency, or altered food intake due to resource constraints.[36] Families are asked whether lack of money in the past 12 months caused them to cut food portions, skip meals, limit the number of foods served, or send children to bed hungry. Families answering affirmatively to five or more of eight such questions are considered hungry. A score of one to four indicates that the family is at risk of hunger due to the presence of at least one food shortage problem.[37] The CCHIP conceptual model (see Figure 12-8) shows the large number of interrelated factors associated with hunger and poverty.

As a result of the ongoing CCHIP surveys, FRAC estimates that approximately 13.6 million children in the United States—under 12 years of age—live in families that must cope with hunger or the risk of hunger. Children in minority groups are particularly at risk for hunger. At least 44 percent of African-American children and 42 percent of Hispanic children under six lived in poverty in 1994.[38]

TABLE 12-8

*Community Childhood Hunger Identification Project's Hunger Scale**

1. Does your household ever run out of money to buy food to make a meal?

2. Do you or members of your household ever eat less than you feel you should because there is not enough money for food?

3. Do you or members of your household ever cut the size of meals or skip meals because there is not enough money for food?

4. Do your children ever eat less than you feel they should because there is not enough money for food?

5. Do you ever cut the size of your children's meals or do they ever skip meals because there is not enough money for food?

6. Do your children ever say they are hungry because there is not enough food in the house?

7. Do you ever rely on a limited number of foods to feed your children because you are running out of money to buy food for a meal?

8. Do any of your children ever go to bed hungry because there is not enough money to buy food?

---

*Hunger criterion: five positive responses out of eight.

Source: Adapted from C. A. Wehler, R. I. Scott, and J. J. Anderson, The Community Childhood Hunger Identification Project: A model of domestic hunger—demonstration project in Seattle, Washington. Reprinted with permission, *Journal of Nutrition Education* 24 (1992): 29S–35S, Society for Nutrition Education.

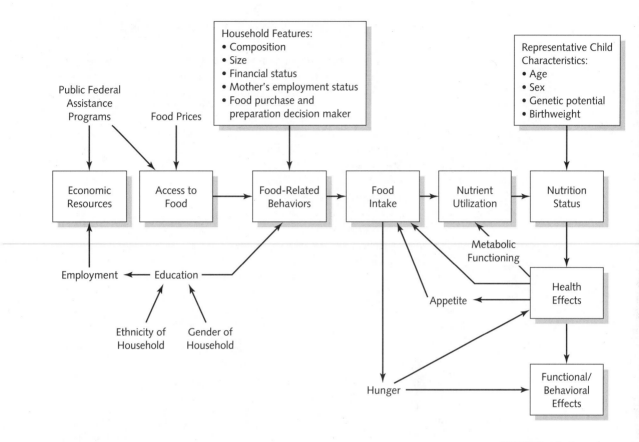

FIGURE 12-8

CCHIP *Conceptual Model of Factors Associated with Hunger and Its Outcomes*

Source: Adapted from C. A. Wehler, R. I. Scott, and J. J. Anderson, The Community Childhood Hunger Identification Project: A model of domestic hunger—demonstration project in Seattle, Washington. Reprinted with permission. *Journal of Nutrition Education* 24 (1992): 32S, Society for Nutrition Education.

The Plight of the Farmers

Changes in the domestic economy are proving adverse for producers as well as for consumers. U.S. farmers today lack significant control over the products they produce, the prices they must pay for supplies, and the prices they receive in return for their commodities. Just prior to 1980, the USDA urged farmers to produce more corn and soybeans for export. Accordingly, farmers borrowed heavily to expand their production capabilities. Since that time, their costs for seed, fertilizer, equipment, and loans have steadily risen while crop prices have declined. Today, thousands of U.S. farmers are hungry, frustrated, and desperate about their debt.[39]

The number of hungry farm families is not known, but agencies that provide aid to the rural poor say the demand for food assistance is increasing. Ironically, farm families generally do not grow fruits, vegetables, and other crops and animals to feed themselves. With modern practices aimed at efficiency, most farmers raise two or three crops—for example, feed corn, sorghum, and wheat—and buy most of their food from the grocery store. As a result, when crop prices drop, farmers struggle to survive under the sagging prices and realize no significant profits. Eventually, the farmers go out of business from lack of profits, just as would happen to any other type of business in the United States.

Beyond Public Assistance: What Can Individuals Do?

Solutions to the hunger problem depend on the willingness of people to take action and work together. Can we realistically hope to end hunger in the United States in the near future? Professor Larry Brown of Tufts University provides our response:

> The chief answer to those who question whether we can eliminate hunger lies in the fact that we virtually did so in our recent past. The programs created by the nation in the late 1960s and early 1970s worked. The evidence indicates that hunger significantly declined in the face of a national commitment expressed through the vehicles of school meals, food stamps, WIC, and elderly feeding programs. By fully utilizing these existing programs, we could again end hunger.
>
> The larger question—the truly complicated one—is how to eliminate the cause of hunger: poverty. The United States pays a high price for poverty, and hunger is only one part of it. From a public health perspective it is the height of folly to permit such a significant risk factor for illness and premature mortality to persist. From a moral perspective the prevalence of poverty in the world's wealthiest nation is yet another matter.
>
> Our nation has the ability to end hunger in a matter of months once we determine to do so. The larger issue is whether we will address poverty and, in doing so, not only eliminate hunger but prevent much untimely disease and death as well.[40]

Regardless of the type and level of involvement a person chooses, each person can make a difference. The government programs described in this chapter need people's support in a number of ways. Individual people can do all of the following:

- Assist in these programs as volunteers. Community nutritionists can educate providers about safe food-handling practices and healthful diets, identify the most effective means of providing the needed nutrients from limited resources, and teach participants how to shop for the most economical nutritious foods.[41]
- Help develop means of informing low-income people about food-related services and programs for which they are eligible.
- Help increase the accessibility of existing programs and services to those who need them.
- Document the needs that exist in their own communities (see the Community Learning Activity at the end of this chapter).
- Join with others in the community who have similar interests. Use the "hunger-free" criteria listed in Table 12-9 as guidelines for implementing comprehensive food assistance networks. The criteria are meant to provide a useful means of evaluating local antihunger networks.
- Conduct or participate in research to document the effectiveness of food assistance programs.[42]
- Document the impact of welfare reform on food security in local communities.
- Learn more about the problem of food insecurity and become familiar with organizations that advocate for sustainable solutions to the problems associated with food insecurity. Use the Internet to contact the agencies listed on page 380.
- Follow current food security legislation; call and write legislators about food insecurity issues; lobby to draw political attention to the need for more job opportunities and a higher minimum wage.

Besides individual actions, all persons who are concerned about the problems of poverty and undernutrition in the United States can exercise their right to affect the political process. Anyone can decide what local, state, and national governments should do to help, and communicate these ideas to elected officials for needed legislative changes. Individuals who volunteer their efforts and express their convictions to improve food assistance programs can also make a difference. Consider the following words, spoken over a hundred years ago:

I am only one,
But still I am one.
I cannot do everything,
But still I can do something;
And because I cannot do everything
I will not refuse to do the something that I can do.[43]

1. Establish a community-based emergency food delivery network.	**TABLE 12-9** *Fourteen Ways to Reduce Hunger in Communities*
2. Assess community hunger problems and evaluate community services. Create strategies for responding to unmet needs.	
3. Establish a group of individuals, including low-income participants, to develop and implement policies and programs to combat hunger and the threat of hunger; monitor responsiveness of existing services; and address underlying causes of hunger.	
4. Participate in federally assisted nutrition programs that are easily accessible to targeted populations.	
5. Integrate public and private resources, including local businesses, to relieve hunger.	
6. Establish an education program that addresses the food needs of the community and the need for increased local citizen participation in activities to alleviate hunger.	
7. Provide information and referral services for accessing both public and private programs and services.	
8. Support programs to provide transportation and assistance in food shopping, where needed.	
9. Identify high-risk populations and target services to meet their needs.	
10. Provide adequate transportation and distribution of food from all resources.	
11. Coordinate food services with parks and recreation programs and other community-based outlets to which residents of the area have easy access.	
12. Improve public transportation to human service agencies and food resources.	
13. Establish nutrition education programs for low-income citizens to enhance their food purchasing and preparation skills and make them aware of the connections between diet and health.	
14. Establish a program for collecting and distributing nutritious foods—either agricultural commodities in farmers' fields or prepared foods that would have been wasted.	

Source: House Select Committee on Hunger, legislation introduced by Tony P. Hall, excerpted in *Seeds*, Sprouts edition, January 1992, p. 3, with permission, © SEEDS Magazine, P.O. Box 6170, Waco, TX 76706.

Internet Resources

Congressional Hunger Center	**www.hungercenter.org/chc**
Foodchain	**www.foodchain.org**
A site for information on model community kitchens, food rescue of prepared foods, and more.	
HungerWeb	**www.brown.edu/Departments/ World_Hunger_Program/index.html**
Contains information and links to research, fieldwork experiences, and hunger-related advocacy, policies, education, and training.	
Second Harvest	**www.secondharvest.org**
Information and links on food security topics; contains results of study on food security—Hunger: The Faces and Facts—which profiles hungry Americans and where they get food.	
Bread for the World Institute	**www.bread.org**
Seeks to educate and motivate concerned citizens for action on policies that affect hungry people.	
InterAction	**www.interaction.org**
Represents over 150 private voluntary agencies.	
USDA Home Page	**www.usda.gov**
Contains information about current food assistance programs.	
USDA Food and Nutrition Research Briefs	**www.nal.usda.gov/fnic/usda/fnrb**
Department of Health and Human Services	**www.hhs.gov**
Census Bureau	**www.census.gov**
National Center for Children in Poverty Columbia University School of Public Health	**cpmcnet.columbia.edu:80/dept/nccp**
Food Research and Action Center	**www.frac.org**
A nonprofit research, public policy, and legal center that serves as the hub of an anti-hunger network of agencies across the country.	
Share Our Strength	**www.foggy.com/SOS/involved.html**
Hosts annual Taste of the Nation fund-raising event.	

Community Learning Activity

Activity 1

Through this activity, you will become familiar with the basic steps for documenting hunger in a community. Your documentation should give you a sense of the food insecurity in your own community. Your hunger documentation may also be useful for informing and/or influencing the decisions of local, state, and federal legislators regarding food security issues.

Four steps are required to provide a basic profile of food security in a community. In this activity, we ask you to focus on Steps 1 and 3.[44]

1. Estimate the size of the group at risk of hunger. Check the most recent U.S. Census of Population and Housing. Track statistics for those listed as "poor" and "near-poor" by age, sex, race, and type of family. These books of statistics from the Bureau of the Census are available in most libraries and on the Internet (www.census.gov).

2. Estimate the amount of help available through government food programs and welfare assistance. For example, if 7500 people are classified as living in poverty-level households, how many of these people receive food stamps? Cash welfare such as TANF and Supplemental Security Income (SSI)? Free or reduced-price meals for their children? WIC benefits? Congregate or home-delivered meals if they are over age 60? Sources of information would be the director or agency staff of the various food assistance and welfare programs.

3. Learn how emergency food needs have changed (increased or decreased) in the community. Talk with the staff of local soup kitchens, food banks, and food pantries to determine whether they are providing more emergency food and if so, how much. Note how many emergency food agencies there are in the community.

4. Determine whether there are indications that overall health status is being affected. You can do this by examining basic health indicators for the community: infant mortality rates, the incidence of low birthweight, the incidence of "failure-to-thrive" among children, and, if possible, the number of hospitalizations in which malnutrition is listed as one of the precipitating causes of treatment. You would need to check vital statistics at local and county health departments and local hospital records for this step. You could also check with the health planning agency for your state to see if a "health status report" has been prepared. Much of this information is documented in these state reports for health planning purposes.

It is recommended that you collect this information for three consecutive years, so that you'll have a sense of whether the situation is improving or getting worse (see Figure 12-9 on the following page).

FIGURE 12-9 *Basic Facts for Documenting Hunger in a Community*

Source: Adapted from Food Research and Action Center, *How to Document Hunger in Your Community* (Washington, D.C.: FRAC, 1983), pp. 5–6.

SAMPLE

for _____, _____ (_____ County)
           City   State

1. Population and Poverty Count

POPULATION	NUMBER IN POVERTY	POOR PLUS NEAR-POOR
_____	_____	_____

Of those below the poverty line:

OVER 65	UNDER 18
_____	_____

	1996	1999
2. Receiving Help from Government		
Food Assistance		
Food Stamps	_____	_____
School Lunch	_____	_____
School Breakfast	_____	_____
WIC	_____	_____
CACFP	_____	_____
Elderly Nutrition Program	_____	_____
Cash Welfare		
TANF (formerly AFDC)	_____	_____
SSI	_____	_____
3. Emergency Assistance		
Number of Providers		
Soup Kitchens	_____	_____
Food Pantries	_____	_____
Food Bank	_____	_____
Other	_____	_____
Serving		
(number of people served or number of meals served)		
4. Health Indicators		
Infant mortality rate	_____	_____
Low-birthweight incidence	_____	_____
Failure-to-thrive incidence	_____	_____
Hospitalizations related to malnutrition	_____	_____

References

1. K. Markandaya, *Nectar in a Sieve*, 2nd American ed. (New York: John Day, 1955), pp. 121–22.
2. P. L. Splett, Federal Food Assistance Programs: A step to food security for many, *Nutrition Today* March/April, 1994, pp. 6–13.
3. The margin definition of food insecurity was adapted from K. Radimer and coauthors, Understanding hunger and developing indicators to assess it in women and children, *Journal of Nutrition Education* 24 (1992): 36S–45S.
4. L. Schwartz-Nobel, *Starving in the Shadow of Plenty* (New York: Putnam, 1981), pp. 35–36.
5. Food Research and Action Committee, *Community Childhood Hunger Identification Project* (Washington, D.C.: Food Research and Action Center, July 1995), pp. 11–13.
6. Physician Task Force on Hunger in America, *Hunger in America: The Growing Epidemic* (Middletown, CT: Wesleyan University Press, 1985), p. 17.

7. N. Kotz, *Hunger in America: The Federal Response* (New York: Field Foundation, 1979), p. 17.

8. Bread for the World Institute, *Hunger 1994: Transforming the Politics of Hunger* (Washington, D.C.: Bread for the World Institute, 1993), pp. 151–56.

9. A. Kendall and E. Kennedy, Position of the American Dietetic Association: Domestic food and nutrition security, *Journal of the American Dietetic Association* 98 (1998): 337–42.

10. M. Nestle and S. Guttmacher, Hunger in the United States: Rationale, methods, and policy implications of state hunger surveys, *Journal of Nutrition Education* 24 (1992): 20S–22S.

11. Position of the American Dietetic Association: Domestic food and nutrition security, 1998.

12. R. Briefel and C. Woteki, Development of food sufficiency questions for the third National Health and Nutrition Examination Survey, *Journal of Nutrition Education* 24 (1992): 24S–28S.

13. J. P. Habicht and L. D. Meyers, Principles for effective surveys of hunger and malnutrition in the United States, *Journal of Nutrition* 121 (1991): 403–7.

14. W. L. Hamilton and coauthors, *Household Food Security in the United States in 1995*, (Washington, D.C.: U.S. Department of Agriculture Food and Consumer Service, 1997); the household food security report is available via www.usda.fov/fcs/measure.htm.

15. Ibid.

16. B. W. Klein, Food security and hunger measures: Promising future for state and local household surveys, *Family Economics and Nutrition Review* 9 (1996): 31–37.

17. *Nutrition Action Themes for the United States: A Report in Response to the International Conference on Nutrition* (Washington, D.C.: U.S. Department of Agriculture Center for Nutrition Policy and Promotion, 1996), CNPP-2 Occasional Paper.

18. The Administration for Children and Families, *HHS Fact Sheet: Work Not Welfare* (Washington, D.C.: ACF Office of Public Affairs, November 1997), pp. 1–3.

19. E. Kennedy, P. M. Morris, and R. Lucas, Welfare reform and nutrition programs: Contemporary budget and policy realities, *Journal of Nutrition Education* 28 (1996): 67–70; D. Rose and M. Nestle, Welfare reform and nutrition education: Alternative strategies to address the challenges of the future, *Journal of Nutrition Education* 29 (1996): 61–66.

20. The discussions of the food assistance programs were adapted from Food and Nutrition Service, *Nutrition Program Facts* (Alexandria, VA: U.S. Department of Agriculture, May, 1998).

21. USDA, *Nutrition Program Facts: Food Stamp Program*, May, 1998, pp. 1–5.

22. Community Nutrition Institute, Farmers' markets growing new sites and new states, *Nutrition Week*, April 24, 1998, pp. 3, 6.

23. Department of Health and Human Services Administration on Aging, The Elderly Nutrition Program Fact Sheet, May 1998, pp. 1–4.

24. USDA, *Nutrition Program Facts: Nutrition Program for the Elderly*, May, 1998, pp. 1–2.

25. Community Nutrition Institute, Nearly 26 million people turned to food banks in 1997, Second Harvest says, *Nutrition Week,*

March 13, 1998, pp. 1–3; Community Nutrition Institute, Cities and charities report an increase demand for emergency food services, *Nutrition Week,* December 19, 1997, p. 1.

26. B. O. Daponte, Food pantry use among low-income households in Allegheny County, Pennsylvania, *Journal of Nutrition Education* 30 (1998): 50–57.

27. United States Department of Agriculture, *Nutrition Program Facts: Food Recovery and Gleaning Fact Sheet,* May, 1998, pp. 1–4.

28. U.S. Department of Agriculture Food Recovery and Gleaning Initiative, *A Citizen's Guide to Food Recovery* (Washington, D.C.: U.S. Department of Agriculture, 1997).

29. Bread for the World Institute, *Hunger 1993: Uprooted People* (Washington, D.C.: Bread for the World Institute, 1992), p. 107.

30. J. L. Wiecha, J. T. Dwyer, and M. Dunn-Strohecker, Nutrition and health services needs among the homeless, *Public Health Reports* 106 (1991): 364–74.

31. The discussion of homelessness and the steps to end it was adapted from Bread for the World Institute, *Hunger 1993,* pp. 108–9.

32. U.S. Conference of Mayors, *A Status Report on Hunger and Homelessness in America's Cities: 1997* (Washington, D.C.: U.S. Conference of Mayors, 1997).

33. Bread for the World Institute, *Hunger 1997: What Governments Can Do,* (Silver Spring, MD: Bread for the World Institute, 1996), pp. 26–35.

34. Bread for the World Institute, *Hunger 1998: Hunger in a Global Economy* (Silver Spring, MD: Bread for the World Institute, 1997), p. 108.

35. Food Research and Action Committee, *Community Childhood Hunger Identification Project.*

36. C. A. Wehler, R. I. Scott, and J. J. Anderson, The Community Childhood Hunger Identification Project: A model of domestic hunger—demonstration project in Seattle, Washington, *Journal of Nutrition Education* 24 (1992): 29S–35S.

37. Ibid., p. 30S.

38. Bread for the World Institute, *Hunger 1998,* p. 108.

39. Select Committee on Hunger, *Farm Crisis: Growing Poverty and Hunger among America's Food Producers* (Washington, D.C.: U.S. Government Printing Office, 1987), p. 163.

40. J. L. Brown and D. Allen, Hunger in America, *Annual Reviews in Public Health* 9 (1988): 503–26.

41. L. Sims and J. Voichick, Our perspective: Nutrition education enhances food assistance programs, *Journal of Nutrition Education* 28 (1996): 83–85; A. B. Joy and C. Doisy, Food stamp nutrition education program: Assisting food stamp recipients to become self-sufficient, *Journal of Nutrition Education* 28 (1996): 123–126.

42. R. Lobosco, A commentary on domestic hunger: A problem we can solve, *Topics in Clinical Nutrition* 9 (1994): 8–12.

43. The quotation is by Edward Everett Hale (1822–1909), *For the Lend-a-Hand Society.*

44. This activity was adapted from *How to Document Hunger in Your Community* (Washington, D.C.: Food Research and Action Center, 1983), pp. 3–21.

Chapter 13

Mothers and Infants: Nutrition Assessment, Services, and Programs

Outline

Learning Objectives

After you have read and studied this chapter, you will be able to:

- List the recommendations for maternal weight gain during pregnancy.
- Explain the relationship of maternal weight gain to infant birthweight and outcome of pregnancy.
- Identify nutritional factors and lifestyle practices that increase health risk during pregnancy.

- Describe the benefits of breastfeeding.
- Describe the purpose, eligibility requirements, and benefits of the federal nutrition programs available to assist low-income women and their children.
- Identify the common nutrition-related problems of infancy.
- Describe current infant feeding recommendations during infancy.

Something To Think About...

All aspects of a society, whether it is relatively static or in a stage of dramatic upheaval, affect the health of every member of the family and of the community, and especially of its most physiologically and culturally vulnerable groups—mothers and young children.

– Cicely D. Williams
– Derrick B. Jelliffe

Introduction

The effects of nutrition extend from one generation to the next, and this is particularly evident during pregnancy. Research has demonstrated that the poor nutrition of a woman during her early pregnancy can impair the health of her *grandchild*, even after that child has become an adult.[1] For example, if a mother's nutrient stores are inadequate early in pregnancy when the placenta is developing, the fetus will develop poorly, no matter how well the mother eats later. After getting such a poor start on life, the female child may grow up poorly equipped to support a normal pregnancy, and she, too, may bear a poorly developed infant.

Infants born of malnourished mothers are more likely than healthy women's infants to become ill, to have birth defects, and to suffer retarded mental or physical development. This remains true even if they later receive abundant, nourishing food. The many growth-retarded Korean orphans adopted by U.S. families after the Korean War, for example, experienced several years of catch-up growth, which did not completely remedy the growth deficits caused by early malnutrition.[2] Malnutrition in the prenatal and early postnatal periods also affects learning ability and behavior. Clearly, it is critical to provide the best nutrition possible at the early stages of life. This chapter focuses on the particular nutrient requirements and nutrition-related problems of pregnancy, lactation, and infancy and examines the nutrition programs and services that target pregnant women and their infants.

Trends in Maternal and Infant Health

The health of a nation is often judged by the health status of its mothers and infants. One of the best indicators of a nation's health, according to epidemiologists, is the **infant mortality rate (IMR).** Although the United States spends more money on health care than most other countries, its IMR of 7.6 is considerably higher than several industrialized countries'—for example, 4.0 for both Japan and Sweden (see Table 13-1). Although the U.S. IMR is at an all-time low, important measures of increased risk of death, such as incidence of low birthweight and receipt of prenatal care, show no recent improvement.[3] In addition, disparities in IMRs persist between ethnic groups and between poor and nonpoor infants, as shown in Figure 13-1. The IMR of 15.1 for black infants remains more than twice as high as the IMR of 6.3 for white infants. The failure to further improve the IMR in the United States has been

TABLE 13-1 A
Comparison of Infant Mortality Rates Worldwide, 1995

Finland	4.0
Japan	4.0
Sweden	4.0
Austria	6.0
Canada	6.0
Denmark	6.0
Germany	6.0
Ireland	6.0
Netherlands	6.0
Norway	6.0
Switzerland	6.0
United Kingdom	6.0
Australia	7.0
France	7.0
Israel	7.0
Italy	7.0
New Zealand	7.0
Belgium	8.0
Spain	8.0
United States	8.0

Source: Adapted from UNICEF, *The State of the World's Children* (Oxford: Oxford University Press, 1997).

FIGURE 13-1 U.S. *Infant Mortality Rates by Race; 1950–1995*

Source: Adapted from Office of Maternal and Child Health, Public Health Service, U.S. Department of Health and Human Services. *Child Health USA '90* (Washington, D.C.: U.S. Department of Health and Human Services, 1990), p. 16, and *Kids Count Data Book* (Baltimore, MD: Annie E. Casey Foundation, 1997).

attributed to the number of infants born with low birthweights. In 1994 the infant mortality rate ranged from a low of 5.0 in Rhode Island to a high of 18.2 in the District of Columbia.[4]

If the pregnant woman does not receive adequate nourishment and does not gain the recommended amount of weight, she may give birth to a baby of **low birthweight (LBW).** Not all small babies are unhealthy, but birthweight and length of gestation are the primary indicators of the infant's future health status. An LBW baby is more likely to experience complications during delivery than a normal-weight baby and has a statistically greater chance of having physical and mental birth defects, contracting diseases, and dying early in life. Low birthweight in full-term infants is a major contributing factor to infant mortality. Clearly, a key to reducing infant mortality is reducing the incidence of LBW babies. To do so, several factors need to be addressed: poverty, minority status, lack of access to health care, inability to pay for health care, poor nutrition, low level of educational achievement, unsanitary living conditions, and unhealthful habits such as smoking, drinking, and drug use. Figure 13-2 depicts the percentage of LBW infants by race in the United States. In 1994 the percentage of births that were low weight ranged from a low of 5.1 percent in New Hampshire to a high of 14.2 percent in the District of Columbia.[5]

Improving the health and nutrition status of pregnant women and infants remains a national challenge. Although **maternal mortality rates** have decreased significantly during the last four decades from a high of 83.3 per 100,000 live births in the 1950s to a low of 6.6 in 1987, black women still have a three times greater risk of dying than white women.[6] A summary of selected objectives to improve maternal and infant health are listed in Table 13-2.

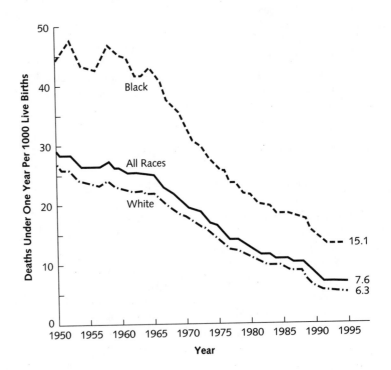

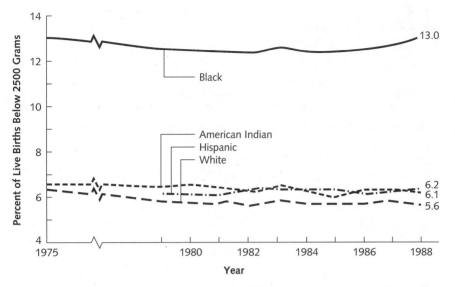

FIGURE 13-2

Percentage of Low-Birthweight Infants by Race in the United States, 1975–1988

Source: Office of Maternal and Child Health, Public Health Service, U.S. Department of Health and Human Services, *Child Health USA '90* (Washington, D.C.: U.S. Department of Health and Human Services, 1990), p. 18.

National Goals for Maternal and Infant Health

Within the past decade a number of research reports, such as the *Surgeon General's Report* and *Healthy People 2000*, have established goals and recommendations designed to improve the nutrition and health status of mothers and infants.[7] In order for the United States to achieve further reductions in infant mortality and eliminate racial and ethnic differences in pregnancy outcome, health care professionals must focus on changing the behaviors and outcomes that affect pregnancy outcomes.[8] For example, health problems and behaviors such as smoking, substance abuse, and poor nutrition need to be addressed prior to conception.[9]

According to the *Healthy People 2000* Midcourse Revisions, progress toward the goals is uneven. Improvement is seen in four areas: (1) there is a decline in infant mortality rates for Hispanics and African Americans, (2) the rate of timely prenatal care improved for women in all race and ethnic groups during the 1990s, (3) there is an increase in breastfeeding by Hispanic, American Indian, and African American women, and (4) cigarette smoking during pregnancy continues to decline. In 1996, 13.6 percent of women smoked during pregnancy compared to 20 percent of pregnant women in 1989.[10]

Healthy Mothers

A number of factors contribute to maternal and infant health.[11] Genetic, environmental, and behavioral factors affect risk and the outcome of pregnancy. A woman's nutrition prior to and throughout pregnancy is crucial to both her health and the growth, development, and health of the infant she conceives. Ideally, a woman starts pregnancy at a healthful weight, with filled nutrient stores and the firmly established habit of eating a balanced and varied diet. In this section, we discuss the nutritional

Infant mortality rate (IMR) Infant deaths under one year of age, expressed as a rate per 1,000 live births.

Maternal mortality rate Number of women's deaths assigned to causes related to pregnancy, expressed as a rate per 100,000 live births.

Low birthweight (LBW) A birthweight of 5½ pounds (2500 grams) or less, used as a predictor of poor health in the newborn and as a probable indicator of poor nutrition status of the mother during and/or before pregnancy. LBW infants are of two different types. Some are premature; they are born early (prior to 38 weeks of gestation) and are the right size for their gestational age. Others have suffered growth failure in the uterus; they may or may not be born early, but they are small for gestational age (small-for-date).

TABLE 13-2 Healthy People 2000 *Objectives to Improve Maternal and Infant Health*

Health Status Objectives

1. Reduce the infant mortality rate to no more than 7 per 1,000 live births. (Baseline: 10.1 per 1,000 live births in 1987.)

   • Among blacks: 11/1,000 live births. (Baseline: 17.9 in 1987.)
   • Among American Indians and Alaska natives: 8.5/1,000 live births. (Baseline: 12.5 in 1984.)
   • Among Hispanics: 8/1,000 live births. (Baseline: 12.9 for Puerto Ricans in 1984.)

2. Reduce the maternal mortality rate to no more than 3.3 per 100,000 live births. (Baseline: 6.6 per 100,000 in 1987.)

3. Reduce the incidence of fetal alcohol syndrome (FAS) to no more than 0.12 per 1,000 live births. (Baseline: 0.22 per 1,000 live births in 1987.) The 1993 incidence of FAS reached 0.67 per 1,000 live births.

Risk Reduction Objectives

1. Reduce low birthweight (LBW) to an incidence of no more than 5% of live births. (Baseline: 6.9% in 1987.) Among blacks: 9% of live births. (Baseline: 12.7% in 1987.) The 1993 rate of LBW increased to 7.2 percent of live births.

2. Increase to at least 85% the proportion of mothers who achieve the minimum recommended weight gain during their pregnancies. (Baseline: 67% of married women in 1987.)

3. Increase to at least 75% the proportion of mothers who breastfeed their babies in the early postpartum period and to at least 50% the proportion who continue breastfeeding until their babies are 5 to 6 months old. (Baseline: 54% at discharge from birth site and 21% at 5 to 6 months in 1988.) The 1995 overall rate of breastfeeding was 60% during the early postpartum period.

4. Increase abstinence from tobacco use by pregnant women to at least 90% and increase abstinence from alcohol, cocaine, and marijuana by pregnant women by at least 20%. (Baseline: 75% of pregnant women abstained from tobacco use in 1985.)

5. Increase calcium intake so at least 50% of pregnant and lactating women consume three or more servings daily of foods rich in calcium. (Baseline: 24% of pregnant and lactating women in 1985–1986.)

6. Reduce iron deficiency to less than 3% among women of childbearing age. (Baseline: 5% for women aged 20 to 44 years in 1976–1980.)

Healthy People 2000 Midcourse Revisions

1. Reduce the incidence of spina bifida and other neural tube defects to 3 per 10,000 live births. (Baseline: 6 cases per 10,000 live births in 1990.) The 1993 rate was 7 per 10,000 live births.

Source: *Healthy People 2000: National Health Promotion and Disease Prevention Objectives* (Washington, D.C.: Department of Health and Human Services, 1990); and C. M. Trahms and J. Powell, Maternal and Infant Nutrition: An inquiry into the issues, *Topics in Clinical Nutrition* 11 (1996): 15.

needs of pregnant women, nutrition-related health problems associated with pregnancy, and the nutrition assessment methods important to this population.

Nutritional Needs of Pregnant Women

For most women, nutrient needs during pregnancy and lactation are higher than at any other time in their adult life and are greater for certain nutrients than for others (see the RDA/DRI Table on the inside front cover). Notice that although nutrient needs are much higher than usual, energy needs are not. An increase of only 15 percent of maintenance calories is recommended to support the metabolic demands of pregnancy and fetal development. The RDA for energy is an additional 300 calories per day during the second and third trimesters.

The Subcommittee on Dietary Intake and Nutrient Supplementation of the Institute of Medicine believes that the nutrient needs of pregnancy are best met by the routine intake of a variety of foods (see Table 13-3). Based on a review of several studies, the subcommittee concluded that protein, folate, iron, zinc, calcium, and vitamins known to be toxic in excess amounts deserved special attention in the diets of pregnant women.[12]

The pregnancy RDA for protein is 60 grams—an additional 10 to 16 grams per day over nonpregnant requirements. Most women are already eating enough protein to cover the increased demand of pregnancy.

The pregnant woman's recommended folic acid intake is 50 percent greater than that of the nonpregnant woman due to the increase in her blood volume and rapid growth of the fetus. Certain studies have shown that folic acid supplements given around the time of conception reduce the recurrence of **neural-tube defects** such as spina bifida in the infants of women who previously have had such births.[13] A woman can obtain the recommended 400 micrograms of folic acid by taking a folic acid supplement daily, eating a breakfast cereal fortified with 100 percent of the daily value of folic acid, or increasing her consumption of foods fortified with folic acid (for example, cereal, bread, rice, or pasta) and foods naturally rich in folates (orange juice and green vegetables).[14]

The RDA for iron during pregnancy is an additional 15 milligrams of iron per day to meet maternal and fetal needs. Iron-deficiency anemia is a common problem among nonpregnant women, and as a result, many women begin pregnancy with diminished iron stores. For this reason, an iron supplement of 30 milligrams of ferrous iron daily during the second and third trimesters is recommended. To facilitate iron absorption from the supplement, it should be taken between meals with vitamin C-rich fruit juices or at bedtime on an empty stomach. Since supplemental doses of iron greater than 30 milligrams can interfere with zinc absorption, pregnant women are encouraged to include good sources of zinc in their daily diets.

The DRI for calcium in pregnancy is the same amount recommended for females of comparable ages. During the last trimester of pregnancy when fetal skeletal growth is maximum and teeth are being formed, the fetus draws approximately 300

The U.S. Public Health Service now recommends that all women of childbearing age consume 400 µg of folic acid daily to reduce the risk of neural tube birth defects. To reduce risk, adequate folic acid consumption must begin before pregnancy because the defects occur within the first month after conception—generally before a women is aware she is pregnant.

Neural-tube defects Any of a number of birth defects in the orderly formation of the neural tube during early gestation. Both the brain and the spinal cord develop from the neural tube; defects result in various central nervous system disorders.

Food	Number of Servings*	
	Nonpregnant Women	**Pregnant or Lactating Women**
Breads/cereals	6 to 11	7 to 11 (7+)
Vegetables	3 to 5	4 to 5 (5+)
Fruits	2 to 4	3 to 4 (4+)
Meat/meat alternates	2 to 3	3 (3+)
Milk/milk products	2	3 to 4 (4+)

TABLE 13-3 Food Guide for Pregnant and Lactating Women

*Numbers in parentheses indicate numbers of servings recommended for the pregnant teenager.

Pregnancy DRI for calcium: 1,000 mg (19 to 50 years); 1,300 mg (14 to 18 years)

milligrams per day from the maternal blood supply. Dairy products are recommended because they are also sources of vitamin D and riboflavin. However, in cases where lactose intolerance is a problem, a calcium supplement should be considered. This is particularly important for women under 30 years of age whose bone mineral density is still increasing. In such cases, a 600-milligram supplement of calcium per day during pregnancy is recommended if the woman normally consumes less than 600 milligrams of calcium a day.

Routine supplementation with vitamins during pregnancy is not advised, and excess intakes of certain vitamins, notably vitamins A and D, can cause fetal malformations.[15] Nutrient supplementation may be appropriate in certain circumstances, however. For example, supplements of vitamin D (10 micrograms per day) and vitamin B_{12} (2 micrograms per day) may be recommended for vegans, or a low-dose multivitamin-mineral supplement beginning in the second trimester may be appropriate for women who do not ordinarily consume an adequate diet or are in high-risk categories, such as women carrying more than one fetus, heavy cigarette smokers, and alcohol and drug abusers. A summary of the Food and Nutrition Board's guidelines is provided in Table 13-4.

Maternal Weight Gain

Normal weight gain and adequate nutrition support the health of the mother and the development of the fetus. The National Academy of Sciences' Committee rec-

TABLE 13-4 *Adapted from Recommendations for Nutrient Supplementation In Pregnancy*

Nutrient	Candidates for Supplementation	Level of Nutrient Supplementation
Iron	All pregnant women (2nd and 3rd trimesters)	30 mg ferrous iron daily
Folic acid	Pregnant women with suspected dietary inadequacy of folate	400 μg/day*
Vitamin D	Complete vegetarians and others with low intake of vitamin D–fortified milk	10 μg/day
Calcium	Women under age 25 whose daily dietary calcium intake is less than 600 mg	600 mg/day
Vitamin B_{12}	Complete vegetarians	2 μg/day
Zinc/copper	Women under treatment with iron for iron deficiency anemia	15 mg Zn/day 2 mg Cu/day
Multivitamin-mineral supplements	Pregnant women with poor diets and for those who are considered high risk: multiple gestation, heavy smokers, alcohol/drug abusers, other	Preparation containing: iron—30 mg zinc—15 mg copper—2 mg calcium—250 mg vitamin B_6—2 mg folate—400 μg* vitamin C—50 mg vitamin D—5 μg

*Centers for Disease Control, Recommendations for the use of folic acid to reduce the number of cases of spina bifida and other neural tube defects, *Morbidity and Mortality Weekly Report* 41 (1991): 1.

Source: Adapted from Institute of Medicine: *Nutrition During Pregnancy: Weight Gain and Nutrient Supplements* (Washington, DC: National Academy Press, 1990).

BMI	Weight Category	Recommended Gain (lbs.)*
< 18.5	Underweight	28–40
18.5–24.9	Normal weight	25–35
25.0–29.9	Overweight	15–25
≥ 30.0	Obesity	≥ 15

TABLE 13-5

Recommended Weight Gain for Pregnant Women Based on Body Mass Index (BMI)

*Teens should strive to gain the maximum pounds in their ranges; short women (less than 62 inches tall) should strive for the minimum. Weight gain varies widely, and these values are suggested only as guidelines for identifying individuals whose weights may be too high or low for health.

Source: Adapted from NHLBI, *Clinical Guidelines on the Identification, Evaluation, and Treatment of Overweight and Obesity in Adults: The Evidence Report,* June 1998, and National Academy of Sciences, Food and Nutrition Board, *Nutrition During Pregnancy* (Washington, D.C.: National Academy Press, 1990).

ommendations for weight gain take into account a mother's prepregnancy weight-for-height or body mass index (BMI), as shown in Table 13-5. The committee recommends that a woman who begins pregnancy at a healthful weight should gain between 25 and 35 pounds. Women pregnant with twins need to gain 35 to 45 pounds; women pregnant with triplets need to gain 45 to 55 pounds. An underweight woman needs to gain between 28 and 40 pounds; an overweight woman, between 15 and 25 pounds. Weight gains at the upper end of the range are recommended for pregnant teenagers and black women because of their increased risk of low weight gains and delivery of LBW infants.

Low weight gain in pregnancy is associated with increased risk of delivering an LBW infant; these infants have high mortality rates.[16] Excessive weight gain in pregnancy increases the risk of complications during labor and delivery, as well as postpartum obesity. Obese women (greater than 135 percent of the standard weight-for-height) also have an increased risk for complications during pregnancy, including hypertension and gestational diabetes.

Weight gain should be lowest during the first trimester—2 to 4 pounds for the trimester—followed by a steady gain of about a pound per week thereafter, as shown in Figure 13-3. If a woman gains more than the recommended amount of weight early in pregnancy, she should not try to diet in the last weeks. Dieting during pregnancy is not recommended. A *sudden* large weight gain, however, may indicate the onset of pregnancy-induced hypertension, discussed later; a woman experiencing this type of weight gain should see her health care provider. See Table 13-6 on page 393 for tips on counseling pregnant women regarding healthy weight gains.

Practices to Avoid

Optimal pregnancy outcome is influenced by maternal nutrient intake, but it can also be affected by maternal use of nonfood substances, excess caffeine, low-calorie diets, tobacco, alcohol, and illicit drugs. **Pica** refers to the craving of nonfood items having little or no nutritional value. Pica of pregnancy typically involves the consumption of dirt, clay, or laundry starch, but episodes of pica have included compulsive ingestion of such things as ice, paper, and coffee grounds.[17] The medical consequences of pica can include malnutrition, as nonfood items replace nutritious

Pica The craving of nonfood items such as clay, ice, and cornstarch. Pica does not appear to be limited to any particular geographical area, race, sex, culture, or social status.

FIGURE 13-3

Desirable Weight Gains During Pregnancy

The woman who is of normal weight prior to pregnancy should gain in the B–C range, 25 to 35 lb during the pregnancy. The underweight woman should gain in the A–B range, 28 to 40 lb. The woman who is overweight prior to pregnancy should gain in the D range, 15 to 25 lb.

Source: Courtesy of National Dairy Council, *Great Beginnings: The Weighting Game Graph* (Rosemont, IL: National Dairy Council, 1991); adapted from Food and Nutrition Board, Institute of Medicine, National Academy of Sciences, *Nutrition During Pregnancy* (Washington, D.C.: National Academy Press, 1990).

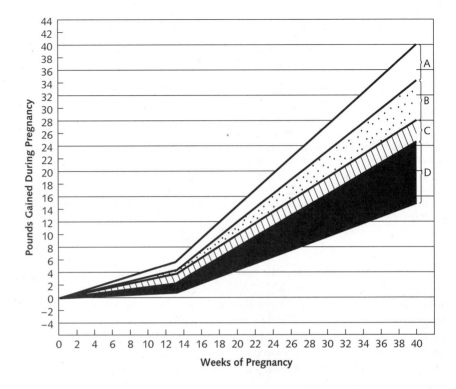

foods in the diet; obesity, from overconsumption of items such as starch; the ingestion of toxic compounds; or intestinal obstruction, from consuming large amounts of clay or starch.

The Food and Drug Administration has advised pregnant women to avoid unnecessary consumption of caffeine because animal studies suggest that it causes birth defects.[18] Studies in humans have generally failed to show that caffeine use has a negative effect on pregnancy outcome, although limited evidence has shown that moderate-to-heavy use may contribute to lower infant birthweight.[19] Women who choose to use caffeine during pregnancy are generally advised to do so in moderation, if at all—the equivalent of a cup of coffee or two 12-ounce cola beverages a day.

Some practices are truly harmful, and their potential impact on pregnancy outcome is too great to risk. Low-carbohydrate or low-calorie diets that cause ketosis deprive the fetus's brain of needed glucose and cause congenital deformity. The invisible effects may be even more serious. For example, carbohydrate metabolism may be rendered permanently defective, or the infant's brain may be permanently damaged. Protein deprivation can cause children's height and head circumference to diminish markedly and irreversibly.

Another harmful maternal practice is smoking, which restricts the blood supply to the growing fetus, thereby limiting the delivery of nutrients and removal of wastes. Smoking stunts growth, thus increasing the risk of premature delivery, low infant birthweight, retarded development, and complications at birth. Smoking is responsible for 20 to 30 percent of all LBW deliveries in the United States.[20] Sudden infant death syndrome (SIDS) has been linked to a mother's smoking during pregnancy as well.[21]

What to Look for If Weight Gain Is Slow or If Weight Loss Occurs:

- Is there a measurement or recording error?

- Is the overall pattern acceptable? Was a lack of gain preceded by a higher than expected gain?

- Was there evidence of edema at the last visit and is it resolved?

- Is nausea, vomiting, or diarrhea a problem?

- Is there a problem with access to food?

- Have psychosocial problems led to poor appetite?

- Does the woman resist weight gain? Is she restricting her energy intake? Does she have an eating disorder?

- If the slow weight gain appears to be a result of self-imposed restriction, does she understand the relationship between her weight gain and her infant's growth and health?

- Is she smoking? How much?

- Is she using alcohol or drugs (especially cocaine or amphetamines)?

- Does her energy expenditure exceed her energy intake?

- Does she have an infection or illness that requires treatment?

What to Look for If Weight Gain Is Very Rapid:

- Is there a measurement or recording error?

- Is the overall pattern acceptable? Was the gain preceded by weight loss or a lower than expected gain?

- Is there evidence of edema?

- Has the woman stopped smoking recently? The advantages of smoking cessation offset any disadvantages associated by gaining some extra weight.

- Are twins a possibility? (A large increase in fundal height may be the earliest sign.)

- Are there signs of gestational diabetes?

- Has there been a dramatic decrease in physical activity without an accompanying decrease in food intake?

- Has the woman greatly increased her food intake? Obtain a diet recall, making special note of high-fat foods. However, rapid weight gain is often accompanied by normal eating patterns, which should be continued. If intake of high-fat or high-sugar foods is excessive, encourage substitutions.

- If serious overeating is occurring, explore why. Does stress, depression, an eating disorder or boredom play a factor? Is there need for special support or a referral?

TABLE 13-6

Points to Consider for Optimal Weight Gain in Pregnancy

Source: B. Worthington Roberts and S. R. Williams, *Nutrition Throughout the Life Cycle*, 3rd ed. (St. Louis: Times Mirror/Mosby, 1996).

During the past 15 years, research has confirmed that excessive consumption of alcohol adversely affects fetal development.[22] Even as few as one or two drinks daily can cause **fetal alcohol syndrome (FAS)**—irreversible brain damage and mental and physical retardation in the fetus. The most severe impact of maternal drinking is likely to be in the first month, before the woman even is sure she is pregnant. This preventable condition (FAS) is estimated to occur in approximately one to two infants per 1,000 live births and is the leading known cause of mental retardation in the United States.[23] Birth defects, low birthweight, and spontaneous abortions

Fetal alcohol syndrome (FAS) The cluster of symptoms seen in an infant or child whose mother consumed excess alcohol during pregnancy, including retarded growth, impaired development of the central nervous system, and facial malformations.

occur more often in pregnancies of women who drink even as little as 2 ounces of alcohol daily during pregnancy. Accumulating evidence that even one drink may be too much has led the surgeon general to take the position that women should stop drinking as soon as they *plan* to become pregnant.[24]

Primary Nutrition-Related Problems of Pregnancy

Common physical problems in pregnancy include "morning" sickness and, later, constipation. The nausea of morning sickness seems unavoidable because it arises from the hormonal changes taking place early in pregnancy, but it can sometimes be alleviated. A suggested strategy is to start the day with a few sips of water and a soda cracker or other bland carbohydrate food to get something into the stomach before getting out of bed.

Later during pregnancy, as the hormones of pregnancy alter her muscle tone and the growing fetus crowds her intestinal organs, an expectant mother may complain of constipation. A high-fiber diet, plentiful fluid intake, and moderate exercise will help to relieve this condition.

More serious problems needing control during pregnancy include hypertension and diabetes:

• **Hypertension.** Preeclampsia and eclampsia are hypertensive conditions induced by pregnancy (see Table 13-7). Preeclampsia is characterized by high blood pressure, protein in the urine, and generalized edema that may cause sudden, large weight gain from retained water.

TABLE 13-7
Pregnancy-Induced Hypertension

Preeclampsia

Hypertension (blood pressure 140/90 mm Hg) after 20th week of pregnancy

Proteinuria

Edema

Severe preeclampsia

One or more of the following:

Systolic blood pressure of 160 mm Hg or diastolic blood pressure of 110 mm Hg

Proteinuria, edema

Blurred vision, headache, altered consciousness

Epigastric pain, impaired liver function

Eclampsia

Preeclampsia with seizure, occurs near time of labor

Transient gestational hypertension

Transient increase in blood pressure during labor or first 24 hours postpartum; no signs or symptoms of preeclampsia

Source: Adapted from National High Blood Pressure Education Program Working Group Report on High Blood Pressure in Pregnancy, *American Journal of Obstetrics and Gynecology* 163 (1990): 1691.

Fluid retention alone, which is quite common in pregnant women, is not sufficient to diagnose preeclampsia.[25] Warning signs of preeclampsia include severe and constant headaches, sudden weight gain (1 pound a day), swelling, dizziness, and blurred vision. Eclampsia, the most severe form of this pregnancy-induced hypertension (PIH), is characterized by convulsions that may lead to coma. Both conditions present serious health risks to mother and fetus. PIH can retard fetal growth and cause the placenta to separate from the uterus, resulting in stillbirth. The former practice of restricting salt and prescribing diuretics for the edema of PIH is not recommended, and pregnant women are advised to use salt to taste. Pregnant women who consume calcium supplements of 1,500 to 2,000 milligrams per day may be at lower risk of PIH.[26]

• **Diabetes.** Infants born to women with diabetes are at greater risk for prematurity, congenital defects, excessively high birthweight, and respiratory distress syndrome.[27] Metabolic control of diabetes before and throughout pregnancy is critical. In some women, pregnancy can alter carbohydrate metabolism and precipitate a condition known as gestational diabetes mellitus. The abnormal blood glucose levels seen during pregnancy return to normal postpartum for about two-thirds of women diagnosed with the condition. Risk factors include age 35 or older, previous history of gestational diabetes, obesity, and a family history of diabetes.

Some women with gestational diabetes have the classic symptoms of diabetes—increased thirst, hunger, urination, weakness—but other women have no warning symptoms of the condition. For this reason, pregnant women at risk for developing gestational diabetes mellitus are screened for gestational diabetes between the 24th and 28th weeks of gestation (see Table 13-8).[28]

Adolescent Pregnancy

More than a million teenagers become pregnant in the United States each year—one out of every five babies is born to a teenager—and more than a tenth of these mothers are under age 15. The complex social, emotional, and physical factors involved make teen pregnancy one of the most challenging situations for nutrition counseling. According to a position paper from the American Dietetic Association, pregnant adolescents are nutritionally at risk and require intervention early and throughout pregnancy.[29] Medical and nutritional risks are particularly high when the teenager is within 2 years of menarche (usually 15 years of age or younger).[30] Risks include higher rates of PIH, iron-deficiency anemia, premature birth, stillbirths, LBW infants, and prolonged labor in pregnant teens than in older women. In addition, mothers under 15 years of age bear more babies who die within the first year than do any other age group.

The increased energy and nutrient demands of pregnancy place adolescent girls, who are already at risk for nutritional problems, at even greater risk. To support the needs of both mother and infant, adolescents are encouraged to strive for pregnancy weight gains at the upper end of the ranges recommended for pregnant women (see Table 13-5). Those who gain between 30 and 35 pounds during pregnancy have lower risks of delivering LBW infants.[31] Adequate nutrition can substantially improve the course and outcome of adolescent pregnancy.[32]

TABLE 13-8 *Screening and Diagnosis of Gestational Diabetes*

Detection and Screening

Screening should be performed between the 24th and 28th weeks of gestation in women meeting one or more of the following criteria:

1. ≥ 25 years of age
2. < 25 years of age and obese
3. Family history of diabetes in first-degree relatives
4. Member of an ethnic/racial group with a high prevalence of diabetes (Hispanic, American Indian, Asian, African American, or Pacific Islander)

Screening Test

1. Administer 50 g oral glucose
2. After 1 hour measure plasma glucose level
3. If plasma glucose > 140 mg/dL, recommend diagnostic glucose tolerance test

Diagnosis: Glucose Tolerance Test (GTT)

1. Measure fasting plasma glucose
2. Administer 100 g oral glucose load
3. Measure plasma glucose at 1, 2, and 3 hours
4. If two or more plasma glucose values exceed the following, a diagnosis of gestational diabetes is made.

Time/Plasma Glucose	Diagnostic of Diabetes Mellitus (mg/dL)
Fasting	105
100 g oral glucose test	
1 hour	190
2 hour	165
3 hour	145

Source: Adapted from American Diabetes Association, Position Statement: Gestational diabetes mellitus, *Diabetes Care* 21 (1998): 560–561.

Nutrition Assessment In Pregnancy

Nutrition assessment and monitoring during pregnancy can be divided into three categories: preconception care, the initial prenatal visit, and subsequent prenatal visits.[33] Since a woman's nutrition status and lifestyle habits prior to pregnancy can influence the outcome of pregnancy, these are important factors to consider prior to pregnancy. Nutritional risk factors that may be present at the start of pregnancy are listed in Table 13-9. Preconception care should ideally be available to all women. It should include nutrition assessment, nutrition counseling, and appropriate supplementation and referral to correct nutritional problems existing prior to conception. Prepregnancy weight-for-height should be categorized using BMI so that appropriate weight gain can be recommended for pregnancy (refer to Table 13-5).

Assessment of pregnant teens follows the general pattern for assessment of older pregnant women (see Table 13-10). Assessment issues include acceptance of the pregnancy, food resources and food preparation facilities, body image, living situation, relationship with the father of the infant, peer relationships, nutrition status, prenatal care, nutrition attitude and knowledge, preparation for child feeding,

Personal Characteristics (Cannot Change)	Personal Habits, Living Situation (Seek to Change)	Chronic Preexisting Maternal Medical Problems (Screen, Treat)	Obstetric History (Screen, Prevent)	Current/Potential Pregnancy-Induced Problems (Screen, Prevent, Treat)
Age (years)	Low socioeconomic status	Hypertension	Low birthweight	Anemia
Adolescent (< 15)	Smoking	Type I diabetes	Macrosomia	Iron deficiency
Older woman (> 35)	Alcohol/drug use	Type II diabetes	Stillbirth	Folate deficiency
Family history	Poor diet	Heart disease	Abortion	PIH
Diabetes	Eating disorders	Pulmonary disease	Fetal anomalies	Gestational diabetes
Heart disease	Malnutrition	Renal disease	High parity	Weight gain
Hypertension	Obesity	Maternal PKU	Multipara	Inadequate
PKU	Underweight	AIDS		Excessive
History of poor	Sedentary lifestyle			
obstetric outcome	Restricted diet			

Source: Adapted with permission from B. Worthington-Roberts and S. R. Williams, *Nutrition in Pregnancy and Lactation*, 5th ed. (St. Louis: Times Mirror/Mosby, 1993) p. 240.

TABLE 13-9
Nutritional Risk Factors in Pregnancy

financial resources, continuation of education, day care, and knowledge and attitudes of infant feeding methods.[34] See Table 13-11 for sample questions to include in a nutrition interview with a pregnant woman.

The nutrition status of all women should be assessed at their initial prenatal visit. This assessment should include the following components:[35]

- **Dietary measures.** Diet history, including food habits, attitudes, and folklore; allergies; use of vitamin and mineral supplements; and lifestyles (for example, substance abuse; existence of pica).
- **Clinical measures.** Obstetric history, including outcome of previous pregnancies, interval between pregnancies, and history of problems during course of previous pregnancies (PIH, gestational diabetes, iron-deficiency anemia, inadequate or excessive weight gain pattern).
- **Anthropometric measures.** Measurement of weight-for-height; the weight should be recorded and plotted on a weight gain grid (refer back to Figure 13-3). Skinfold measures are not recommended for routine assessment.[36]
- **Laboratory values.** Screening for anemia by hematocrit and/or hemoglobin. A urine analysis for ketones, glucose, and protein spillage may be ordered for women with gestational diabetes or preeclampsia.

During each subsequent prenatal visit, weight gain should be monitored and the pattern of weight gain evaluated. Screening for anemia should be repeated on at least one other occasion during pregnancy. Routine assessment of dietary practices is recommended for all women so that their need for an improved diet or vitamin or mineral supplementation can be evaluated. The nutrition status of women in high-risk categories should be reevaluated at each visit.

Poor dietary practices should be improved by appropriate interventions. These may include general nutrition education, individualized diet counseling, and referral

TABLE 13-10

Protocol for Nutrition Assessment in Pregnancy

Initial Evaluation

Review clinical data:
Height and weight
Gynecological age (adolescents)
Physical signs of health
Expected delivery date

Review laboratory data:
Hematocrit or hemoglobin
Urinalysis

Assess attitude about and acceptance of prepregnancy weight and feelings about weight gain during pregnancy.
Assess intake patterns using dietary methodology best suited to client and professional.
Make preliminary assessment of food resources and refer to supportive agencies if necessary.
Check for nausea and vomiting and suggest possible remedy.
Discuss supplemental vitamins and minerals.
Make initial plan that sets priorities for issues.
Determine client's understanding of relation between nutrition and health.

Second Visit

Check on referrals to other agencies.
Discuss results of initial evaluation and suggest any changes necessary in dietary patterns (use printed materials as appropriate).
Do any further investigations when necessary:
Laboratory studies
For specific diagnosis of anemia:

- Protoporphyrin heme or serum ferritin
- Serum or red cell folate
- Serum vitamin B_{12}

to food assistance programs (for instance, Special Supplemental Nutrition Program for Women, Infants, and Children [WIC] and The Food Stamp Program) or to programs that promote improved food acquisition or preparation practices (such as the Expanded Food and Nutrition Education Program—EFNEP).

Healthy Babies

The growth of infants directly reflects their nutritional well-being and is the major indicator of their nutrition status. A baby grows faster during the first year of life than ever again; its birthweight doubles during the first 4 to 6 months and triples by the end of the first year. Adequate nutrition during infancy is critical to support this rapid rate of growth and development. Clearly, from the point of view of nutrition, the first year is the most important year of a person's life. This section provides a brief overview of nutrient requirements, current recommendations and health objectives for feeding healthy infants, and the relationship between infant feeding and selected pediatric nutrition issues.

TABLE 13-10

Continued

Second Visit—continued

Further probing of dietary habits if necessary
Monitor weight gain; discuss projected weight gain for following visit and total for gestation.
Assess and address issues affecting nutrition status in order of priority for the individual:

- Activity level
- Appetite changes
- Pica, food cravings, and aversions
- Allergies/food intolerances
- Supplementation practices

Subsequent Visits

Monitor and support appropriate weight gain; include discussion of fitness and encourage safe
exercise.
Continue to address issues affecting nutrition status.
Check for heartburn, small food-intake capacity, and elimination problems; suggest dietary
interventions.
Begin preliminary discussion and comparison of advantages/disadvantages of breastfeeding and
formula feeding.

Final Prenatal Visit(s)

Discuss infant feeding.
If breastfeeding is chosen, provide preliminary guidance about breastfeeding practices.
If formula feeding is chosen, discuss product selection and preparation; define important details
about feeding techniques.

Postpartum Visits

Help client understand safe methods of managing weight following delivery.
Review infant feeding practices and infant growth; provide assistance when problems are
identified.

Source: Adapted with permission from L. K. Mahan and J. M. Rees, *Nutrition in Adolescence* (St.
Louis: Times Mirror/Mosby, 1984).

Nutrient Needs and Growth Status in Infancy

The infant's rapid growth and metabolism demand an adequate supply of all essential nutrients. Because of their small size, infants need smaller total amounts of the nutrients than adults do, but based on body weight, infants need over twice as much of many of the nutrients. After 6 months, energy needs increase less rapidly as the growth rate begins to slow down, but some of the energy saved by slower growth is spent on increased activity.

Anthropometric Measures in Infancy

Anthropometric measurements that are routinely obtained in the examination of infants include length, weight, and head circumference. These measures assess physical size and growth.

Infants should be weighed nude, using a table model beam scale that allows the infant to lie or sit. Length should be measured in the recumbent position on a

TABLE 13-11

Questions for a Client-Focused Nutrition Interview in Pregnancy

What you eat and some of the lifestyle choices you make can affect your nutrition and health now and in the future. Your nutrition can also have an important effect on your baby's health. Please answer these questions by circling the answers that apply to you.

Eating Behavior

1. Are you frequently bothered by any of the following? (circle all that apply)

   Nausea   Vomiting   Heartburn   Constipation

2. Do you skip meals at least 3 times a week?   No   Yes

3. Do you try to limit the amount or kind of food you eat to control your weight?   No   Yes

4. Are you on a special diet now?   No   Yes

5. Do you avoid any foods now for health or religious reasons?   No   Yes

Food Resources

6. Do you have a working stove?   No   Yes

   Do you have a working refrigerator?   No   Yes

7. Do you sometimes run out of food before you are able to buy more?   No   Yes

8. Can you afford to eat the way you should?   No   Yes

9. Are you receiving any food assistance now?   No   Yes
   (Circle all that apply):

   Food stamps   School breakfast   School lunch
   WIC   Donated food/commodities   CSFP

10. Do you feel you need help in obtaining food?   No   Yes

Food and Drink

11. Which of these did you drink yesterday?
    (Circle all that apply):

    Soft drinks   Coffee   Tea   Fruit drink
    Orange juice   Grapefruit juice   Other juices   Milk
    Kool-Aid   Beer   Wine   Alcoholic drinks
    Water   Other beverages (list) _____

measuring board that has a fixed headboard and movable footboard attached at right angles to the surface.

Head circumference measures can confirm that growth is proceeding normally or help detect protein-energy malnutrition (PEM) and evaluate the extent of its impact on brain size. To measure head circumference, a nonstretchable tape is placed around the largest part of the infant's head: just above the eyebrow ridges, just above the point where the ears attach, and around the occipital prominence at the back of the head.[37] The head circumference percentile should be similar to the infant's weight and length percentiles.

The National Center for Health Statistics (NCHS) growth charts, such as those shown in Figure 13-4, are used to analyze measures of growth status in infants. A single plotting is used to assess how an infant ranks in comparison to other infants of the same age and sex in the United States. If a measurement is less than the 10th percentile, it should be checked for accuracy and further evaluated. Height-for-age less than the 5th percentile reflects chronic undernutrition. Weight-for-height (length)

TABLE 13-11

Continued

Food and Drink—continued

12. Which of these foods did you eat yesterday?
    (Circle all that apply):

Cheese	Pizza	Macaroni and cheese
Yogurt	Cereal with milk	

Other foods made with cheese (such as tacos, enchiladas, lasagna, cheeseburgers)

Corn	Potatoes	Sweet potatoes	Green salad
Carrots	Collard greens	Spinach	Turnip greens
Broccoli	Green beans	Green peas	Other vegetables
Apples	Bananas	Berries	Grapefruit
Melon	Oranges	Peaches	Other fruit
Meat	Fish	Chicken	Eggs
Peanut butter	Nuts	Seeds	Dried beans
Cold cuts	Hot dog	Bacon	
Cake	Cookies	Doughnut	Sausage
Chips	French fries	Pastry	

Other deep-fried foods, such as fried chicken or egg rolls

Bread	Rolls	Rice	
Noodles	Spaghetti	Tortillas	Cereal

Were any of these whole grain?   No   Yes

13. Is the way you ate yesterday the way you usually eat?   No   Yes

Lifestyle

14. Do you exercise for at least 30 minutes on a regular basis (3 times a week
    or more)?   No   Yes

15. Do you ever smoke cigarettes or use smokeless tobacco?   No   Yes

16. Do you ever drink beer, wine, liquor, or any other alcoholic beverages?   No   Yes

17. Which of these do you take?
    (Circle all that apply):
    Prescribed drugs or medications
    Over-the-counter products (such as aspirin, Tylenol, antacids, or vitamins)
    Street drugs (such as marijuana, speed, downers, crack, or heroin)

Source: Reprinted with permission from *Nutrition During Pregnancy and Lactation: An Implementation Guide.* Copyright 1992 by the National Academy of Sciences. Courtesy of the National Academy Press, Washington, D.C.

less than the 5th percentile may reflect acute malnutrition. Excessive weight-for-length (above the 95th percentile) indicates overweight. An infant whose weight-for-length falls below the 5th percentile is classified as **failure-to-thrive,** and those below the 10th percentile are suspect for failure-to-thrive. These infants require further evaluation and care. Under normal conditions, the growth rate usually varies within two percentiles. Greater variation indicates the possibility of inadequate nutrition and needs to be evaluated further.

Failure-to-thrive Inadequate weight gain of infants.

Breastfeeding: Promotion and Recommendations

Breastfeeding offers both emotional and physical health advantages. Emotional bonding is facilitated by many events and behaviors of mother and infant during the early months and years; one of the first can be breastfeeding.

FIGURE 13-4 *Examples of NCHS Growth Charts*

The percentile of an infant's weight, height, or head circumference is the point at which the age of the infant on the axis and the measurement along the abscissa intersect. To plot a weight measurement on a percentile graph:

1. Locate the infant's age along the horizontal axis on the bottom or top of the chart.

2. Locate the infant's weight in pounds or kilograms along the vertical axis on the lower left or right side of the chart.

3. Mark the chart where the age and weight lines intersect, and read off the percentile.

4. Assess length, height, or head circumference using the appropriate chart and the same procedure.

The two charts shown here are used for girls from birth to 36 months. The first chart gives percentiles for length and weight for age; the other, percentiles for head circumference for age and weight for length.

Check out the National Center for Health Statistics website (www.cdc.gov/nchswww) for updated versions of these growth charts.

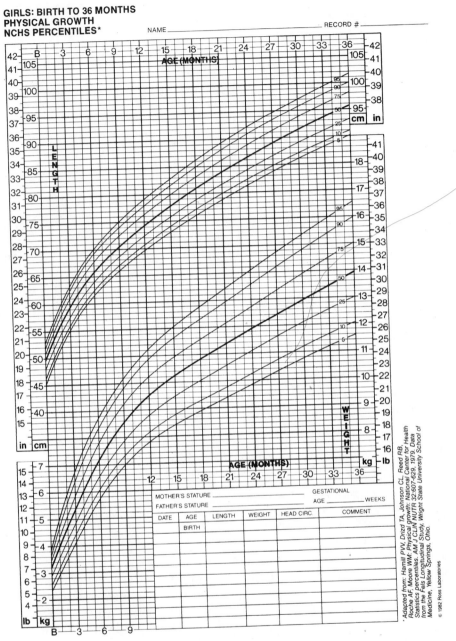

GIRLS: BIRTH TO 36 MONTHS
PHYSICAL GROWTH
NCHS PERCENTILES*

During the first 2 or 3 days of lactation, the breasts produce colostrum, a premilk substance containing antibodies and white cells from the mother's blood. Colostrum and breast milk both contain the bifidus factor that favors the growth of the "friendly" bacteria *Lactobacillus bifidus* in the infant's digestive tract so that other, harmful bacteria cannot grow there. Breast milk also contains the powerful antibacterial agent, lactoferrin, and other factors including several enzymes, several

**GIRLS: BIRTH TO 36 MONTHS
PHYSICAL GROWTH
NCHS PERCENTILES***

NAME _____ RECORD # _____

FIGURE 13-4

Continued

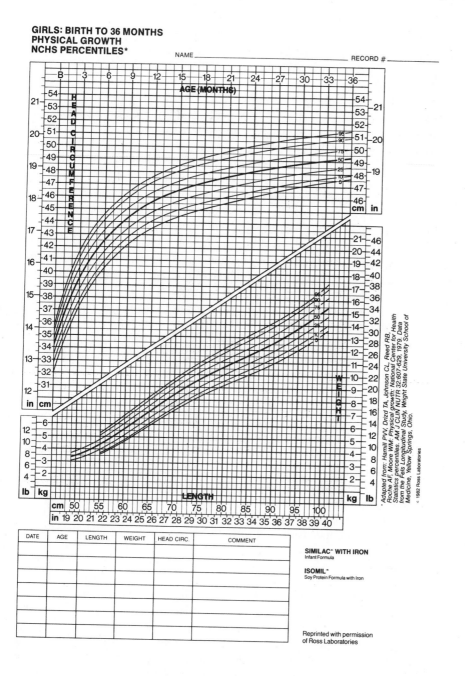

DATE	AGE	LENGTH	WEIGHT	HEAD CIRC.	COMMENT

SIMILAC® WITH IRON
Infant Formula

ISOMIL®
Soy Protein Formula with Iron

Reprinted with permission
of Ross Laboratories

*Adapted from: Hamill PVV, Drizd TA, Johnson CL, Reed RB, Roche AF, Moore WM: Physical growth: National Center for Health Statistics percentiles. AM J CLIN NUTR 32:607-629, 1979. Data from the Fels Longitudinal Study, Wright State University School of Medicine, Yellow Springs, Ohio.

© 1982 Ross Laboratories

hormones (including thyroid hormone and prostaglandins), and lipids that help protect the infant against infection.

Breast milk is also tailor-made to meet the nutrient needs of the young infant. Breastfed infants usually require no nutrient supplements except for vitamin D and fluoride. At 4 to 6 months, infants may require an iron supplement, depending on food intake.

Breastfeeding provides other benefits as well. It protects against allergy development during the vulnerable first few weeks, the act of suckling favors normal tooth and jaw alignment, and breastfed babies are less likely to be obese because they are less likely to be overfed. A woman who wants to breastfeed can derive justification and satisfaction from all these advantages.

These attributes, along with the convenience and lower cost of breastfeeding, have led many organizations and medical experts to encourage breastfeeding for all normal full-term infants. Breastfeeding in the United States declined after World War II to a low of about 25 percent of infants in 1970. Breastfeeding rates more than doubled in the late 1970s and early 1980s and then declined slightly in the late 1980s. By 1994, 57.4 percent of mothers initiated breastfeeding, and 19.7 percent were still breastfeeding babies who were 6 months old (see Figure 13-5). The *Healthy People 2000* goal for breastfeeding is to increase the incidence of breastfeeding to 75 percent at discharge from the hospital and 50 percent at 5 to 6 months.[39] Analysis of data from the Ross Laboratories Mothers Survey indicates that breastfeeding rates continue to be the highest among women who are older, well educated, relatively affluent, and/or living in the western United States (see Figure 13-5). Among those least likely to breastfeed are women who are low income, black, less than age 20, and/or living in the southeastern United States.[40]

A number of barriers to achieving the nation's health objective of increasing the incidence of breastfeeding have been identified. These include lack of knowledge, an absence of work policies and facilities that support lactating women (for example, extended maternity leave, part-time employment, facilities for pumping breast milk or breastfeeding, and on-site child care); the portrayal of bottle feeding rather than breastfeeding as the norm in American society; and the lack of breastfeeding incentives and support for low-income women.[41]

One example of a successful approach to increasing breastfeeding rates in low-income, urban populations is the peer counseling method promoted by the La Leche League International.[42] The WIC program has initiated breastfeeding promotion projects among low-income women using the peer counselor method.[43] With this approach, peer support counselors are trained to provide culturally appropriate interventions for initiating and maintaining breastfeeding in their communities. The peer counselor is paired with an expectant or new mother for individual assistance and informal discussions in their neighborhood.[44] Results of similar grassroots approaches to breastfeeding promotion are encouraging, with both the rate and duration of breastfeeding among the women in the programs showing significant increases.[45]

Focus groups can be useful in designing breastfeeding promotion projects.[46] For example, the project "Best Start: Breastfeeding for Healthy Mothers, Healthy Babies" was organized in the southeastern United States by a coalition of nutrition and public health officials concerned about declining rates of breastfeeding in the region.[47] Focus groups were especially helpful in exploring topics related to breastfeeding among participants who were ambivalent or undecided. Research findings pointed to the need for a carefully coordinated campaign that utilized a combination of strategies to improve the image of breastfeeding and help women overcome the barriers they perceived to breastfeeding. Based on information generated from the focus group interviews, guidelines were formulated for program development (see the boxed insert on page 406 for the guidelines).

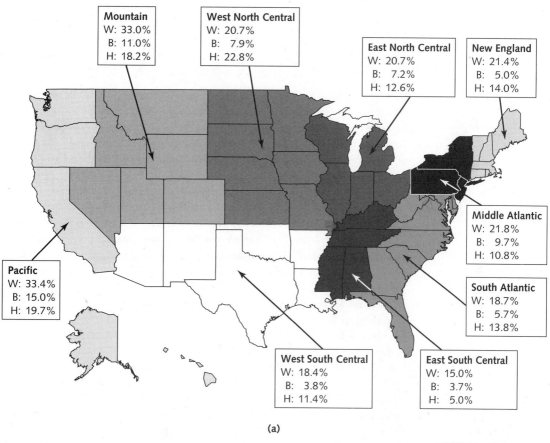

Mountain
W: 33.0%
B: 11.0%
H: 18.2%

West North Central
W: 20.7%
B: 7.9%
H: 22.8%

East North Central
W: 20.7%
B: 7.2%
H: 12.6%

New England
W: 21.4%
B: 5.0%
H: 14.0%

Pacific
W: 33.4%
B: 15.0%
H: 19.7%

Middle Atlantic
W: 21.8%
B: 9.7%
H: 10.8%

South Atlantic
W: 18.7%
B: 5.7%
H: 13.8%

West South Central
W: 18.4%
B: 3.8%
H: 11.4%

East South Central
W: 15.0%
B: 3.7%
H: 5.0%

(a)

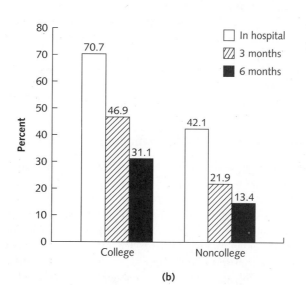

(b)

FIGURE 13-5

Percentage of Mothers Breastfeeding

(a) Breastfeeding rates at 5 to 6 months postpartum, by census region and ethnic background. *W,* white; *B,* black; *H,* Hispanic.

Source: B. Worthington Roberts and S. R. Williams, *Nutrition Throughout the Life Cycle,* 3rd ed. (St. Louis: Times Mirror/Mosby, 1996); and Institute of Medicine, National Academy of Sciences: *Nutrition During Lactation,* Washington, D.C., 1991, National Academy Press.

(b) Percentage breastfeeding by education.

Source: Office of Maternal and Child Health, Public Health Service. U.S. Department of Health and Human Services. *Child Health USA '89* (Washington, D.C.: Department of Health and Human Services, 1989).

Using Focus Group Findings

Using comments made by women participating in focus groups, planners of the Best Start breastfeeding promotion project formulated the following guidelines:

- **Campaign tone.** The tone should be emotional to reflect the strong feelings women attach to their aspirations for their children and themselves as mothers.
- **Message design.** Educational messages should be succinct and easily understood to counteract the mistaken belief that breastfeeding is complicated and requires major lifestyle changes. Promotional messages should emphasize confidence and the pride breastfeeding mothers gain from nursing.
- **Spokespersons.** Celebrities are not perceived as credible sources of advice on infant feeding. Also, most focus group respondents find it difficult to identify with wealthy women featured on many of the pamphlets, posters, and other breastfeeding promotion materials used in health departments. Whenever possible, women featured in the materials should be of the same economic, ethnic, and age groups as those targeted or should not have a clear class affiliation. Visual images in print and broadcast materials should communicate modernity and confidence.
- **Educational approaches.** Educational strategies and materials need to be redesigned so that they no longer reinforce women's perceptions of breastfeeding as difficult. The emphasis on being healthy and relaxed and following special dietary guidelines needs to be replaced with reassurance that most women produce sufficient quantities of highly nutritious breast milk despite variations in diet, stress levels, and health status.
- **Professional training.** Motivational and training materials are needed to counter professionals' mistaken belief that their economically disadvantaged clients are not interested in breastfeeding and do not value health professionals' advice. Counseling strategies need to be redesigned to address the special needs of low-income women. Of special concern are recommendations for overcoming clients' lack of confidence and enabling them to realize their aspirations as women and mothers.
- **Program activities/components.** A variety of mutually reinforcing activities are needed to reach social network members who influence women's infant feeding decisions and create hospital, home, and community environments conducive to lactation.

Source: Adapted from C. Bryant and D. Bailey, The use of focus group research in program development. Unpublished manuscript, Lexington, KY, 1989.

Other Infant Feeding Recommendations

Like the breastfeeding mother, the mother who offers formula to her baby has reasons for making her choice, and her feelings should be honored. Infant formulas are manufactured to approximate the nutrient composition of breast milk. The immunologic protection of breast milk, however, cannot be duplicated.

Many mothers breastfeed at first and then wean within the first one to six months. When a woman chooses to wean her infant during the first six months of life, it is imperative that she shift to *formula*, not to plain milk of any kind—whole, low-fat, or nonfat. Only formula contains enough iron (to name but one of many factors) to support normal development in the baby's first months of life. National and international standards have been set for the nutrient content of infant formulas.

For infants with special problems, many variations of infant formulas are available. Special formulas based on soy protein are available for infants allergic to milk protein, and formulas with the lactose replaced can be used for infants with lactose intolerance. Whole cow's milk is not recommended during the first year of life, according to the American Academy of Pediatrics (AAP).[48]

Solid foods should normally be added to a breastfed baby's diet when the baby is between six months and one year of age. A baby who is formula fed might be started on solid foods between 4 and 6 months, depending on readiness.[49]

Solids should not be introduced too early, because infants are more likely to develop allergies to them in the early months. But all babies are different, and the program of additions should depend on the individual baby, not on any rigid schedule. Table 13-12 presents a suggested sequence for feeding infants.

Primary Nutrition-Related Problems of Infancy

Iron deficiency and food allergies are two of the most significant nutrition-related problems of infants.

Iron Deficiency. Iron deficiency remains a prevalent nutritional problem in infancy, although it has declined in recent years largely because of the increasing use of iron-fortified formulas.[50] The use of cow's milk earlier than recommended in infancy can cause iron deficiency because of its poor iron content and the potential to cause gastrointestinal blood loss in susceptible infants.[51] Other factors contributing to iron deficiency in infancy include breastfeeding for more than six months without providing supplemental iron, feeding infant formula not fortified

TABLE 13-12 *First Foods for the Infant*

Age (Months)	Addition
0 to 4	Feed breast milk or infant formula
4 to 6	Iron-fortified rice cereal, followed by other cereals (for iron; baby can swallow and can digest starch now)*
	Pureed vegetables and/or fruits and their juices, one by one (perhaps vegetables before fruits so that the baby will learn to like their less sweet flavors)†
6 to 8	Mashed vegetables and fruits and their juices, infant breads and crackers
8 to 10	Protein foods (soft cheeses, yogurt, tofu, mashed cooked beans, finely chopped meat, fish, chicken, egg yolk), toast, teething crackers (for emerging teeth), soft-cooked vegetables and fruit
10 to 12	Whole egg (allergies are less likely now), whole milk (at 1 year)

*Later, other cereals can be introduced, but they should still be iron-fortified varieties.
†All baby juices are fortified with vitamin C. Orange juice may cause allergies; apple juice may be a better juice to feed first. Dilute juices with water and offer in a cup to prevent nursing bottle syndrome.

Source: Adapted from Committee on Nutrition, American Academy of Pediatrics, *Pediatric Nutrition Handbook*, 3rd ed., ed. L. A. Barness (Elk Grove Village, IL: American Academy of Pediatrics, 1993), pp. 23–33.

with iron, the infant's rapid rate of growth, low birthweight, and low socioeconomic status. To prevent iron deficiency, the AAP recommends that infants be fed breast milk or iron-fortified formula for the first year of life, with appropriate foods added between the ages of 4 and 6 months as shown in Table 13-12

Food Allergies. Genetics is probably the most significant factor affecting an infant's susceptibility to food allergies.[52] Nevertheless, food allergies are much less prevalent in breastfed babies than in formula-fed infants. At-risk infants can be identified from elevated cord blood levels of Immunoglobulin E (IgE) or by a family history. Breast milk is recommended for infants allergic to cow's milk protein and is preferable to soy or goat's milk formulas, because infants are sometimes intolerant of these proteins as well. To reduce the risk of food sensitivity or allergic reactions to other foods, new foods should be introduced singly to facilitate prompt detection of allergies. For example, if a cereal causes irritability due to skin rash, digestive upset, or respiratory discomfort, discontinue its use before going on to the next food. Several days should elapse between the introduction of each new food to allow time for clinical symptoms to appear.

Domestic Maternal and Infant Nutrition Programs

Nutrition plays a vital role in the outcome of pregnancy as well as the growth and development of infants. A stable base of essential programs and services is required to meet maternal and infant health care needs. This section describes the nutrition programs and related services available to meet the demands of the vulnerable life-cycle periods of pregnancy and infancy (see Table 13-13).

Nutrition Programs of the U.S. Department of Agriculture

Several programs of the U.S. Department of Agriculture (USDA) directly or indirectly provide nutrition support during pregnancy and infancy.

Food Stamp Program. The Food Stamp Program (FSP) is designed to improve the diets of people with low incomes by providing coupons to cover part or all of their household's food budget. Chapter 12 provided an overview of this program. As explained previously, program participants can use the food stamp coupons to buy food in any retail store that has been approved by the Food and Nutrition Service to accept and redeem them. For many economically disadvantaged women, the FSP is the major means by which they are able to purchase adequate diets for their families. The Select Panel for the Promotion of Child Health reported that food stamp users purchase more nutritious foods per dollar spent on food than eligible households that do not participate in the program.[53]

WIC Farmers' Market Nutrition Program. The WIC Farmers' Market Nutrition Program (FMNP) was created to provide fresh, nutritious fruits and vegetables to WIC participants, and to expand the awareness and use of farmers' markets by consumers. Eligible recipients receive FMNP coupons. A study reported by the Food Research and Action Center showed that WIC women who received FMNP coupons increased both fruit and vegetable consumption by about 5 percent com-

The Program Spotlight on page 412 provides information about the WIC Program.

Program	Participants	Benefits
Food Stamps	Anyone with income < 130% of poverty level	Increased ability to purchase food
WIC*	Pregnant women, postpartum and lactating women, infants, and children up to 5 years of age	Vouchers to purchase healthy foods or direct food supplements, nutrition education, and referral to health services
FMNP	Persons eligible for WIC	Increased fruit and vegetable consumption
Commodity Supplemental Foods	Pregnant and postpartum women, infants, and children < 6 years of age with incomes ≤ 185% of poverty level	Monthly food package of fruits, vegetables, meats, infant formula, beans, and other available foods
EFNEP	Persons with incomes ≤ 125% of poverty level who have children under 19 years of age	Nutrition education
Medicaid	Anyone with income < 133% of poverty level	Complete health care
Healthy Start (Medicaid)	Pregnant women with incomes < 185% of poverty level; certain high risk pregnancies	Prenatal and postpartum care
EPSDT (HEALTH CHEK)	Infants, children, and adolescents up to 18 years of age	Health screening: dental checks, health education, hearing, vision

TABLE 13-13 *Federal Nutrition and Health Care Programs That Assist Mothers and Their Infants*

*WIC = Special Supplemental Nutrition Program for Women, Infants, and Children; FMNP = Farmers' Market Nutrition Program; EFNEP = Expanded Food and Nutrition Education Program; EPSDT = Early Periodic Screening, Diagnosis, and Treatment.

Source: Adapted from M. K. Mitchell, *Nutrition Across the Life Span* (Philadelphia, PA: W. B. Saunders, 1997).

pared with WIC women who did not receive coupons. The FMNP participants also patronized farmers' markets more than nonrecipients, even after they were no longer eligible for WIC.[54]

Commodity Supplemental Food Program. The Commodity Supplemental Food Program (CSFP) is a direct food distribution program providing supplemental foods and nutrition education. The CSFP provides supplemental foods to infants and children and to pregnant, postpartum, and breastfeeding women with low incomes who are vulnerable to malnutrition and live in approved project areas. Recipients may not participate in both WIC and CSFP. The USDA purchases the foods for distribution through state agencies on a monthly basis.

Expanded Food and Nutrition Education Program (EFNEP). The Expanded Food and Nutrition Education Program (EFNEP) is one of two federally funded programs designed specifically for nutrition education. (The other is the Nutrition Education and Training [NET] program, described in Chapter 14.) The EFNEP is directed at low-income families and is administered by the USDA Extension Service. It was authorized in 1968 to provide food and nutrition education to homemakers with young children. The program is implemented by trained nutrition aides from the local community under the supervision of county Cooperative Extension Home Economists.

These paraprofessionals work to develop a one-to-one relationship with disadvantaged homemakers enrolled in the program. In recent years, multimedia strategies have been tested to expand the scope of nutrition education for this population.[55]

Nutrition Programs of the U.S. Department of Health and Human Services

The U.S. Department of Health and Human Services (DHHS) also sponsors several programs that are concerned with health and nutrition status during pregnancy and infancy.

The state programs for Children with Special Health Care Needs (CSHCN) were formerly known as Crippled Children's Services (CC).

Title V Maternal and Child Health Program. Enacted in 1935, Title V of the Social Security Act is the only federal program concerned exclusively with the health of mothers, infants, and children. It provides federal support to the states to enhance their ability to "promote, improve, and deliver" maternal, infant, and child health (MCH) care and programs for children with special health care needs (CSHCN), especially in rural areas and regions experiencing severe economic stress. The aim of Congress in passing this legislation was to improve the health of mothers, infants, and children in areas where the need was greatest.

The states are allocated Title V MCH funds to be used for (1) services and programs to reduce infant mortality and improve child and maternal health; and (2) services, programs, and facilities to locate, diagnose, and treat children who have special health care needs (for example, chronic medical conditions) or are at risk of physical or developmental disabilities.[56] The Title V MCH program provides for nutrition assessment, dietary counseling, nutrition education, and referral to food assistance programs for infants, preschool and school-aged children, children with special health care needs, adolescents, and women of childbearing age. Title V helps create healthy communities by working with local groups to identify and address their local health needs ranging from teen pregnancy to low immunization rates. It also supports training in nutrition for health and nutrition professionals who are involved in developing nutrition services.[57]

On several occasions, significant amendments have been added to the original law:

- **1963.** A series of amendments authorized grants to state and local health departments for Maternal and Infant Care (MIC) projects aimed mainly at reducing mental retardation and infant mortality through improved prenatal, perinatal, and postpartum care.
- **1965.** Amendments authorized special project grants to provide comprehensive health services to children and youth through state and local health agencies and other public or nonprofit organizations.
- **1967.** An amendment provided for special grants for dental services for children. Other amendments linked CSHCN programs and Medicaid in the provision of Early Periodic Screening, Diagnosis, and Treatment (EPSDT), described below.

The Title V MCH program is administered federally by the Bureau of Maternal and Child Health and Resources Development in the Health Resources and

Services Administration of the Public Health Service. Administration of the MCH program is the responsibility of the MCH unit within each state's health agency. Most of the CSHCN programs are also administered through state health agencies; some, however, are delivered through other state agencies, such as welfare departments, social service agencies, or state universities. Under the law, each state is required to operate a "program of projects" in each of five areas: (1) maternity and infant care, (2) intensive infant care, (3) family planning, (4) health care for children and youth, and (5) dental care for children.[58]

States have varying degrees of control over local use of Title V funds. Most state MCH funds are used to support well-child checkups, immunization programs, vision and hearing screenings, and school health services, as well as other programs. CSHCN program funds are used to provide direct services to children with special health care needs through local clinics and/or fee-for-service arrangements with physicians in private practice. Regardless of the method of delivering CSHCN programs, a multidisciplinary approach to providing health care is used by nearly all states.[59]

The Select Panel for the Promotion of Child Health reported that Title V program efforts have resulted in significant improvements in maternal and child health. The program is believed to have contributed to the decline in infant and maternal mortality, the reduction of disability in handicapped children, and a general improvement in the health status of children.[60]

Medicaid and EPSDT. Congress created the Medicaid program in 1965 to ensure financial access to health care for the economically disadvantaged. It was enacted through Title XIX of the Social Security Act. The Medicaid program is a state-administered entitlement program built upon the welfare model. It is constructed as a medical assistance program that reimburses providers for specific services delivered to eligible recipients. Under amendments enacted in 1967, the states are required to provide EPSDT (HEALTH CHEK) as a mandatory Medicaid service. As outlined by Congress, the purpose of EPSDT is to improve the health status of children from low-income families by providing health services not typically found under the current Medicaid program. EPSDT requires an assessment of the nutrition status of eligible children and their referral for treatment.[61] Whereas Medicaid is mainly a provider payment program for medical services, EPSDT regulations stipulate that states must develop protocols for identifying eligible children, informing them of the EPSDT program, and ensuring that referral, preventive, and treatment services are made available to participants.

The federal administration of Medicaid is the responsibility of the Health Care Financing Administration (HCFA) within the DHHS. The HCFA's Office of Child Health administers EPSDT at the federal level.

Medicaid and EPSDT have been credited with increasing the access of low-income women, infants, and children to health care services. EPSDT, by virtue of its aggressive preventive strategy, has been effective in improving child health.[62] However, strict eligibility requirements and federal statutory policies have to some extent limited the ability of these programs to reach their full potential.[63]

Continued on page 418.

Program Spotlight *The WIC Program*

In 1969, the White House Conference on Food, Nutrition, and Health recommended that special attention be given to the nutritional needs of pregnant and breastfeeding women, infants, and preschool children. As a result, WIC was authorized in 1972 by PL 92-433 as an amendment to the Child Nutrition Act of 1966. The legislation states that the WIC program is to "serve as an adjunct to good health care, during critical times of growth and development." To encourage earlier and more frequent utilization of health services, federal regulations mandate that local agencies may qualify as WIC sponsors only if they can make health care services available to WIC enrollees.[1]

The WIC program is based on two assumptions. One is that inadequate nutritional intakes and health behaviors of low-income women, infants, and children make them vulnerable to adverse health outcomes. The other is that nutrition intervention at critical periods of growth and development will prevent health problems and improve the health status of participants.

WIC is federally funded but administered by the states. Cash grants are made to authorized agencies of each state and to officially recognized Indian tribes or councils, which then provide WIC services through local service sites. Priority for the creation of local programs is given to areas whose populations need benefits most, based on high rates of infant mortality, low birthweight, and low income.

WIC began as a 2-year pilot project for each of fiscal years 1973 and 1974 to provide supplemental foods to infants, children up to age five, and pregnant, breastfeeding, and non-breastfeeding postpartum women who qualify financially and are considered by competent professionals to be at nutritional risk because of inadequate nutrition and inadequate income.[2] Competent professionals include physicians, nutritionists, nurses, and other health officials. Financial eligibility is determined by income (between 100 percent and 185 percent of the poverty line or below) or by participation in the Food Stamp Program or Medicaid. Nutritional risks, determined by a health professional, may include one of three types: medically based risks (anemia, underweight, obesity, maternal age, history of high-risk pregnancies, HIV infection); diet-based risks (inadequate dietary pattern, gastrointestinal disorders, renal or cardiorespiratory disorders); or conditions that make the applicant predisposed to medically-based or diet-based risks, such as alcoholism or drug addiction (see Table 13-14).

TABLE 13-14 *Nutritional Risk Criteria for the Special Supplemental Nutrition Program for Women, Infants, and Children*

To be considered for the WIC program, an applicant must exhibit at least one of the nutritional risk factors listed below:

Women

Conditions Complicating the Prenatal and/or Postpartum Periods:
- low hematocrit or hemoglobin
- insufficient or excessive prenatal weight gain
- insufficient or excessive pregravid or postpartum weight
- excessive use of alcohol, drugs, or tobacco
- inadequate dietary status (as assessed by WIC standards)

General Obstetrical Risks:
- younger than 18
- multifetal gestation
- closely spaced pregnancies
- high parity and young age

History of:
- preterm delivery
- low birthweight infant
- infant with congenital defect
- miscarriage or stillbirth
- neonatal death
- gestational diabetes

Nutrition-Related Risk Conditions:
- eating disorders
- gastrointestinal disorders
- chronic or pregnancy-induced hypertension
- diabetes
- infectious diseases
- depression
- cancer
- renal disease
- thyroid disorders
- homelessness
- migrancy

Infants and Children

Low birthweight or preterm infants

Abnormal pattern of growth: short stature, underweight, overweight

Failure-to-thrive

Inadequate growth

Low head circumference

Inadequate dietary status (as assessed by WIC standards)

Low hemotocrit or hemoglobin (after 6 months of age)

Elevated blood lead levels

Food allergies/intolerances (as specified by WIC)

Fetal alcohol syndrome

Inborn errors of metabolism

Source: USDA, Nutrition Program Facts: WIC, May 1998; and www.usda.gov/fcs/wic.

The WIC program serves both a remedial and a preventive role. Services provided include the following:

- Food packages or vouchers for food packages to provide specific nutrients (protein, calcium, iron, and vitamins A and C) known to be lacking in the diets of the target population. WIC foods include iron-fortified infant formula and infant cereal, iron-fortified breakfast cereal, vitamin C–rich fruit or vegetable juice, eggs, milk, cheese, and peanut butter or dried beans and peas.
- Nutrition education including individual nutrition counseling and group nutrition classes[3]

A sample WIC Coupon.

- Referral to health care services including breastfeeding support, immunizations, prenatal care, family planning, or substance abuse treatment

The provision of nutritious supplemental foods to needy pregnant women is expected to improve the outcome of pregnancy. For infants and children, the food supplements are intended to reduce the incidence of anemia and to improve physical and mental development. Some states distribute food directly, but most provide vouchers or checks that participants can use at authorized food stores. The food voucher lists the quantities of specific foods, including brand names, that can be purchased with the voucher or check.

The combination of supplementary food, nutrition education, and preventive health care distinguishes WIC from other federal food assistance programs. Some of the potential impacts of program participation are summarized in Figure 13-6. Program benefits include improved dietary quality, more efficient food purchasing, better use of health services, and improved maternal, fetal, and child health and development.[4]

The provision of WIC benefits can improve the outcome of pregnancy. Everyone benefits: Mother, infant, and child.

Continued

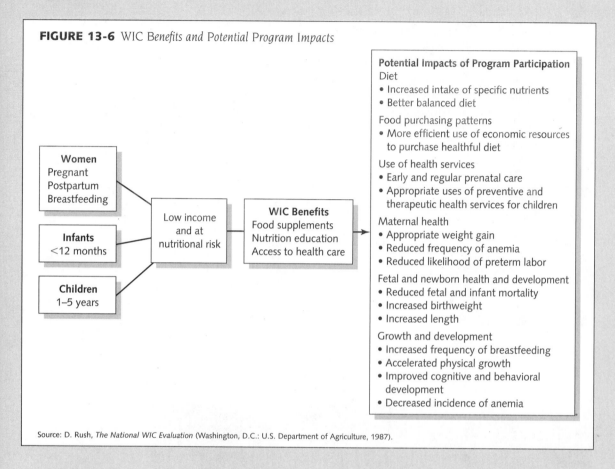

FIGURE 13-6 WIC Benefits and Potential Program Impacts

Women
Pregnant
Postpartum
Breastfeeding

Infants
<12 months

Children
1–5 years

Low income
and at
nutritional risk

WIC Benefits
Food supplements
Nutrition education
Access to health care

Potential Impacts of Program Participation
Diet
• Increased intake of specific nutrients
• Better balanced diet

Food purchasing patterns
• More efficient use of economic resources
  to purchase healthful diet

Use of health services
• Early and regular prenatal care
• Appropriate uses of preventive and
  therapeutic health services for children

Maternal health
• Appropriate weight gain
• Reduced frequency of anemia
• Reduced likelihood of preterm labor

Fetal and newborn health and development
• Reduced fetal and infant mortality
• Increased birthweight
• Increased length

Growth and development
• Increased frequency of breastfeeding
• Accelerated physical growth
• Improved cognitive and behavioral
  development
• Decreased incidence of anemia

Source: D. Rush, *The National WIC Evaluation* (Washington, D.C.: U.S. Department of Agriculture, 1987).

WIC has been described as one of the most efficient programs undertaken by the federal government. Over the years, the program has expanded significantly; the 1.9 million women and children served in 1980 at a cost of $725 million had grown by 1997 to more than 7.4 million women, infants, and children served at a cost of $3.7 billion. The dramatic growth in WIC since 1972 has caused policy makers to focus attention on quantifying the program's benefits.[5] To date, the program's proven benefits include the following:

• A General Accounting Office review of previous WIC studies concluded that WIC reduces the incidence of low birthweight (< 2,500 grams) and very-low birthweight (< 1,500 grams) rates by 25 percent and 44 percent, respectively.[6]

• The 1986 National WIC Evaluation released by the USDA found that WIC has contributed to a reduction

of 20 to 33 percent in fetal mortality and that the head size of infants whose mothers received WIC during pregnancy increased measurably.[7]

• In 1990, a USDA study of the health records of more than 100,000 mothers and babies found that the program increased birthweight significantly.

• Women who participate in WIC have longer pregnancies leading to fewer premature births. This not only benefits the infants, but it saves millions of dollars in Medicaid bills that would otherwise have been required for neonatal intensive care. It is estimated that every dollar spent on WIC for pregnant women can save as much as $4.21 in Medicaid costs.[8] (See the cost-benefit analysis of the WIC Program provided in the boxed insert on page 416.)

• A study conducted in 5 states—Florida, North Carolina, Minnesota, Texas, and South Carolina—

found that each dollar spent on WIC participants prenatally resulted in a $1.77 to $3.90 savings per participant in Medicaid costs.[9]

- A Yale University School of Medicine study found a remarkable decrease in the prevalence of anemia among low-income children in New Haven since the early 1970s. The researchers concluded, "The marked improvement can most probably be attributed to the nutritional supplementation with iron-fortified foods provided by the WIC program."[10]

- WIC also appears to lead to better mental performance. WIC children whose mothers participated in WIC during pregnancy had better vocabulary and memory test scores than nonparticipants.

- WIC participation can lead to improved breastfeeding rates.[11]

- WIC participation is associated with regular use of health care services; WIC children are more likely than not to receive some form of immunization against infectious diseases.

Unlike most of the other food assistance programs, WIC is not an entitlement program and therefore can serve only as many people as its annual appropriation from Congress permits.[12] Under federal regulations, once a local agency has reached its maximum caseload, vacancies must be filled in the following order of priority to assure that program resources are allocated to those at greatest nutritional risk:

1. Pregnant women, breastfeeding women, and infants determined to be at nutritional risk by a blood test, anthropometric measures, or other documentation of a nutrition-related medical condition

2. Infants, up to 6 months of age, whose mothers were at nutritional risk during pregnancy

3. Children at nutritional risk as determined by a blood test, anthropometric measures, or documentation of a nutrition-related medical condition

4. Pregnant or breastfeeding women and infants at nutritional risk because of an inadequate dietary pattern

5. Children at nutritional risk because of an inadequate dietary pattern

6. Nonbreastfeeding, postpartum women at nutritional risk

7. Persons certified for WIC solely because of homelessness or migrancy

Because of the caps placed on allocated federal funds, WIC currently reaches about 60 percent of eligible persons. Barriers to participation in WIC include limited government funding, lack of transportation, insufficient time or money to travel to the clinic, insufficient outreach to potential WIC recipients, the absence or expense of child care, inconvenient clinic hours, and understaffing of WIC facilities.[13]

Overall, research evaluating WIC's effectiveness finds positive outcomes for WIC participants compared to nonparticipants.[14] The following measures would help lower the barriers to program participation and improve the nutrition status and health care of women, infants, and children even further:

- Increase congressional funding for the program to enable more eligible women, infants, and children to receive WIC benefits.

- Improve the coordination between WIC and programs providing or financing maternal and child health care services.

- Implement outreach activities to increase access to WIC program benefits by all who are eligible.[15]

References

1. Select Panel for the Promotion of Child Health, WIC: The Special Supplemental Food Program for Women, Infants, and Children, in *Better Health for Our Children: A National Strategy*, 2 vols. (Washington, D.C.: U.S. Department of Health and Human Services, 1981), 2:57–68.

2. A. L. Owen and G. M. Owen, Twenty years of WIC: A review of some effects of the program, *Journal of the American Dietetic Association* 97 (1997): 777–82.

3. Food and Nutrition Service, United States Department of Agriculture, *Infant Nutrition and Feeding: A Reference Handbook for Nutrition and Health Counselors in the WIC and CSF Programs* (Washington, D.C.: USDA, 1993).

4. D. Rush and coauthors, The national WIC evaluation: Evaluation of the Special Supplemental Food Program for Women, Infants, and Children, *American Journal of Clinical Nutrition* 48 (1988): 389–93.

5. *Early Intervention: Federal Investments Like WIC Can Produce Savings*, Publication No. 92-18 (Washington, D.C.: US General Accounting Office, 1992); S. Avruch and A. P. Cackley, Savings achieved by giving WIC benefits to women prenatally, *Public Health Reports* 110 (1995): 27–34; and M. Kotelchuck and coauthors, WIC participation and pregnancy outcomes: Massachusetts statewide evaluation project, *American Journal of Public Health* 74 (1994): 1086–89.

Continued

6. Avruch and Cackley, Savings achieved by giving WIC benefits to women prenatally; *Early Intervention: Federal Investments Like WIC Can Produce Savings* (Washington, D.C.: General Accounting Office, 1992), GAO Publication No. 92-18.
7. Rush, The national WIC evaluation, p. 412.
8. Mathematica Policy Research, *The Savings in Medicaid Costs for Newborns and Their Mothers from Prenatal Participation in the WIC Program*, Report prepared for the U.S. Department of Agriculture, Food and Nutrition Service, Office of Analysis and Evaluation, Washington, D.C., 1990.
9. B. Devaney and A. Schirm, *Infant Mortality Among Medicaid Newborns in Five States: The Effects of Prenatal WIC Participation* (Princeton, N.J.: Mathematica Policy Research, 1993).
10. WIC reform means more mothers and children served, *Nutrition Forum Newsletter* (February 1988): 9–11.
11. Community Nutrition Institute, Breastfeeding rates rise significantly among at risk groups, WIC moms, *Nutrition Week* 27 (21) (1997): 4–6; D. L. Montgomery and P. L. Splett, Economic benefit of breastfeeding infants enrolled in WIC, *Journal of the American Dietetic Association* 97 (1997): 379–85; and Community Nutrition Institute, Food assistance: A variety of practices may lower the costs of WIC, *Nutrition Week* 27 (39) (1997): 4–5.
12. Food Research and Action Center, *WIC Works: Let's Make It Work for Everyone* (Washington, D.C.: Food Research and Action Center, 1993), pp. 1–19.
13. Community Nutrition Institute, Working women's access to WIC benefits, *Nutrition Week* 27 (41) (1997): 4–6.
14. P. A. Buescher and coauthors, Prenatal WIC participation can reduce low birthweight and newborn medical costs: A cost-benefit analysis of WIC participation in North Carolina, *Journal of the American Dietetic Association* 93 (1993): 163–66.
15. Community Nutrition Institute, WIC on wheels winnebago serves a sprawling Vegas, *Nutrition Week* 27 (29) (1997): 6.

 ## A Cost-Benefit Analysis of the WIC Program

A cost-benefit analysis helps managers gauge how well a program is meeting its clients' needs and provides data that can be used to influence policy makers. Let's consider an evaluation of prenatal participation in the Special Supplemental Nutrition Program for Women, Infants, and Children (WIC). This cost-benefit analysis was undertaken in North Carolina in 1988.[1] It involved the following steps:[2]

1. **Identify the primary client.** The primary client was the Division of Maternal and Child Health, Department of Environment, Health, and Natural Resources, Raleigh, North Carolina.

2. **Specify the purpose of the evaluation.** The cost-benefit analysis was designed to assess the effect of participation in the prenatal WIC program on low birthweight and Medicaid costs for the medical care of infants born in 1988.

3. **Specify the objectives of the evaluation.** The objectives for the evaluation of WIC prenatal participation in North Carolina in 1988 were to determine the following:
   - The birthweight of all infants born to WIC mothers
   - The birthweight of a random sample of infants born to non-WIC mothers
   - The cost per client of participating in the WIC program
   - The type and cost of hospital claims for newborn care paid by Medicaid

4. **Calculate a dollar value for each benefit of the program.** The direct benefits to participants in the WIC program included food and nutrition counseling. Indirect benefits included the birth of infants weighing more than 2,500 grams and reduced costs to Medicaid for the medical management of newborn care. (The dollar value assigned to these benefits was not reported in the published analysis.)

5. **Calculate the costs associated with the program.** Both personnel and material costs were considered. In this analysis, the direct costs included the total value of food vouchers

redeemed through the WIC program; administrative costs, estimated at an average of $170 per woman; and the newborn medical care costs paid by Medicaid. Medicaid covered such costs for newborn care as physician services, medications, and inpatient or outpatient care. The direct costs are summarized in Table 13-15, which shows the Medicaid costs for infants whose mothers received WIC prenatal care, Medicaid costs for infants whose mothers did not participate in WIC, and the estimated WIC program costs.

6. **Calculate the total cost per client.** The average cost of the program was $179 per white woman, $164 per black woman, and $170 for both groups.

7. **Calculate the benefit-to-cost ratio and/or the net savings.** The benefit-to-cost ratio, also shown in Table 13-15, was calculated by subtracting the Medicaid costs for the WIC group (column 1) from the Medicaid costs for the non-WIC group (column 2) and dividing by the WIC costs (column 3). The net savings was the actual dollar difference between the total benefits and the total costs. The estimated net savings was $343 for whites and $615 for blacks (column 1 subtracted from column 2).

The cost-benefit analysis revealed that the savings in Medicaid costs outweighed the costs of WIC services. The benefit-to-cost ratio was 1.92 for white women and 3.75 for black women, meaning that for each dollar spent on WIC, Medicaid saved $1.92 on whites and $3.75 on blacks. In addition, women who received WIC prenatal care gave birth to fewer infants with low and very low birthweights than women who did not participate in WIC.

A cost-benefit analysis undertaken by another state with a different sample of women and infants would likely produce slightly different results in program costs and the benefit-to-cost ratio. Even so, the finding of fiscal evaluations can be used to convince policy makers that money allocated for the WIC program is well spent and that cost savings can be achieved with nutrition intervention.

References

1. The description of the cost-benefit analysis of the WIC program was adapted from P. A. Buescher and coauthors, Prenatal WIC participation can reduce low birthweight and newborn medical costs: A cost-benefit analysis of WIC participation in North Carolina, *Journal of the American Dietetic Association* 93 (1993): 163–66.

2. The procedure for conducting a cost-benefit analysis was adapted from the Ross Roundtable Report, *Benefits of Nutrition Services: A Costing and Marketing Approach* (Columbus, OH: Ross Laboratories, 1987), pp. 24–29.

TABLE 13-15 *Average Costs to Medicaid and Average Costs of* WIC *Services*

| | Medicaid Costs | | WIC Costs* | Benefit-to-Cost Ratio |
	WIC	Non-WIC		
White	$1778	$2121	$179	1.92
Black	1902	2517	164	3.75
Total	1856	2350	170	2.91

*WIC costs include administrative and food costs.

Source: Copyright The American Dietetic Association. Reprinted by permission from *Journal of the American Dietetic Association*, Vol. 93: 1993, p. 166.

Community Health Centers. The Community Health Centers program was initiated by the Office of Economic Opportunity in 1966 and authorized by the Public Health Service Act. It is designed to provide health services and related training in medically underserviced areas. The primary program focus is on comprehensive primary care services through community health centers, including migrant health centers, Appalachian Health Demonstration projects, the Rural Health Initiative project, and the Urban Health Initiative project.[64] Preventive services are also offered through Community Health Centers, including well-child care, nutrition assessment, and health education. The program is administered federally by the Bureau of Community Health Services within the Health Resources and Services Administration of the Public Health Service.[65]

Looking Ahead:
Improving the Health of Mothers and Infants

Many of the existing health care programs do not have nutrition counseling or education available within their own sites. Often, these programs refer their clients to the WIC program for nutrition services. The heavy caseloads in many WIC programs, however, limit the amount of personalized nutrition counseling they can offer. Clearly, more must be done to assure that quality nutrition counseling is available and accessible for pregnant women and their infants.

A few states have been successful in providing reimbursable nutrition counseling services to maternal and child health programs. The South Carolina High Risk Channeling Project provides nutrition services, reimbursable by Medicaid, to all participants. The Kentucky Department for Health Services uses some of its MCH Block Grant funds to hire public health nutritionists specifically to provide nutrition counseling for high-risk clients in the local maternal and child health care programs.[66]

Some voluntary health organizations such as local chapters of the La Leche League International and the March of Dimes Birth Defects Foundation offer classes that are helpful to particular groups or can provide appropriate nutrition education materials helpful to the community nutritionist working with maternal-child populations (see the list of Internet Resources on page 419).

The increasing numbers of working women, including those planning families or with young infants, along with the growth in worksite health promotion programs have implications for community nutritionists in providing nutrition education and related services to this population. Some worksites have included components designed for pregnant and lactating women (for example, breastfeeding promotion and prenatal education programs) in their overall health promotion programs.[67]

Where adolescents are concerned, a variety of comprehensive community programs exist to assist the pregnant teen with her educational, social, medical, and nutritional needs. Most of these programs include three common components: (1) early and consistent prenatal care, (2) continuing education on a classroom basis, and (3) counseling on an individual or group basis.[68] The importance of these programs in improving maternal and child health is well recognized. Because teen pregnancy is usually unplanned, efforts should be made to improve the nutrition and health status of all adolescents through nutrition education and counseling in the classroom as well as in physicians' offices.

To assure that all pregnant women have access to satisfactory prenatal services in the future, efforts are needed to convince policy makers of the importance of the following recommendations:[69]

- Food supplementation and nutrition education should be available to all pregnant women with low incomes.
- Additional federal funds should be appropriated to WIC to make it available to all pregnant low-income women.
- Nutrition counseling and education should be provided to all pregnant women whose care is financed by Medicaid.
- State Medicaid programs should be required to include nutrition counseling and education as reimbursable services.
- Federal and state funds should be provided to health department clinics and community health centers to allow for employment of public health nutritionists to offer nutrition counseling to all pregnant women who attend these facilities.
- Health insurance policies should include prenatal nutrition counseling as a reimbursable service for all pregnant women living in the United States.

 ## Internet Resources

Promotion of Mother's Milk, Inc. (ProMom)   **www.promom.org**
   ProMom is a nonprofit organization dedicated to increasing public awareness and public acceptance of breastfeeding. Click on the Breastfeeding Gallery (fine art, photography, and clip art), the Breastfeeding Advocacy Page, Myths about Breastfeeding, and Articles and Essays for information about breastfeeding.

Infant Feeding Action Coalition (INFACT) Canada   **www.infactcanada.ca**
   INFACT is a nonprofit, voluntary organization that promotes maternal and infant health by promoting breastfeeding and fostering appropriate mother and infant nutrition. This site explains the problems caused by declines in breastfeeding.

La Leche League   **www.lalecheleague.org**
   La Leche League International is an international, nonprofit organization dedicated to providing encouragement to women who wish to breastfeed. This site contains a broad range of links that provides access to educational, informational and support programs offered by the league. An overview and history of La Leche, information on services provided to women and health care professionals, answers to FAQs, and an opportunity to search for a local La Leche leader are included.

Continued

Internet Resources—continued

Mayo Clinic Health Oasis: Children's Health Center
   www.mayohealth.org/mayo/common/htm/pregpg.htm

Maintained by the Mayo Clinic, this user-friendly site features reference articles, a due date calculator, advice about sex during pregnancy, and parenting information. The opportunity to ask questions of clinic physicians and links to related sites benefit visitors.

U.S. Food and Drug Administration, Center for Food Safety and Applied Nutrition: Information for Women Who Are Pregnant
   vm.cfsan.fda.gov/~dms/wh-preg.html

Links-only site offers sensible information from the Center for Food Safety and Applied Nutrition of the U.S. Food and Drug Administration. Topics include eating for two, folate and neural tube defects, vitamin A and birth defects, healthy eating during pregnancy, breastfeeding, and infant formula.

WIC Program
   www.usda.gov/fcs/wic.htm

Read about the findings of the WIC Infant Feeding Practices study and the WIC National Breastfeeding Promotion Project at this site. Plus, just released: WIC Policy Memorandum specifying how the WIC program will determine nutrition risk eligibility beginning April 1, 1999.

WIC Program Database of Nutrition Education Materials
   www.nal.usda.gov/fnic

This database describes the various materials produced by WIC Program directors and staff across the nation.

World Alliance for Breastfeeding Action
   www.elogica.com.br/waba

International links to information about breastfeeding are featured at this site. Other topics include a description of the World Alliance for Breast Feeding Action, the Lactational Amenorrhea Method (LAM) page, and breastfeeding resources.

Community Learning Activity

Activity 1

As mentioned in the chapter, WIC is not an entitlement program. Because of the caps placed on allocated federal funds, WIC currently reaches about 60 percent of eligible persons. Describe the benefits of participation in the WIC program by searching the relevant nutrition literature. Next, draft a letter to your legislator, requesting increased funding for WIC to make it a fully funded program.

Activity 2

Prepare a list of local health and nutrition resources for participants of the local WIC agency in your area. Include sample printed materials and a reference list of agencies or organizations providing information and services to pregnant or lactating women and their infants and children; include phone numbers and on-line addresses when available.

References

1. E. Hackman and coauthors, Maternal birthweight and subsequent pregnancy outcome, *Journal of the American Medical Association* 250 (1983): 2016–19.

2. N. M. Lien, K. K. Meyer, and M. Winick, Early malnutrition and "late" adoption: A study of the effects of the development of Korean orphans adopted into American families, *American Journal of Clinical Nutrition* 30 (1977): 1734–39.

3. U.S. Department of Health and Human Services, Public Health Service, *Healthy People 2000: National Health Promotion and Disease Prevention Objectives*, DHHS Pub. No. 91-50212 (Washington, D.C.: U.S. Department of Health and Human Services, 1990), pp. 366–90.

4. Annie E. Casey Foundation, *Kids Count Data Book* (Baltimore, MD: Annie E. Casey Foundation, 1997).

5. Ibid.

6. R. W. Rochat and coauthors, Maternal mortality in the United States: Report from the maternal mortality collaborative, *Obstetrics and Gynecology* 72 (1988): 91–97.

7. U.S. Department of Health and Human Services, Public Health Service, *The Surgeon General's Report on Nutrition and Health*, DHHS Pub. No. 88-50210 (Washington, D.C.: U.S. Government Printing Office, 1988), pp. 540–93.

8. Food and Nutrition Board, Institute of Medicine, National Academy of Sciences, *Nutrition During Pregnancy* (Washington, D.C.: National Academy Press, 1990); Food and Nutrition Board, Institute of Medicine, National Academy of Sciences, *Nutrition During Lactation* (Washington, D.C.: National Academy Press, 1991).

9. Predictors of poor maternal weight gain from baseline anthropometric, psychosocial, and demographic information in a Hispanic population, *Journal of the American Dietetic Association* 97 (1997): 1264–68.

10. National Center for Health Statistics, *Healthy People 2000 Review* (Hyattsville, MD: Public Health Service, 1997).

11. C. M. Trahms and J. Powell, Maternal and infant nutrition: An inquiry into the issues, *Topics in Clinical Nutrition* 11 (1996): 10–17.

12. Nutrition during pregnancy and lactation, *Dairy Council Digest* 62 (1991): 13–18.

13. Folate supplements prevent recurrence of neural tube defects, *Nutrition Reviews* 50 (1992): 22–26.

14. Centers for Disease Control and Prevention, Use of folic acid-containing supplements among women of childbearing age—United States, 1997, *Journal of the American Medical Association* 279 (1998): 1430.

15. Food and Nutrition Board, *Nutrition During Pregnancy*; C. W. Suitor and J. D. Gardner, Supplement use among a culturally diverse group of low-income pregnant women, *Journal of the American Dietetic Association* 90 (1990): 268.

16. J. E. Brown and coauthors, Development of a prenatal weight gain intervention program using social marketing methods, *Journal of Nutrition Education* 24 (1992): 21–29.

17. A. J. Rainville, Pica practices of pregnant women are associated with lower maternal hemoglobin level at delivery, *Journal of the American Dietetic Association* 98 (1998): 293–96.

18. B. Worthington-Roberts, Nutritional support of successful reproduction: An update, *Journal of Nutrition Education* 19 (1987): 1–10.

19. Food and Nutrition Board, *Nutrition During Pregnancy*, pp. 397–99.

20. U.S. Department of Health and Human Services, *Healthy People 2000.*

21. M. G. Bulterys, S. Greenland, and J. F. Kraus, Chronic fetal hypoxia and sudden infant death syndrome: Interaction between maternal smoking and low hematocrit during pregnancy, *Pediatrics* 86 (1990): 535–40.

22. B. Worthington-Roberts and R. M. Pitkin, Women's nutrition for optimal reproductive health, in *Call to Action: Better Nutrition for Mothers, Children, and Families*, ed. C. O. Sharbaugh (Washington, D.C.: National Center for Education in Maternal and Child Health, 1991), pp. 113–136.

23. Worthington-Roberts and Pitkin, Women's nutrition for optimal reproductive health, p. 129.

24. U.S. Department of Health and Human Services, *Surgeon General's Report.*

25. J. C. King and J. Weininger, Pregnancy and lactation, in *Present Knowledge in Nutrition*, 6th ed., ed. M. L. Brown

(Washington, D.C.: International Life Sciences Institute—Nutrition Foundation, 1990), pp. 314–19.

26. J. M. Belizan and coauthors, Calcium supplementation to prevent hypertensive disorders of pregnancy, *New England Journal of Medicine* 325 (1991): 1399–1405.

27. U.S. Department of Health and Human Services, *Surgeon General's Report*, pp. 56–57.

28. American Diabetes Association, Position Statement: Gestational Diabetes Mellitus, *Diabetes Care* 21 (1998): S60–S61.

29. Position of the American Dietetic Association: Nutrition management of adolescent pregnancy, *Journal of the American Dietetic Association* 89 (1989): 104–9.

30. M. Story and I. Alton, Nutrition and the pregnant adolescent, *Contemporary Nutrition* 17 (1992): 1–2.

31. M. L. Hediger and coauthors, Rate and amount of weight gain during adolescent pregnancy: Associations with maternal weight-for-height and birth weight, *American Journal of Clinical Nutrition* 52 (1990): 793–99.

32. M. S. Bergman, Improving birth outcomes with nutrition intervention, *Topics in Clinical Nutrition* 13 (1997): 74–79.

33. Worthington-Roberts and Pitkin, Women's nutrition, pp. 132–33.

34. Position of the American Dietetic Association: Nutrition management of adolescent pregnancy, pp. 106–7.

35. B. Worthington-Roberts and L. Klerman, Maternal nutrition, in *New Perspectives in Prenatal Care*, ed. I. R. Merkatz, J. E. Thompson, and R. Goldenberg (New York: Elsevier Science, 1990).

36. Ibid.

37. The description of head circumference measurement was taken from Whitney, Cataldo, and Rolfes, *Understanding Normal and Clinical Nutrition*, p. 616.

38. American Dietetic Association, Position of the American Dietetic Association: Promotion of breast-feeding, *Journal of the American Dietetic Association* 97 (1997): 662–65; M. A. Murtaugh, Optimal breastfeeding duration, *Journal of the American Dietetic Association* 97 (1997): 1252–54.

39. U.S. Department of Health and Human Services, *Healthy People 2000*, pp. 379–80.

40. Ibid.

41. J. B. Sciacca and coauthors, Influences on breastfeeding by lower-income women: An incentive-based, partner-supported educational program, *Journal of the American Dietetic Association* 95 (1995): 323–27.

42. La Leche League International, Breastfeeding Peer Counselor Program, personal communication, November 1993.

43. Community Nutrition Institute, USDA kicks off campaign to promote breastfeeding in nine state WIC agencies, *Nutrition Week* 27 (30) (1997): 1, 6.

44. Food and Nutrition Services, *Promoting Breastfeeding in WIC: A Compendium of Practical Approaches* (Arlington, VA: U.S. Department of Agriculture, 1988), pp. 1–6.

45. C. R. Tuttle and K. G. Dewey, Potential cost savings for Medicaid, AFDC, Food Stamps, and WIC programs associated with increasing breast-feeding among low-income Hmong women in California, *Journal of the American Dietetic Association* 96 (1996): 885–90.

46. B. Dobson, Community-based coalition building for breastfeeding promotion and support, *Nutrition Education for the Public Networking News*, Autumn 1996, 3, 5, 12.

47. The discussion of focus group findings was taken from C. Bryant and D. Bailey, The use of focus group research in program development. Unpublished manuscript, Lexington, KY, 1989.

48. American Academy of Pediatrics, Committee on Nutrition, The use of whole cow's milk in infancy, *Pediatrics* 89 (1992): 1105.

49. D. B. Johnson, Nutrition in infancy: Evolving views on recommendations, *Nutrition Today* 32 (1997): 63–68.

50. E. E. Ziegler, S. J. Fomon, and S. E. Nelson, Cow milk feeding in infancy: Further observations on blood loss from the gastrointestinal tract, *Journal of Pediatrics* 116 (1990): 11.

51. M. C. Holst, Developmental and behavioral effects of iron deficiency anemia in infants, *Nutrition Today* 33 (1998): 27–36.

52. J. S. Forsyth and coauthors, Relation of infant diet to childhood health: Seven year followup of cohort of children in Dundee infant feeding study, *British Medical Journal*, January 3, 1998, pp. 27–32.

53. Select Panel for the Promotion of Child Health, *Better Health for Our Children: A National Strategy*, 2 vols. (Washington, D.C.: U.S. Department of Health and Human Services, 1981), 1:168.

54. Community Nutrition Institute, Farmers' market program to grow with new funds, *Nutrition Week* 28 (2) (1998): 2–3.

55. Ibid., 1:156.

56. Ibid., 2:18, and C. Garza and C. Cowell, Infant nutrition, in *Call to Action: Better Nutrition for Mothers, Children, and Families*, ed. C. Sharbaugh (Washington, D.C.: National Center for Education in Maternal and Child Health, 1991), pp. 135–150.

57. U.S. Department of Health and Human Services, *Surgeon General's Report*, p. 544.

58. Select Panel for the Promotion of Child Health, *Better Health*, 2:17–22.

59. Ibid., 2:20.

60. Ibid., 2:21.

61. U.S. Department of Health and Human Services, *Surgeon General's Report*, p. 544.

62. J. J. Haas and coauthors, The effect of providing health coverage to poor uninsured pregnant women in Massachusetts, *Journal of the American Medical Association* 269 (1993): 87–92; K. A. Johnson, What assures good outcomes in Medicaid-financed prenatal care? *Public Health Reports* 112 (1997): 133–35.

63. Select Panel for the Promotion of Child Health, *Better Health*, 2:44.

64. Ibid., 1:165.

65. Ibid., 2:101.

66. R. E. Brennan and M. N. Traylor, Components of nutrition services, in *Call to Action: Better Nutrition for Mothers, Children, and Families,* ed. C. Sharbaugh (Washington, D.C.: National Center for Education in Maternal and Child Health, 1991), pp. 243–55.

67. B. Barber-Madden and coauthors, Nutrition for pregnant and lactating women: Implication for worksite health promotion, *Journal of Nutrition Education* 18 (1986): S72–S75.

68. Worthington-Roberts and Williams, *Nutrition in Pregnancy and Lactation,* pp. 239–40.

69. The list of recommendations was adapted from Worthington-Roberts and Pitkin, Women's nutrition, p. 133.

Chapter 14

Children and Adolescents: Nutrition Services and Programs

Outline

Learning Objectives

After you have read and studied this chapter, you will be able to:

- Describe three nutritional problems currently experienced by U.S. children and adolescents.
- Specify four *Healthy People 2000* nutrition objectives for children and adolescents.
- Describe five components of an interdisciplinary assessment of children with special health care needs.
- Discuss four food assistance programs aimed at improving the health and nutritional status of children, including their purposes, types of assistance offered to clients, and eligibility requirements.

Something To Think About...

Children are one-third of our population and all of our future. – *Select Panel for the Promotion of Child Health, 1981*

Introduction

The health profile of children and adolescents in the United States has changed dramatically over the past century. Widespread immunization, improved sanitation, public education on nutrition and health, and the discovery of antibiotics have dramatically reduced the rates of child morbidity and mortality due to infectious diseases. Unfortunately, the status of this group today is far from satisfactory, and new perils have arisen in the past few decades: motor vehicle accidents; violence due to suicide, homicide, and abuse; sexually transmitted diseases; substance abuse; and exposure to environmental pollutants.[1] Despite advances in clinical and preventive medicine, children and adolescents in the United States have significant health and nutritional concerns that deserve attention; and there is a sense that policy makers and health practitioners are still falling short of doing what most people believe is necessary to promote the health of all children in the United States.

This chapter reviews the current nutrition-related problems of children and adolescents, and it examines the nutrition programs that target children and teenagers. These programs have the common objective of improving child and adolescent nutrition and, ultimately, enhancing health.

For our purposes, children are generally categorized as ages 1 to 12 years and adolescents as ages 13 to 19 years. Other age categories appear in this chapter, however, because the literature is not entirely consistent on the ages that constitute childhood and adolescence.

National Nutrition Objectives

More than 20 years ago, delegates to the 1970 White House Conference on Children outlined a multidisciplinary strategy for addressing the health, social, physical, educational, and environmental needs of children. Concerned about the widespread neglect of children and the lack of a comprehensive system for delivering health and other services, the delegates called for "a reordering of priorities at all levels of American society so that children and families come first."[2] In the nutrition arena, the delegates recommended that existing food programs be expanded and that all children receive quality nutrition education in school. A separate White House Conference on Youth held in 1971 likewise called for improved nutrition education and services for all young people in need.[3]

Child advocates hoped that the White House Conferences would bring a renewed commitment to improving the health and nutritional status of children (among other objectives). While the conferences may not have fulfilled their ambitious agendas entirely, they did set the stage for major policy developments. In 1979, the Surgeon General issued the first federal report, *Healthy People: The Surgeon General's Report on Health Promotion and Disease Prevention*, which called for increased efforts to reduce death and disabilities from preventable diseases in the general population and in certain population subgroups, including children and adolescents.[4]

In 1990, the U.S. Department of Health and Human Services (DHHS) updated the health and nutrition objectives for children and adolescents in its publication *Healthy People 2000: National Health Promotion and Disease Prevention Objectives*.[5] The most recent publication in the *Healthy People 2000 Review* series, released in 1997, lists the priority areas for children and adolescents. Physical activity and fitness, nutrition, and dental health are among the priority concerns. Several nutrition objectives are shown in Table 14-1. Progress has been made in reaching some goals for the year 2000. The prevalence of growth retardation among low-income

TABLE 14-1 *Major Healthy People 2000 Nutrition Objectives for Children and Adolescents*

Objective Number	2000 Objective	Attainment Status
2.3	Reduce overweight to a prevalence of no more than 15 percent among adolescents aged 12–19.	Prevalence of overweight in 1994 = 24 percent (an increase of 3 percent from 21 percent in 1988–1991).
2.4	Reduce growth retardation* among low-income children aged 5 and younger to less than 10 percent.	Baseline = 11 percent. Prevalence of growth retardation in this age group in 1995 = 8 percent.
2.5	Reduce dietary fat intake to an average of 30 percent of calories or less among people aged 2 and older.	Baseline = 36 percent. Average percentage of calories from fat in 1994 = 34 percent.
	Reduce average saturated fat intake to less than 10 percent of calories among people aged 2 and older.	Baseline = 13 percent. Average percentage of calories from saturated fat in 1994 = 12 percent.
2.8	Increase calcium intake so at least 50 percent of people aged 11–24 consume an average of three or more daily servings of foods rich in calcium, and at least 75 percent of children aged 2–10 consume an average of two or more servings daily.	Data for people aged 11–24: Baseline = 20 percent. Percentage who consume 3 or more servings daily in 1994 = 17 percent. Data for children aged 2–10: Baseline = 48 percent. Percentage who consume 2 or more servings daily in 1994 = 42 percent.

*Growth retardation is defined as height-for-age below the 5th percentile of children in the National Center for Health Statistics' reference population derived from the 1971–1974 National Health and Nutrition Examination Survey (NHANES I).

Source: National Center for Health Statistics, *Healthy People 2000 Review, 1997* (Hyattsville, MD: Public Health Service, 1997). These figures were obtained from the on-line publication at www.cdc.gov/nchswww.

children, for example, has decreased for all races combined and for some Hispanics and Asian/Pacific Islanders. However, the prevalence of overweight among adolescents of all ethnic groups has increased.[6] Additional information about the *Healthy People 2000* goals for children and adolescents is given in the next section.

Nutrition-Related Problems of Children and Adolescents

Children

Childhood is a critical time in human development. Children typically grow taller by 2 to 3 inches and heavier by 5 or more pounds each year between the age of one and adolescence.[7] They master fine motor skills (including those related to eating and drinking), become increasingly independent, and learn to express themselves appropriately. Nutrition plays a critical role in the development and growth of children. The most common nutrition-related problems among U.S. children include undernutrition (in some populations), overweight and obesity, iron-deficiency anemia, dental caries, high blood cholesterol concentrations, and blood lead levels:

- **Undernutrition.** Undernutrition is a problem for some children in the United States, especially those from low-income and migrant families or certain ethnic and racial minority groups (for example, blacks, Asians).[8] Children in foster care, many of whom live in poverty, and homeless children are also at risk for undernutrition.[9] About 5.5 million U.S. children—nearly one-quarter of all children under the age of 6—lived in poverty in 1996. Although this figure represents a decrease from the peak poverty level for children in 1993, it is more than double the rates for adults and the elderly.[10]

- **Overweight and obesity.** Overnutrition is one of the most widespread nutrition disorders among children in the United States. No generally accepted definition of overweight and obesity for children and adolescents exists, although a definition of weight status based on weight and height (for example, the body mass index or BMI) is reasonably precise and practical in a clinical setting. In children, overweight is defined as having a body mass index between the 85th and 94th percentiles; obesity is defined as a BMI above the 95th percentile.[11]

Prevalence estimates for childhood obesity range from 10 to 30 percent, depending on how obesity is defined.[12] Data from the National Health and Nutrition Examination Survey (NHANES) indicate that the prevalence of obesity among 6- to 11-year-old children increased 54 percent in the past 15 to 20 years.[13] Data from the Hispanic HANES indicate that between 20 and 25 percent of Mexican-American girls and boys aged 2 to 5 years are moderately overweight or obese.[14] American Indian children have the highest prevalence of overweight and obesity among all ethnic groups, with about 9 to 20 percent of American Indian preschool children and 25 to 78 percent of school-age children being obese.[15] Childhood obesity is associated with hyperinsulinemia, hypertriglyceridemia, and reduced HDL-cholesterol concentrations and is considered a risk factor for obesity in adulthood.[16] According to one recent study, the strongest predictors of overweight and overfatness in fifth graders was their prior status in third grade.[17] Second to prior obesity, the strongest predictor of subsequent obesity in children is television viewing.[18]

Refer to Appendix L for the physical growth charts for children and information on assessing children's growth status.

- **Iron deficiency and iron-deficiency anemia.** Iron deficiency is one of the most common nutritional deficiencies in the United States and, indeed, in the world. To be considered iron deficient, a child (or any individual) must have an abnormal value for two of the following three indicators of iron status: transferrin saturation, serum ferritin, and free erythrocyte protoporphyrin. Iron-deficiency anemia is defined as having iron deficiency and a low hemoglobin concentration. The cutoff values for laboratory tests of iron status in children and adolescents are given in Tables 14-2 and 14-3.

    The year 2000 target is to reduce iron deficiency to less than 3 percent for children aged 1 to 4 years.[19] Prevalence estimates of impaired iron status indicate that 9 percent of children aged 12 to 36 months are iron deficient—a figure considerably higher than the target level—and 3 percent have iron-deficiency anemia. Roughly 2 to 3 percent of older children are iron deficient, and less than 1 percent have iron-deficiency anemia. These estimates were based on data from the Third National Health and Nutrition Examination Survey (NHANES III).[20]

    The prevalence of iron deficiency is higher among children living at or below the poverty level than among those living above the poverty level.[21] Data from the Pediatric Nutrition Surveillance System survey (PedNSS), which monitors the general health and nutrition status of low-income U.S. children who participate in the public health programs, reveal a prevalence of anemia of 20 to 30 percent in the PedNSS population. This figure is much higher than the national average, reflecting the greater risk for iron-deficiency anemia among low-income children.[22] Black, Mexican-American, and Southeast Asian children are more likely to be iron deficient than children of other ethnic groups.[23] The prevalence rate of iron deficiency among Alaska Native children under 5 years of age was 27 percent in 1994.[24]

TABLE 14-2 *Cutoff Values for Laboratory Tests of Iron Deficiency in Children and Adolescents**

Sex and Age (yr)	Transferrin Saturation (%)[†]	Serum Ferritin (µg/L)	Free Erythrocyte Protoporphyrin (µmol/L RBC[‡])
Both sexes			
1–2	< 10	< 10	> 1.42
3–5	< 12	< 10	> 1.24
6–11	< 14	< 12	> 1.24
Female			
12–15	< 14	< 12	> 1.24
16–19	< 15	< 12	> 1.24
Male			
12–15	< 14	< 12	> 1.24
16–19	< 15	< 12	> 1.24

*Based on data from the Third National Health and Nutrition Examination Survey (NHANES III).
†The following percentiles of the NHANES III reference population were used as cutoff values: Transferrin saturation, 12th percentile; serum ferritin, 5th percentile; free erythrocyte protoporphyrin, 5th percentile.
‡RBC = Red blood cells.
Source: Adapted from A. C. Looker and coauthors, Prevalence of iron deficiency in the United States, *Journal of the American Medical Association* 277 (1997): 973–76.

Sex and Age(yr)	Hemoglobin Concentration		Hematocrit	
	Traditional unit (g/dL)	SI unit[†] (g/L)	Traditional unit (< %)	SI unit[‡] (Fraction of 1.00)
Children				
1–2	11.0	110	32.9	0.33
2– < 5	11.1	111	33.0	0.33
5– < 8	11.5	115	34.5	0.34
8– < 12	11.9	119	35.4	0.35
Men				
12– < 15	12.5	125	37.3	0.37
15– < 18	13.3	133	39.7	0.40
> 18	13.5	135	39.9	0.40
Nonpregnant women				
12– < 15	11.8	118	35.7	0.36
15– < 18	12.0	120	35.9	0.36
> 18	12.0	120	35.7	0.36

TABLE 14-3 *Cutoff Values for Anemia for Children and Non-pregnant Adolescents**

*Based on the 5th percentile from the Third National Health and Nutrition Examination Survey (NHANES III).

[†]For hemoglobin, to convert the traditional unit to the SI unit, multiply by 10.0.

[‡]For hematocrit, to convert the traditional unit to the SI unit, multiply by 0.01.

Source: U.S. Department of Health and Human Services, Centers for Disease Control and Prevention, Recommendations to prevent and control iron deficiency in the United States, *Morbidity and Mortality Weekly Report* 47, No. RR-3 (1998): 1–29.

- **Dental caries.** Dental caries affect 98 percent of all U.S. children, making this condition a widespread public health problem.[25] Data collected in 1986–1987 by the Centers for Disease Control and Prevention (CDC) indicate that about 34 percent of 9-year-old children and 58 percent of 12-year-old children have dental caries, and the number of decayed, missing, and filled permanent teeth increases steadily with age.[26] In 1980, the average child had at least one carious lesion in a permanent tooth by the age of 8 years. By the age of 12 years, the average child had 4 carious lesions; by 17 years of age, 11 lesions. Fortunately, the incidence of dental caries in children has decreased by as much as 30 to 50 percent over the last two decades owing, in part, to fluoridation of public drinking water, improved dental hygiene, and the use of fluoride in toothpastes and mouthwashes.[27]
- **High blood cholesterol.** There is considerable evidence that atherosclerosis begins in childhood and that this process is related to high blood cholesterol levels. In fact, data from the Bogalusa Heart Study, a long-term epidemiologic study of cardiovascular disease risk factors in a white and black population of children and young adults, found that elevated LDL-cholesterol concentrations in children persist into adulthood and increase cardiovascular disease risk.[28]

The Expert Panel on Blood Cholesterol Levels in Children and Adolescents of the National Cholesterol Education Program (NCEP) classifies a total blood cholesterol level of ≥ 200 mg/dL (≥ 5.17 mmol/L) or an LDL-cholesterol level of ≥ 130 mg/dL (≥ 3.36 mmol/L) as high, when associated with a family or parental history of hypercholesterolemia. Refer to

TABLE 14-4 *Classification of Total and LDL-Cholesterol Levels in Children and Adolescents from Families with Hypercholesterolemia or Premature Cardiovascular Disease*

Category	Total Cholesterol (mmol/L)	Total Cholesterol (mg/dL)	LDL-Cholesterol (mmol/L)	LDL-Cholesterol (mg/dL)
Acceptable	< 4.40	< 170	< 2.84	< 110
Borderline	4.40–5.16	170–199	2.84–3.35	110–129
High	≥ 5.17	≥ 200	≥ 3.36	≥ 130

Source: National Cholesterol Education Program, *Report of the Expert Panel on Blood Cholesterol Levels in Children and Adolescents* (Washington, D.C.: U.S. Department of Health and Human Services, Public Health Service, National Institutes of Health, NIH Pub. No. 91-2732, 1991), p. 5. To convert milligrams (mg) of cholesterol per deciliter (dL) of blood to millimoles (mmol) per liter (L), multiply by a factor of 0.02586.

Table 14-4 for the NCEP classification of total and LDL-cholesterol levels in children and adolescents. The panel recommends that children and youths with high blood cholesterol be evaluated further and receive diet therapy, if appropriate. Drug therapy is recommended only for children over 10 years of age whose blood cholesterol level has not responded to an adequate trial of diet therapy lasting from six months to one year. When compared with children in other countries, children and adolescents in the United States have higher blood cholesterol levels and higher dietary intakes of saturated fat and cholesterol.[29]

• **Blood lead level.** Concern about the link between high blood lead levels and lowered intelligence in children led to a large-scale public health effort to eliminate or reduce lead in gasoline, food cans, drinking water, and house paint. Today, small children are most likely to be exposed to lead if they live in houses built before 1950 that were painted with lead-based paints. Such children may eat paint chips containing lead or other lead-contaminated objects, thus increasing their blood lead levels. An elevated blood lead level is defined as a blood lead level high enough to warrant further medical evaluation for the possibility of adverse mental, behavioral, physical, or biochemical effects. The Centers for Disease Control and Prevention recommends that children with blood lead levels exceeding 10 µg/dL be evaluated and referred for treatment.

Overall, the percentage of children aged 1 to 5 years with elevated blood lead levels decreased more than 80 percent between 1976 and 1994. Even so, as many as 890,000 children in the United States have blood lead levels ≥10 µg/dL. Data from the Third National Health and Nutrition Examination Survey found that children who live in poor families are more likely to have high blood lead levels than children of the same age living in high-income families. Black children are more likely than children of other race and ethnic groups to have high blood lead levels. Continued efforts are needed to reduce household lead hazards and prevent lead exposure in children.[30]

Children with Special Health Care Needs

Several terms and classifications have been used in describing the population with special needs, including developmental disabilities, developmental social needs, handicapping conditions, chronic disorders, and chronic illnesses.[31] This book uses the term "special health care needs," which encompasses the definition of developmental disabilities given in the Developmental Disabilities Assistance and Bill of

Rights Act of 1990 (PL 101-496) and applies to a broad spectrum of handicapping conditions and mental and physical impairments.*

Children with special health care needs are those "who have or are at increased risk for a chronic physical, developmental, behavioral, or emotional condition and who also require health and related services of a type or amount beyond that required by children generally." This definition was developed by the federal Maternal and Child Health Bureau's Division of Services for Children with Special Health Care Needs to help state programs plan and develop systems of care for all children with special health care needs.[32]

An estimated 4.4 million U.S. children experience some form of disability that limits their participation in school, play, and social activities.[33] Children with special health care needs are at increased nutritional risk because of feeding problems, metabolic aberrations, drug/nutrient interactions, decreased mobility, and alterations in growth patterns. An interdisciplinary approach to managing children with special health care needs is recommended. The interdisciplinary team may include a physician, nurse, psychologist, dentist, dietitian, school administrator, food service director, special education staff, and occupational, physical, and speech therapists. The child and his or her parents or caregivers are important team members whose insights and observations help the team determine which foods and feeding strategies work.[34]

The interdisciplinary team evaluation of the health and nutritional status of children with special needs includes the dietary history, medical history, anthropometric measurements, clinical assessment, and feeding assessment.[35] These components are outlined in Table 14-5. The nutritional assessment forms the basis for the development of a nutrition plan for the toddler or school-aged child and his or her family. The nutrition plan describes the child's food preferences, the family's beliefs and values about foods, and the child's mealtime behavior, nutritional needs, and ability to feed him- or herself. It includes feeding goals for the child—for example, "David will learn to chew solid food" or "Karen will learn to hold a spoon in her right hand"—and provides detailed information about the child's special diet and any equipment required during feeding. Some examples of special utensils that help children feed themselves are shown in Figure 14-1 on page 434.[36] The primary feeding goal is to help the child learn to feed him- or herself.

---

*The Developmental Disabilities Assistance and Bill of Rights Act of 1990 states: "The term 'developmental disability' means a severe chronic disability of a person 5 years of age or older which is attributable to a mental or physical impairment or combination of mental and physical impairments; is manifested before the person attains age twenty-two; is likely to continue indefinitely; results in substantial functional limitations in three or more of the following areas of major life activity: (i) self-care, (ii) receptive and expressive language, (iii) learning, (iv) mobility, (v) self-direction, (vi) capacity for independent living, and (vii) economic sufficiency; and reflects the person's need for a combination and sequence of special, interdisciplinary, or generic care, treatment, or other services which are of lifelong or extended duration and are individually planned and coordinated, except that such term, when applied to infants and young children means individuals from birth to age 5, inclusive, who have substantial developmental delay or specific congenital or acquired conditions with a high probability of resulting in developmental disabilities if services are not provided." (As cited in *Journal of the American Dietetic Association* 92 [1992]: 613.)

TABLE 14-5 *Inter-disciplinary* Assessment of *Children with Special Health Care Needs*	**Dietary history** Methods of feeding (e.g., breast, bottle, gavage, spoon, cup); gavage, bottle, and breast weaning Type of food, order of introduction, and child's age at introduction of solid foods and present intake of solid foods Foods that cause aspiration or gagging Current and past special diet orders, including texture modifications Use of nutritional supplements and vitamin/mineral supplements Food intolerances/allergies Current and past medications and noted side effects Pica or inappropriate food consumption Appetite, regularity of meals Fluid preferences and intake pattern Fad, cultural, or religious diet preferences Nutrition knowledge and attitude of caregiver regarding meal planning and diet of child Use of food as reinforcer/pacifier Where and with whom food is consumed Participation in school/child care breakfast and/or lunch programs **Medical history** Chronic/metabolic disorders Records of any substance abuse Incidence of vomiting, nausea, diarrhea, or constipation Acute or recurring illnesses and/or infections Obvious manifestation of nutritional deficiencies Physical activity

Adolescents

Adolescence is a time of change. Between the ages of about 10 to 18 years in girls and 12 to 20 years in boys, there are marked changes in physical, intellectual, and emotional growth and development. The maturation process is initiated and controlled by a variety of hormones, including growth hormone, prolactin, estrogen, testosterone, and the thyroid hormones, among others.[37] Many aspects of the maturation process are influenced by dietary intake and nutrition status.

Most adolescents in the United States are perceived as "healthy." Nevertheless, many U.S. adolescents experience a variety of health and nutritional problems, some related to their risk-seeking behaviors and inability to deal with abstract notions such as "good health" and the link between current behaviors and long-term health. In general, the nutritional health of U.S. teenagers is better today than ever before. Overt nutrient deficiencies, with the exception of iron deficiency, are

TABLE 14-5 *Continued*

Medical history—continued

Dental screening

Socioeconomic data, including home environment, adequacy of resources, educational status of primary caregiver, social/economic habits, parenting skills, support system for caregiver, and influential sources of information

Anthropometric assessment

Height/length

Weight

Mid-arm muscle circumference

Triceps skinfold

Feeding assessment

Neurological dysfunction or mechanical obstructions reflected in inefficient sucking and swallowing, bite reflex, strong or hypoactive gag reflex, tongue thrust, protrusion reflex, poor lip control, impaired chewing, malocclusion and high arched palate, and lack of head and trunk control

Dysphagia caused by gross congenital anatomical defects of the tongue, palate, mandible, pharynx, larynx, esophagus, and thorax; neuromuscular causes, such as delayed maturation, cerebral palsy, muscular dystrophy, various syndromes, and bacterial and viral infections; and acute infective conditions, such as stomatitis

Particular problems associated with feeding difficulties, such as delayed introduction of solid foods, delays in self-feeding, mealtime tantrums or disruptive behavior, coercive or manipulative interactive patterns between child and caregiver, and inappropriate uses of rewards/punishment for eating/mealtime

Clinical assessment

Skeleton, musculature, and general appearance

Hair, skin, eyes, nails, face

Lips, teeth, gums, and tongue

Source: L. A.. Wodarski, An interdisciplinary nutrition assessment and intervention protocol for children with disabilities. Copyright The American Dietetic Association. Reprinted by permission from JOURNAL OF THE AMERICAN DIETETIC ASSOCIATION, Vol. 90: 1990, p. 1564.

not the public health problems they once were, although low calcium intakes during the peak bone-building years increases the risk of osteoporosis in later life, especially among females. Teenage girls, for example, consumed only 57 percent of the Dietary Reference Intake (DRI) for calcium, whereas boys consumed 89 percent of the DRI for calcium, according to data from the 1996 Continuing Survey of Food Intakes by Individuals.[38] Specific nutrition-related problems among U.S. adolescents include undernutrition, overweight and obesity, iron-deficiency anemia, high blood cholesterol levels, dental caries, and eating disorders:

• **Undernutrition.** Some groups of adolescents are at risk for reduced energy and food intakes. Adolescents from low-income families and those who have run away from home or abuse

FIGURE 14-1

Examples of Utensils for Children with Special Health Care Needs

Modified utensils help special needs children feed themselves. A large, firm handle makes it easier for children to grasp the utensil. Bent handles help the child move foods from the bowl or plate to his or her mouth. Utensils should be unbreakable and coated with plastic, in case a child bites down on it.

Source: *Project CHANCE: A Guide to Feeding Young Children with Special Needs* (Phoenix: Arizona Department of Health Services, Office of Nutrition Services, 1995), p. 57. Used with permission.

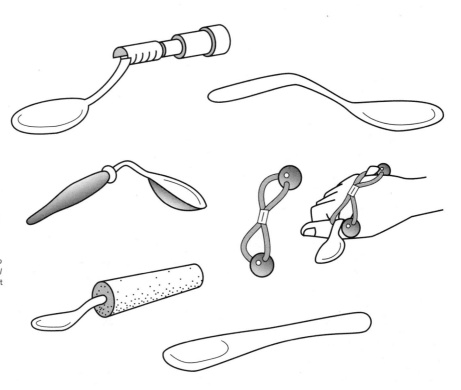

alcohol or drugs are at risk nutritionally. In 1987, 15.9 percent of youth aged 14 to 21 years lived below the poverty line. Black or Hispanic teenagers are nearly three times more likely to live in poverty than are white youth. Juveniles who live on the street tend to have a host of health problems, including substance abuse and malnutrition. Irregular meal patterns, combined with a high rate of use of substances such as alcohol, marijuana, cocaine, and amphetamines, contribute to low nutrient and energy intakes among street youth.[39] Chronic dieters are also at risk. Thirteen percent of the 17,354 females in grades 7 through 12 who were interviewed in the Minnesota Adolescent Survey reported being chronic dieters, defined as always on a diet or having been on a diet for 10 of the previous 12 months.[40]

- **Overweight and obesity.** One in five adolescents (21 percent) in the United States is overweight. The prevalence of overweight in this group has more than doubled in the past 30 years, with most of the increase occurring during the 1970s.[41] Some subgroups of this population are at particular risk. American Indian teenagers of both sexes have higher BMIs than teenagers of the general population,[42] and about 20 percent of Hispanic girls aged 12 to 18 years exceed the 85th percentile of U.S. standards for BMI. In general, Hispanic children living in the United States carry significantly greater amounts of body fat than do their non-Hispanic counterparts.[43]

Given that childhood obesity can lead to adult obesity and that approximately 25 to 30 percent of the adult population is overweight, obesity among teenagers is cause for concern. Genetic susceptibility to obesity, lifestyle, family eating patterns, lack of positive role models, and inactivity all contribute to overweight and obesity in this population. Obese teenagers are also at risk for hypertension and diabetes mellitus, among other disorders.[44] In light of current weight trends, it is not likely that the health objective to reduce overweight among adolescents to a prevalence of no more than 15 percent by the year 2000 will be met.

• **Iron deficiency and iron-deficiency anemia.** Iron needs increase during adolescence, especially for females. In boys, the requirements for absorbed iron increase from about 1.0 to 2.5 milligrams/day. This increase reflects the expanding blood volume and rise in hemoglobin concentration that accompanies sexual maturation. (After the adolescent growth spurt, the need for iron falls off.)[45] Adolescent girls and women have small obligatory iron losses of about 0.8 milligrams/day. Once menses begins, usually around age 13, they have an additional iron loss of about 0.6 milligrams/day when averaged over 1 month. Thus, the total daily iron loss of menstruating women is about 1.4 milligrams. Some adolescents and women have daily iron losses as high as 2.4 milligrams. An oft-quoted average daily iron loss or iron requirement for women is 1.5 milligrams/day.[46] Whereas most males have an adequate iron intake during adolescence, many females do not. Data from the 1996 Continuing Survey of Food Intakes by Individuals indicate that females between 12 and 19 years of age have iron intakes below the RDA.[47]

• **High blood cholesterol levels.** Teenagers have many of the same risk factors for high blood cholesterol as adults: family history of coronary heart disease; diets high in total fat, saturated fat, and cholesterol; hypertension; low activity levels; and smoking. Although the process of atherosclerosis is not completely understood, it is believed that the fatty streaks that develop in young people progress to the fibrous plaques of adulthood.[48]

• **Dental caries.** Dental caries are a significant public health problem among teenagers, with about 78 percent of adolescents having one or more caries in permanent or primary teeth.[49] Dental caries are more prevalent among adolescents than among children. A National Dental Research Survey found that only 22 percent of 15-year-olds were caries-free, compared with

Successful nutrition education activities help children and teenagers focus on their interests and the relationship of good eating and physical activity to health.

56 percent of 10-year-olds. Some population subgroups, such as American Indians, Alaska Natives, blacks, and Hispanics, are at greater risk of dental caries than other groups because they lack access to or do not avail themselves of dental services.[50]

• **Eating disorders.** Eating disorders have become serious health problems in recent years. The most common eating disorders are anorexia nervosa and bulimia nervosa. A constellation of individual, familial, sociocultural, and biological factors contribute to these disorders, which threaten physical health and psychological well-being. Some individuals are more predisposed to develop an eating disorder than others. For example, about 90 percent of people with eating disorders are female. Most are Caucasian, with few cases seen among blacks and other minority groups. Finally, most individuals who develop eating disorders are adolescents or young adults, who typically begin experiencing food-related and self image problems between the ages of 14 and 30 years. Because these syndromes are surrounded by secrecy, their prevalence is not known with certainty, although it has increased dramatically within the past two decades. Anorexia nervosa and bulimia nervosa may affect about 3 percent of all teenage girls in the United States. These types of eating disorders are often seen in adolescent athletes, many of whom compete in sports such as gymnastics, wrestling, distance running, diving, horse racing, and swimming that demand a rigid control of body weight.[51]

Domestic Child and Adolescent Nutrition Programs

Federal programs addressing the nutritional needs of children and adolescents have existed since the beginning of this century. The Children's Bureau, at one time part of the U.S. Department of Labor, issued dietary advice to parents and teachers and conducted nutrition surveys of low-income children. School feeding programs, augmented by the financial support of local school districts, philanthropic organizations, and private donors, likewise began in the early 1900s. Federal involvement in school food programs increased during the 1930s with the passage of an amendment to the Agricultural Act of 1933, which established a fund to purchase surplus agricultural commodities for donation to needy families and child nutrition programs, including school lunch programs. The Social Security Act of 1935 authorized grants to the states for health services for children and mothers and formulated a national program of nutrition services that included assessment, counseling, referral, and follow-up for this population.

In 1946, the National School Lunch Act, which authorized permanent grants-in-aid to the states, was enacted. To receive cash and commodity assistance under the statute, states had to operate school lunch programs on a nonprofit basis, serve free or reduced-price lunches for needy children, and provide lunches that met certain federal standards. Twenty years later, in 1966, the Child Nutrition Act was passed. It expanded federal efforts to improve child nutrition by establishing numerous programs of year-round food assistance to children of all ages. These programs led to the authorization of the Special Supplemental Food Program for Women, Infants and Children (WIC) in 1972. (This program was later renamed the Special Supplemental Nutrition Program for Women, Infants and Children.) Many of the programs and policies developed during the 1960s and 1970s still exist today, although most have been modified over the years. This section describes the major domestic nutrition programs for children and adolescents.[52]

Nutrition Programs of the U.S. Department of Agriculture

The Food and Nutrition Service was established in 1969 to administer the food assistance programs of the U.S. Department of Agriculture (USDA). The primary aim of this agency is to make food assistance available to people who need it. Other goals include improving the eating habits of U.S. children and stabilizing farm prices through the distribution of surplus foods. In this section, we examine three of the agency's largest food assistance programs for children, the School Breakfast Program, the National School Lunch Program, and the Summer Food Service Program for Children. Refer to Table 14-6 for a brief description of the programs designed specifically for children and to Table 14-7 for an outline of programs that benefit children by assisting their families. These programs were also described in Chapters 12 and 13.

• **School Breakfast Program.** The School Breakfast Program was authorized by the Child Nutrition Act of 1966 (as amended, PL 89-642, PL 94-105, PL 95-627). The program was initially introduced as a two-year pilot project in 1966 and then was made permanent and extended to all public and nonprofit private schools of high school grade or under in 1975. The program is administered in the same manner as the School Lunch Program. Through **formula grants,** the program helps states provide a nutritious, nonprofit breakfast for students. Participating schools must follow standard meal patterns in which breakfasts provide one-fourth of the RDA over time. The breakfast can be either hot or cold, but it must include the foods shown in Table 14-8. All students attending a school where the program is offered are eligible to participate. Breakfast is served free or at a reduced price to students from households with incomes at or below the income eligibility guidelines. In October, 1997, more than 7.3 million children participated in breakfast programs at a cost of $152 million.

> **Formula grants** A type of funding mechanism in which the funding agency distributes funds to states on the basis of a "formula" that takes into account any number of factors, such as the number of breakfasts or lunches served to eligible children, the number of breakfasts or lunches served free or at a reduced price, and the national average payment for the program.

A universal school breakfast program demonstration project was included recently in legislation passed by the U.S. House of Representatives. The demonstration project is planned for elementary schools in five states. Children in these schools will receive a free breakfast, regardless of income. The continued interest in the universal breakfast program echoes the philosophy of Representative Joseph Kennedy (D-MA), who remarked, "Breakfast is as important a part of the school day as buses, books, or blackboards."[53]

• **National School Lunch Program.** School lunch programs began informally and unofficially in 1935 with the donation of commodity foods to schools. The National School Lunch Program became official in 1946 with the passage of the National School Lunch Act (as amended, PL 79-396). Through formula grants, the program helps states make lunches available to schoolchildren and encourages the consumption of domestic agricultural commodities. The program is administered at the federal level by the USDA's Food and Nutrition Service, at the state level through state agencies, and at the local level through school boards. More than 92,000 schools and residential child care institutions take part in this program. Read the Program Spotlight for a detailed description of the school lunch program.

• **Summer Food Service Program for Children.** This entitlement program was created by Congress in 1968.[54] It provides funds for any eligible sponsoring organization such as public

TABLE 14-6 USDA *Food Assistance Programs Specifically for Children*

Program	Purpose(s)	Type of Assistance	Eligibility Requirements of Program Participants
National School Lunch Program	Assist states in making the school lunch program available to students and encourage the domestic consumption of nutritious agricultural commodities. Public and nonprofit private schools of high school grade and under, public and private nonprofit residential child care institutions (except Job Corp Centers), residential summer camps that participate in the Summer Food Service Program for Children, and private foster homes are eligible to participate	Formula grants	1. All students attending schools where the lunch program is operating may participate. 2. Lunch is served free to students who are determined by local school authorities to live in households with incomes at or below 130 percent of the federal poverty level. 3. Lunch is served at a reduced price to students who live in households with incomes between 130 percent and 185 percent of the poverty level. 4. Students from families with incomes over 185 percent of the poverty level pay full price for lunch.
School Breakfast Program	Assist states in providing a nutritious, nonprofit breakfast for school students. Eligible schools and residential child care facilities are the same as for the National School Lunch Program	Formula grants	Eligibility requirements are the same as for the National School Lunch Program.
Special Milk Program for Children	Provide subsidies to schools and institutions to encourage the consumption of fluid milk by children. Any public or private nonprofit school or child care institution (for example, nursery school, child-care center) of high school grade or under (except Job Corps Centers) may participate on request if it does not participate in a meal service program authorized under the National School Lunch Act or the Child Nutrition Act of 1966	Formula grants	All students attending schools and institutions in which the program is operating may participate.
Summer Food Service Program for Children	Assist states in conducting nonprofit food service programs for low-income children during the summer months and at other approved times, when area schools are closed for vacation	Formula grants	Homeless children and children attending public or private nonprofit schools and residential camps or participating in the National Youth Sports Program

Source: *Catalog of Federal Domestic Assistance Programs*, June, 1996, available at family.info.gov/cfda/index.htm.

Program	Purpose(s)
Child and Adult Care Food Program	Assist states in initiating, maintaining, and expanding nonprofit food service programs for children and elderly or impaired adults in nonresidential day care facilities. New legislation includes a provision to allow after-school programs operated by community groups to serve snacks to teenagers aged 12–18 years in low-income areas.[†]
Commodity Supplemental Food Program	Improve the health and nutritional status of low-income pregnant, postpartum and breastfeeding women, infants, and children up to 6 years of age, and elderly persons through the donation of supplemental foods
Emergency Food Assistance Program (Food Commodities)	Make food commodities available to states for distribution to needy persons such as the unemployed, welfare recipients, or low-income individuals
Food Commodities for Soup Kitchens	Improve the diets of the homeless
Food Distribution (popularly called the Food Donation Program)	Improve the diets of school and preschool children and other groups and increase the market for domestically produced foods acquired under surplus removal or price support operations
Food Distribution Program on Indian Reservations	Improve the diets of needy persons in households on or near Indian reservations and to increase the market for domestically produced foods acquired under surplus removal or price support operations
Food Stamp Program	Improve the diets of low-income households by increasing their ability to purchase foods
Homeless Children Nutrition Program	Assist state, city, county, and local governments, other public institutions, and private nonprofit organizations in providing food service throughout the year to homeless children under the age of 6 in emergency shelters
Nutrition Assistance for Puerto Rico	Improve the diets of needy persons and families living in Puerto Rico through a cash grant alternative to the Food Stamp Program
Special Supplemental Nutrition Program for Women, Infants, and Children	Provide, at no cost, supplemental nutritious foods, nutrition education, and referrals to health care to low-income pregnant, breastfeeding and postpartum women, infants, and children to age 5 who are determined to be at nutritional risk
WIC Farmers' Market Nutrition Program	Provide fresh, nutritious unprepared foods such as fruits and vegetables from farmers' markets to low-income women, infants, and children; expand the awareness and use of farmers' markets; and increase sales at farmers' markets

TABLE 14-7 *Other USDA Food Assistance Programs that Benefit Children**

*Refer to Table 12-5 on pages 360–363 for a complete description of the programs shown in this table.

[†]Community Nutrition Institute, *Nutrition Week,* June 5, 1998, p. 1.

Source: *Catalog of Federal Domestic Assistance Programs,* June, 1996, available at family.info.gov/cfda/index.htm.

TABLE 14-8 *Foods Required for Breakfasts Provided Under the School Breakfast Program*

- ½ pt of fluid milk as a beverage, on cereal, or both—one serving; AND

- ½ c serving fruit OR ½ c full-strength fruit or vegetable juice—one serving; AND

- bread or bread-alternate (one slice whole-grain or enriched bread, or an equivalent serving of cornbread, biscuits, rolls, muffins, etc.; or ¾ c or 1 oz serving of cereal)—two servings; OR

- meat or meat-alternate (one serving of protein-rich foods such as an egg; or a 1 oz serving of meat, poultry, fish, or cheese; or 2 tbsp of peanut butter)—two servings; OR

- one bread AND one meat

Source: *Fact Sheets on the Federal Food Assistance Programs* (Washington, D.C.: Food Research and Action Center, May 1993), pp. 11–12. Used with permission from the Food Research and Action Center.

and private nonprofit school food authorities, residential camps, youth sports camps, nonprofit organizations, and units of local, municipal, county, tribal, or state government for the purpose of serving nutritious meals to needy children. The main target group of this program is children from households with incomes below 185 percent of the federal poverty line who would otherwise not have access to nutritious meals during the summer months when school is not in session.

State agencies receive funds from USDA to administer the program and are reimbursed either on a per-meal basis or for the actual cost of running the program, whichever is less. The per-meal reimbursement rate is divided into administrative and operating costs, as shown in Table 14-9. Administrative costs are those incurred in the management of the program, including office expenses, salaries, and insurance. Operating costs are those incurred in food preparation and distribution, transportation of food in rural areas, and program activities for children participating in the program.

State agencies reimburse eligible sponsors for meals served to children at feeding sites. Sponsors may be reimbursed for lunch and either breakfast or a snack per day. Youth sports and residential camps may be reimbursed for up to four meals per day (breakfast, lunch, supper, snack). There are two types of eligible feeding sites:

"Open sites" are located in areas where 50 percent or more of the children come from families whose household income is below 185 percent of the federal poverty level. Sponsors of open site programs may be reimbursed for all eligible meals served, regardless of the income of each individual child.

"Enrolled sites" are those where 50 percent of the children *attending* the program come from families whose income is under 185 percent of the federal poverty line. Sponsors of "enrolled sites" must document the incomes of participating children. They may be reimbursed for all eligible meals served to children participating in the program.

The average daily attendance in the Summer Food Service Program for Children was about 2.22 million children in 1996. Participation rates vary among states, largely because of differences in the quality of leadership in state agencies, the level of cooperation between state agencies and sponsors, and state budget priorities. Even state geography affects participation rate, with mainly rural states having lower rates than mainly urban states.

| Meal Type | Operating Costs | | Administrative Costs | | |
			Urban or Vended*	or	Rural or Self-preparation†
Breakfast	$1.19	+	$0.0925	or	$0.1175
Lunch	2.08	+	0.18	or	0.2175
Supper	2.08	+	0.18	or	0.2175
Snack	0.48	+	0.475	or	0.06

TABLE 14-9 *Reimbursement Rates for Sponsors of the Summer Food Service Program,* 1998

*A vended program is one in which the sponsor contracts with another food service operator to prepare meals.
†A self-preparation site is one in which the sponsor prepares its own meals.

Source: *Hunger Doesn't Take a Vacation: A Status Report on the Summer Food Service Program for Children,* 5th ed. (Washington, D.C.: Food Research and Action Center, 1997), p. 16, and "Summer Food Service Program—Maximum per Meal Reimbursement Rates, Effective January 1, 1998—December 31, 1998," available on the U.S. Department of Agriculture's Web site at www.usda.gov/fcs.

Nutrition Programs of the U.S. Department of Health and Human Services

Recall from Chapter 13 that several DHHS programs include a child care component: Title V Maternal and Child Health Program; Medicaid and the Early and Periodic Screening, Diagnosis, and Treatment (EPSDT) programs; and the primary care programs of Community Health Centers. Another DHHS program that benefits children is the Head Start Program.

Initiated in 1965 by the Office of Economic Opportunity, Head Start was authorized by the Economic Opportunity and Community Partnership Act of 1967 (PL 93-644, as amended in 1974). The program is coordinated by the Administration for Children and Families within the DHHS. It is one of the most successful federal government programs for child development. Head Start provides children from low-income families with comprehensive education, social, health, and nutrition services. Eligible children range in age from birth to the age at which they begin school. Parental involvement in the program planning and operation is emphasized. Head Start projects provide meals and snacks as well as nutrition assessment and education for children and their parents. The nutrition services are meant to complement the health and education components of the program. In FY1998, about 830,000 preschool children from low-income families were enrolled. The Select Panel for the Promotion of Child Health reported that Head Start has been shown to improve children's health. Children participating in Head Start have a lower incidence of anemia, receive more immunizations, and have better nutrition and improved overall health compared with children who do not participate.[55]

Programs for Children with Special Health Care Needs

In the early 1970s, the U.S. courts began to accord greater recognition to the rights of children with disabilities. In 1975, the growing advocacy for children with special needs resulted in the passage of PL 94-142, the Education for All Handicapped Children Act. This statute requires states to provide a "free appropriate public education" for all children aged 3 to 21 years with disabilities. For those who require special education because of their impairments, states must provide individualized instruction designed to meet their special needs and any related or supportive services that will enable the children to benefit from the education program. Related services include school health services, counseling services, and parent counseling and training.[56]

In 1986, Congress passed the Education of the Handicapped Act Amendments (PL 99-457) to provide added incentives to the states to expand public education for children between 3 and 5 years of age.[57] This legislation mandates the provision of comprehensive services, including nutrition services, to children with special health care needs, using a community-based approach that focuses on the family.[58] The legislation also recognizes nutritionists as the health professionals qualified to provide developmental services to children with special health care needs.

Surveys reveal some deficiencies in delivering these services. For example, in the southeastern and southwestern United States in 1983 and 1984, only seven full-time equivalent nutrition positions were in the State Programs for Children with Special Health Care Needs for more than 200,000 children. A survey of these programs conducted in 1986 found that fewer than 25 state programs had full-time nutrition consultants.[59] In some instances, children with special health care needs are served by dietitians or nutritionists in WIC programs, Head Start, public health departments, private clinics, university-based programs, or group homes.

Fortunately, nutrition care for these children has improved in recent years because of new legislation, increased involvement of various agencies, improved delivery of community-based programs, and better home health care. The dietetics community is committed to helping families and communities provide needed nutrition services for children with special health care needs. The American Dietetic Association outlined its position on the importance of providing nutrition services to persons with developmental disabilities, as follows:

- The American Dietetic Association encourages and supports nutrition services which are coordinated, interdisciplinary, family centered, and community based for children with special health care needs.[60]
- It is the position of The American Dietetic Association that persons with developmental disabilities should receive comprehensive nutrition services as part of all health care, vocational, and educational programs.[61]

Nutrition Education Programs

Nutrition education strategies aimed at children or their caregivers are found in both the public and private sectors. They strive to improve eating patterns among children. Numerous government and commercial sites on the Internet provide databases of educational materials and resources and, in some cases, offer interactive games and puzzles for children. Refer to the boxed insert on page 447 for a list of government and commercial Web sites related to children and adolescents. Public and private sector nutrition education initiatives are described beginning on page 445.

| Program Spotlight | *National School Lunch Program* |

When Congress passed the National School Lunch Act in 1946, it had the following policy objectives:

It is hereby declared to be the policy of Congress, as a measure of national security, to safeguard the health and well-being of the Nation's children and to encourage the domestic consumption of nutritious agricultural commodities and other food, by assisting the states, through grants-in-aid and other means in providing an adequate supply of foods and other facilities for the establishment, maintenance, operation, and expansion of non-profit school lunch programs.[1]

The statute requires schools to serve lunches free or at a reduced price to students *who are determined by local authorities* to be unable to pay full price because they come from low-income families.[2] The income guidelines for children participating in child nutrition programs are shown in Chapter 12 (Table 12-7 on page 370).

Schools that elect to participate in the program get cash subsidies and donated commodities from the USDA for each meal served. In return, the lunches they serve must meet federal requirements and must be offered free or at reduced price to eligible children. Specifically, the lunches must include the following:

- 8 ounces fluid milk
- 2 ounces protein (meat, fish, cheese, two eggs, four tablespoons peanut butter, or one cup dried beans or peas)
- 3/4 cup serving consisting of two or more vegetables or fruits or both (juice can meet half of this requirement)
- 8 servings bread, pasta or grain per week

The law requires school districts to prepare and publish a statement of the criteria used for free and reduced-price lunches. The officials responsible for determining a child's eligibility must be identified and the procedural steps used in reaching eligibility decisions specified. A system for appeals in individual cases must exist. The names of children who receive free or reduced-price lunches cannot be "published, posted, or announced in any manner to other children," nor can children who participate in the program be required to use a separate lunchroom, lunchtime, serving line, cafeteria entrance, or medium of exchange.[3]

Problems with Participation

The passage of the National School Lunch Act brought federal recognition to the importance of nutrition to learning and growth. Senator George McGovern, chairman of the Senate Select Committee on Nutrition and Human Needs, observed in 1970 that "the need for adequate food during the school day stands as an obvious one. A child cannot learn if he is hungry—that is a simple and undeniable fact. Hunger makes him restless, lethargic, and physically ill. Without at least one nutritious meal during his 4- to 7-hour stay at the school, no child can benefit from the tremendous educational opportunities offered in American schools today."[4]

Almost from the very beginning, however, there were problems with participation in the program. A survey conducted in 1967 by Their Daily Bread, a women's organization, showed that two out of three children did not participate in the program.[5] Another survey carried out in 1968 confirmed this finding. Of some 50 million eligible schoolchildren, only 18 million children (36 percent) took part in the program. Only about 55 percent of schools participated in the program in 1968.

A number of reasons were cited for the low participation rates during the late 1960s. First, many parents objected to the manner in which school officials determined which children were eligible. Parents were ashamed to sign an affidavit verifying their low income and thus effectively prevented their children from participating. To make matters worse, schools sometimes failed to protect the identity of low-income families. Also, the lack of uniform national standards for eligibility led to inequities. Without uniform guidelines, for example, a child's deportment or attendance record could influence the decision of local officials. Second, some schools refused to participate because they perceived the administrative tasks as being too demanding. Many communities chose not to participate and, instead, elected to reinforce the concept of the local school where children went home for lunch. Finally, some states diverted school lunch funds meant for needy children to other foodservice purposes or to programs for middle-income children.

Some of these concerns have been addressed over the last two decades. For example, in 1989 the law was changed so that schools can "directly certify as eligible" those children whose families participate in food stamp or Aid for Families with Dependent Children programs. Efforts to

Continued

reduce paperwork and the demands on staff time have resulted in increased program participation. In 1997, more than 95,000 schools and residential child care institutions offered a school lunch. This figure represents about 99 percent of all public schools. About 58 percent of all public schoolchildren participate in the program. More than 26 million children got their lunch through the program in FY1997.

Does the Program Work?

In 1977, the U.S. General Accounting Office considered whether the school lunch program was working. Although the program had been operating for more than 30 years, no major studies of its effectiveness had been undertaken. The results of one study, published in 1984, showed that participating children from families at all income levels had superior nutrient intakes at lunch and higher total daily intakes than children who did not participate. In addition, the quality of the food was higher for program participants than for children who purchased their lunch from other vendors or brought their lunch from home. These findings have been confirmed by other researchers.[6]

Although evaluations have consistently shown that the National School Lunch Program makes a significant contribution to the daily nutrient and food intakes of participating children, there are concerns that school lunch meals are not as healthy as they might be. A 1993 USDA survey of 545 schools and 3,350 students found that 38 percent of calories in the average USDA school lunch came from fat, 15 percent of calories came from saturated fat, and the typical lunch contained 1,479 milligrams of sodium. These intakes exceeded the USDA's recommendation that children consume no more than 30 percent of calories from fat, 10 percent of calories from saturated fat, and 800 milligrams of sodium at lunch. On the positive side, the survey found that students who consumed the USDA school lunch exceeded the target goal of one-third the RDA for all vitamins and minerals. The typical school lunch also provided 95 percent of the RDA for protein. These findings led the USDA to call for a reduction in the amount of fat, saturated fat, and sodium in school lunches.[7]

Team Nutrition: Making a Good Program Better

In 1995, the USDA launched its School Meals Initiative for Healthy Children to "improve the health and education of children through better nutrition." The initiative's overall aim is to provide meals that are consistent with the Dietary Guidelines for Americans and other current scientific recommendations for children at school. It is implemented through a national program called "Team Nutrition," which aims to improve child health and nutrition by developing creative public and private partnerships through the media, schools, businesses, families, and the community. Program elements include a Team Nutrition Supporter Network of more than 200 organizations; spokestoons Timon and Pumbaa from Disney's *The Lion King,* which appear in public service announcements and other promotional materials; Pilot Communities, which serve as model demonstration projects; Team Nutrition Schools, a national network of local schools that share ideas and strategies; and Training Grants, which provide competitive funding for states to help local school districts. The Healthy School Meals Resource System is an information system for food service professionals available in print form, computer disk, and on the Internet. The Team Nutrition logo is shown in Figure 14-2.[8]

FIGURE 14-2 *The Team Nutrition Logo*

Toward a Universal School Lunch Program

In 1970, Senator McGovern and other policy makers argued that schools had the responsibility to provide food during the day in much the same way as they provide classroom facilities, free bus rides, and free textbooks. Their idea was to move the nation toward a universal school lunch program that would be available to all schoolchildren, regardless of income level. In 1975, Senator Hubert Humphrey introduced S. 3123, a bill to establish such a program. The proposal recognized schools as the proper delivery system for ensuring that all children have proper and adequate nourishment during the school day.[9]

Although no universal program exists today, nutritionists remain interested in the concept, according to the Food

Research and Action Center. Indeed, members of the American School Food Service Association lobbied Congress to pass a bill that would allow all students to receive a free breakfast and lunch, regardless of their family income. The Universal Student Nutrition Act of 1993 (HR 11), introduced by Representative George Miller (D-CA), would give schools the option of offering free meals to all students. Under this "universal" system, schools would not have to verify family income, thus reducing administrative work. If the bill is passed by Congress, the universal program would not be implemented fully until July 2000.[10]

References

1. The quotation was cited in J. E. Austin and C. Hitt, *Nutrition Intervention in the United States* (Cambridge, Mass.: Ballinger, 1979), p. 93.
2. *1993 Catalog of Federal Domestic Assistance* (Washington, D.C.: U.S. General Services Administration, 1993); and *Fact Sheets on the Federal Food Assistance Programs* (Washington, D.C.: Food Research and Action Center, May 1993); and Nutrition Program Facts, National School Lunch Program, available at www.usda.gov/fcs.
3. Hearings before the Select Committee on Nutrition and Human Needs of the United States Senate, Part 8—Review of the National School Lunch Act (Washington, D.C.: U.S. Government Printing Office, October 13, 1970), pp. 2166–67.
4. Austin and Hitt, *Nutrition Intervention*, p. 91.
5. The description of the survey by Their Daily Bread was taken from the Senate Hearings, Part 8, p. 2121.
6. The description of the effectiveness of the School Lunch Program was adapted from Chapter III: Impact of hunger and malnutrition on student achievement, *School Food Service Research Review* 13 (1989): 17–21.
7. USDA uses school food report to blast meals' excessive fat, sodium, *Nutrition Week* 23 (October 29, 1993): 1, 3.
8. Information about the USDA's School Meals Initiative for Healthy Children and Team Nutrition program was adapted from the following Web site: http://schoolmeals.nal.usda.gov:8001.
9. Austin and Hitt, *Nutrition Intervention*, p. 119.
10. Meeting in Washington, school food advocates push for universal meals, *Nutrition Week* 23 (February 26, 1993): 1.

Nutrition Education in the Public Sector. Public nutrition education programs include the Expanded Food and Nutrition Education Program, described in Chapter 13, and the Nutrition Education and Training program, which is described here. A new federal initiative is the *YourSELF* program, also described below.

• **Nutrition Education and Training.** The Nutrition Education and Training (NET) program was established in 1977 by PL 95-166, an amendment to the Child Nutrition Act of 1966; it is administered by the USDA's Food and Nutrition Service. NET's purpose is to disseminate scientifically valid information about diet and health to children participating or eligible to participate in school lunch and related child nutrition programs. NET funds are distributed through formula grants to state educational agencies for use by local schools and districts in providing nutrition-related training for teachers and school food service personnel. NET funds are also used to conduct nutrition education activities that use both the classroom and the school's food service facility. The NET program is designed to help children learn to apply the basic principles of healthful eating and good nutrition to their daily lives and to acquire a firm understanding of the relationship of good nutrition to good health. In 1995, about 101,000 school food service personnel, 117,000 teachers, and 56,000 schools participated in NET-funded programs.

• *YourSELF.* YourSELF is one of the first federal initiatives on nutrition and physical activity. It is designed to speak directly to adolescents in the 7th and 8th grades. Released in 1998 by USDA, the *YourSELF* Middle School Nutrition Education Kit contains materials for health education, home economics, or family living classes that teach students to make healthy eating choices and become physically active. The kit includes the *yourSELF* magazine, a student workbook, teacher's guide, reproducible masters, and ideas for linking classroom and cafeteria activities.[62] A portion of one of the *YourSELF* handouts is shown in Figure 14-3.

Nutrition Education in the Private Sector. Some voluntary, nonprofit health organizations have developed nutrition education programs or materials for children or their caregivers. The American Heart Association (AHA), for example, has a variety of Schoolsite Program modules for children in kindergarten through grade 12. Each module comes with a teacher's guide, learning and kickoff activities, videotape, and handouts for students, all geared to the appropriate grade level.[63] The American Cancer Society developed "Changing the Course," a comprehensive nutrition education program for students in kindergarten through grade 12. Changing the Course uses a video, together with handouts and overheads, to help students develop good eating habits and reduce cancer risk by eating more fruits and

FIGURE 14-3 A *Nutrition Education Activity from USDA's YourSELF Program for 7th and 8th Graders*

Source: U.S. Department of Agriculture, *YourSELF Middle School Nutrition Education Kit,* 1998, available at www.usda.gov/fcs.

Let's Eat!

Treat your brain to this puzzle. When you're through, give your brain a break: eat something! Answers page 8.

```
T D I N H R B E A N S D I O A E A B
C E R E A L A U T N B M I D A T N E
E R E G G S W C O C N I T R P P O T
P I Z Z A F Y T M H T L I O P O S F
E R N W A O E W A E W K Y T L R E E
S O C P A S T A T D I E O L E K T A
D H E E A I A A O D E Z G I R C E S
H U G R A P E S E E A H U O A H B E
M U F F I N D U S R H E R Y O O P E
Y B R O C C O L I C N A T H N P A R
D N I T I O M T E H E G T B H V N E
P I A E C R A C K E R S B U F E C C
K I W I D A I U H E U L T R S C A R
O O S D A N E E T S A H P R Y S K I
S O R H R G E F T E C P N I R E E I
D G O E H E L A S A G N A T I N S D
P E P P E R S T E W N A S O A E E S
T S P E A R O N V E O N S I A O E D
```

How many words can you find? Hidden within this puzzle are:

THREE MILK GROUP FOODS

THREE MEAT GROUP FOODS

FIVE FRUIT GROUP FOODS

THREE VEGETABLE GROUP FOODS

FIVE BREAD GROUP FOODS

THREE COMBINATION FOODS

Internet Resources

Government and University Web Sites

Administration for Children and Families	www.acf.dhhs.gov
Adolescence Directory On-Line	education.indiana.edu/cas/adol/adol.html
Head Start Bureau	www.acf.dhhs.gov/programs/hsb
National Clearinghouse on Families & Youth	www.ncfy.com
Nutrition Expedition	Fscn.che.umn.edu/nutrexp/default.html
USDA's School Meals Initiative for Healthy Children/Team Nutrition	schoolmeals.nal.usda.gov:8001

Disability Web Sites

Cornucopia of Disability Information (CODI)	codi.buffalo.edu
Government Resources—General Disability Information	mirconnect.com/feds/general.html
National Council on Disability	www.ncd.gov
National Information Center for Children and Youth with Disabilities	www.nichcy.org

Other Web Sites

Kids Count	www.aecf.org/aeckids.htm
Kids Food Cyberclub	www.kidsfood.org
KidsHealth	www.kidshealth.org
Nutrition on the Web—for teens	hyperion.advanced.org/10991
Produce for Better Health Foundation	www.5aday.com
Slack's Pediatric Internet Directory	www.slackinc.com/child/pednet-x.htm

vegetables and less fat.[64] A variety of books, pamphlets, and brochures written for children or their caregivers are available from the American Dietetic Association, the American Academy of Pediatrics, and other organizations.

Trade organizations likewise direct some of their nutrition education strategies to children. Brochures or booklets that describe good child feeding practices, outline tips for nutritious snacks, or answer questions about foods or ingredients are available from the National Cattlemen's Beef Association, the National Dairy Council, and the Sugar Association, among others. Puzzles, posters, and games designed to interest children in foods and nutrition can be purchased from such companies as NCES (Nutrition Counseling and Education Services) and Nasco Nutrition Teaching Aids.

Keeping Children and Adolescents Healthy

Programs and services designed to keep children and adolescents healthy can have a lasting effect on the nation's public health. Healthy children and adolescents mean healthy communities. Children and youth who learn good eating habits, exercise regularly, refrain from smoking, learn to manage stress, and develop a strong sense of self are less likely to turn to drugs and alcohol to solve problems and more likely to know how to live constructively.

Successful, effective programs or services for children and youth recognize the stresses of life in the late twentieth century and the mixed media messages about products and values. Programs and services founded on basic health promotion principles must consider the urgent health issues facing today's young people: suicide, child abuse, teen pregnancy, eating disorders, substance abuse, and others. Positive nutrition messages support and expand on other health promotion concepts.

What types of programs work? Health educators have found that programming for children and adolescents succeeds when it is fun and informative. Programs work best when they are geared to a specific health or nutritional objective, such as weight loss, improved eating patterns, or increased activity levels. For instance, Wings of America, a youth program for Indians based in Santa Fe, New Mexico, has sparked a running revival among American Indian children and teenagers. Wings teams help Indian children develop pride, self-esteem, and cultural identity, in addition to promoting fitness.[65] Involving children and adolescents in the planning and implementation of a program increases its effectiveness, as does using peer support to help the participants make decisions about their health. Effective programs also employ trained staff who work well with these populations. The youth development approach offers a strategy for linking young people with their communities and involving them in designing and implementing programs and services.[66]

In the final analysis, developing programs and services that meet the needs of our nation's children and youth means recognizing that as today's children grow, they have to perform increasingly complex tasks in a world of constant technological and environmental change. Improving the health of today's young people will enhance the quality of their lives in the immediate future and expand their potential for contributing to our nation's future as adults. In promoting the health and well-being of children and youth, we are recognizing "that children matter for themselves, that childhood has its own intrinsic value, and that society has an obligation to enhance the lives of children today."[67]

Community Learning Activity

Activity 1

Let's assume that you are responsible for evaluating either the School Breakfast Program or Head Start (choose one).

1. What is the first step you would take in evaluating the program's effectiveness? Why did you begin at this point?

2. Describe the outcomes and indicators you would select to determine whether the program meets the needs of its clients. Why did you select these particular outcomes and indicators?

3. Let's assume that the program is *not* meeting the needs of its participants or the target population. What steps would you take to correct the situation as it relates to the following areas?
   a. Program delivery
   b. Policy

4. Describe one marketing strategy you would undertake to enhance program participation.

References

1. P. A. Colón and A. R. Colón, The health of America's children, in *Caring for America's Children*, ed. F. J. Macchiarola and A. Gartner (New York: Academy of Political Science, 1989), pp. 45–57.

2. Report to the President, *White House Conference on Children* (Washington, D.C., 1970), p. 11.

3. U.S. Department of Health and Human Services, Public Health Service, *The Surgeon General's Report on Nutrition and Health*, DHHS Pub. No. 88-50210 (Washington, D.C.: U.S. Government Printing Office, 1988), pp. 539–93.

4. U.S. Department of Health and Human Services, *Health United States 1991 and Prevention Profile* (Hyattsville, MD: U.S. Department of Health and Human Services, 1992), pp. 78–80 and 93–98; and U.S. Department of Health and Human Services, Public Health Service, *Healthy People: The Surgeon General's Report on Health Promotion and Disease Prevention*, PHS Pub. No. 79-55071 (Washington, D.C.: U.S. Government Printing Office, 1979).

5. U.S. Department of Health and Human Services, Public Health Service, *Healthy People 2000: National Health Promotion and Disease Prevention Objectives* (Washington, D.C.: U.S. Government Printing Office, 1990), pp. 114–34.

6. National Center for Health Statistics. *Healthy People 2000 Review, 1997* (Hyattsville, MD: Public Health Service, 1997), pp. 34–46. Data were taken from the on-line publication at www.cdc.gov/nchswww.

7. The general descriptions of child growth and development and nutrition assessment methods were adapted from E. N. Whitney and S. R. Rolfes, *Understanding Nutrition*, 6th ed. (St. Paul, MN: West, 1993), pp. 514–30 and E-1–E-37.

8. R. Yip and coauthors, Pediatric Nutrition Surveillance System—United States, 1980–1991, *Morbidity and Mortality Weekly Report* 41/No. SS-7 (November 27, 1992): 1–24.

9. P. C. DuRousseau and coauthors, Children in foster care: Are they at nutritional risk? *Journal of the American Dietetic Association* 91 (1991): 83–85; and M. L. Taylor and S. A. Koblinsky, Dietary intake and growth status of young homeless children, *Journal of the American Dietetic Association* 93 (1993): 464–66.

10. Community Nutrition Institute, Nearly six million young children live in poverty, *Nutrition Week* 28 (March 20, 1998): 2.

11. The definitions of overweight and obesity were taken from R. P. Troiano and K. M. Flegal, Overweight children and adolescents: Description, epidemiology, and demographics, *Pediatrics* 101 (1998): 497–504; S. C. Wilkins and coauthors, Family functioning is related to overweight in children, *Journal of the American Dietetic Association* 98 (1998): 572–74; A. Must, G. E. Dallal, and W. H. Dietz, Reference data for obesity: 85th and 95th percentiles of body mass index (wt/ht^2) and triceps skinfold thickness, *American Journal of Clinical Nutrition* 53 (1991): 839–46; and A. Must, G. E. Dallal, and W. H. Dietz, Reference data for obesity: 85th and 95th percentiles of body mass index (wt/ht^2)—A correction, *American Journal of Clinical Nutrition* 54 (1991): 773.

12. J. Dwyer and J. Arent, Child nutrition, in *Call to Action: Better Nutrition for Mothers, Children, and Families*, ed. C. O. Sharbaugh (Washington, D.C.: National Center for Education in Maternal and Child Health, 1991), pp. 151–68.

13. G. Kolata, Obese children: A growing problem, *Science* 232 (1986): 20–21.

14. I. G. Pawson and coauthors, Prevalence of overweight and obesity in US Hispanic populations, *American Journal of Clinical Nutrition* 53 (1991): 1522S–28S.

15. M. Story and coauthors, Nutritional concerns in American Indian and Alaska Native children: Transitions and future directions, *Journal of the American Dietetic Association* 98 (1998): 170–76.

16. M. Knip and O. Nuutinen, Long-term effects of weight reduction on serum lipids and plasma insulin in obese children, *American Journal of Clinical Nutrition* 57 (1993): 490–93; and W. H. Dietz, Jr., Prevention of childhood obesity, *Pediatric Clinics of North America* 33 (1986): 823–33.

17. J. T. Dwyer and coauthors, Predictors of overweight and overfatness in a multiethnic pediatric population, *American Journal of Clinical Nutrition* 67 (1998): 602–10.

18. R. A. Behrens and M. E. Longe, *Hospital-Based Health Promotion Programs for Children and Youth* (Chicago: American Hospital Publishing, 1987), pp. 16–17.

19. *Healthy People 2000 Review,* 1997, p. 45.

20. A. C. Looker and coauthors, Prevalence of iron deficiency in the United States, *Journal of the American Medical Association* 277 (1997): 973–76.

21. Centers for Disease Control and Prevention, Recommendations to prevent and control iron deficiency in the United States, *Morbidity and Mortality Weekly Report* 47, No. RR-3 (1998): 1–29.

22. Yip and coauthors, Pediatric Nutrition Surveillance System.

23. J. D. Sargent and coauthors, Iron deficiency in Massachusetts communities: Socioeconomic and demographic risk factors among children, *American Journal of Public Health* 86 (1996): 544–50.

24. *Healthy People 2000 Review,* 1997, p. 38.

25. Dwyer and Arent, Child nutrition.

26. U.S. Department of Health and Human Services, *Health United States 1991,* p. 78.

27. U.S. Department of Health and Human Services, Public Health Service, *Surgeon General's Report,* pp. 345–80.

28. W. Bao and coauthors, Usefulness of childhood low-density lipoprotein cholesterol level in predicting adult dyslipidemia and other cardiovascular risks: The Bogalusa Heart Study, *Archives of Internal Medicine* 156 (1996): 1315–320.

29. National Cholesterol Education Program, *Report of the Expert Panel on Blood Cholesterol Levels in Children and Adolescents,* NIH Pub. No. 91-2732 (Washington, D.C.: U.S. Department of Health and Human Services, Public Health Service, National Institutes of Health, 1991), pp. 1–22; and American Academy of Pediatrics, Committee on Nutrition, Cholesterol in childhood, *Pediatrics* 101 (1998): 141–47.

30. The discussion of high blood lead levels was adapted from U.S. Department of Health and Human Services, Centers for Disease Control and Prevention, *Health, United States, 1998,* DHHS Pub. No. 98-1232 (Hyattsville, MD: U.S. Government Printing Office, 1998), pp. 62–63, available on-line at www.cdc.gov/nchswww; Interagency Forum on Child and Family Statistics, *America's Children: Key National Indicators of Well-Being, 1998,* pp. 1–3 of "Special Features," available on-line at www.childstats.gov; E. W. Manheimer and E. K. Silbergeld, Critique of CDC's retreat from recommending universal lead screening for children, *Public Health Reports* 113 (1998): 38–46; and N. M. Tips, H. Falk, and R. J. Jackson, CDC's lead screening guidance: A systematic approach to more effective screening, *Public Health Reports* 113 (1998): 47–51.

31. ADA Reports, Position of The American Dietetic Association: Nutrition services for children with special health care needs, *Journal of the American Dietetic Association* 89 (1989): 1133–37.

32. M. McPherson and coauthors, A new definition of children with special health care needs, *Pediatrics* 102 (1998): 137–40.

33. P. W. Newacheck and N. Halfon, Prevalence and impact of disabling chronic conditions in childhood, *American Journal of Public Health* 88 (1998): 610–17.

34. Minnesota Department of Children, Families and Learning, *Nutrition Management for Children with Special Needs* (St. Paul: Minnesota Department of Children, Families and Learning, 1996).

35. L. A. Wodarski, An interdisciplinary nutrition assessment and intervention protocol for children with disabilities, *Journal of the American Dietetic Association* 90 (1990): 1563–68.

36. Arizona Department of Health Services, Office of Nutrition Services, *Project CHANCE: A Guide to Feeding Young Children with Special Needs* (Phoenix: Arizona Department of Health Services, 1995).

37. C. N. Bianculli, Physical growth and development in adolescents, in *Health of Adolescents and Youths in the Americas,* Scientific Pub. No. 489 (Washington, D.C.: Pan American Health Organization, 1985), pp. 45–50.

38. L. W. Green and D. Horton, Adolescent health: Issues and challenges, in *Promoting Adolescent Health: A Dialog on Research and Practice,* ed. T. J. Coates, A. C. Petersen, and C. Perry (New York: Academic Press, 1982), pp. 23–43; data on the calcium intakes of adolescents were taken from U.S. Department of Agriculture, Agricultural Research Service, *Data tables: Results from USDA's 1996 Continuing Survey of Food Intakes by Individuals and 1996 Diet and Health Knowledge Survey* (On-line), ARS Food Surveys Research Group, available under "Releases" at www.barc.usda.gov/bhnrc/foodsurvey/home.htm.

39. D. J. Sherman, The neglected health care needs of street youth, *Public Health Reports* 107 (1992): 433–40.

40. M. Story and coauthors, Adolescent nutrition: Trends and critical issues for the 1990s, in *Call to Action: Better Nutrition for Mothers, Children, and Families,* ed. C. O. Sharbaugh (Washington, D.C.: National Center for Education in Maternal and Child Health, 1991), pp. 169–89.

41. Centers for Disease Control and Prevention, Guidelines for school health programs to promote lifelong healthy eating, *Morbidity and Mortality Weekly Report* 45, No. RR-9 (1996): 1–36.

42. Federation of American Societies for Experimental Biology, Life Sciences Research Office, *Third Report on Nutrition Monitoring in the United States,* vol. 1 (Washington, D.C.: U.S. Government Printing Office, 1995), p. 175.

43. I. G. Pawson and coauthors, Prevalence of overweight and obesity in US Hispanic populations, *American Journal of Clinical Nutrition* 53 (1991): 1522S–28S.

44. U.S. Department of Health and Human Services, Public Health Service, *Surgeon General's Report,* pp. 539–93.

45. L. Brabin and B. J. Brabin, The cost of successful adolescent growth and development in girls in relation to iron and vitamin A status, *American Journal of Clinical Nutrition* 55 (1992): 955–58.

46. National Cattlemen's Beef Association, Iron in Human Nutrition. (Chicago: National Cattlemen's Beef Association, 1998), p. 24; and L. Brabin and B. J. Brabin, The cost of successful adolescent growth and development in girls in

relation to iron and vitamin A status, *American Journal of Clinical Nutrition* 55 (1992): 955-958; L. Hallberg and L. Rossander-Hultén, Iron requirements in menstruating women, *American Journal of Clinical Nutrition* 54 (1991): 1047–58.

47. U.S. Department of Agriculture, *Data tables: Results from USDA's 1996 Continuing Survey of Food Intakes by Individuals and 1996 Diet and Health Knowledge Survey* (On-line), available under "Releases" at www.barc.usda.gov/bhnrc/foodsurveyhome.htm.

48. National Cholesterol Education Program, *Report of the Expert Panel.*

49. U.S. Department of Health and Human Services, Public Health Service, *Healthy People 2000*, pp. 571–78.

50. Story and coauthors, Adolescent nutrition.

51. R. A. Thompson and R. T. Sherman, *Helping Athletes with Eating Disorders* (Champaign, IL: Human Kinetics, 1993).

52. Information about specific nutrition programs, including participating levels, legislation, and participant eligibility, was adapted from *Catalog of Federal Domestic Assistance Programs*, June, 1996, available on-line at family.info.gov/cfda/index.htm.

53. The quote by Representative Kennedy was taken from Community Nutrition Institute, Free breakfasts debated, *Nutrition Week* 28 (March 13, 1998): 2; other information about the school breakfast and lunch programs was adapted from Community Nutrition Institute, Full House panel approves child nutrition bill with a universal breakfast pilot, *Nutrition Week* 28 (June 5, 1998): 1, and October Food Stamp rolls remain under 21 million, *Nutrition Week* 28 (January 30, 1998): 8.

54. Food Research and Action Center, *Hunger Doesn't Take a Vacation: A Status Report on the Summer Food Service Program for Children*, 5th ed. (Washington, D.C.: Food Research and Action Center, 1997).

55. Select Panel for the Promotion of Child Health, *Better Health for Our Children: A National Strategy*, vols. 1 and 2,

DHHS Pub. No. 79-55071 (Washington, D.C.: U.S. Department of Health and Human Services, 1981); and the Web site of the Administration for Children and Families at www.acf.dhhs.gov.

56. Ibid., vol. 2, pp. 69–72.

57. ADA Reports, Nutrition services for children with special health care needs.

58. M. T. Baer and coauthors, Children with special health care needs, in *Call to Action: Better Nutrition for Mothers, Children, and Families*, ed. C. O. Sharbaugh (Washington, D.C.: National Center for Education in Maternal and Child Health, 1991), pp. 191–208.

59. S. Farnan, Role of nutrition in Crippled Children's Service agencies, *Topics in Clinical Nutrition* 3 (1988): 33–42; and Final Report, *Nutrition Programming for the Chronically Ill/Handicapped Child* (Birmingham, AL: Sparks Center for Developmental and Learning Disorders, University of Alabama, and Child Development Center, University of Tennessee Center for the Health Sciences, 1986).

60. ADA Reports, Nutrition services for children with special health care needs.

61. ADA Reports, Position of The American Dietetic Association: Nutrition in comprehensive program planning for persons with developmental disabilities, *Journal of the American Dietetic Association* 92 (1992): 613–15.

62. Information about USDA's *YourSELF* program was obtained from the USDA's Web site at www.usda.gov/fcs.

63. American Heart Association Schoolsite Program (Dallas: National Center, American Heart Association).

64. American Cancer Society "Changing the Course" Program (Minneapolis: Minnesota Division, American Cancer Society).

65. J. Brooke, Indians revive a running tradition, *New York Times*, August 2, 1998, p. 16.

66. National Clearinghouse on Families and Youth, *Reconnecting Youth and Community: A Youth Development Approach*, July, 1996, available at www.ncfy.com.

67. Select Panel for the Promotion of Child Health, vol. 1, p. 2.

Chapter 15

Growing Older: Nutrition Assessment, Services, and Programs

Outline

Learning Objectives

After you have read and studied this chapter, you will be able to:

- Describe the potential impact of the graying of America on health care services.
- List national goals for health promotion for adults.
- Identify factors influencing the nutrition status of older adults.
- Describe the components of a nutrition assessment of older adults.
- Describe the purpose and function of the Nutrition Screening Initiative.
- Describe community nutrition programs that are intended to provide nutrition assistance to older adults.

Something To Think About...

How far you go in life depends on your being tender with the young, compassionate with the aged, sympathetic with the striving, and tolerant of the weak and the strong. Because someday in life you will have been all of these.

— *George Washington Carver*

Introduction

It is easy to get the impression from mortality statistics that people are living longer and longer lives, but this is not the case. Certainly, the *average* age at death (life expectancy) has changed dramatically. Life expectancy at birth is now 76 years—most men die at a little past 72, and most women die at about 79.[1] On the other hand, the *maximum* age at which people die—that is, the *maximum life span*—has changed very little. It seems that the aging phenomenon cuts off life at a rather fixed point in time.

To what extent is aging inevitable? Apparently, aging is an inevitable, natural process programmed into our genes at conception. Nevertheless, we can adopt lifestyle habits, such as consuming a healthful diet, exercising, and paying attention to our work and recreational environments, that will slow the process within the natural limits set by heredity. Clearly, good nutrition can retard and ease the aging process in many significant ways. However, no potions, foods, or pills will prolong youth. People who claim to have found the fountain of youth have been selling its waters for centuries, but products purporting to prevent aging profit only the sellers, not the buyers.

One approach to the prevention of aging has been to study other cultures in the hope of finding an extremely long-lived people and then learning their secrets of long life. The views of the experts can best be summed up by saying that disease can *shorten* people's lives and that poor nutrition practices make diseases more likely to occur. Thus, by postponing and slowing disease processes, optimal nutrition can help to prolong life up to the maximum life span—but cannot extend it further.[2] This chapter focuses on the diseases that seem to come with age, their risk factors, and the nutrition assessment of older adults; it also examines the programs and services that target older adults for health promotion and disease prevention. We begin with a look at the demographic trends characteristic of this segment of the population and the national nutrition and health goals for improving the quality of life of Americans as they age.

Demographic Trends and Aging

The number of elderly (aged 65 years and older) in the United States will double by 2030 to more than 70 million people. In 1996, the elderly accounted for 12.8 percent of the population, and this proportion is expected to rise to approximately 14 percent in 2010 and to nearly 22 percent by 2030. Nearly 12 percent of the population will be over age 74 by 2030.[3] The increased growth in the elderly population in the United States is illustrated in Figure 15-1.

Technically, the life span is the oldest documented age to which a member of a given species is known to have survived. For humans, this is about 115 years.

Researchers and marketing analysts have struggled to find a useful way to segment—and, therefore, target—the elderly market. One frequently used segmentation divides the elderly into the "mature," aged 55–64; the "young-old," aged 65–74; and the "old-old," over age 74.

Both the baby boom that took place between 1946 and 1964 and improved life expectancy are important contributors to the growing elderly population in the United States. Baby boomers will increase the numbers of the older middle-aged (ages 46 through 63) until 2010, when they will begin to swell the ranks of the retired population.

Life expectancy has risen as a result of better prenatal and postnatal care and improved means of combating disease in older adults. For example, the death rate from heart disease began to decline in the 1960s and continues to fall today. Over half of the drop is attributed to a decline in smoking and fewer people with high blood pressure or high blood cholesterol. As Figure 15-2 illustrates, life expectancy at birth has increased dramatically for both men and women and for both whites and nonwhites since 1900. By 2030, minorities will represent 25 percent of the population of older adults (up from 13 percent in 1990), with the largest increases in Hispanics and Asians.[4]

Policy makers and others concerned with meeting the health needs of older adults are alert to the implications of these demographic changes because the elderly tend to consume a large amount of total health care and long-term care resources. Consider that persons aged 65 years and over represent 12.8 percent of the total population today—some 33.9 million men and women—yet they account for more than:[5]

- 30 percent of the country's health care costs,
- 25 percent of the prescription drugs used,
- 15 percent of all visits to doctors' offices, and
- 40 percent of all days in acute-care hospitals.

With the "graying of America," these health care demands can only increase. As Joseph A. Califano, a former secretary of health and human services, testified before the Committee on a National Research Agenda on Aging:

> The aging of America will challenge all our political, retirement, and social service systems. As never before, it will test our commitment to decent human values. Nowhere is the aging of America freighted with more risk and opportunity than in the area of health care.[6]

FIGURE 15-1 *The Aging of the Population*

Source: U.S. Census Bureau

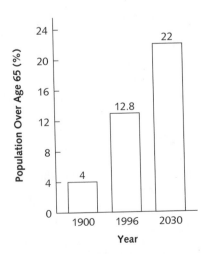

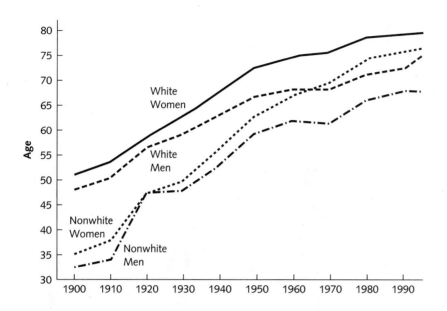

FIGURE 15-2 *Life Expectancies at Birth, by Race and Sex, for the United States, 1900–1990*

Source: National Center for Health Statistics, *Health United States and Prevention Profile, 1991* (Washington, D.C.: U.S. Department of Health and Human Services, 1991).

Healthy Adults

An individual's current health profile is substantially determined by behavioral risk factors. The leading causes of death for adults aged 25 through 64 are cancer, heart disease, stroke, injuries, chronic lung disease, and liver disease; all have been associated with behavioral risk factors. Thus, many adults today would benefit from changes in their lifestyle behaviors. For example, incorporating exercise and a balanced, low-fat diet into one's lifestyle can contribute to weight loss and to controlling three important risk factors for heart disease—high fat intake, overweight, and a sedentary lifestyle.

National Goals for Health Promotion

The most important goal of health promotion and disease prevention for adults as they age is maintaining health and functional independence. Many of the health problems associated with the later years are preventable or can be controlled.[7] For example, changing certain risk behaviors into healthy ones can improve the quality of life for older persons and lessen their risk of disability. Improvements in diet and nutrition status and weight control can enhance the health of older adults as well as help control risk factors for disease in middle-aged and younger adults. The *Surgeon General's Report on Nutrition and Health* demonstrates how the same set of dietary recommendations can both promote general health and help prevent a broad spectrum of chronic diseases.[8] As Figure 15-3 shows, these basic dietary recommendations reduce the risk of a variety of chronic diseases or their complications. The accompanying box presents the key national health objectives for the year 2000 that focus on improving the health of adults as they age.[9]

Efforts at health promotion and disease prevention are conducted at several levels. *Primary* prevention efforts seek to prevent the occurrence of a disease in susceptible

FIGURE 15-3 *Consistency of Recommendations for Reducing the Risk of Chronic Diseases or Their Complications*

Source: Adapted from J. M. McGinnis and M. Nestle, The Surgeon General's Report on Nutrition and Health: Policy implications and implementation strategies, *American Journal of Clinical Nutrition* 49 (1989): 26. © American Society for Clinical Nutrition.

Change Diet ➡	Reduce Fats	Control Calories	Increase Starch and Fiber	Reduce Sodium	Control Alcohol
Reduce Risk ⬇					
Heart disease	🍎	🍎		🍎	
Cancer	🍎	🍎	🍎		🍎
Stroke	🍎	🍎		🍎	🍎
Diabetes	🍎	🍎	🍎		
Gastrointestinal Diseases*	🍎	🍎	🍎		🍎

*Gastrointestinal diseases affected by dietary factors are primarily gallbladder disease (fat and energy), diverticular disease (fiber), and cirrhosis of the liver (alcohol).

people by decreasing their risks for the disease, typically through behavior modification. Examples include the promotion of healthful behaviors such as stopping smoking, improved dietary behaviors, and regular physical activity. The goal of *secondary* prevention is to detect and treat a disease in its early stages. Health fairs offering blood pressure and cholesterol screening are an example of this type of prevention. *Tertiary* prevention seeks to minimize the disabling effects of a disease once it occurs. Rehabilitative services such as physical and occupational therapies are an example of this level of prevention. Thus, health promotion activities are designed to help adults of all ages change their eating patterns and other behaviors to reduce the risk of chronic disease.

Understanding Baby Boomers

Baby boomers, or the approximately 77 million individuals who were born between 1946 and 1964, represent almost one-third of the U.S. population. By virtue of their large numbers, baby boomers are a driving force for current and future trends. An understanding of their preferences, character, lifestyle, and location is and will continue to be critical to health promotion programs and services. In 1991, the first of this generation turned 45, and in 2029, the last of the baby boomers will turn 65.

Although there are several subsets of baby boomers, some general characteristics can be noted:[10]

- Boomers have the power to change the marketplace. Because of their numbers and affluence, they are able to drive trends, especially as they age. Of importance

Healthy People 2000 Objectives

Health Status Objectives

- Reduce coronary heart disease deaths to no more than 100/100,000 people. (Baseline: 135/100,000 in 1987.)

  Healthy People Progress Report. Positive trend: Overall death rate for coronary heart disease has declined by 49 percent. In 1987, CHD death rate for persons aged 65 years and older was 2,075 per 100,000; in 1993, the death rate was 1,891 per 100,000.

- Reverse the rise in cancer deaths to achieve a rate of no more than 130/100,000. (Baseline: 133/100,000 in 1987.)

  Healthy People Progress Report. Goal met: In 1995, the 2000 target for total cancer death rates was achieved with 130 reported cases per 100,000.

- Reduce overweight to a prevalence of no more than 20 percent among people aged 20 and older. (Baseline: 26 percent for people aged 20 through 74 in 1976–1980; for men, 24 percent; for women, 27 percent.)

  Healthy People Progress Report. Negative trend: 34 percent of persons aged 20 and older were obese in 1988–1991.

- Reduce to no more than 90/1000 people the proportion of all people aged 65 and older who have difficulty performing two or more personal care activities, thereby preserving independence.* (Baseline: 111/1000 in 1984–1985.)

  Healthy People Progress Report. Little improvement: In 1992, more than half of the elderly (53.9 percent) reported having at least one disability which limited them from carrying out one or more activities of daily living and instrumental activities of daily living.

Risk Reduction Objectives

- Increase to at least 30 percent the proportion of people aged 6 and older who engage regularly, preferably daily, in light to moderate physical activity for at least 30 minutes per day. (Baseline: 22 percent of people over 18 years were active for at least 30 minutes five or more times per week in 1985.)

  Healthy People Progress Report. Modest improvement: In 1991, 24 percent of people over age 18 were active five or more times per week.

- Reduce dietary fat intake to an average of 30 percent of calories or less and average saturated fat intake to less than 10 percent of calories among people aged 2 years and older. (Baseline: 36 percent of calories from total fat and 13 percent from saturated fat for people aged 20 through 74 in 1976–1980.)

  Healthy People Progress Report. Positive trend: From 1988 to 1991, dietary fat intake averaged 34 percent of calories and saturated fat intake averaged 12 percent of calories.

- Increase complex carbohydrate and fiber-containing foods in the diets of adults to 5 or more daily servings for vegetables (including legumes) and fruits, and to 6 or more daily servings for grain products. (Baseline: 2½ servings of vegetables and fruits and 3 servings of grain products for women aged 19 through 50 in 1985.)

  Healthy People Progress Report. Positive trend: In 1996, people 2 years of age and older consumed an average of 4.7 servings of vegetables and 6.9 servings of grain products.

Continued

---

*Personal care activities are bathing, dressing, using the toilet, getting in or out of bed or chair, and eating.

Healthy People 2000 Objectives—continued

- Increase to at least 50 percent the proportion of overweight people aged 12 and older who have adopted sound dietary practices combined with regular physical activity to attain an appropriate body weight. (Baseline: 30 percent of overweight women and 25 percent of overweight men for people aged 18 and older in 1985.)
  Healthy People Progress Report. Negative trend: In 1993, only 19 percent of overweight women and 17 percent of overweight men had adopted sound dietary practices combined with regular physical activity to lose weight.
- Increase to at least 85 percent the proportion of people 18 years of age or older who use food labels to make nutritious food selections. (Baseline: 74 percent used labels in 1988.)
  Healthy People Progress Report. Slight positive trend: From 1988 to 1991, 75 percent of people aged 18 years and older used food labels.

Services and Protection Objectives

- Increase to at least 90 percent the proportion of restaurants and institutional foodservice operations that offer identifiable low-fat, low-calorie food choices, consistent with the *Dietary Guidelines for Americans.* (Baseline: About 70 percent of fast-food and family restaurant chains with 350 or more units had at least one low-fat, low-calorie item on their menu in 1989.)
  Healthy People Progress Report. Positive trend: 75 percent of fast food and family restaurant chains offered heart-healthy, nutritious items on their menus.
- Increase to at least 80 percent the receipt of home foodservices by people aged 65 and older who have difficulty in preparing their own meals or are otherwise in need of home-delivered meals.
  Healthy People Progress Report. No improvement: A survey by USDA in 1991 showed that only 7 percent of eligible older persons were receiving home-delivered meals.
- Increase to at least 50 percent the proportion of worksites with 50 or more employees that offer nutrition education and/or weight management programs for employees. (Baseline: 17 percent offered nutrition education activities, and 15 percent offered weight control activities in 1985.)
  Healthy People Progress Report. Positive trend: In 1992, 31 percent of worksites offered nutrition education activities and 24 percent offered weight control activities.

Source: Adapted from *Healthy People 2000: National Health Promotion and Disease Prevention* (Washington, D.C.: U.S. Department of Health and Human Services, Public Health Service, 1990), pp. 588–89, 611–13, and National Center for Health Statistics, *Healthy People 2000 Review*, 1996 (Hyattsville, MD: Public Health Service, 1996).

to community nutritionists, these consumers are generally concerned about what is healthful and convenient.
- Boomers make decisions based on personal beliefs and want to be empowered. They prefer health programs that offer information and options in a learner-involved format, such as a supermarket tour.

- Boomers are constantly pressed for time as they juggle careers, child care, home responsibilities, and leisure activities. Programs need to be practical and convenient and should be presented in an understandable format.
- Boomers look for value and quality in their investments and are becoming thriftier with age. They seek information on how to relate market choices to value.
- Boomers will not age gracefully. Programs must be upbeat and dynamic for these on-the-go consumers.
- Boomers like nostalgia. Nostalgia can be used to reinforce nutrition messages—for example, modifying traditional family recipes and holiday menus to reflect current nutritional advice.

Nutrition Education Programs

Nutrition education strategies aimed at adults are found in both the public and private sectors. They strive to increase nutrition knowledge and skills and improve eating patterns among adults of all ages. Nutrition education for health promotion is generally based on the *Dietary Guidelines for Americans* and the Food Guide Pyramid.[11] The nutritional goals of these educational tools are to help consumers select diets that provide an appropriate amount of energy to maintain a healthful weight; meet the recommended intakes for all nutrients without depending on supplements; are varied in types of fat and moderate in total fat, caloric sweeteners, sodium, cholesterol, and alcohol; and are adequate in complex carbohydrate and fiber.

Public nutrition education programs include the Expanded Food and Nutrition Education Program, described in Chapter 13, and the Food and Drug Administration (FDA) and U.S. Department of Agriculture (USDA) public education campaign on the food label in cooperation with other federal, state, and local agencies (for example, state cooperative extension). For example, the Michigan State University Cooperative Extension offers a workbook for culturally diverse low-income consumers on how to use the food label to manage fat intake.

A primary challenge facing nutrition educators is to improve nutrition education strategies to reduce the major risk factors for coronary heart disease and cancer, the leading causes of death among adults.[12] Public efforts are now under way. The National Heart, Lung, and Blood Institute initiated the guidelines of the National Cholesterol Education Program (NCEP) to help prevent heart disease by reducing saturated fat and cholesterol in the American diet.[13] The National Cancer Institute (NCI) designed its *5 a Day for Better Health* program to increase per capita fruit and vegetable consumption. The NCI's 5 a Day program promotes a simple nutrition message that physicians, nurses, community nutritionists, and other health care professionals can reinforce to their clients: *Eat five or more servings of fruits and vegetables every day for better health.*[14]

A number of trade and professional organizations are likewise directing some of their nutrition education strategies to help adults understand the role of nutrition in health promotion and disease prevention. Brochures, booklets, newsletters, and videos that offer simple ways to trim dietary fat, interpret nutrition labels, implement the Dietary Guidelines, or improve shopping skills are available from the American Dietetic Association, American Association for Retired Persons, the National Dairy Council, General Mills, and the Produce Marketing Association, among others.[15] For example, Kraft General Foods' *A Matter of Balance: Using the New Food Labels* is a consumer brochure that uses the themes of balance and moderation to explain how to use the new food label to plan a healthful diet.

In many communities, food retailers and food service establishments provide point-of-purchase information and literature to their customers.[16] Other communities offer seminars and grocery store tours to help consumers understand the food label.[17]

Health Promotion Programs

Evidence continues to indicate that adults of all ages need to modify their current eating patterns and other behaviors to reduce the risk of chronic diseases. However, as mentioned in the discussion of social marketing in Chapter 10, behavior changes can be very difficult to make. An important characteristic of community nutrition interventions is that they can reach people in many different contexts of their daily lives. Supportive social environments can help individuals change their behavior.[18] For this reason, community- and employer-based programs for health promotion are expanding and are facilitating lifestyle changes. For example, many employers now provide worksite health promotion programs offering classes and activities for smoking cessation, weight loss, and stress management. These programs vary widely—some are simple and inexpensive (distribution of health information pamphlets), whereas others are more complex (comprehensive risk factor screening and intensive follow-up counseling).[19] In general, worksite health promotion efforts can be classified under four main areas:[20]

- **Policies.** Smoking, alcohol and other drugs, and AIDS/HIV infection
- **Screenings.** Health risk/health status, cancer, high blood pressure, and cholesterol
- **Information or activities.** Individual counseling, group classes, workshops, lectures, special events, and resource materials such as posters, brochures, pamphlets, and videos. Topics typically covered include cancer, high blood pressure, cholesterol, smoking, exercise and fitness, nutrition, weight control, prenatal care, medical self-care, mental health, stress management, alcohol and other drugs, AIDS/HIV infection, sexually transmitted infections (STIs), job hazards and injury prevention, back care, and off-the-job accidents.
- **Facilities or services.** Nutrition, physical fitness, alcohol and other drugs, and stress management

> The national health promotion and disease prevention objectives for the nation include the target that by the year 2000, at least 85 percent of all workplaces with 50 or more employees will offer employee health promotion programs.

Figure 15-4 shows the prevalence of worksites with 50 or more employees offering information or activities in 17 subject areas. According to the U.S. Department of Health and Human Services (DHHS), two out of three worksites with at least 50 employees offer some form of health promotion programming.[21] Successful nutrition promotion programs range from introducing heart-healthy menus into company cafeterias to reducing blood cholesterol levels through screening and intervention.[22] At the world headquarters for Coca-Cola in Atlanta, Georgia, the *HealthWorks Program* focuses on comprehensive health promotion. The nutrition theme is carried through group weight-loss classes, cooking demonstrations and taste tests, and individualized diet instruction available to all employees, spouses, children, and retirees. The *HealthWorks* cafeteria line features healthful entrees, and the annual worldwide recipe contest for healthful cafeteria recipes encourages customer involvement and results in new menu selections.[23]

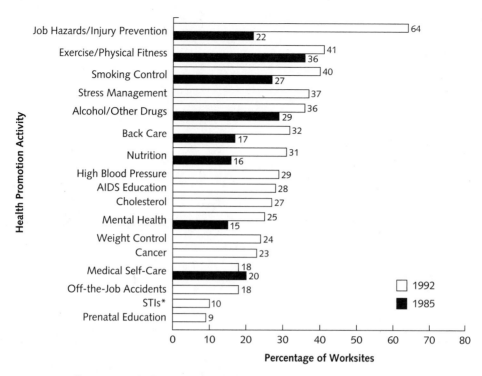

*STIs = sexually transmitted infections

Current efforts to help young adults identify their familial risk factors for chronic disease conditions and programs that tout the benefits of lifelong healthful eating and regular physical activity should enable the older adults of tomorrow to enjoy a productive and satisfying life well into advanced age.

Aging and Nutrition Status

Growing old is often associated with frailty, sickness, and a loss of vitality. Although the aging in our society do experience chronic illness and associated disabilities, this population is very heterogeneous: older people vary greatly in their social, economic, and lifestyle situations, functional capacity, and physical conditions.[24] Each person ages at a different rate, sometimes making chronological age different from biological age. Most older persons live at home (see Figure 15-5), are fully independent, and have lives of good quality.[25] Only 4 percent of older adults live in nursing homes.[26] Older persons who have problems with the **activities of daily living (ADLs)** are known as the frail elderly. Because they depend on others to perform these essential activities, they are likely to be at risk for malnutrition.[27]

Primary Nutrition-Related Problems of Aging

Although aging is not completely understood, we know that it involves progressive changes in every body tissue and organ: the brain, heart, lungs, digestive tract, and

FIGURE 15-4 *Worksites Offering Health Promotion Information or Activities, 1985 and 1992* The Office of Disease Prevention and Health Promotion (ODPHP) of the U.S. Public Health Service conducted a 1992 survey to measure the growth of worksite health promotion activities since its first survey in 1985. This figure shows the prevalence of worksites with information or activities in 17 subject areas in 1992 and comparisons with 1985 in eight areas.

Source: Reprinted with permission from U.S. Department of Health and Human Services, 1992 National Survey of Worksite Health Promotion Activities: Summary, *American Journal of Health Promotion* 7 (1993): 454.

FIGURE 15-5 *Living Arrangements of Older Adults by Age and Sex*

Source: Adapted from E. L. Schneider and J. M. Guralank, The aging of America: Impact of health-care costs, *Journal of the American Medical Association* 263 (1990): 2335–40.

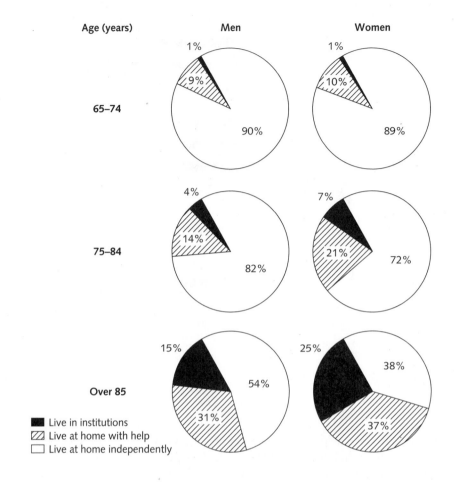

Age (years)

	Men	Women
65–74	1%, 9%, 90%	1%, 10%, 89%
75–84	4%, 14%, 82%	7%, 21%, 72%
Over 85	15%, 31%, 54%	25%, 37%, 38%

■ Live in institutions
▨ Live at home with help
☐ Live at home independently

ADLs Activities of daily living include bathing, dressing, grooming, transferring from bed or chair, going to the bathroom, and feeding oneself.

bones (see Table 15-1). After age 35, functional capacity declines in almost every organ system. Such changes affect nutrition status: some, including oral problems, interfere with nutrient intake; others affect absorption, storage, and utilization of nutrients; and still others increase the excretion of, and need for, specific nutrients. Examples of various conditions associated with aging that can affect nutrition status include sensory impairments, altered endocrine, gastrointestinal, and cardiovascular functions, and changes in the renal and musculoskeletal systems. For example, as many as 30 percent of persons older than 50 years may experience hypochlorhydria, which can interfere with their ability to absorb vitamin B_{12} effectively from protein-rich foods. In addition, there may be increased needs in the elderly for other vitamins, particularly, vitamin D and vitamins B_6 and B_{12}. Both genetic and environmental factors contribute to these declines.[28] Many of the changes are inevitable, but a healthful lifestyle that combines moderation with adequate intakes of all essential nutrients can forestall degeneration and improve the quality of life into the later years.

As a person gets older, the chances of suffering a chronic illness or functional impairment are greater. Among the diseases that befall some people in later life are heart disease, hypertension, cancer, diverticulosis, osteoporosis, dementia, diabetes, and gum disease. More than 60 percent of people over age 65 have high blood pres-

Cardiovascular System	Change (%)
Cardiac output	↓ 30
Maximum heart rate	↓ 25
Respiratory System	
Vital capacity	↓ 40
Maximum O_2 uptake	↓ 60
Musculoskeletal	
Muscle mass	↓ 30
Hand grip, flexibility	↓ 30
Bone mineralization	↓ 20–30
Renal function	↓ 40
Nervous System	
Conduction velocity	↓ 15
Taste and smell	↓ 90
Metabolism	
Basal metabolic rate	↓ 15

TABLE 15-1 *Changes in Biological Function Between the Ages of 30 and 70*

Source: Adapted from E. M. Berry, Undernutrition in the elderly: A physiological and pathological process, in H. Munro and A. Schlierf, eds., *Nutrition of the Elderly,* Nestle Workshop Series Vol. 29 (New York: Vervey/Raven Press, 1992).

sure, and approximately 30 percent have heart disease. Chronic conditions contributing to **disability** include arthritis, heart disease, strokes, disorders of vision and hearing, nutritional deficiencies, and oral-dental problems (see Table 15-2). Dementia (especially Alzheimer's disease) is a major contributor to disability and placement in nursing homes for those over age 75.[29] Malnutrition can occur secondary to these conditions, as noted in Table 15-3.[30] Many of these conditions require special diets that can further compromise nutrition status in the older adult. Also, there are differences in disease prevalence among racial and ethnic groups. For this reason, nutrition interventions designed to reduce disease risks must be sensitive to ethnic or cultural differences and preferences. Minority-group elderly are more likely to have malnutrition secondary to chronic diseases such as heart disease, renal disease, diabetes mellitus, obesity, and certain cancers. Many times, they have less access to health care, and have decreased quality of life and increased mortality.[31]

Polypharmacy, or the use of multiple drugs, is problematic for many older adults. The average older person receives more than 13 prescriptions a year and may take as many as six drugs at a time.[32] Cardiac drugs are most widely used by the elderly, followed by drugs to treat arthritis, psychic disorders, and respiratory and gastrointestinal conditions. Long-term use of a variety of drugs increases the risk of drug-nutrient interactions. Individuals with impaired nutrition status and poor dietary intakes are at the highest risk.[33]

Individually or in combination, the social, economic, psychological, cultural, and environmental factors associated with aging may interact with the physiological changes and further affect nutrition status in older adults. These interactions are illustrated in Figure 15-6 on page 466.[34]

Disability Any restriction on or impairment in performing an activity in the manner or within the range considered normal for a human being.

Polypharmacy The taking of three or more medications regularly; occurs in one-third of those over 65 years.

See Appendix L for a listing of commonly used drugs and their nutritional side-effects.

TABLE 15-2 *Distribution of Most Common Chronic Conditions That Cause Disability in Older Persons*

Condition	Adults over 65 years Affected (%)
Arthritis	50
Hypertension	36
Heart disease	32
Hearing impairment	29
Cataracts	17
Orthopedic impairment	16
Chronic sinusitis	15
Diabetes	10
Visual impairment	9
Tinnitus	9

Source: U.S. Department of Health and Human Services, *A Profile of Older Americans: 1997* (Washington, D.C.: U.S. Department of Health and Human Services, 1997).

Nutrition Policy Recommendations for Health Promotion for Older Adults

The first major effort to develop national policies to reduce chronic disease risk factors in older adults took place during the Surgeon General's Workshop on Health Promotion and Aging.[35] The workshop issued several major nutrition policy recommendations for promoting the health of older adults that are related to dietary guidance and nutrition services:

1. **Dietary guidance for older adults should:**
   - Target messages to the special concerns of this population.
   - Emphasize the need for balance between food intake to both meet nutrient needs and maintain a healthful weight.
   - Promote the consumption of more fruits, vegetables, and whole-grain products, as well as the choice of lean meats and low-fat or fat-free dairy foods.
   - Address drug-nutrient interactions.
   - Become integrated into the training of physicians, dietitians, and other health professionals.
2. **Nutrition services should be:**
   - Structured to include nutrition assessment and guidance.
   - Incorporated into institution-, community-, or home-based health care programs for older adults.
   - Tailored to the needs of older persons who are homebound, live in isolation, or are chronically ill.
   - Provided by credentialed nutrition professionals.
3. **Food manufacturers should:**
   - Develop easy-to-prepare, tasty food products that are nutrient dense.
   - Use food labels set in larger type that provide sufficient information to guide older consumers.

Disease or Condition	Effects on Nutrition Status
Atherosclerosis	May increase difficulties in regulating fluid balances if caused by congestive heart failure. If the individual is incapacitated, energy needs decrease.
Cancer	Weight loss, lack of appetite, and secondary malnutrition are common.
Dental and oral disease	May alter the ability to chew and thus reduce dietary intake. Increased likelihood of choking and aspiration.
Depression and dementia	Increased or decreased food intakes are common. A person with dementia may have decreased ability to get food, or the appetite may be very small or very great. Judgment and balance in meal planning are generally absent.
Diabetes mellitus (Type 1)	If untreated, increased risk of undernutrition results; increased risk of other diet-related diseases, such as hyperlipidemia, and decreased resistance to infections.
Diabetes mellitus (Type 2)	Increased risk of other diet-related diseases such as hyperlipidemia; weight loss is needed if obesity is present.
End-stage kidney disease	Alters fluid and electrolyte needs; uremia may alter appetite and increase risk of malnutrition. Infections and low-grade fever may increase energy output and weight loss.
Gastrointestinal disorders	Increased risk of malabsorption of nutrients and consequent undernutrition.
High blood pressure	Hyper- or hypokalemia can be increased by dietary means; weight gain may exacerbate high blood pressure.
Osteoarthritis	Makes motion difficult, including those activities related to purchasing, serving, eating, and cleaning up after meals. Predisposes people to a sedentary lifestyle and may give rise to obesity. Drug-nutrient interactions are common.
Osteoporosis	Limits the ability to purchase and prepare food if mobility is affected. If severe scoliosis is present, the appetite may be altered.
Smoking	Smoking may alter weight status. Alters serum levels of some nutrients such as ascorbic acid and carotenes. Chronic smoking gives rise to emphysema and chronic obstructive pulmonary diseases (COPD), which makes it difficult to eat owing to breathing problems.
Stroke	May alter abilities in the cognitive and motor realms related to food and eating. If the individual is incapacitated, his or her energy needs decrease.

TABLE 15-3 *Malnutrition That Is Secondary to Disease, Physiologic State, or Medication Use*

Source: Adapted from Institute of Medicine, *The Second Fifty Years: Promoting Health and Preventing Disability* (Washington, D.C.: National Academy Press, 1992). pp. 168–69.

4. **The federal government should:**
   - Disseminate information about model programs that have successfully delivered nutrition education and services to older adults.
   - Require its sponsored programs to address the calorie and nutrient needs of older participants.

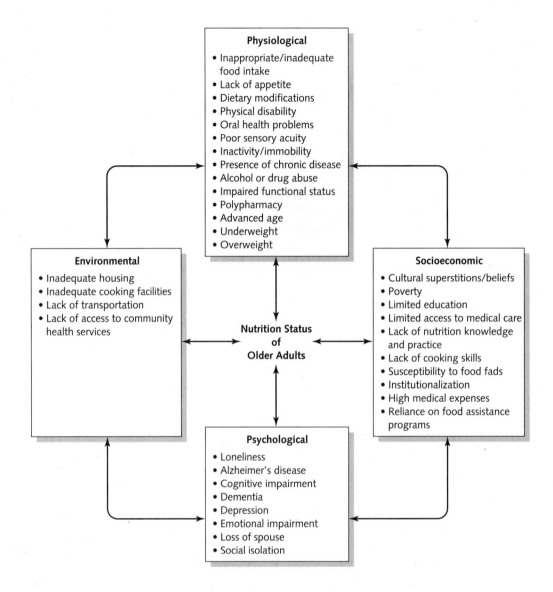

Physiological

- Inappropriate/inadequate food intake
- Lack of appetite
- Dietary modifications
- Physical disability
- Oral health problems
- Poor sensory acuity
- Inactivity/immobility
- Presence of chronic disease
- Alcohol or drug abuse
- Impaired functional status
- Polypharmacy
- Advanced age
- Underweight
- Overweight

Environmental

- Inadequate housing
- Inadequate cooking facilities
- Lack of transportation
- Lack of access to community health services

Nutrition Status of Older Adults

Socioeconomic

- Cultural superstitions/beliefs
- Poverty
- Limited education
- Limited access to medical care
- Lack of nutrition knowledge and practice
- Lack of cooking skills
- Susceptibility to food fads
- Institutionalization
- High medical expenses
- Reliance on food assistance programs

Psychological

- Loneliness
- Alzheimer's disease
- Cognitive impairment
- Dementia
- Depression
- Emotional impairment
- Loss of spouse
- Social isolation

FIGURE 15-6 *Risk Factors Influencing the Nutrition Status of Older Adults*

- Promote research and surveillance programs to improve nutrition status.
- Develop a system for appropriate funding for nutrition services delivered to older adults.

The Surgeon General's Workshop on Health Promotion and Aging recommended that a nutrition assessment be completed on all older adults admitted to health care institutions or community-based health services.[36] This recommendation is similar to those proposed by the American Dietetic Association (ADA) in its position paper on nutrition and aging. The ADA recommends that nutrition services (including nutrition assessment and monitoring), therapeutic interventions as needed, and nutrition counseling and education be included throughout the continuum of health care services for older adults.[37] A discussion of the guidelines and tools for nutrition screening and assessment of older adults follows in the next two sections.

Evaluation of Nutrition Status

Up to one-quarter of all elderly patients and one-half of all hospitalized elderly may be suffering from malnutrition.[38] In addition, a national survey in 1990 found that one-third of noninstitutionalized Americans over the age of 65 live alone, 45 percent take multiple prescription drugs that can interfere with appetite and nutrient absorption, 30 percent skip meals almost daily, and 25 percent have annual incomes under $10,000—all factors placing elderly people at potential for nutritional risk.[39] Identifying older adults at nutritional risk is an important first step in maintaining quality of life and functional status.[40]

Nutrition Screening

The American Dietetic Association, the American Academy of Family Physicians, and the National Council on Aging have collaborated since 1990 on an effort to promote nutrition screening and early intervention as part of routine health care. Its focus is on the elderly, one of the largest population groups in the United States at risk of poor nutrition.

The Nutrition Screening Initiative has identified a number of specific risk factors and indicators of poor nutrition status in older adults (see Figure 15-6 and Table 15-4). Some of the risk factors shown in Figure 15-6 increase the risks for dietary inadequacy, excess, or imbalance and involve social, economic, and lifestyle factors rather than physical health problems. Others indicate risks of malnutrition secondary to disease and/or treatment modalities rather than those caused primarily by lack of food.[41]

A key premise of the Nutrition Screening Initiative is that nutrition status is a "vital sign"—as vital to health assessment as blood pressure and pulse rate.[42] The Nutrition Screening Initiative has developed a 10-question self-assessment "checklist" that can be distributed at any office or agency for the elderly (see Figure 15-7). Individuals identify factors placing them at nutritional risk to arrive at their score. This checklist addresses disease, eating status, tooth loss or mouth pain, economic hardship, reduced social contact, multiple medications, involuntary weight loss or gain, and need for assistance with self-care. The word DETERMINE is used as a mnemonic device with the checklist and helps provide basic nutrition information (see Figure 15-8 on page 471). Each letter in DETERMINE stands for a risk factor.[43] As Figure 15-9 shows, people identified as being at risk should be followed up with more in-depth screening of risk factors and assessment of nutrition status (Level I or Level II screens) by a health professional (see Table 15-5 on page 470).

The Level I screen is a basic nutrition screen designed for social service and health professionals to use to identify older adults who may need medical attention or nutrition services. This screen includes determinations of height and weight and charts to convert these measures to body mass index (BMI). The Level I screen also evaluates eating habits and provides a brief review of socioeconomic and functional status.

TABLE 15-4 *Major Indicators of Poor Nutrition Status in Older Adults*

- Significant weight loss over time:
    5.0% or more of prior body weight in 1 month.
    7.5% or more of body weight in 3 months.
    10.0% or more of body weight in 6 months or involuntary weight loss.
- Significantly low or high weight for height: 20% below or above desirable weight for height at a given age.
- Significant reduction in serum albumin: Serum albumin of less than 3.5 g/dL.
- Significant change in functional status: Change from "independent" to "dependent" in two of the ADLs or one of the nutrition-related IADLs.*
- Significant and sustained inappropriate food intake:
    Failure to consume the recommended minimum from one or more of the food groups suggested in the Food Guide Pyramid, or a sufficient variety of foods for a period of 3 months or more.
    Excessive consumption of fat, saturated fat, and/or alcohol (alcohol: > 1 oz/day, women; > 2 oz/day, men).
- Significant reduction in mid-arm circumference: To less than 10th percentile (NHANES standards).
- Significant increase or decrease in triceps skinfolds: To less than 10th percentile or more than 95th percentile (NHANES standards).
- Significant obesity: More than 120% of desirable weight or body mass index over 30 or triceps skinfolds above 95th percentile (NHANES standards).
- Other nutrition-related disorders: Presence of osteoporosis, osteomalacia, folic acid deficiency, or vitamin B_{12} deficiency.

*IADLs are instrumental activities of daily living (e.g., meal preparation and financial management).

Source: Reprinted with permission by the Nutrition Screening Initiative, a project of the American Academy of Family Physicians, The American Dietetic Association, and the National Council on Aging, Inc. and funded in part by a grant from Ross Laboratories, a division of Abbott Laboratories.

Examples of the Nutrition Screening Initiative's in-depth assessments are provided in Appendix L.

The Level II screen provides more specific diagnostic information on nutrition status. It is designed for health and medical professionals to use with older adults who have a potentially serious medical or nutritional problem. This in-depth screening tool focuses on the components of nutrition assessment: anthropometric indicators (for instance, weight, height, body composition); clinical indicators (such as oral health and general physical exam), biochemical indicators (for example, serum albumin, serum cholesterol, hemoglobin, plasma glucose), and dietary indicators (including dietary history). The Level II screen also assesses chronic medication use and the living environment of the individual (assistance, facilities, support systems, safety), cognitive status, emotional status, and functional status.[44]

Nutrition Assessment

Periodic nutrition assessment is useful for identifying and tracking elderly persons at nutritional risk.[45] The components of nutrition assessment for the elderly are listed in Table 15-6 on page 473 along with indicators to determine risk. A geriatric nutrition assessment includes the elements discussed in the bulleted list that follows:[46]

- **Anthropometric measures.** Height, weight, and skinfold measures are affected by aging. Height decreases over time due to changes in the integrity of the skeletal system as a result of bone loss. Measurements of height are sometimes difficult to obtain because of poor posture or the inability to stand erect unassisted. In such cases, a recumbent anthropometric measure such as knee-to-heel height can be used as an alternative measure of stature.[47]
- **Clinical assessment.** The clinical assessment should evaluate the condition of hair, skin, nails, musculature, eyes, mucosa, and other physical attributes. An oral examination is useful for determining the condition of the mouth and teeth, the need for dentures or condition of existing dentures, and oral lesions. An assessment of the client's ability to chew, swallow, and self-feed is recommended.[48]

The Warning Signs of poor nutritional health are often overlooked. Use this checklist to find out if you or someone you know is at nutritional risk.

Read the statements below. Circle the number in the yes column for those that apply to you or someone you know. For each yes answer, score the number in the box. Total your nutritional score.

DETERMINE YOUR NUTRITIONAL HEALTH

	YES
I have an illness or condition that made me change the kind and/or amount of food I eat.	2
I eat fewer than 2 meals per day.	3
I eat few fruits or vegetables, or milk products.	2
I have 3 or more drinks of beer, liquor or wine almost every day.	2
I have tooth or mouth problems that make it hard for me to eat.	2
I don't always have enough money to buy the food I need.	4
I eat alone most of the time.	1
I take 3 or more different prescribed or over-the-counter drugs a day.	1
Without wanting to, I have lost or gained 10 pounds in the last 6 months.	2
I am not always physically able to shop, cook and/or feed myself.	2
TOTAL	

Total Your Nutritional Score. If it's —

0-2   **Good!** Recheck your nutritional score in 6 months.

3-5   **You are at moderate nutritional risk.**
See what can be done to improve your eating habits and lifestyle. Your office on aging, senior nutrition program, senior citizens center or health department can help. Recheck your nutritional score in 3 months.

6 or more **You are at high nutritional risk.** Bring this checklist the next time you see your doctor, dietitian or other qualified health or social service professional. Talk with them about any problems you may have. Ask for help to improve your nutritional health.

The Nutrition Screening Initiative • 1010 Wisconsin Avenue, NW • Suite 800 • Washington, DC 20007

These materials developed and distributed by the Nutrition Screening Initiative, a project of:

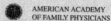

AMERICAN ACADEMY
OF FAMILY PHYSICIANS

THE AMERICAN
DIETETIC ASSOCIATION

NATIONAL COUNCIL
ON THE AGING, INC.

Remember that warning signs suggest risk, but do not represent diagnosis of any condition. Turn the page to learn more about the Warnings Signs of poor nutritional health.

sponsored in part through a grant from Ross Products Division of Abbott Laboratories.

FIGURE 15-7 *Checklist to Determine Your Nutritional Health*

Source: Reprinted with permission by the Nutrition Screening Initiative, a project of the American Academy of Family Physicians, The American Dietetic Association, and the National Council on Aging, Inc. and funded in part by a grant from Ross Laboratories, a division of Abbott Laboratories.

TABLE 15-5 *Nutrition Screening Initiative: Level I and Level II Screens*

	Level I Screen	Level II Screen
Primary User	Social workers and health care professionals	Physicians and other qualified health care professionals
Data Evaluation	Height Weight Dietary data Daily food intake Living environment Functional status	Height Weight Dietary data Daily food intake Living environment Functional status Laboratory and anthropometric data Clinical features Mental/cognitive status Medication use

Source: Adapted from Nutrition Screening Initiative, *Nutrition Screening Manual for Professionals Caring for Older Americans* (Washington, D.C.: Nutrition Screening Initiative, 1991).

Activities of Daily Living (ADL)

Bathing
Dressing
Grooming
Eating
Toileting
Transferring from bed
  or chair
Walking
Getting outside

Instrumental Activities of Daily Living (IADL)

Food preparation
Use of the telephone
Housekeeping
Laundry
Use of transportation
Responsibility for
  medication
Managing money
Shopping

- **Biochemical assessment.** Biochemical parameters are affected by the aging process, as well as by polypharmacy, chronic disease, and hydration status. However, serial measures of blood parameters can be useful in evaluating nutritional risk. Serum albumin is generally used to assess visceral protein status in the elderly. Low serum albumin levels are associated with increased morbidity and mortality in the elderly. Measurements of serum cholesterol, hemoglobin, blood glucose, and antigen-recall skin tests are also included.

- **Dietary assessment.** A detailed record of current food consumption and a history of changes in eating habits over time are needed to assess diet adequacy. The evaluation should detect persons who avoid certain food groups, adhere to unusual dietary practices, or consume excessive or insufficient amounts of essential nutrients. The adequacy of fluid intake should be assessed as well.

- **Functional assessment.** Functional assessment measures changes in the basic functions necessary to maintain independent living. ADLs refer to self-care activities (such as bathing, dressing, feeding). Instrumental ADLs, or IADLs, which require a higher level of functioning, include activities such as meal preparation, financial management, and housekeeping. Many of these activities (for example, shopping, cooking, self-feeding) are closely related to adequate nutrition status.

- **Medication assessment.** Note the types and doses of various prescription and over-the-counter drugs. Evaluate the individual's drug intake for possible nutrient-drug interactions that could affect the absorption, metabolism, and requirements for specific nutrients. Also identify any drugs that may depress the appetite or alter the perception of taste.

- **Social assessment.** Financial resources, living arrangements, and social support network, including availability of caregivers, should be evaluated as part of the nutrition assessment since these factors can directly impact a person's nutrition status. Poverty (that is, annual income of less than $8,000 per person) and social isolation particularly impair the nutrition status of many older adults, as noted perceptively by a professor of psychiatry:

> It is not what the older person eats but with whom that will be the deciding factor in proper care for him. The oft-repeated complaint of the older patient that he has little incentive to prepare food for only himself is not merely a statement of fact but also a rebuke to the questioner for failing to perceive his isolation and aloneness and to realize

The Nutrition Checklist is based on the Warning Signs described below. Use the word <u>DETERMINE</u> to remind you of the Warning Signs.

DISEASE

Any disease, illness or chronic condition which causes you to change the way you eat, or makes it hard for you to eat, puts your nutritional health at risk. Four out of five adults have chronic diseases that are affected by diet. Confusion or memory loss that keeps getting worse is estimated to affect one out of five or more of older adults. This can make it hard to remember what, when or if you've eaten. Feeling sad or depressed, which happens to about one in eight older adults, can cause big changes in appetite, digestion, energy level, weight and well-being.

EATING POORLY

Eating too little and eating too much both lead to poor health. Eating the same foods day after day or not eating fruit, vegetables, and milk products daily will also cause poor nutritional health. One in five adults skip meals daily. Only 13% of adults eat the minimum amount of fruit and vegetables needed. One in four older adults drink too much alcohol. Many health problems become worse if you drink more than one or two alcoholic beverages per day.

TOOTH LOSS/ MOUTH PAIN

A healthy mouth, teeth and gums are needed to eat. Missing, loose or rotten teeth or dentures which don't fit well or cause mouth sores make it hard to eat.

ECONOMIC HARDSHIP

As many as 40% of older Americans have incomes of less than $6,000 per year. Having less--or choosing to spend less--than $25-30 per week for food makes it very hard to get the foods you need to stay healthy.

REDUCED SOCIAL CONTACT

One-third of all older people live alone. Being with people daily has a positive effect on morale, well-being and eating.

MULTIPLE MEDICINES

Many older Americans must take medicines for health problems. Almost half of older Americans take multiple medicines daily. Growing old may change the way we respond to drugs. The more medicines you take, the greater the chance for side effects such as increased or decreased appetite, change in taste, constipation, weakness, drowsiness, diarrhea, nausea, and others. Vitamins or minerals when taken in large doses act like drugs and can cause harm. Alert your doctor to everything you take.

INVOLUNTARY WEIGHT LOSS/GAIN

Losing or gaining a lot of weight when you are not trying to do so is an important warning sign that must not be ignored. Being overweight or underweight also increases your chance of poor health.

NEEDS ASSISTANCE IN SELF CARE

Although most older people are able to eat, one of every five have trouble walking, shopping, buying and cooking food, especially as they get older.

ELDER YEARS ABOVE AGE 80

Most older people lead full and productive lives. But as age increases, risk of frailty and health problems increase. Checking your nutritional health regularly makes good sense.

 The Nutrition Screening Initiative • 1010 Wisconsin Avenue, NW • Suite 800 • Washington, DC 20007
The Nutrition Screening Initiative is funded in part by a grant from Ross Products Division of Abbott Laboratories.

FIGURE 15-8 *The Warning Signals Nutrition Checklist*

Source: Reprinted with permission by the Nutrition Screening Initiative, a project of the American Academy of Family Physicians, The American Dietetic Association, and the National Council on Aging, Inc. and funded in part by a grant from Ross Laboratories, a division of Abbott Laboratories.

that food . . . for one's self lacks the condiment of another's presence which can transform the simplest fare to the ceremonial act with all its shared meaning.[49]

Appendix L contains forms to assess mental status in older adults.

Community-Based Programs and Services

Until the early 1970s, nutrition services for older adults, with the exception of food stamps, were found primarily in hospitals and long-term care facilities. Efforts were then made to expand nutrition services from the hospitals to include communities and homes.[50]

FIGURE 15-9 *A Practical Approach to Nutrition Screening*

Source: Reprinted with permission from Nutrition Screening Initiative, *Incorporating Nutrition Screening and Interventions Into Medical Practice*, 1994, p. 28.

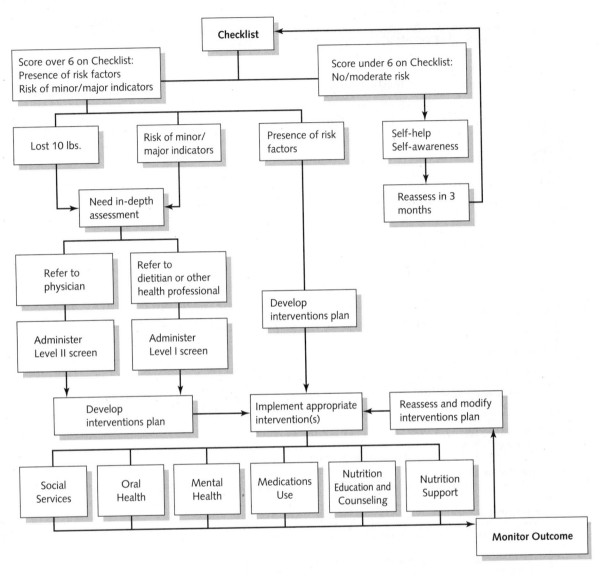

Risk Factors

- Inappropriate food intake
- Poverty
- Social isolation
- Dependency/disability
- Acute/chronic diseases or conditions
- Chronic medications use
- Advanced age (80+)

Major Indicators

- Weight loss of 10 lbs.+
- Under-/overweight
- Serum albumin below 3.5 g/dL
- Change in functional status
- Inappropriate food intake
- Mid-arm muscle circumference <10th percentile
- Triceps skinfold <10th percentile or >95th percentile
- Obesity
- Nutrition-related disorders
    Osteoporosis
    Osteomalacia
    Folate deficiency
    B_{12} deficiency

Minor Indicators

- Alcoholism
- Cognitive impairment
- Chronic renal insufficiency
- Multiple concurrent medications
- Malabsorption syndromes
- Anorexia, nausea, dysphagia
- Change in bowel habit
- Fatigue, apathy, memory loss
- Poor oral/dental status
- Dehydration
- Poorly healing wounds
- Loss of subcutaneous fat or muscle mass
- Fluid retention
- Reduced iron, ascorbic acid, zinc

Nutrition Assessment Component	Indicator of Risk
Anthropometric Measurements	
Weight Height	Measured weight-for-height 80% or below or 120% or above midpoint of medium frame (Metropolitan Life Tables)
	Reported usual body weight more than 5% less than actual weight
	Involuntary weight loss or gain of more than 10 lb
Body mass index (BMI)	Underweight: BMI < 16
	Overweight: BMI > 25
Body composition	Mid-arm muscle circumference 20% or more below NHANES standards
	Triceps skinfold 40% or less, or 190% or more above NHANES standards
	Waist-to-hip ratio:
	> 1.0 for men
	> 0.8 for women
Clinical Assessment	
Individual/family medical history	Use of tobacco, alcohol, drugs
	Chronic illness or disability
	Blood pressure > 140/90 mm Hg
Oral cavity exam	Lesions in mouth, bleeding gums, mobile teeth, ill-fitting dentures, need for dentures
Physical examination	Signs and symptoms of possible diet-related problems
Biochemical Assessment	
Total blood cholesterol	Elevated:
	> 240 mg/dL on 2 occasions
	> 200 mg/dL plus risk factors
LDL-C	Elevated: > 130 mg/dL
HDL-C	Low: < 35 mg/dL
Hemoglobin	Risk of anemia:
	Males: 45–64 yr: 13.2 g/dL; 65+ yr: 13.6 g/dL
	Females: 45–64 yr: 11.8 g/dL; 65+ yr: 11.9 g/dL
Serum albumin	Low: < 3.5 g/dL
Blood glucose	Outside normal range of 70–110 mg/100 ml
Dietary Assessment	
3-day food diary Food frequency 24-hour recall	Deficient or excessive intakes of calories and/or nutrients
Functional Assessment	
Activities of daily living (ADLs) Instrumental activities of daily living (IADLs)	A change from independence to dependence (needs assistance most of the time) with 2 ADLs or 1 nutrition-related IADL
Medication Assessment	Use of 3 or more prescribed medications; daily use of over-the-counter medications (e.g., aspirin, antacids); nutrition quackery

TABLE 15-6 *Nutrition Assessment for Older Adults*

Source: Adapted with permission by the Nutrition Screening Initiative, a project of the American Academy of Family Physicians, The American Dietetic Association, and the National Council on Aging, Inc. and funded in part by a grant from Ross Laboratories, a division of Abbott Laboratories.

In response to the socioeconomic problems that trouble many older adults and may lead to malnutrition—low income, inadequate facilities for preparing food, lack of transportation, and inability to afford dental care, among others—federal, state, and local agencies have mandated nutrition programs for the elderly.[51] Older adults' need for nutrition services depends on their level of independence, which can be depicted as a continuum, as in Figure 15-10.[52] Currently, community nutrition programs support the functional independence of older individuals in ambulatory care centers, adult day care centers, hospices, and home settings. Community nutritionists need to be familiar with organizations and programs providing nutrition and other health-related services to older adults.[53] A summary of the nutrition programs for older adults is provided in Table 15-7. Check out the resources available on-line from the organizations listed in the boxed insert.

FIGURE 15-10
Overview of Community Nutrition Programs for Older Adults

Source: Adapted from H. T. Philips and S. H. Gaylord, eds., *Aging and Public Health.* Copyright Springer Publishing Company, Inc., New York 10012. Used by permission.

General Assistance Programs

The Supplemental Security Income (SSI) Program improves the financial plight of the very poor directly by increasing a person's or family's income to the defined poverty level. This sometimes helps older people retain their independence.

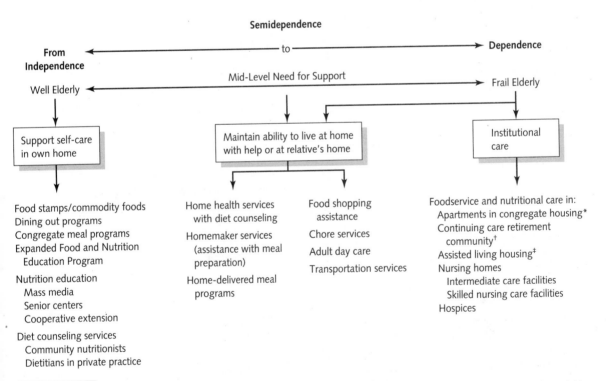

*Apartments in congregate housing: allows older persons to maintain a private apartment but makes supportive services easily available (transportation, common dining room, other personal services).

†Continuing-care retirement community: offers full spectrum of services as needed—from meal service only to assisted living services to skilled nursing care.

‡Assisted living housing: offers elderly people more supportive services and supervision than congregate housing (for example, help with ADL, meals provided in common dining room, emergency assistance available 24-hours/day).

Program	Type of Intervention	Funding Source	Eligible/Available Services	Percentage of Older Population Served
Elderly Nutrition Program	Congregate and home-delivered meals, therapeutic diets	DHHS AoA	Meals; transportation; shopping assistance; limited nutrition education, information, and referral	10% to 13% of population aged 60+ years
Food Stamps	Income subsidy	USDA	Coupons for food purchases or electronic benefits transfer (EBT)	50% to 80% of the eligible low-income elderly
Meals-on-Wheels America	Direct food delivery	Private	Home meals that complement weekday congregate programs; attention to needs of home-bound elderly	< 1%
Adult Day Care	Meal program, supervised day care	USDA	Day care programs for older Americans	58,000 adults per month
Medicare/ Medicaid	Third-party payment system	DHHS HCFA SSA	Covers medical and related services provided by participating hospitals, HMOs, private medical practices, ambulatory centers, rehabilitation and skilled nursing facilities, home health agencies, and hospice programs. Eligible nutrition services vary depending on the setting of care and the deemed medical necessity. Home meals, enteral/parenteral nutrition, and weight reduction are particularly limited.	Virtually all people over age 65 years are eligible (but many eligible people reportedly lack coverage by either program).

AoA = U.S. Department of Health and Human Services, Administration on Aging.
DHHS = U.S. Department of Health and Human Services.
HCFA = U.S. Department of Health and Human Services, Health Care Financing Administration.
SSA = Social Security Administration.
USDA =U.S. Department of Agriculture.

Source: Adapted from *Geriatric Nutrition: The Health Professional's Handbook* by R. Chernoff, pp. 428–29, with permission of Aspen Publishers, Inc. © 1991; and Food and Nutrition Service, *Food Program Facts* (Washington, D.C.: United States Department of Agriculture, 1998).

Another system of financial support to older Americans is the third-party reimbursement system, discussed earlier in Chapter 3. Third-party payers (for example, Medicare, Medicaid, Blue Cross/Blue Shield) sometimes reimburse the costs of receiving health-related services, including such nutrition services as nutrition screening, assessment, and counseling and enteral or parenteral nutrition support.[54] Generally, the nutrition service must be deemed "medically necessary." Whether a service is reimbursable and the extent of reimbursement varies from state to state and from case to case.

TABLE 15-7 *Nutrition Programs for Older Adults*

◎ Internet Resources

Administration on Aging	**www.aoa.dhhs.gov**
National Institute on Aging	**www.nih.gov/nia**
Look for information on the National Institute on Aging, research, funding, and training through links provided here.	
Shape Up America BMI Chart	**www.shapeup.org/sua/bmi/chart.htm**
Alzheimer's Association	**www.alz.org**
Visit the Alzheimer's Association's homepage to learn about the organization, the latest in the news, caregiver resources, public policy, and conferences and events.	
Jean Mayer USDA, Human Nutrition Research Center on Aging	**www.hnrc.tufts.edu**
The Jean Mayer USDA, Human Nutrition Research Center on Aging studies the effects of human nutrition on health. Links provided take visitors to an introduction of the center, research programs, human studies programs, and scientific publications.	
National Institute on Aging's Alzheimer's Disease Education and Referral Center	**www.alzheimers.org**
National Evaluation of the Elderly Nutrition Program, 1993–1995	**www.aoa.dhhs.gov/aoa/pages/nutreval.html**
National Policy and Resource Center on Nutrition and Aging	**www.aoa.dhhs.gov/aoa/dir/186.html**
American Association of Retired Persons (AARP)	**www.aarp.org**
American Geriatrics Society	**www.americangeriatrics.org**
North American Menopause Society	**www.menopause.org**
National Center for Health Statistics	**www.cdc.gov/nchswww**
Weight-Control Information Network	**www.niddk.nih.gov/health/nutrit/win.htm**
National Osteoporosis Foundation	**www.nof.org**
American Heart Association	**www.americanheart.org**
American Institute for Cancer Research	**www.aicr.org**
National Heart, Lung, and Blood Institute	**www.nhlbi.nih.gov/nhlbi/nhlbi.htm**
American Diabetes Association	**www.diabetes.org**
Women's Health Initiative	**www.nih.gov**

Social work agencies can provide older adults with information about appropriate nutrition resources in the community, such as congregate meal sites or Meals-on-Wheels programs.* On physician referral, Home Health Services, offered through local private and public organizations, provide home health aides to assist older people with shopping, housekeeping, and food preparation.

Nutrition Programs of the U.S. Department of Agriculture

The Food Stamp Program was not designed specifically for older people, but it can nevertheless help older adults in need of financial assistance. The Food Stamp Program, which is administered by the USDA, enables qualifying people to obtain coupons that they can use to buy food at grocery stores or to pay for meals at "Dining Out Programs" sometimes offered by restaurants. Currently, about 7 percent of the Food Stamp Program's participants are elderly persons (age 60 or over). Reasons for nonparticipation by the elderly include the "stigma" of receiving assistance, confusing paperwork, and a lack of public information about eligibility requirements. In an effort to reach more eligible older adults, California and Wisconsin provide those over 65 years of age with a cash equivalent to their entitled food stamp benefit as part of their SSI check.

The USDA also contributes to the elderly nutrition programs through its Nutrition Program for the Elderly (NPE). The NPE provides cash and commodity foods to local senior citizen centers for use in the Congregate and Home-Delivered Meals Programs of the U.S. Department of Health and Human Services (DHHS). The USDA also sponsors meal programs for the Adult Day Care Centers operating in many communities through its Child and Adult Day Care Program. Adult day care facilities care for seniors while their care providers are away from the home. USDA's Commodity Supplemental Food Program is also available in a limited number of states and can provide monthly food packages to persons 60 years of age and older.

The Food Stamp Program is discussed in Chapter 12's Program Spotlight.

Nutrition Programs of the U.S. Department of Health and Human Services

The Older Americans Act (OAA) of 1965 was amended in 1972 (PL 92-258) to establish the federal Elderly Nutrition Program. This formula-grant program distributes funds under Title III-C (formerly known as Title VII) of the OAA. Funds are given to state agencies on aging, which coordinate community services for the elderly based on the number of elderly residing in each state. This chapter's Program Spotlight provides an overview of the congregate and home-delivered meals programs.

Private Sector Nutrition Assistance Programs

In some communities, food banks enable older people on limited incomes to buy good food for less money. A food bank project buys industry "irregulars"—products that have been mislabeled, underweighted, redesigned, or mispackaged and would therefore ordinarily be thrown away. Nothing is wrong with this food; the industry

* The *Eldercare Locator* (800/677-1116) is a new service available nationwide that provides the name and telephone number for information resources on aging in local areas.

The social atmosphere can be as valuable as the foods served at congregate meal sites.

can credit it as a donation, and the buyer (often a food-preparing site) can obtain the food for a small handling fee and make it available at a greatly reduced price.

The Meals-on-Wheels America program is a similar but separate program to the federal home-delivered meals program. Meals-on-Wheels is a national project operated under the auspices of local volunteer groups. Its purpose is to help fill gaps in services provided by the federal meals programs by reaching older adults in communities not fully serviced by the Elderly Nutrition Program. [55] The preparation and delivery of the meals provided by Meals-on-Wheels is usually integrated into existing programs. In some communities, the Meals-on-Wheels program provides weekend and holiday meals in addition to the standard five luncheon meals.

Nutrition Education and Health Promotion Programs for Older Adults

Nationwide, about one-third of all seniors live in rural areas—communities with populations of 2,500 or less.[56] Few nutrition education efforts for these seniors are available, although some new programs have been designed to target this large audience. In Florida, the Area Agency on Aging of Central Florida has teamed with Florida Cooperative Extension to provide food labeling educational materials for the elderly and training programs for volunteers working with the elderly.

The Harvest Health at Home—Eating for the Second Fifty Years (HHH) project in North Dakota, which was designed using social learning theory, reaches rural communities through a series of newsletters.[57] The focus is on improving eating behaviors with an emphasis on preventive nutrition messages such as decreasing fat intake and increasing fiber in the diet. Part of the intervention's success is due to the practical suggestions included in the newsletters (such as lists comparing brand names); such measures are generally valuable and effective when working with elders. The HHH intervention has also been successful in collaborating with other health professionals; for example, the newsletter publicizes the schedule of routine screenings sponsored by local health departments and clinics.

Focus group interviews show that seniors are interested in changing their eating behavior.[58] Including practical activities in programs can help motivate these changes. For example, at the White Crane Senior Center in Chicago, a combination health care and wellness center founded by seniors, monthly cooking classes provide an opportunity to modify and taste new recipes and try new foods.

The supermarket can serve as a forum to promote healthful diets to older persons. At some supermarkets, dietitians interact with older consumers through store tours for people on special diets and in-store cooking classes showing how to prepare meals with foods that help lower the risks for chronic diseases. Retailers now recognize the value of providing customers with point-of-purchase information about nutrition and health. Supermarket programs for older adults can increase product sales, attract new customers, and contribute to customer loyalty. Many quality nutrition programs exist that can be implemented with a minimum of cost and effort. The Food Marketing Institute's *To Your Health!* program offers a supermarket kit that provides program planners with tools and ideas for promoting healthful eating and activity habits for older shoppers.[59]

Whatever the setting, nutrition programs for seniors can be designed for cost-effectiveness by considering the following elements which have been found to contribute to the cost-effectiveness of past interventions involving seniors:[60]

- Begin nutrition programs with a personalized approach, such as a self-assessment of nutrition status or behaviors and subsequent comparison with recommendations.
- Use a behavioral approach that combines self-assessment with self-management techniques (for example, goal setting or social support).
- Allow for active participation in the program (for example, hands-on cooking classes and small group discussions).
- Pay attention to motivators and reinforcements (for example, ease of food preparation, opportunities for social interaction).
- Empower participants by enhancing personal choice and self-control of health-related behaviors.
- Target specific subgroups of older adults. Needs and interests differ by age, income, and health status.
- Be sensitive to age-related physical changes as necessary. For example, consider the visual and hearing capabilities of the audience.

Finally, as discussed in Chapter 11, remember to plan for the evaluation of the program. Clarify and document program outcomes in order to measure the impact of the intervention. Who benefits? How does the impact vary by type of person served? Is the program cost-effective?

For adults over 50, health promotion efforts seek to preserve independence, productivity, and personal fulfillment.[61] The premise of health promotion is that individuals can enjoy benefits from healthful behaviors at any age. To this end, most states now offer community wellness centers for seniors that include services at all levels of prevention.[62] Resource people for these efforts at the local level include public health nutritionists employed by county health departments, consultant registered dietitians working with local nursing homes and community hospitals, and county extension service home economists.[63] Community groups and churches also offer support and self-help groups.

Program Spotlight   *The Elderly Nutrition Program*

With the graying of America, increased attention is being given to delivering cost-effective nutrition and health-related services to older persons in the community. The Elderly Nutrition Program (ENP) is intended to improve older people's nutrition status and enable them to avoid medical problems, continue living in communities of their own choice, and stay out of institutions.* Its specific goals are to provide the following:

- Low-cost, nutritious meals
- Opportunities for social interaction
- Nutrition education and shopping assistance
- Counseling and referral to other social and rehabilitation services
- Transportation services

The current ENP legislation makes one hot noon meal available five days a week, supplying a third of the RDA (see Table 15-8). There is no cost for meals, but participants sometimes make voluntary contributions.

One aspect of ENP is the Congregate Meals Program. Administrators try to select sites for congregate meals that will be accessible to as many of the eligible elderly as possible. The congregate meal sites are often community centers, senior citizen centers, religious facilities, schools, extended care facilities, or elderly housing complexes. Through the Home-Delivered Meals Program, meals are delivered to those who are homebound either permanently or temporarily. The home-delivery program—often referred to as "meals-on-wheels"—ensures nutrition, but its recipients miss out on the social benefit of the congregate meal sites; every effort is made to persuade them to come to the shared meals, if they can. The DHHS's Administration on Aging administers these programs, while the states, usually in conjunction with local county and city agencies, have responsibility for the daily operation and administration of these feeding programs.

All persons over 60 years of age and their spouses (regardless of age) are eligible to receive meals from these

* The ENP is authorized under the Older Americans Act of 1972 to provide grants to promote the delivery of nutrition services in local communities: (1) Under Title III, Grants for State and Community Programs on Aging, grants are made to the 655 Area Agencies on Aging, and (2) Under Title VI (added in 1978), Grants for Native Americans, grants are made to 221 Tribal Organizations representing American Indians, Alaska Natives, and Native Hawaiians.

TABLE 15-8 *Title III-C Meal Pattern*

Food Type	Recommended Portion Size
Meat or meat alternate	3 oz, cooked portion
Vegetables and fruits	Two ½ c portions*
Enriched white or whole-grain bread or alternate	1 serving (one slice bread or equivalent)
Butter or margarine	1 tsp
Milk	8 oz milk or calcium equivalent
Dessert	1 serving

*A Vitamin C-rich fruit or vegetable is to be served each day; a vitamin A-rich fruit or vegetable is to be served at least three times per week.

Source: U.S. Department of Health and Human Services.

programs, regardless of their income level. However, priority is given to low-income minorities, older persons with the greatest economic or social need, and extremely old individuals.[1] Since American Indians, Alaska Natives, and Native Hawaiians tend to have lower life expectancies and higher rates of illness at younger ages, Title VI allows Tribal Organizations to set the age at which older people can participate in the program. Since 1972, these programs have grown significantly, accounting for annual federal funding of about $486.4 million in 1998 and representing some 244 million meals served each year—125 million congregate meals and 119 million home-delivered meals. For every federal dollar spent, the ENP collects more than $2 from state, local, and private donations. Funding assists with food purchasing and preparation, facilities, and transportation for persons otherwise unable to participate.

By the early 1980s, it became apparent that Title III funds were insufficient to provide meals to the increasing numbers of frail elderly in the community in need of home-delivered meals. Home-delivered meals are usually more costly than congregate site meals. Some of this increased demand by homebound elderly for meals was a result of the initiation of the prospective payment system for hospitals, causing more of the elderly to be discharged early from hospitals. Certain states then initiated state-funded home-delivered meals programs for those who could not be served by the federal meals programs. One such program is the Supplemental Nutrition Assistance Program (SNAP) in New York state.[2]

Current evaluations of home-delivered meals programs ask whether the most needy elderly are going unserved and

TABLE 15-9 *Ten Nutritional Risk Factors Diagnostic for the Need for Assistance Among the Elderly*

1. Consumption of fewer than 8 main meals (hot or cold) per week.
2. Drinking of very little milk (less than half a pint per day).
3. Little or no intake of fruits or vegetables.
4. Wastage of food, even if supplied hot and ready to eat.
5. Long periods of the day without food or beverages.
6. Depression or loneliness.
7. Unexpected weight change (gain or loss).
8. Shopping difficulties.
9. Low income.
10. Presence of disabilities (including alcoholism).

Source: L. Davies, Nutrition and the elderly: Identifying those at risk, *Proceedings of the Nutrition Society* 43 (1984): 299. © 1984, reprinted with the permission of Cambridge University Press.

who should be given priority for receiving food assistance—those who lack access to food because of social or economic disabilities, or those with medical disabilities.[3] Criteria of nutritional risk are needed in order to assign priority status among the elderly experiencing food insecurity. One such assessment tool was designed in Great Britain so that social workers would be able to identify nutritional risk factors related to poverty, frailty, and loss of coping skills among the homebound elderly (see Table 15-9).[4] In the United States, eligibility criteria for home-delivered meals varies with whether the program is a federally, state, or locally operated program.

In an effort to reduce the cost of providing home-delivered meals, some states have initiated "luncheon clubs." These clubs have permitted several seniors living in close proximity to one another and receiving home-delivered meals to congregate in neighbors' homes. Only one meal delivery stop is therefore required and the seniors benefit from the social interaction.[5]

A recent, 2 year congressionally-mandated evaluation of the programs for congregate and home-delivered meals generally shows that the programs improve the dietary intake and nutrition status of their clients (see Figure 15-11).[6] Participants generally have greater diversity in their diets and higher intakes of essential nutrients, and they are less likely to report food insecurity than nonparticipants.[7] Other benefits come as a result of screening and the referrals generated by such programs. Additional benefits

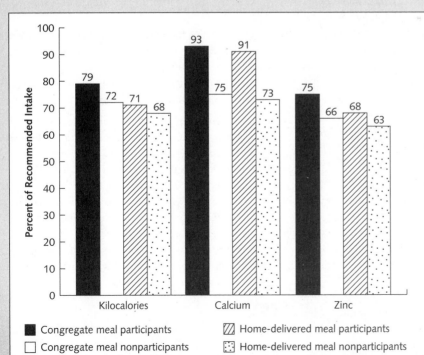

FIGURE 15-11 *Intakes of Selected Nutrients by Participants and Nonparticipants of the Congregate Meal and Home-Delivered Meal Programs*

Meal participants have a higher calorie intake and reach a higher percentage of the recommended intakes for most nutrients than nonparticipants similar in age and socioeconomic status. Homebound elderly have lower nutrient intakes than elderly persons who can leave their home.

Source: Data from M. Ponza, J. Ohls, and B. Millen, *Serving Elders at Risk: The Older Americans Act Nutrition Programs: National Evaluation of the Elderly Nutrition Program 1993–1995, Vol. I, Title III Evaluation Findings* (Princeton, NJ: Mathematica Policy Research, 1996).

Continued

are derived from the activities associated with the congregate meals services: diet counseling, exercise, adult education, and other classes and activities. Participants benefit, too, from the opportunity for improved socialization.[8]

However, despite these positive outcomes, deficiencies in the meals programs are noted in the lack of provision of weekend or evening meals to those who cannot get food or cannot cook, lack of special diets for persons with special dietary needs (for instance., persons with diabetes), and delivery of meals that are too high in fat or lacking in nutrients that the meal participant is unlikely to obtain from the food they get for themselves (such as folate).[9] In addition, 41 percent of Title III ENP service providers have waiting lists for home-delivered meals, suggesting a significant unmet need for these meals.[10]

The special needs of the homebound elderly need to be given greater priority.[11] The dietary intake of the homebound elderly might improve if more than one meal per day were provided, if the meal furnished to the client were prepared with greater percentages of the DRI or RDA, and if meals were provided 7 days per week rather than 5.[12]

Although Title III nutrition programs are required to provide nutrition education to their clients, these efforts are usually limited to the congregate meal sites.[13] With the exception of a limited amount of printed material, virtually no education is provided to the staff who purchase and prepare these meals.[14] Greater emphasis on nutrition education for both the helper and the client receiving home-delivered meals is warranted.

Amendments to the Older Americans Act (OAA) include recommendations believed necessary to improve the services in the OAA and permit the Administration on Aging to make significant changes in OAA programs. Highlights of these changes include the following:[15]

- Nutrition education will be provided to nutrition program participants at least on a semi-annual basis.
- Each state must establish and administer the nutrition programs under the advice of dietitians.
- Criteria for nonfinancial eligibility to receive nutrition services will be developed.
- States must ensure that meals comply with the Dietary Guidelines for Americans and provide one-third of the RDA for one meal, two-thirds of the RDA for two meals, and 100 percent of the RDA if three meals are served.

In addition, whereas Title III administers the congregate and home-delivered nutrition services, a more recent Title IIIf authorizes the provision of services related to health promotion and disease prevention. Health promotion and disease prevention services as defined in Title IIIf can include these elements:[16]

- Health risk appraisals
- Routine health screening
- Health promotion programs (for example, substance abuse, weight loss, smoking cessation)
- Programs that offer physical fitness, group exercise, music, art, and dance movement therapy
- Home injury control services, including screening of high-risk home environments and injury prevention programs
- Screening for the prevention of depression, coordination of community mental health services, provision of educational activities, and referral to psychiatric and psychological services
- Educational programs about preventive health services covered under Medicare
- Medication management programs and screening
- Information concerning diagnosis, prevention, treatment, rehabilitation of age-related diseases like osteoporosis and cardiovascular disease

Some states are using the new Title IIIf funds to incorporate the Nutrition Screening Initiative into the services that they already provide to elders.

The Elderly Nutrition Program faces certain challenges on the horizon, including the following:

- Changing demographics will likely increase the demand for program services—particularly for home-delivered meals. The number of persons 85 years and older is expected to double by 2030; this group is less likely to live independently due to disabling conditions.
- Changes in the present health care system will affect the ENP as more people are discharged early and in need of community health services.
- Depending on changes in public policy and funding, the ENP may be challenged to meet increased demand at a time of decreasing federal funding.

References

1. R. Voelker, Federal program nourishes poor elderly, *Journal of the American Medical Association* 278 (1997): 1301.

2. D. A. Roe, Development and current status of home-delivered meals programs in the United States: Who is served? *Nutrition Reviews* 48 (1990): 181–85.

3. N. S. Wellman and coauthors, Elder insecurities: Poverty, hunger, and malnutrition, *Journal of the American Dietetic Association* 97 (1997): S120–S122.

4. Roe, Development and current status of home-delivered meals programs, p. 183.

5. S. Riggs, Serving the seventies: Meal programs for the elderly, together or at home, *Restaurants and Institutions* 95 (1984): 193–96.

6. M. R. Neyman, S. Zidenberg-Cherr, and R. B. McDonald, Effect of participation in congregate-site meal programs on nutritional status of the healthy elderly, *Journal of the American Dietetic Association* 96 (1996): 475; F. M. Torres-Gil, J. L. Lloyd, and J. Carlin, Role of elderly nutrition in home and community-based care, *Perspectives in Applied Nutrition* 2 (1995): 9; and M. Ponza and coauthors, *Serving Elders at Risk: The Older Americans Act Nutrition Programs 1993–1995, Vol. 1, Title III Evaluation Findings* (Princeton, NJ: Mathematica Policy Research, 1996), available on the Web site of the National Policy and Resource Center on Nutrition and Aging, Florida State University; access at www.aoa.dhhs.gov/aoa/pages/nutreval.html.

7. D. L. Edwards and coauthors, Home-delivered meals benefit the diabetic elderly, *Journal of the American Dietetic Association* 93 (1993); 585–87; D. L. MacLellan, Contribution of home-delivered meals to the dietary intake of the elderly, *Journal of Nutrition for the Elderly* 16 (1997):

23–27; and E. Fogler-Levitt and coauthors, Utilization of home-delivered meals by recipients 75 years of age or older, *Journal of the American Dietetic Association* 95 (1995): 552.

8. Institute of Medicine, *The Second Fifty Years*, p. 182.

9. Roe, Development and current status of home-delivered meals programs, p. 185; L. Y. Yamaguchi and coauthors, Improvement in nutrient intake by elderly meals-on-wheels participants receiving a liquid nutrition supplement, *Nutrition Today* 33 (1998): 37–44.

10. *Serving Elders at Risk: The Older Americans Act Nutrition Programs*, 1996.

11. A. Balsam, J. Carlin, and B. Rogers, Weekend home-delivered meals in Elderly Nutrition Program, *Journal of the American Dietetic Association* 92 (1992): 1125–26.

12. D. A. Stevens, L. E. Grivetti, and R. B. McDonald, Nutrient intake of urban and rural elderly receiving home-delivered meals, *Journal of the American Dietetic Association* 92 (1992): 714–18; and O. Walden and coauthors, The provision of weekend home-delivered meals by state and a pilot study indicating the need for weekend home-delivered meals, *Journal of Nutrition for the Elderly* 8 (1988): 31–43.

13. *Nutrition Services for the Elderly*, Hearing Pub. No. Y4, Ag 4/2 N 95, (Washington, D.C.: U.S. House of Representatives Committee on Aging, June 10, 1988).

14. Stevens, Grivetti, and McDonald, Nutrient intake of urban and rural elderly.

15. Legislative highlights: Older Americans Act passes, *Journal of the American Dietetic Association* 92 (1992): 1458.

16. *Gerontological Nutritionists Newsletter*, Winter 1994, p. 7.

Looking Ahead: And Then We Were Old

As a nation, we tend to value the future more than the present, putting off enjoying today so that we will have money, prestige, or time to have fun tomorrow. The elderly feel this loss of future. The present is their time for leisure and enjoyment, but often they have no experience in using leisure time.

The solution is to begin to prepare for old age early in life, both psychologically and nutritionally (see Figure 15-12). Preparation for this period should, of course, include financial planning, but other lifelong habits should be developed as well. Each adult needs to learn to reach out to others to forestall the loneliness that will otherwise ensue. Adults need to develop some skills or activities that they can continue into their later years—volunteer work with organizations, reading, games, hobbies, or intellectual pursuits—and that will give meaning to their lives. Each adult needs to develop the habit of adjusting to change, especially when it comes without consent, so that it will not be seen as a loss of control over one's life. The goal is to arrive at maturity with as healthy a mind and body as possible; this means cultivating good nutrition status and maintaining a program of daily exercise.

In general, the ability of the elderly to function well varies from person to person and depends on several factors. The "life advantages" listed on the next page seem to contribute to good physical and mental health in later years.[64]

AGING WELL

Stress Busters

- Relax
- Go for a walk
- Breathe deeply
- Think positively

Emotional Well-Being

- Reduce stress
- Learn relaxation techniques
- Cultivate a garden
- Laugh often
- Take time for spiritual growth
- Adopt and love a pet
- Take time off

Social Health

- Be socially active
- Volunteer for a special cause
- Make new friends
- Enroll in lifelong learning

Nutritional Health

- Choose nutrient-dense foods
- Eat at least 5 fruits and vegetables every day
- Drink plenty of water
- Keep fat intake to a minimum
- Get adequate fiber

Physical Health

- Be physically active
- Get adequate sleep
- Challenge your mental skills
- Do aerobic and strength-training exercises at least 3 times a week
- Stretch for flexibility

Lifelong Habits for Health

- Cherish your personal values and goals
- Develop good communication skills
- Balance diet and exercise to maintain a healthful weight
- Practice preventive health care
- Develop skills and hobbies to enjoy for a lifetime

- Manage time
- Learn from mistakes
- Nurture relationships with family and friends
- Enjoy, respect, and protect nature
- Accept change as inevitable
- Plan ahead for financial security

FIGURE 15-12 *The Aging Well Pyramid*

The time to prepare for old age is early in life. Practice the items found at the base of the pyramid to achieve an optimal sense of well-being. Use the inner four compartments of the pyramid to create a balance among all aspects of your life: nutrition, physical activity, social health, and emotional well-being. Use the tip of the pyramid to manage everyday stresses such as traffic gridlock, exams, and work deadlines.

- Genetic potential for extended longevity. Some persons seem to have inherited a reduced susceptibility to degenerative diseases.
- A continued desire for new knowledge and new experiences. Some studies suggest that "active" minds, ever involved in learning new things, may be more resistant to decline.
- Socialization, intimacy, and family integrity. Older persons thrive in situations where love, understanding, shared responsibility, and mutual respect are nurtured.
- Adherence to a prudent diet while avoiding excesses of food energy, fat, cholesterol, and sodium. A prudent diet with adequate intakes of all essential nutrients has a positive impact on health and weight management.
- Avoidance of substance abuse
- Acceptable living arrangements
- Financial independence
- Access to health care, including a family physician, health clinic, public health nursing service providing home health care, dentist, podiatrist, physical therapist, pharmacist, and community nutritionist

Everyone knows older people who have maintained many contacts—through relatives, church, synagogue, or fraternal orders—and have not allowed themselves to drift into isolation. Upon analysis, you will find that their favorable environment came through a lifetime of effort. These people spent their entire lives reaching out to others and practicing the art of weaving others into their own lives. Likewise, a lifetime of effort is required for good nutrition status in the later years. A person who has eaten a wide variety of foods, stayed trim, and remained physically active will be best able to withstand the assaults of change.

Community Learning Activity

Activity 1

In this activity, we ask that you become familiar with an older adult's circumstances and needs. The older adult you choose to interview can be a relative, neighbor, associate, or other individual over the age of 65. Ask your interviewee to complete the checklist developed by the Nutrition Screening Initiative (see Figure 15-7). Next, depending on the results you obtain, refer to the schematic in Figure 15-9 and determine your appropriate next step. Use the Level I screen to further evaluate risk factors for poor nutrition status in this individual (see Appendix L). Finally, what nutrition interventions and referrals, if any, would you recommend to this person (for instance, nutrition education, food stamps, congregate or home-delivered meals programs, dental care, physical activity and exercise programs, homemaker or home health aide assistance, and socialization activities)?

References

1. National Center for Health Statistics, *Health United States, 1997* (Washington, D.C.: U.S. Department of Health and Human Services, December 1997).

2. A. E. Harper, Nutrition, aging, and longevity, *American Journal of Clinical Nutrition* (supplement) 36 (October 1982): 737–49.

3. Administration on Aging, *Profile of Older Americans: 1997* (Hyattsville, MD: U.S. Department of Health and Human Services, 1997).

4. U.S. Census Bureau, Sixty-five plus in the United States, Statistical Brief, June 6, 1998: 1–7; www.census.gov/ftp/pub/socdemo/www/agebrief.html.

5. Administration on Aging, *Profile of Older Americans, 1997.*

6. Institute of Medicine, *Extending Life, Enhancing Life: A National Research Agenda on Aging* (Washington, D.C.: National Academy Press, 1991), p. 1.

7. *Healthy People 2000: National Health Promotion and Disease Prevention* (Washington, D.C.: U.S. Department of Health and Human Services, Public Health Service, 1990), pp. 579–92.

8. J. M. McGinnis and M. Nestle, The Surgeon General's Report on Nutrition and Health: Policy implications and implementation strategies, *American Journal of Clinical Nutrition* 49 (1989): 23–28.

9. *Healthy People 2000*, pp. 588–89, 611–13; P. Fishman, *Healthy People 2000:* What progress toward better nutrition, *Geriatrics* 51 (1996): 38.

10. *The Boomer Report*, vol. II, nos. 4, 5, 6, and 7, 1990; and S. T. Borra, Food and nutrition education for baby boomers: Challenges and opportunities, *Nutrition News*, 54(2) 1991: 5–6.

11. J. V. White and coauthors, Beyond nutrition screening: A systems approach to nutrition intervention, *Journal of the American Dietetic Association* 93 (1993): 405–7.

12. B. G. Janas, C. A. Bisogni, and C. C. Campbell, Conceptual model for dietary change to lower serum cholesterol, *Journal of Nutrition Education* 25 (1993): 186–92; T. Byers, Dietary trends in the United States: Relevance to cancer prevention, *Cancer* 72 (1993): 1015–18.

13. Report of the National Cholesterol Education Program Expert Panel on Detection, Evaluation, and Treatment of High Blood Cholesterol in Adults, *Archives of Internal Medicine* 148 (1988): 36–69.

14. S. Havas and coauthors, 5 a day for better health: Nine community research projects to increase fruit and vegetable consumption, *Public Health Reports* 110 (1995): 68–79.

15. K. Lancaster and coauthors, Evaluation of a nutrition newsletter by older adults, *Journal of Nutrition Education* 29 (1997): 145–51.

16. A. Eldridge and coauthors, Development and evaluation of a labeling program for low-fat foods in a discount department store foodservice area, *Journal of Nutrition Education* 29 (1997): 159–61.

17. Food Marketing Institute, *Supermarket Consumer Affairs, 1997: Directory of Nutrition and Health Programs* (Washington, D.C.: Food Marketing Institute, 1997); and T. Hammonds, Nutrition professionals and supermarkets: A strategic alliance to promote consumer health and food safety, Keynote presentation, 80th Annual Meeting & Exhibition, American Dietetic Association, October 27, 1997.

18. A. Worick and M. Petersons, Weight-loss contests at the worksite: Results of repeat participation, *Journal of the American Dietetic Association* 93 (1993): 680–81.

19. R. W. Jeffery, The healthy worker project, *American Journal of Public Health* 83 (1993): 395–401; J. C. Erfurt, A. Foote, and M. A. Heirich, Worksite wellness programs: Incremental comparisons of screening and referral alone, health education, follow-up counseling, and plant organization, *American Journal of Health Promotion* 5 (1991): 438–48; and The American Dietetic Association and U.S. Public Health Service, *Worksite Nutrition: A Guide to Planning, Implementation, and Evaluation*, 2nd ed. (Chicago: American Dietetic Association, 1993).

20. The list of classes of health promotion activities is from U.S. Department of Health and Human Services, 1992 National Survey of Worksite Health Promotion Activities: Summary, *American Journal of Health Promotion* 7 (1993): 452–64.

21. Jeffery, The healthy worker project, p. 395.

22. K. Richmond, Introducing heart healthy foods in a company cafeteria, *Journal of Nutrition Education* 18 (1986): S63; G. S. Peterson and coauthors, Strategies for lowering cholesterol at the worksite, *Journal of Nutrition Education* 18 (1986): S54; and H. Quigley, L. L. Bean cholesterol reduction program, *Journal of Nutrition Education* 18 (1986): S58.

23. The American Dietetic Association and U.S. Public Health Service, *Worksite Nutrition: A Guide to Planning, Implementation, and Evaluation*, pp. 72–74.

24. Institute of Medicine, *The Second Fifty Years: Promoting Health and Preventing Disability* (Washington, D.C.: National Academy Press, 1992), pp. 1–21.

25. Institute of Medicine, *Extending Life*.

26. J. E. Kerstetter, B. A. Holthausen, and P. A. Fitz, Malnutrition in the institutionalized older adult, *Journal of the American Dietetic Association* 92 (1992): 1109–16.

27. J. T. Dwyer, J. J. Gallo, and W. Reichel, Assessing nutritional status in elderly patients, *American Family Physician* 47 (1993): 613–20.

28. E. L. Smith, P. E. Smith, and C. Gilligan, Diet, exercise, and chronic disease patterns in older adults, *Nutrition Reviews* 46 (1998): 52–61.

29. Institute of Medicine, *Extending Life*, pp. 1–39.

30. NHANES III health data relevant for aging nation, *Journal of the American Medical Association* 277 (1997): 100–2.

31. M. Bernard, V. Lampley-Dallas, and L. Smith, Common health problems among minority elders, *Journal of the American Dietetic Association* 97 (1997): 771–76; and M. L. Lopez and coauthors, Building educational partnerships to serve Latinos in central California, *Journal of Family and Consumer Sciences*, Summer 1997.

32. Kerstetter, Holthausen, and Fitz, Malnutrition, p. 1113.

33. F. G. Abdellah and S. R. Moore, eds., *Surgeon General's Workshop on Health Promotion and Aging: Proceedings* (Washington, D.C.: Office of the Surgeon General, 1988), pp. G1–G19.

34. M. T. Fanelli and M. Kaufman, Nutrition and older adults, in *Aging and Public Health*, ed. H. T. Philips and S. H. Gaylord (New York: Springer, 1985), pp. 70–100.

35. M. Nestle and J. A. Gilbride, Nutrition policies for health promotion in older adults: Education priorities for the 1990s, *Journal of Nutrition Education* 22 (1990): 316.

36. *Surgeon General's Workshop: Health Promotion and Aging* (Washington, D.C.: U.S. Department of Health and Human Services, 1988).

37. Position of the American Dietetic Association: Nutrition, aging, and the continuum of care, *Journal of the American Dietetic Association* 96 (1996): 1048–52.

38. Administration on Aging, *Food and Nutrition for Life: Malnutrition and Older Americans* (Washington, D.C.: Department of Health and Human Services, 1994).

39. *Nutrition Screening Initiative Survey* (Washington, D.C.: Peter D. Hart Research Associates, February 1990); and President's page: The Nutrition Screening Initiative—An emerging force in public policy, *Journal of the American Dietetic Association* 93 (1993): 822.

40. K. Gray-Donald, The frail elderly: Meeting the nutritional challenges, *Journal of the American Dietetic Association* 95 (1995): 538–40; and C. Ryan and M. Shea, Recognizing depression in older adults: The role of the dietitian, *Journal of the American Dietetic Association* 96 (1996): 1042–44.

41. J. T. Dwyer, *Screening Older Americans' Nutritional Health: Current Practices and Possibilities* (Washington, D.C.: Nutrition Screening Initiative, 1991), pp. 1–23.

42. M. A. Hess, President's page: ADA as an advocate for older Americans, *Journal of the American Dietetic Association* 91 (1991): 847–49.

43. J. V. White and coauthors, Nutrition Screening Initiative: Development and implementation of the public awareness checklist and screening tools, *Journal of the American Dietetic Association* 92 (1992): 163–67.

44. The discussion of tools used by the Nutrition Screening Initiative was adapted from White and coauthors, Nutrition Screening Initiative, pp. 163–67; J. V. White and coauthors,

Consensus of the Nutrition Screening Initiative: Risk factors and indicators of poor nutritional status in older Americans, *Journal of the American Dietetic Association* 91 (1991): 783–87; and J. V. White, Risk factors for poor nutritional status in older Americans, *American Family Physician* 44 (1991): 2087–97.

45. N. J. Stiles and coauthors, A geriatric nutrition clinic: Addressing the nutritional needs of the elderly through an interdisciplinary team, *Journal of Nutrition for the Elderly* 15 (1996): 33–41.

46. The discussion of assessment in the elderly was adapted from P. P. Barry and M. Ibarra, Multidimensional assessment of the elderly, *Hospital Practice* 25 (1990): 117–28; Mobarhan and Trumbore, Nutritional problems of the elderly; C. O. Mitchell, Nutritional assessment of the elderly, *Clinics in Applied Nutrition* 1 (1991): 76–88; and R. Chernoff, Physiological aging and nutrition status, *Nutrition in Clinical Practice* 5 (1990): 8–13.

47. W. C. Chumlea, A. F. Roche, and M. L. Steinbaugh, Estimating stature from knee height for persons 60 to 90 years of age, *Journal of the American Geriatric Society* 33 (1985): 116–20.

48. Mitchell, Nutritional assessment of the elderly.

49. J. Weinberg, Psychologic implications of the nutritional needs of the elderly, *Journal of the American Dietetic Association* 60 (1972): 293–96.

50. U.S. Department of Health and Human Services, *The Surgeon General's Report*, pp. 598–99.

51. W. S. Wolfe, Understanding food insecurity in the elderly: A conceptual framework, *Journal of Nutrition Education* 28 (1996): 92–100.

52. Fanelli and Kaufman, Nutrition and older adults, pp. 88–89.

53. S. Saffel-Shrier and B. M. Athas, Effective provision of comprehensive nutrition case management for the elderly, *Journal of the American Dietetic Association* 93 (1993): 439–444.

54. R. Chernoff, *Geriatric Nutrition: The Health Professional's Handbook* (Gaithersburg, MD: Aspen, 1991), pp. 436–42.

55. *Meals-on-Wheels America: More Meals for the Homebound Through Public/Private Partnerships—A Technical Assistant Guide* (New York: New York City Department for the Aging, 1989).

56. G. L. Klein and coauthors, Nutrition and health for older persons in rural America: A managed care model, *Journal of the American Dietetic Association* 97 (1997): 885–88.

57. E. D'urso-Fischer and coauthors, Reaching out to the elderly, *Journal of Nutrition Education* 23 (1991): 20–25.

58. Ibid., pp. 20–21.

59. T. Hammonds, Nutrition professionals and supermarkets, 1997.

60. L. K. Maloney and S. L. White, Nutrition education for older adults, *Journal of Nutrition Education* 27 (1995): 339–46.

61. M. G. Ory, Considerations in the development of age-sensitive indicators for assessing health promotion, *Health Promotion* 3 (1988): 139–49.

62. S. Maloney, Healthy older people, in *Surgeon General's Workshop on Health Promotion and Aging* (Washington, D.C.: U.S. Department of Health and Human Services, 1988).

63. Fanelli and Kaufman, Nutrition and older adults, p. 95.

64. The list of advantages is adapted from D. A Roe, *Geriatric Nutrition*, 3rd ed. (Upper Saddle River, NJ: Prentice Hall, 1992), pp. 1–9.

Chapter 16

Nurturing Global Awareness: Community Nutrition with an International Perspective

Outline

Learning Objectives

After you have read and studied this chapter, you will be able to:

- Describe the current status of world food security.
- List causes of world food insecurity.
- Give reasons why women and children are particularly at risk with regard to hunger.
- Describe the purpose and goals of recent international food policy initiatives.

- Give examples of current international nutrition intervention programs.
- Describe the global public health issues related to global food insecurity that will continue to challenge policy makers and program designers into the twenty-first century.
- List actions that individuals might take to eliminate global food insecurity.

Something To Think About...

The world does not exist in a series of separately functioning compartments. The last lines from the chapter on poverty in the State of the World Report sum this up succinctly: "Sheets of rain washing off denuded watersheds flood exclusive neighborhoods as surely as slums. Potentially valuable medicines lost with the extinction of rain forest species are as unavailable to the rich in their private hospitals as they are to the poor in rural clinics. And the carbon dioxide released as landless migrants burn plots in the Amazon or the Congo warms the globe as surely as do the fumes from automobiles and factory smoke-stacks in Los Angeles or Milan."

Poverty, food security, environment, nutrition—they are all inescapably interlinked. Those of you who have the knowledge to embrace issues of nutrition education also have the power to ameliorate the human predicament. And there is no objective in this world more worth pursuing than improving the human condition. – *Stephen Lewis*

Introduction

All people need food. Regardless of race, religion, sex, or nationality, our bodies experience similarly the effects of food insecurity and its companion malnutrition—listlessness, weakness, failure to thrive, stunted growth, mental retardation, muscle wastage, scurvy, anemia, rickets, osteoporosis, goiter, tooth decay, blindness, and a host of other effects, including death.[1] Apathy and shortened attention span are two of a number of behavioral symptoms that are often mistaken for laziness, lack of intelligence, or mental illness in undernourished people.

The Food and Agriculture Organization of the United Nations (FAO) estimates that of the more than 5.7 billion people in the world, at least 840 million people—20 percent of the developing world's population, or one in seven people worldwide—suffer from chronic, severe undernutrition, consuming too little food each day to meet even minimum energy requirements (see Figure 16-1).[2] Over 200 million children under the age of 5 still suffer from basic protein and energy deficiencies.[3] Some 2 billion people, mostly women and children, are deficient in one or more of three major micronutrients: iron, iodine, and vitamin A.[4] By the year 2030, the world's population will have grown by another 3 billion people, thus further stretching the world's food resources.[5]

Mapping Poverty and Undernutrition

Food insecurity was once viewed as a problem of overpopulation and inadequate food production, but now many people recognize it as a problem of poverty. Food is *available* but not *accessible* to the poor who have neither land nor money. In 1978, Robert McNamara, then-president of the World Bank, gave what stands as the classic

FIGURE 16-1 A
*Regional Distribution of
the World's Chronically
Undernourished
Populations*

Source: Food and Agriculture
Organization of the United Nations,
The Sixth World Food Survey
(Rome: FAO, 1996).

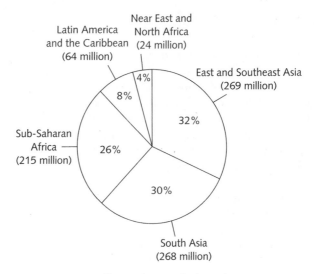

840 million undernourished people

Food security is defined as access by all people at all times to enough food for an active and healthy life. Food security has two aspects: ensuring that adequate food supplies are available and ensuring that households whose members suffer from undernutrition have the ability to acquire food, either by producing it themselves or by being able to purchase it.

To qualify as chronically and severely undernourished by FAO standards, a person must consume fewer than the calories required to perform the basic physiological functions and light physical activity. This minimum is usually within the range of 1,700 to 1,960 kilocalories/day.

Low-income food deficit countries (LIFDCs): FAO defines these countries as nations that have been net importers of food for over a three-year period and whose net income per person is below the eligibility threshold ($1,395 in 1996) to receive financial assistance from the **World Bank.**

The **World Bank** is a group of international financial institutions owned by the governments of more than 150 nations. The bank provides loans for economic development.

description of *absolute* poverty: "A condition of life so limited by malnutrition, illiteracy, disease, squalid surroundings, high infant mortality, and low life expectancy as to be beneath any reasonable definition of human decency (see Table 16-1)."[6]

Trends in the total numbers of people estimated to be undernourished are given in Figure 16-2. The present estimate of 1.2 billion people living in poverty translates to 23 percent of the world's population. Living standards declined during the past two decades owing in part to accelerated rates of population growth and environmental decline but also as a result of lower export earnings, rising inflation, and higher interest rates on foreign debts.[7] In other words, the poor earned less and paid more.

According to the FAO, widespread chronic hunger is centered in more than 80 nations known as the **low-income food deficit countries (LIFDCs).** These developing countries can neither produce enough food to feed their populations fully nor earn enough foreign exchange to import food to cover their food deficits.[8] In 1996, 82 countries fit this description, half of them in Africa (see Table 16-2).[9]

Those who live with chronic poverty must constantly face unsafe drinking water, intestinal parasites, insufficient food, a low-protein diet, stunted growth, low birthweights, illiteracy, disease, shortened life spans, and death. In *Quiet Violence: View from a Bangladesh Village*, Hartman and Boyce provide a good introduction to life in the villages of the developing world. The lives of these villagers are more difficult than anything we have ever known, and yet their hopes and dreams are not unlike our own. They exhibit resourcefulness, hard work, and dignity in the midst of circumstances that require a persistence and personal strength that most of us will never need to call upon in our lifetimes. Hari, one of the landless laborers in the village, reflects on his life just days before his death: "Between the mortar and the pestle, the chili cannot last. We poor are like chilies—each year we are ground down, and soon there will be nothing left."[10]

Poverty is much more than an economic condition and exists for many reasons, including overpopulation, the greed of others, unemployment, and the lack of pro-

Region	GNP	Infant Mortality Rate (IMR)	Life Expectancy	Literacy Rate	Safe Water Supply (%)	Children Under 5 Mortality Rate (U5MR)	Population per Doctor
General Differences*							
Low-income food deficit countries	$301	120	51	40	50	200	17,000
Other developing countries	$684	67	63	70	70	120	2,700
Developed countries	$19,542	7	76	98	100	20	520

	IMR	Undernourished Children (%)	U5MR
Regional Differences			
Sub-saharan Africa	106	30	175
South Asia†	82	58	121
Southeast Asia‡	42	24	55
Latin America and Caribbean	38	12	47
Middle East and North Africa	46	25	60

*Notice that the poorer nations have higher infant and children-under-5 mortality rates, shorter life expectancies, lower literacy rates, and fewer doctors available than richer nations. In short, the quality of life suffers from poverty.
†South Asia consists of seven countries: Afghanistan, Bangladesh, Bhutan, India, Nepal, Pakistan, and Sri Lanka.
‡Southeast Asia includes Cambodia, China, Fiji, Hong Kong, Indonesia, North Korea, South Korea, Laos, Malaysia, Mongolia, Myanmar, Papua New Guinea, the Philippines, Singapore, Taiwan, Thailand, and Vietnam.
Source: Data from Bread for the World Institute, *Hunger 1998: Hunger in a Global Economy* (Silver Spring, MD: Bread for the World Institute, 1997).

ductive resources such as land, tools, and credit.[11] Consequently, if we are to provide adequate nutrition for all the earth's hungry people, we must transform the economic, political, and social structures that both limit food production, distribution, and consumption and create a gap between rich and poor.[12]

TABLE 16-1 *The Gap Between Developed and Developing Countries*

Malnutrition and Health Worldwide

Worldwide, about 40,000 to 50,000 people die each day as a result of undernutrition. Millions of children die each year from the parasitic and infectious diseases associated with poverty: dysentery, whooping cough, measles, tuberculosis, cholera, and malaria (see Table 16-3). These diseases interact with poor nutrition to form a vicious cycle in which the outcome for many is death, as shown in Figure 16-3. **UNICEF** estimates that malnutrition and disease claim the lives of 250,000 children *every week* at the rate of one every two seconds.[13]

UNICEF The United Nations International Children's Emergency Fund, now referred to as the United Nations Children's Fund.

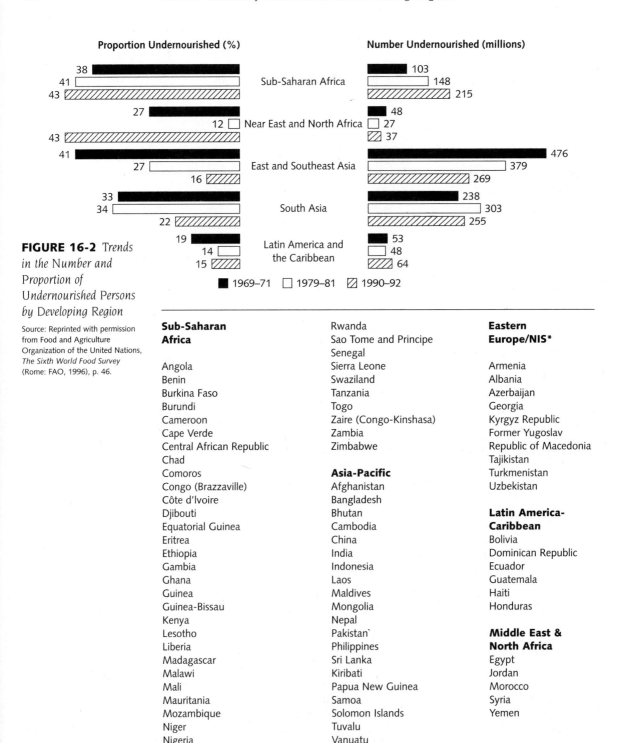

FIGURE 16-2 *Trends in the Number and Proportion of Undernourished Persons by Developing Region*

Source: Reprinted with permission from Food and Agriculture Organization of the United Nations, *The Sixth World Food Survey* (Rome: FAO, 1996), p. 46.

Sub-Saharan Africa

Angola
Benin
Burkina Faso
Burundi
Cameroon
Cape Verde
Central African Republic
Chad
Comoros
Congo (Brazzaville)
Côte d'Ivoire
Djibouti
Equatorial Guinea
Eritrea
Ethiopia
Gambia
Ghana
Guinea
Guinea-Bissau
Kenya
Lesotho
Liberia
Madagascar
Malawi
Mali
Mauritania
Mozambique
Niger
Nigeria
Rwanda
Sao Tome and Principe
Senegal
Sierra Leone
Swaziland
Tanzania
Togo
Zaire (Congo-Kinshasa)
Zambia
Zimbabwe

Asia-Pacific

Afghanistan
Bangladesh
Bhutan
Cambodia
China
India
Indonesia
Laos
Maldives
Mongolia
Nepal
Pakistan`
Philippines
Sri Lanka
Kiribati
Papua New Guinea
Samoa
Solomon Islands
Tuvalu
Vanuatu

Eastern Europe/NIS*

Armenia
Albania
Azerbaijan
Georgia
Kyrgyz Republic
Former Yugoslav Republic of Macedonia
Tajikistan
Turkmenistan
Uzbekistan

Latin America-Caribbean

Bolivia
Dominican Republic
Ecuador
Guatemala
Haiti
Honduras

Middle East & North Africa

Egypt
Jordan
Morocco
Syria
Yemen

TABLE 16-2 *Low-Income Food Deficit Countries*

*NIS = the newly independent states of the former Soviet Union.

Source: Bread for the World Institute, *Hunger 1998: Hunger in a Global Economy* (Silver Spring, MD: Bread for the World Institute, 1997), p. 10.

Causes of Death	Percentage
Acute respiratory infection (ARI-mostly pneumonia)	27.6
ARI-measles	3.7
ARI-pertussis	2.0
Pertussis	0.8
Measles	1.7
Diarrhea	23.3
Diarrhea measles	1.4
Malaria	6.2
Birth asphyxia	6.7
Neonatal tetanus	4.3
Congenital anomalies	3.5
Birth trauma	3.3
Prematurity	3.3
Neonatal sepsis and meningitis	2.3
Tuberculosis	2.3
Accidents	1.6
All other causes	6.0

TABLE 16-3 *Estimated Causes of Death in Developing Countries for Children Under 5 Years*

Note: Since 1985 the number of deaths due to ARI has decreased; neonatal tetanus has dropped; neonatal sepsis and meningitis have increased; measles and pertussis deaths have decreased; neonatal and perinatal deaths have increased slightly, to 30 percent from 27 percent; vaccine-preventable deaths have decreased from 27 percent to 16 percent.

Source: C. D. Williams, N. Baumslag, and D. B. Jelliffe, *Mother and Child Health: Delivering the Services,* 3rd ed. (New York: Oxford University Press, 1994), p. 6.

UNICEF estimates that most child malnutrition in the developing world could be eliminated with the expenditure of an additional $25 billion a year (see Figure 16-4 to put this sum of money in perspective). This amount would cover the cost of the resources needed to control the major childhood diseases, halve the rate of child malnutrition, bring clean water and safe sanitation to all communities, make family planning services universally available, and provide almost every child with at least a basic education.[14]

Hungry people receive such small quantities of food that they develop multiple nutrient deficiencies. Their undernutrition may result from lack of food energy or from a lack of both food energy and protein—protein-energy malnutrition (PEM).

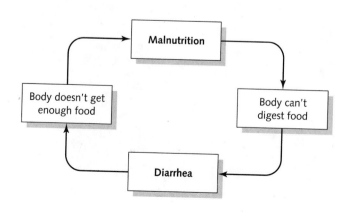

FIGURE 16-3 *The Vicious Circle of Malnutrition*

Source: UNICEF.

FIGURE 16-4

Affording the Cost of Solving the World Food Problem

This $25 billion is UNICEF's estimate of the extra resources required to control the major childhood diseases, halve child malnutrition, reduce child deaths by 4 million a year, bring safe water and sanitation to all communities, provide basic education for children, and make family planning universally available.

Source: UNICEF, derived from various sources.

It is no longer possible to say that the task of meeting basic human needs is too vast or expensive. With present knowledge, the task could be accomplished within a decade and at a cost of an extra $25 billion per year. Some comparisons:

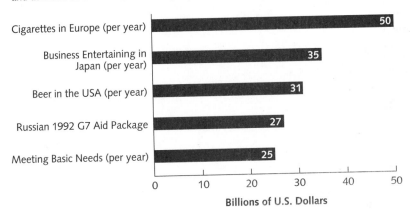

Such distinctions can easily be made on paper, and the extremes are evident in individuals; but for the most part, the differences blur.

Protein-energy malnutrition (PEM) is the most widespread form of malnutrition in the world today, affecting over 500 million children. Children who are thin for

Infants can be the first to show the signs of undernutrition due to their high nutrient needs. No famine, no flood, no earthquake, no war has ever claimed the lives of 250,000 children in a single week. Yet malnutrition and disease claim that number of child victims every week.

their height may be suffering from acute PEM (recent severe food restriction), whereas children who are short for their age may be suffering from chronic PEM (long-term food deprivation). PEM includes the classifications of **kwashiorkor**, a protein deficiency disease, **marasmus**, a deficiency disease caused by inadequate food intake, and the states in which these two extremes overlap.* Children suffering from PEM are likely to develop infections, nutrient deficiencies, and diarrhea (see Table 16-4).

Worldwide, three micronutrient deficiencies are of particular concern: vitamin A deficiency, the world's most common cause of preventable child blindness and vision impairment; iron-deficiency anemia; and iodine deficiency, which causes high levels of goiter and child retardation.[15]

- **Vitamin A deficiency.** Some 13 million children below six years of age have xerophthalmia. Of these, an estimated 500,000 children become partially or totally blind as a result of insufficient vitamin A in the diet. Vitamin A deficiency is also associated with other forms of malnutrition, infection, diarrhea, and a high rate of mortality.
- **Iron deficiency.** Iron-deficiency anemia is estimated to affect some 2.1 billion people. Iron deficiency in infancy and early childhood is associated with decreased cognitive abilities and resistance to disease.
- **Iodine deficiency.** Iodine deficiency, the major preventable cause of mental retardation worldwide, is a risk factor for both physical and mental retardation in about 1 billion people. About 655 million people worldwide—especially in mountainous regions—are estimated to have goiter, and over 3 million suffer overt cretinism.

The malnutrition that comes from living with food insecurity is one of the major factors influencing life expectancy. According to the *1997 World Population Data Sheet*, life expectancy at birth in the United States and Canada is now about 77 years.[16] Worldwide, life expectancy averages about 65 years, but in Africa it is approximately 50, and in the small African country of Sierra Leone it is the lowest of all—only 37.

Hunger and malnutrition can be found in people of all ages, sexes, and nationalities. Even so, these problems hit some groups harder than others.

Children at Risk

When nutrient needs are high (as in times of rapid growth), the risk of undernutrition increases. If family food is limited, pregnant and lactating women, infants, and children are the first to show the signs of undernutrition. Effects of food insecurity can be devastating to this group of the population.

As Chapter 13 pointed out, to support normal fetal growth and development, women must gain adequate weight during pregnancy. Healthy women in developed

Protein-energy malnutrition (PEM) The world's most widespread malnutrition problem, includes **kwashiorkor** (a deficiency disease caused by inadequate protein intake), **marasmus** (a deficiency disease caused by inadequate food intake—starvation), and the states in which they overlap.

This chapter's Program Spotlight highlights the international activities of the Vitamin A Field Support Project (VITAL).

---

*The exact cause of kwashiorkor remains uncertain. Some research suggests that kwashiorkor may involve more than protein deficiency and develops when malnourished children eat moldy grains or peanuts. The mold *Aspergillus flavus*, commonly found in hot, humid areas, produces a potent aflatoxin that inhibits protein synthesis. (R. G. Hendickse, Kwashiorkor: The hypothesis that incriminates aflatoxins, *Pediatrics* 88 [1991]: 376–79.)

TABLE 16-4

Characteristic Features of Marasmus and Kwashiorkor

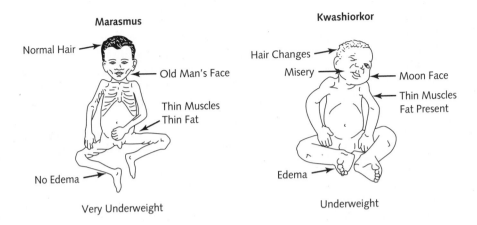

Features	Marasmus	Kwashiorkor
Essential features		
Edema	None*	Lower legs, sometimes face, or generalized*
Wasting	Gross loss of subcutaneous fat, "all skin and bone"*	Less obvious; sometimes fat, blubbery
Muscle wasting	Severe*	Sometimes
Growth retardation in terms of body weight	Severe*	Less than in marasmus
Mental changes	Usually none	Usually present
Variable features		
Appetite	Usually good	Usually poor
Diarrhea	Often (past or present)	Often (past or present)
Skin changes	Usually none	Often, diffuse depigmentation; occasional, "flaky-paint"* or "enamel" dermatosis
Hair changes	Texture may be modified but usually no dispigmentation	Often sparse—straight and silky, dispigmentation—greyish or reddish
Moon face	None	Often
Hepatic enlargement	None	Frequent, although it is not observed in some areas
Biochemistry pathology		
Serum albumin	Normal or slightly decreased	Low*
Urinary urea per g of creatinine	Normal or decreased	Low*
Urinary hydroxy-proline index	Low	Low
Serum-free amino acid ratio	Normal	Elevated*
Anemia	May be observed	Common; iron or folate deficiency may be associated
Liver biopsy	Normal or atrophic*	Fatty infiltration*

*The most characteristic or useful distinguishing features.

Source: C. D. Williams, N. Baumslag, and D. B. Jelliffe, *Mother and Child Health: Delivering the Services*, 3rd ed. (New York: Oxford University Press, 1994), p. 156.

countries gain an average of about 27 pounds. Studies among poor women show a weight gain of only 11 to 15 pounds.[17] In some areas, women may have caloric deficits of up to 42 percent and do not gain any weight at all during pregnancy.[18] As a consequence, they give birth to babies with low birthweights.

Birthweight is a potent indicator of an infant's future health status. A low-birthweight baby (less than 5½ pounds or 2,500 grams) is more likely to experience complications during delivery than a normal baby and has a statistically greater-than-normal chance of having physical and mental birth defects, contracting diseases, and dying early in life. Worldwide, low birthweight contributes to more than half of the deaths of children under 5 years of age. Low-birthweight infants suffering undernutrition after their births incur even greater risks. They are more likely to get sick, to fail to obtain nourishment by sucking, and to be unable to win their mothers' attention by energetic, vigorous cries and other healthy behavior. They can become apathetic, neglected babies, which compounds the original malnutrition problems.

Until the middle of the twentieth century, in most **developing countries,** babies were breastfed for their first year of life—with supplements of other milk and cereal gruel added to their diets after the first several months. Today, the percentage of infants who are exclusively breastfed to the age of 4 months has dropped to below 10 percent.[19] A number of factors contributed to this unfortunate decline, including the aggressive promotion and sale of infant formula to new mothers; the encouragement of bottlefeeding by health care practitioners, who send mothers home from the hospital with free samples of formula after delivery of the newborn; and the global pattern of urbanization and accompanying loss of cultural ties supporting breastfeeding combined with more women working outside the home.[20] Overall, the World Health Organization (WHO) estimates that more than a million children's lives could be saved each year if all mothers gave their babies nothing but breast milk for the first 4 to 6 months of life.[21]

Breastfeeding permits infants in many developing countries to achieve weight and height gains equal to those of children in developed countries until about 6 months of age, but then the majority of these children fall behind in growth and development because inadequate supplementary foods are added to their diets. Visitors to developing countries may underestimate childhood undernutrition because they do not realize that the children they think are three or four years old are actually eight or nine years old. Failure of children to grow is a warning of the extreme malnutrition that may soon follow (see Figure 16-5).

Replacing breast milk with infant formula in environments and economic circumstances that make it impossible to feed formula safely may lead to infant undernutrition. Breast milk, the recommended food for infants, is sterile and contains antibodies that enhance an infant's resistance to disease. In the absence of sterilization and refrigeration, formula in bottles is an ideal breeding ground for bacteria. More than 1.9 billion people in developing countries do not have access to safe drinking water.[22] Thus, feeding infants formula prepared with contaminated water often causes infections leading to diarrhea, dehydration, and failure to absorb nutrients. In countries where poor sanitation is prevalent, breastfeeding should take priority over feeding formula. Failure to breastfeed an infant who lives in a house without indoor plumbing incurs twice the risk of perinatal mortality as breastfeeding an infant who lives in a house with good sanitation.[23]

Developing countries
Countries with low per capita incomes and low standards of living. Such countries make up most of Africa, Latin America, and Asia.

Height-for-age: a measure of linear growth; a measure below the normal range reflects growth stunting.

Weight-for-height: a sensitive measure of acute malnutrition; a low weight-for-height is described as wasting.

Weight-for-age: a synthesis of height-for-age and weight-for-height; a low weight-for-age—underweight—reflects both stunting and wasting.

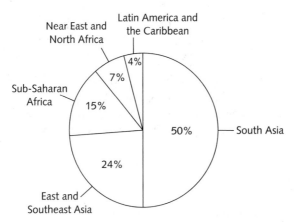

LOW WEIGHT-FOR-AGE

180 million children

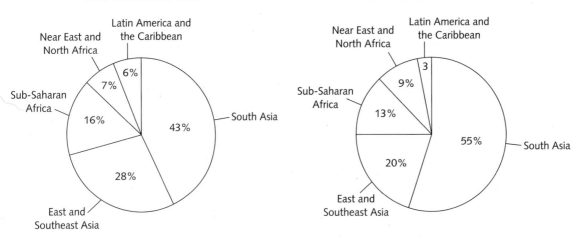

LOW HEIGHT-FOR-AGE

215 million children

LOW WEIGHT-FOR-HEIGHT

50 million children

FIGURE 16-5 *Regional Distribution of Children with Insufficient Growth*

Source: The worldwide magnitude of protein-energy malnutrition: An overview from the WHO Global Database on Child Growth and Malnutrition, *WHO Bulletin* 71 (1996): 703.

Even if infants are protected by breastfeeding at first, they must be weaned. The **weaning period** is one of the most dangerous times for children in developing countries for a number of reasons. For one, newly weaned infants often receive nutrient-poor diluted cereals or starchy root crops. For another, infants' foods are often prepared with contaminated water, making infection almost inevitable. Attitudes toward food may also affect nutrition. In some areas of India, for example, a child may be forbidden to eat curds and fruit because they are "cold" or bananas because they "cause convulsions."[24]

Mortality statistics reflect the hazards to infants and children. The infant mortality rate ranges from about 17 (Costa Rica) to over 162 (Afghanistan) in the poorest of the developing countries (see Table 16-1). The death rate for children from 1 to 4 years of age is no more favorable; it ranges from 20 to 30 times higher in developing countries than in developed countries.[25] Maternal mortality rates are equally shocking, as shown in Figure 16-6.

UNICEF regards the **under-5 mortality rate (U5MR)** as the single best indicator of children's overall health and well-being.[26] UNICEF argues that this rate reflects the overall resources a country directs at children:

> The U5MR reflects the nutritional health and the health knowledge of mothers; the level of immunization and ORT (oral rehydration therapy) use; the availability of maternal and health services (including prenatal care); income and food availability in the family; the availability of clean water and safe sanitation; and the overall safety of the child's environment.[27]

Women at Risk

Women are more susceptible than men to food insecurity and undernutrition for a number of reasons. In addition to their increased nutrient needs during childbearing years, many women in developing countries are responsible, even during their pregnancies, for most of the physical labor required to procure food for their families. The poor nutrition of some women results from both their family's lack of access to food and unequal distribution of food within the family itself. A woman will feed her husband, children, and other family members first, eating only what-

Weaning period The time during which an infant's diet is changed from breast milk to other nourishment.

Under-five mortality rate (U5MR) The number of children who die before the age of 5 for every 1,000 live births.

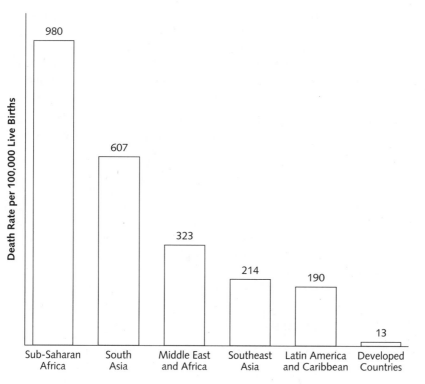

FIGURE 16-6

Maternal Mortality in Developing Regions Compared with Average for Developed Countries

Source: Data from Bread for the World Institute, *Hunger 1998: Hunger in a Global Economy* (Silver Spring, MD: Bread for the World Institute, 1997).

ever is left. Furthermore, each time she becomes pregnant, her body's nutrient reserves are drained.

Social beliefs may also limit women's food intakes. In the Indian Punjab, the director of a program aimed at relieving undernutrition in local villages found that an undernutrition rate of 10 to 15 percent persisted even after a major effort to provide supplementary foods to families. The majority of those affected were young girls, who were unable to demand their share and were regarded by other family members as not deserving a fair share.[28]

Food Insecurity in Developing Countries

World hunger is more extreme than domestic hunger. In fact, most people would find it hard to imagine the severity of poverty in the developing world:

> Many hundreds of millions of people in the poorest countries are preoccupied solely with survival and elementary needs. For them, work is frequently not available or pay is low, and conditions barely tolerable. Homes are constructed of impermanent materials and have neither piped water nor sanitation. Electricity is a luxury. Health services are thinly spread, and in rural areas only rarely within walking distance. Permanent insecurity is the condition of the poor. . . . In the wealthy countries, ordinary men and women face genuine economic problems. . . . But they rarely face anything resembling the total deprivation found in the poor countries.[29]

Table 16-5 reveals the human dimensions of the world hunger problem; Table 16-6 shows some of its many causes. World hunger is a problem of supply and demand, inappropriate technology, environmental abuse, demographic distribution, unequal access to resources, extremes in dietary patterns, and unjust economic systems. Two generalizations and an important question are suggested by these tables:

- The underlying causes of global hunger and poverty are complex and interrelated.
- Hunger is a product of poverty resulting from the ways in which governments and businesses manage national and international economies.
- The question then is, Why are people poor?

Poverty contributes to hunger in many important ways. Oftentimes, people who are poor are powerless to change their situation because they have little access to vital resources such as education, training, food, health services, and other vehicles of change.[30] The roots of hunger and poverty, like those of many other current problems, can be found in numerous historical and natural developments, including colonialism, economic institutions, corporate systems, population pressure, resource distribution, and agricultural technology.

The Role of Colonialism

The colonial era led to hunger and malnutrition for millions of people in developing countries. Although no longer called colonialism, much of this same activity

TABLE 16-5 *The Realities of Hunger*

- The United Nations reports that there are 840 million malnourished people in the world.

- Each year, 15 to 20 million people die of hunger-related causes, including diseases brought on by lowered resistance due to malnutrition. Of every four of these, three are children.

- Over 40% of all deaths in poor countries occur among children under 5 years old.

- The United Nations (UNICEF) states that 17 million children died last year from preventable diseases—one every two seconds, 40,000 a day. (A vaccination immunizing one child against a major disease costs 7 cents.) At least 500,000 children are partially or totally blinded each year simply through lack of vitamin A.

- More than 2 billion people in poor countries suffer from chronic anemia.

- Every day, the world produces 2 lb of grain for every man, woman, and child on earth. This is enough to provide everyone with 3,000 kcal/day, well above the average need of 2,300 kcal.

- A person born in the rich world will consume 30 times as much food as a person born in the poor world.

- The poor countries have nearly 75% of the world's population, but consume only about 15% of the world's available energy.

- One-eighth of the world's population lives on an income of less than $300 a year—many use 80 to 90% of that income to obtain food.

- Of the 5.7 billion people on earth, more than 1 billion drink contaminated water. Water-related disease claims 25 million lives a year. Of these, 15 million are under 5 years of age.

- An estimated 1 billion adults can't read or write. In many countries, half of the population over 15 is illiterate. Two-thirds of these are women.

Source: Adapted from *World Hunger: Facts,* available from Oxfam America, 26 West St., Boston, MA 02111; Office on Global Education, Church World Service, 2115 N. Charles St., Baltimore, MD 21218.

still continues today. The African experience provides a good example of the colonial process.

Britain, the Netherlands, Germany, France, and other nations originally colonized the African continent largely to gain a source of raw materials for industrial use. Accordingly, the colonial powers created a governing infrastructure designed merely to move Africa's minerals, metals, cash crops, and wealth to Europe. They provided few opportunities for education, disrupted traditional family structures and community organization, and greatly diminished the ability of the African people to produce their own food.

Before the Europeans arrived, small landowners throughout much of Africa had cleared forests to grow beans, grains, or vegetables for their own use. With colonialism, wealthy Africans and foreign investors took over the fertile farmland and forced the rural poor onto marginal lands that could produce adequate food only with irrigation and fertilizer, which were beyond the means of the poor. The fertile lands were used to grow cotton, sesame, sugar, cocoa, coffee, tea, tobacco, and livestock for export. As more raw materials were exported, more food and manufactured products had to be imported. Imported goods cost money—also beyond the reach of the poor.

TABLE 16-6 *Causes of the World Food Problem*

Worldwide Problems

- Natural catastrophes—drought, heavy rains and flooding, crop failures.
- Environmental degradation—soil erosion, inadequate water resources, deforestation.
- Food supply-and-demand imbalances.
- Inadequate food reserves.
- Warfare and civil disturbances.
- Migration—refugees.
- Culturally based food prejudices.
- Declining ecological conditions in agricultural regions.

Problems of the Developing World

- Underdevelopment.
- Excessive population growth.
- Lack of economic incentives—farmers using inappropriate methods and laboring on land they may lose or can never hope to own.
- Parents lacking knowledge of basic nutrition for their children.
- Insufficient government attention to the rural sector.

Problems of the Industrialized World

- Excessive use of natural resources.
- Pollution.
- Inefficient animal protein diets.
- Inadequate research in science and technology to meet basic human needs.
- Excessive government bureaucracy.
- Loss of farmland to competing uses.

Problems Linking the Industrialized and Developing Worlds

- Unequal access to resources.
- Inadequate transfer of research and technology.
- Lack of development planning.
- Insufficient food aid.
- Excessive food aid that undermines local initiatives for food production.
- Politics of food aid and nutrition education.
- Inappropriate technological research.
- Inappropriate role of multinational corporations.
- Insufficient emphasis on agricultural development for self-sufficiency.

Source: Adapted from C. G. Knight and R. P. Wilcox, *Triumph or Triage? The World Food Problem in Geographical Perspective*, Resource Paper no. 75–3 (Washington, D.C.: Association of American Geographers, 1976), p. 4.

Per capita production of grains for food use has declined for the last 20 years in Africa, while sugar cane production has doubled and tea production has quadrupled. The country of Chad recently harvested a record cotton crop in the same year that it experienced an epidemic of famine. Sixty percent of the gross national product of Ghana, Sudan, Somalia, Ethiopia, Zambia, and Malawi is derived from cash crops that finance both luxury imported goods for the minority and international debts. Huge amounts of soybeans and grains are fed to livestock to produce protein foods the poor cannot afford to purchase.

International Trade and Debt

Over the years, developing countries have seen the prices of imported fuels and manufactured items rise much faster than the prices they receive for their export goods, such as bananas, coffee, and various raw materials, on the international market. The combination of high import costs and low export profits often pushes a developing country into accelerating international debt that sometimes leads to bankruptcy.

Debt and trade are closely related to the progress a country can make toward achieving adequate diets for its people. As import prices increase relative to export prices, a country's money moves abroad to pay for the imports. With more and more of its money abroad, the country is forced to borrow money, usually at high interest rates, to continue functioning at home. Many of its financial resources must then go to pay the interest on the borrowed money, thus draining the economy further. Creditor nations may not demand much, or any, capital back, but they do require that interest be paid each year, and the interest can consume most of a country's gross national product. Large and growing debts can slow or halt a nation's attempt to deal effectively with its problems of local food insecurity. As more and more of its financial resources are used to pay off interest on the country's trade debts, less and less money is available to deal with food insecurity at home.[31] Each year, the debt crisis worsens and leads to further problems with hunger.

Currently, the developed and developing countries in the United Nations are discussing ways to lighten this burden of international debt. These discussions are known as the **North-South Dialogue** and represent what is referred to as a **New International Economic Order (NIEO)**. The NIEO proposals include rescheduling interest payments or even totally forgiving the debts of developing countries. As Stephen Lewis, formerly Canada's ambassador and permanent representative to the United Nations, noted: "It would scarcely be felt if we [the North] forgave the entire African debt. Industrial capitalism lost on the stock exchange, in one day in October 1987, 10 times the African debt."[32]

Little consensus has been reached on the merits of these proposals. Some fear the loss of profits that the lender nations and multinational banks will sustain. Others believe that the loans were ill-advised to begin with and that the lenders have collected sufficient interest already, often many times the original value of the loans. In any case, the NIEO may have identified one of the key steps in solving debtor countries' hunger problems. Once the tremendous financial drain caused by their international debt is eliminated, the countries can allocate more financial resources to economic development that will lead to less hunger. Undoubtedly, the debate between the developing and developed world will continue. These North-South discussions are likely to be as crucial to the future of both sides as the arms control negotiations between the East and West were during the Cold War.

The Role of Multinational Corporations

National economic policies based on the export of cash crops, such as coffee, often have a negative effect on household food security. The competition between cash crops and food crops for farmland provides a classic example of the plight of the poor. Typically, the fertile farmlands are controlled by large landowners and **multinational corporations** who hire indigenous people for below-subsistence wages to grow crops to be exported for profit, leaving little fertile land for the local farmers to use to grow food. The local people work hard cultivating cash crops for others, not food crops for themselves. The money they earn is not even enough to

North-South Dialogue Discussions between the developed countries and the developing countries aimed at achieving global economic change and addressing the economic imbalance in international trade experienced by the developing countries. (The more developed countries, with the exceptions of New Zealand and Australia, are located geographically to the north of most of the developing countries.)

New International Economic Order (NIEO) Proposals made by the developing countries, requesting structural changes in the international economic system.

Multinational corporations Transnational companies (TNC) with direct investments and/or operative facilities in more than one country. U.S. oil and food companies are examples.

buy the products they help produce. As a result, imported foods—bananas, beef, cocoa, coconuts, coffee, pineapples, sugar, tea, winter tomatoes, and the like—fill the grocery stores of developed countries, while the poor who labored to grow these foods have less food and fewer resources than when they farmed the land for their own use. Additional cropland is diverted for nonfood, cash crops such as tobacco, rubber, and cotton. These practices have also had an adverse effect on many U.S. farmers. The foreign cash crops often undersell the same U.S.-grown produce. The U.S. farmers cannot compete against these lower-priced imported foods and may be forced out of business.

Export-oriented agriculture thus uses the labor, land, capital, and technology that is needed to help local families produce their own food. For example, the resources used to produce bananas for export could be reallocated to provide food for the local people. Some have suggested that the developed countries could help alleviate the world food problem not by *giving* more food aid to the poor countries, but by *taking* less food away from them.[33]

Countless examples can be cited to illustrate how natural resources are diverted from producing food for domestic consumption to producing luxury crops for those who can afford them:

- Africa is a net *exporter* of barley, beans, peanuts, fresh vegetables, and cattle (not to mention luxury crops such as coffee and cocoa), yet it has a higher incidence of protein-energy malnutrition among young children than any other continent.
- Over half of the U.S. supply of several winter and early spring vegetables comes from Mexico where infant deaths associated with poor nutrition are common.
- Half of the agricultural land in Central America produces food for export, while in several Central American countries the poorest 50 percent of the population eat only half the protein they need.[34]

Besides diverting acreage away from the traditional staples of the local diet, some multinational corporations may also contribute to hunger through their marketing techniques. Their advertisements lead many consumers with limited incomes to associate products like cola beverages, cigarettes, infant formulas, and snack foods with Western culture and prosperity. A poor family's nutrition status suffers when its tight budget is pinched further by the purchase of such goods. Even worse is the inappropriate use of infant formula. Use of formula leads a mother to wean her infant. Then, when her money runs out and she cannot afford to buy more formula, her breast milk has ceased to flow, and she cannot resume breastfeeding. All too often, the result is malnutrition, sickness, and death of the infant.

The United Nations has commissioned several studies in the hopes of establishing an international code of conduct for multinational corporations.[35] These corporations could increase the credit and capital available to the developing world; and these resources, if properly used, could help to eliminate food insecurity. The multinational corporations also possess the scientific knowledge and organizational skills needed to help develop improved food and agricultural systems. Experience shows, however, that sustained outside pressure may have to be applied to some of these corporations to help ensure that human needs do not become subordinate to political and financial gains. U.S. consumers can often influence these multinational corporations because they are shareholders in them.

The Role of Overpopulation

The current world population is approximately 5.7 billion, and the United Nations projects 6 billion by the year 2000. The earth may not be able to support this many people adequately.[36] The world's present population is certainly of concern as is the projected increase in that population. Nevertheless, population is only one aspect of the world food problem. Poverty seems to be at the root of both problems—hunger and overpopulation.

Three major factors affect population growth: birth rates, death rates, and standards of living. Low-income countries have high birth rates, high death rates, and low standards of living.

When people's standard of living rises, giving them better access to health care, family planning, and education, the death rate falls. In time, the birth rate also falls. As the standard of living continues to improve, the family earns sufficient income to risk having smaller numbers of children. A family depends on its children to cultivate the land, secure food and water, and provide for the adults in their old age. Under conditions of ongoing poverty, parents will choose to have many children to ensure that some will survive to adulthood. Children represent the "social security" of the poor. Improvements in economic status help relieve the need for this "insurance" and so help reduce the birth rate. The relationships between the infant mortality rate and the population growth rate reveal that hunger and poverty reflect both the level of national development and the people's sense of security.[37]

In many countries where economic growth has occurred and all groups share resources relatively equally, the rate of population growth has decreased. Examples include Costa Rica, Sri Lanka, Taiwan, and Malaysia. In countries where economic growth has occurred but the resources are unevenly distributed, population growth has remained high. Examples include Brazil, Mexico, the Philippines, and Thailand, where a large family continues to be a major economic asset for the poor.[38]

As the world's population continues to grow, it threatens the world's capacity to produce adequate food in the future. The activity of billions of human beings on the earth's limited surface is seriously and adversely affecting our planet: wiping out many varieties of plant life, using up our freshwater supplies, and destroying the protective ozone layer that shields life from the sun's damaging rays—in short, overstraining the earth's ability to support life.[39] Population control is one of the most pressing needs of this time in history. Until the nations of the world resolve the population problem, they must all deal with its effects and make efforts to support the life of the populations that currently exist.

The transition of population growth rates from a slow-growth stage (high birth rates and high death rates) through a rapid-growth stage (high birth rates and low death rates) to a low-growth stage (low birth rates and low death rates) is known as the **demographic transition**.

Distribution of Resources

Land reform—giving people a meaningful opportunity to produce food for local consumption, for example—can combine with population control to increase everyone's assets. Some background information is important to understanding this relationship:

- Much of the world's agriculture is primitive. More than 50 percent of all food consumed in the world is still produced by hand.
- In many countries, up to 90 percent of the population live on rural land.

- Most governments dictate the day-to-day lives of their people, and their policies may not be equitable.
- Securing enough food on a day-to-day basis is a problem for as many as a billion human beings.
- Even the best land in many parts of the world does not support the growing of food, even by those who can afford fertilizer and irrigation.[40]

If you give a man a fish, he will eat for a day. If you teach him to fish, he will eat for a lifetime.

Resources are distributed unequally not only between the rich and the poor within nations, but also between rich and poor nations. But if the wealthy nations simply give aid to the poor nations, the poor nations will be weakened further. Instead, the wealthy nations must foster self-reliance in the poor nations. Doing so will initially require some economic sacrifices by the wealthy nations but will ultimately benefit large numbers of hungry people.

Poor nations must be allowed to increase their agricultural productivity. Much is involved, but to put it simply, poor nations must gain greater access to five things simultaneously: land, capital, water, technology, and knowledge (see Figure 16-7).[41] Equally important, each nation must make improving the condition of all its people a political priority. International food aid may be required temporarily during the development period, but eventually this aid will be less and less necessary.

FIGURE 16-7 *Breaking Out of the Cycle of Despair*

Source: Adapted from Bread for the World Institute, *Hunger 1996: Causes of Hunger* (Silver Spring, MD: Bread for the World Institute, 1995), p. 91.

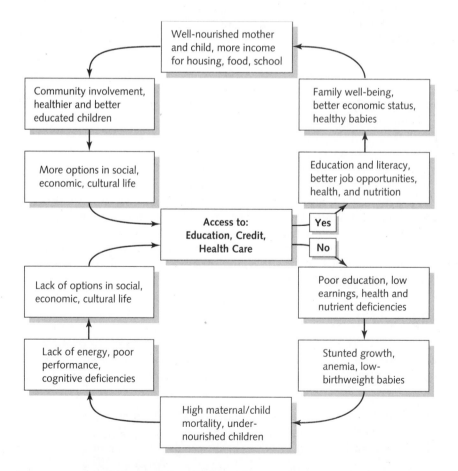

Agricultural Technology

Governments can learn from recent history the importance of developing local agricultural technology. A major effort made in the 1960s and 1970s—the **green revolution**—demonstrated both the potential for increased grain production in Asia and the necessity of considering local conditions. The industrial world made an effort to bring its agricultural technology to the developing countries, but the high-yielding strains of wheat and rice that were selected required irrigation, chemical fertilizers, and pesticides—all costly and beyond the economic means of many of the farmers in the developing world.

International research centers need to examine the conditions of developing countries and orient their research toward **appropriate technology**—labor-intensive rather than energy-intensive agricultural methods. Instead of transplanting industrial technology into the developing countries, small, efficient farms and local structures for marketing, credit, transportation, food storage, and agricultural education should be developed.

For example, labor-intensive technology, such as the use of manual grinders for grains, is appropriate in some places because it makes the best use of human, financial, and natural resources. A manual grinder can process 20 pounds of grain per hour, replacing the mortar and pestle, which can grind a maximum of only 3 pounds in the same time.[42] The specific technology that is appropriate for use varies from situation to situation.

Green revolution The development and widespread adoption of high-yielding strains of wheat and rice in developing countries. The term *green revolution* is also used to describe almost any package of modern agricultural technology delivered to developing countries.

Appropriate technology A technology that utilizes locally abundant resources in preference to locally scarce resources. Developing countries usually have a large labor force and little capital; the appropriate technology would therefore be labor-intensive.

A Need for Sustainable Development

Environmental concerns must be taken more seriously as well. The amount of land available for crop production is as important as the condition of the soil and the availability of water. Soil erosion is now accelerating on every continent at a rate that threatens the world's ability to continue feeding itself. Erosion of soil has always occurred; it is a natural process. But in the past, processes that build up the soil—such as the growth of trees—have compensated for erosion.

Where forests have already been converted to farmland and there are no trees, farmers can practice crop rotation, alternating soil-devouring crops with soil-building crops. An acre of soil planted one year in corn, the next in wheat, and the next in clover loses 2.7 tons of topsoil each year, but if it is planted only in corn three years in a row, it will lose 19.7 tons a year. When farmers must choose whether to make three times as much money planting corn year after year or to rotate crops and go bankrupt, many choose the short-term profits. Ruin may not follow immediately, but it will follow.[43]

There is a growing recognition that governments need to encourage and support efforts at sustainable development. **Sustainable development** is defined as the successful management of agricultural resources to satisfy changing human needs while maintaining or enhancing the natural resource base and avoiding environmental degradation.

Consider that:

Poverty drives ecological deterioration when desperate people overexploit their resource space, sacrificing the future to salvage the present. The cruel logic of short-term needs forces landless families to raise plots in the rain forest, plow steep slopes, and shorten

Sustainable development Development that meets the needs of the present without compromising the ability of future generations to meet their own needs.

fallow periods. Ecological decline, in turn, perpetuates poverty as degraded ecosystems offer diminishing yields to their poor inhabitants. A self-defeating spiral of economic deprivation and ecological degradation takes hold.[44]

Sustainable development includes the reduction of poverty and food insecurity in environmentally-friendly ways. It includes the following four interrelated objectives:[45]

1. Expand economic opportunities for low-income people, to increase their income and productivity in ways that are economically, environmentally, and socially viable in the long-term.
2. Meet basic human needs for food, clean water, shelter, health care services, and education.
3. Protect and enhance the natural environment by managing natural resources that respect the needs of present and future generations.
4. Promote democratic participation by all people in economic and political decisions that impact their lives.

People-Centered Development

We have used the word "developing" in this chapter to classify certain countries, but what is development? According to Oxfam America, a nonprofit international agency that funds self-help development and disaster relief projects worldwide, development enables people to meet their essential needs, extends beyond food aid and emergency relief, reverses the process of impoverishment, enhances democracy, and makes possible a balance between populations and resources. It also improves the well-being and status of women, respects local cultures, sustains the natural environment, measures progress in human, not just monetary terms, and involves change, not just charity. Finally, development requires the empowerment of the poor and promotes the interests of the majority of people worldwide, in the global North as well as the South.[46]

Consider, for example, the history of development in Sri Lanka.[47] This island nation off the southeastern coast of India is the size of Ireland with a population of 14 million people. Sri Lanka, which achieved independence in 1948 and developed welfare and education programs in the 1960s and 1970s, now boasts a literacy rate of 85 percent and an infant mortality rate that has dropped from 141 to 37 deaths per 1,000 live births. Despite these improvements, the national average income of $270 remains one of the lowest in the world.

Many of the villagers live in mud huts with little more than 100 square feet of space; light enters only from the open doorway. Several children, parents, and grandparents live in each home. Food procurement and preparation head the list of endless chores each day. Children must be cared for, rice must be dried and pounded into flour, chilies must be dried and ground for curry, water must be collected from the well when it's working, the animals must be grazed, the garden needs to be cultivated, and laundry must be carried to the reservoir. Parents often spend time working away from the home as well. In the poorer villages, less time is needed for meal preparation—there are fewer meals.

In an effort to improve these conditions by increasing average income, the current government has shifted its emphasis away from human welfare programs. To stimulate the economy, it has created free trade zones and maintains a cheap labor force to attract multinational corporations whose products are chiefly for export. Tourism is being widely developed on the tropical island to generate national income. These monies are largely spent to provide imported food and luxury goods for those who can afford them. Thus, the *average* income of Sri Lanka is rising, but only because the rich are becoming richer; the poor remain poor.

Contrast this consumer-oriented, high-technology, import-oriented, plantation-crop form of development with an alternative grass roots development movement called the *Sarvodaya Shramadana*, which is active in about 30 percent of the 23,000 villages on the island. Believing national development should begin with the villages, Sarvodaya helps people organize themselves to deal directly with their own problems and sponsors programs in health, education, agriculture, and local industry. Villagers have identified 10 basic needs: fuel, housing, basic health care, minimum requirements of clothing, a balanced and adequate diet, communication facilities (including roads), an adequate supply of clean water, education relevant to their lives, a clean beautiful environment, and a spiritual and cultural life. Villagers are asked to describe projects that would fulfill their basic needs, and then set priorities. They are expected to furnish the maximum amount of their own human and material resources. The remaining input is supplied by other able Sarvodaya villages and district and national Sarvodaya headquarters. Training is provided as necessary in community organization, health and preschool education, and practical income-generating skills. Resources from the national headquarters include savings and loan programs, health and nutrition monitoring programs, community and marketing cooperatives, and appropriate agricultural technology.

Sarvodaya demonstrates that development cannot be measured by gross national product or the quantity of the community harvest. Instead, development should serve the people. A preschool may be struck by lightning and burn the day after it is built, but what matters most is that the people saw the need for a school and labored together to build it. Harsha Navarante, a Sarvodaya district director, summarizes the movement: "When the politician goes into a village plagued by poor roads he announces, 'The roads are terrible! Change the government! Vote for me!' Sarvodaya's message is that the roads are terrible, therefore the villagers must join together to fix them. We build the road and the road builds us."

The Sri Lankan experience highlights the need for ongoing community involvement and participation in project development and implementation. In Tanzania, the Iringa Nutrition Program involves community members in the assessment and analysis of problems and decisions about appropriate actions.[48] The program is designed to increase people's awareness of malnutrition and thus improve their capacity to take action. Fundamental to the program are the United Nations' child survival activities; these include the regular quarterly weighing of children under five years of age in the villages with discussion of the results by the village health committees. From this analysis, a set of appropriate interventions for solving problems can be identified. Recent evaluations indicate that this process has contributed to significant decreases in infant and child malnutrition and mortality.[49]

The cornerstone of true development was best expressed decades ago by Mahatma Gandhi: "Whenever you are in doubt . . . apply the following test. Recall the face of the poorest and the weakest man whom you may have seen, and ask yourself if the

Sarvodaya means awakening, or the well-being of all; *Shramadana* is the sharing of one's time, thought, and labor for the welfare of all.

step you contemplate is going to be of any use to him. Will he gain anything by it? Will it restore him to a control over his own life and destiny?"[50]

Nutrition and Development

The first global International Conference on Nutrition (ICN), held in December 1992 in Rome, Italy, was organized by the food and health agencies of the United Nations (FAO and WHO). The intergovernmental conference focused world attention on nutritional and diet-related problems, especially among the poor and other vulnerable groups, in hopes of mobilizing governments, United Nations organizations, nongovernmental organizations, local communities, the private sector, and individuals in the fight against hunger and malnutrition.[51]

The ICN reflects not only the recognition of the problems surrounding food insecurity but also the realization that solutions are possible, especially for children. The United Nations views a healthy nutritious diet as a basic human right—one that the FAO and WHO are pledged to secure. However, achieving improved nutritional well-being worldwide requires broad action on many issues, including the following, which were explored at the ICN:

- Ensuring that the poor and malnourished have adequate access to food
- Preventing and controlling infectious diseases by providing clean water, basic sanitation, and effective health care
- Promoting healthy diets and lifestyles
- Protecting consumers through improved food quality and safety
- Preventing micronutrient deficiencies
- Assessing, analyzing, and global monitoring of nutrition status of populations at risk
- Incorporating nutrition objectives into development policies and programs

The ICN helped focus world attention for the first time on the role of nutrition in achieving sustainable development. Nutrition and health are now seen as instruments or tools of economic development as well as goals. The inclusion of nutrition objectives in growth and development policies holds the promise of increasing the productivity and earning power of people worldwide. Well-nourished people are more productive, are sick less often, and earn higher incomes.[52] More recently, the 1996 World Food Summit, convened by the Food and Agriculture Organization, served to focus the world's attention on the continuing problem of food insecurity. World leaders from 186 nations adopted 7 commitments as part of the World Food Summit Plan of Action (see the box on the next page).[53] The overall objective of these commitments is to achieve the goal of reducing by half the number of undernourished people by no later than the year 2015.[54]

Agenda for Action

Although the problem of world hunger may seem overwhelming, it can be broken down into many small, local problems. Significant strides can then be made toward

The World Food Summit Plan of Action, 1996

Commitment 1: We will ensure an enabling political, social and economic environment designed to create the best conditions for the eradication of poverty and for durable peace, based on full and equal participation of women and men, which is most conducive to achieving sustainable food security for all.

Commitment 2: We will implement policies aimed at eradicating poverty and inequality and improving physical and economic access by all, at all times, to sufficient, nutritionally adequate and safe food and its effective utilization.

Commitment 3: We will pursue participatory and sustainable food, agriculture, fisheries, forestry and rural development policies and practices in high and low potential areas, which are essential to adequate and reliable food supplies at the household, national, regional and global levels, and combat pests, drought and desertification, considering the multifunctional character of agriculture.

Commitment 4: We will strive to ensure that food, agricultural trade and overall trade policies are conducive to fostering food security for all through a fair and market-oriented world trade system.

Commitment 5: We will endeavor to prevent and be prepared for natural disasters and man-made emergencies and to meet transitory and emergency food requirements in ways that encourage recovery, rehabilitation, development and a capacity to satisfy future needs.

Commitment 6: We will promote optimal allocation and use of public and private investments to foster human resources, sustainable food, agriculture, fisheries and forestry systems and rural development, in high and low potential areas.

Commitment 7: We will implement, monitor and follow-up this Plan of Action at all levels in cooperation with the international community.

---

solving them at the local level. Even if the problem of poverty itself is not immediately or fully solved, progress is possible. For example, infants and children need not be raised in middle-class homes to be protected from malnutrition. Slight modifications of the children's diets can be immensely beneficial.

Focus on Children

Children are the group most strongly affected by poverty, malnutrition, and food insecurity and its related effects on the environment.[55] However, there is hopeful news for children in developing countries. **GOBI**, a child survival plan set forth by UNICEF in 1983, has made outstanding progress in cutting the number of hunger-related child deaths. GOBI is an acronym formed from four simple, but profoundly important, elements of UNICEF's "Child Survival" campaign: growth charts, oral rehydration therapy (ORT), breast milk, and immunization.

The use of growth monitoring to determine the adequacy of child feeding requires a worldwide education campaign. A mother can learn to weigh her child every month and chart the child's growth on specially designed paper. She can learn to detect the early stages of hidden malnutrition that can leave a child irreparably retarded in mind and body. Then at least she will know she needs to take steps to remedy the malnutrition—if she can.

GOBI An acronym formed from the elements of UNICEF's Child Survival campaign—Growth charts, Oral rehydration therapy, Breast milk, and Immunization.

**Oral rehydration therapy
(ORT)** The treatment of
dehydration (usually due to
diarrhea caused by
infectious disease) with an
oral solution; as developed
by UNICEF, ORT is
intended to enable a
mother to mix a simple
solution for her child from
substances that she has at
home.

Most children who die of malnutrition do not starve to death—they die because
their health has been compromised by dehydration from infections causing diarrhea.
Until recently, there was no easy way of stopping the infection-diarrhea cycle and saving their lives; now, the spread of **oral rehydration therapy (ORT)** is preventing an
estimated 1 million dehydration deaths each year.[56] ORT involves the administration
of a simple solution that mothers can make themselves, using locally available ingredients; the solution increases a body's ability to absorb fluids 25-fold.[57] International
development groups also provide mothers with packets of premeasured salt and sugar
to be mixed with *boiled* water in rural and urban areas.* A safe, sanitary supply of drinking water is a prerequisite for the success of the ORT program. Contaminated water
perpetuates the infection-diarrhea cycle.

The promotion of breastfeeding among mothers in developing countries has
many benefits. Breast milk is hygienic, readily available, and nutritionally sound,
and it provides infants with immunologic protection specific for their environment.
In the developing world, its advantages over formula feeding can mean the difference between life and death.

An important contributor to children's malnutrition in developing countries is the
high bulk and low energy content of the available foods. The diet may be based on
grains, such as wheat, rice, millet, sorghum, and corn, as well as starchy root crops,
such as the cassava, sweet potato, plantain, and banana. These may be supplemented
with legumes (peas or beans), but rarely with animal proteins. Infants have small
stomachs, and most cannot eat enough of these staples (grains or root crops) to meet
their daily energy and protein requirements. They need to be fed more nutrient-dense
foods during the weaning period. The most promising weaning foods are usually concentrated mixtures of grain and locally available **pulses**—that is, peas or beans—
which are both nourishing and inexpensive.[58] Mothers are advised to continue
breastfeeding while they introduce weaning foods.

Pulses A term used for
legumes, especially those
that serve as staples in the
diets of developing
countries.

As for immunizations (the *I* of GOBI), they could prevent most of the 5 million
deaths each year from measles, diphtheria, tetanus, whooping cough, poliomyelitis,
and tuberculosis. However, adequate protein nutrition is necessary for vaccinations
to be effective; otherwise the body may use the vaccine itself as a source of protein.
It used to be difficult to keep vaccines stable in their long journeys from laboratory
to remote villages. Now, however, the discovery of a new measles vaccine that does
not require refrigeration has made universal measles immunization for young children possible, and many countries are reporting coverage rates of 80 to 90 percent.
The immunization achievements of the last 2 decades are credited with preventing
approximately 3 million deaths a year and the protection of many millions more
from disease, malnutrition, blindness, deafness, and polio.[59]

The first World Summit for Children in history was convened by UNICEF in
September 1990, bringing together representatives of 159 nations for the purpose of
making a renewed commitment to ending child deaths and child malnutrition on
today's scale by the year 2000. Significantly, *nutrition* was mentioned for the first
time in world history as an internationally recognized human right.[60] The overall

---

* Oral rehydration solutions are easily prepared from common ingredients including: ⅓ to ⅔ tsp table salt,
¼ tsp sodium bicarbonate, ⅓ tsp potassium chloride, and 3½ tbsp sugar in one liter of boiled water. Several
commercial preparations are also available. (F. J. Zeman, *Clinical Nutrition and Dietetics*, 2nd ed. [New
York: Macmillan, 1991], p. 241.)

goal of ending child deaths and malnutrition was broken down into 24 specific targets in a Plan of Action agreed upon by the countries in attendance (see the box on the next page). An immediate result of this summit has been an increase in the number of governments actively adopting the child survival strategies of UNICEF and WHO. UNICEF's goals for nutrition and food security to be reached by the year 2000 include the following:

- A 50 percent reduction in the 1990 levels of moderate to severe malnutrition among children under 5 years old; this in itself would save the lives of 100 million children by the year 2000.
- A 50 percent reduction in the 1990 levels of low-birthweight infants.
- The virtual elimination of blindness and other consequences of vitamin A deficiency.

Strategies devised to achieve these goals by the year 2000 include universal immunization; oral rehydration therapy; a massive effort to promote breastfeeding as the ideal food for at least the first 4 to 6 months of an infant's life; an attack on malnutrition involving nutrition surveillance that focuses on growth monitoring and weighing of infants at least once every month for the first 18 months of life; and nutrition and literacy education that will empower women in developing countries and lead to a reduction in nutrition-related diseases among vulnerable children.[61]

Progress on Meeting the World Summit for Children Goals

According to WHO, one-third of children younger than 5 years are undernourished.[62] Of these, 70 percent live in Asia, 16 percent in Africa, and 3 percent in Latin America. Countries can be grouped by their level of commitment to reducing child malnutrition. In Chile, Egypt, Swaziland, and Malaysia, fewer than 10 percent of children are malnourished. Thailand, Zimbabwe, and Tanzania are likely to reach the goal of halving childhood malnutrition. Some more populous countries—

Vulnerability comes with age: Infants and children are disproportionately vulnerable to hunger. Breastfeeding offers the infant a safe and adequate supply of nutrients.

Plan of Action from the World Summit for Children

The following is a partial list of goals, to be attained by the year 2000, which were adopted by the World Summit for Children on September 30, 1990.

Overall Goals, 1990–2000

- A one-third reduction in under-five death rates (or a reduction to below 70 per 1,000 live births—whichever is lower)
- A halving of maternal mortality rates
- A halving of severe and moderate malnutrition among the under-5s
- Safe water and sanitation for all families
- Basic education for all children and completion of primary education by at least 80 percent
- A halving of the adult illiteracy rate and the achievement of equal educational opportunity for males and females

Protection for Girls and Women

- Family planning education and services to be made available to all couples to empower them to prevent unwanted pregnancies and births. Such services should be adapted to each country's cultural, religious, and social traditions.
- All women to have access to prenatal care, a trained attendant during childbirth, and facilities for high-risk pregnancies and emergencies
- Universal recognition of the special health and nutritional needs of females during early childhood, adolescence, pregnancy, and lactation

Nutrition

- A reduction in the incidence of low birthweight to less than 10 percent
- A one-third reduction in iron-deficiency anemia among women, and virtual elimination of vitamin A deficiency and iodine deficiency disorders
- All families to know the importance of supporting women in the task of exclusive breastfeeding for the first 4 to 6 months of a child's life
- Growth monitoring and promotion to be institutionalized in all countries
- Dissemination of knowledge to enable all families to ensure food security

Child Health

- The eradication of polio, a 90 percent reduction in measles cases, and a 95 percent reduction in measles deaths, compared with preimmunization levels
- Achievement and maintenance of at least 90 percent immunization coverage of 1-year-old children and universal tetanus immunization for women in the childbearing years
- A halving of child deaths caused by diarrhea and a 25 percent reduction in the incidence of diarrheal diseases
- A one-third reduction in child deaths caused by acute respiratory infections

Source: Adapted from UNICEF, *The State of the World's Children 1993* (London: Oxford University Press, 1993), p. 32.

China, India, and Bangladesh—have designed effective nutrition strategies, but lack sufficient resources for implementation. Other countries, such as Ethiopia, Nigeria, and Nepal, are without strategies due to lack of political consensus on the causes of the malnutrition problem. Lastly, some countries are unlikely to meet the Summit goals because of war or civil strife (for example, Angola, Haiti, Rwanda, and Sudan).

Large scale efforts in certain areas of Africa and India show promise for halving the malnutrition rates for children in those regions. Despite economic hardship, Tanzania's Iringa nutrition program, mentioned earlier, has more than halved the rate of severe malnutrition in 3 years. Some progress has been made on vitamin and mineral deficiency goals. Many countries, including Bangladesh, Brazil, India, Malawi, and the Philippines, now use immunization systems to deliver vitamin A capsules. However, there are still 3 million deaths each year from diseases that could have been prevented with vaccination. Nearly 80 developing countries have now banned the free or subsidized distribution of infant formula to new mothers in hospitals and clinics. Although oral rehydration therapy now saves the lives of 1 million children annually, another 3 million children still die of dehydration from diarrheal disease each year. More improvement is needed.

Focus on Women

Women make up 50 percent of the world's population. With their children, they represent the majority of those living in poverty. Thus, any solution to the problems of poverty and hunger is incomplete and even hopeless if it fails to address the role of women in developing countries (see Table 16-7).

In many countries, over 90 percent of the population live in rural areas. Women living in rural poverty endure oppressive conditions. They are often overworked and underfed, yet they are expected to carry most of the burden of their family's survival. In many cultures, they are the last to get food, although they spend long hours each day procuring water and firewood and pounding grain by hand (see Figure 16-8). In many countries, women in rural areas not only are the primary food producers, but also are responsible for child care and food preparation. Often they have to work as harvesters on other people's lands as well. Husbands are frequently absent from their homes—not by choice, but because the changing global economy has forced many men to leave home to find paying jobs. They have gone to look for work in the cities or to find employment growing export crops on distant commercial farms.

Development projects are often large in scale and highly technological, but they frequently overlook women's needs. Typically, only men have access to education and training programs (see Figure 16-8). Yet women play a vital role in the nutrition of their nation's people. Their nutrition during pregnancy and lactation determines the future health of their children. If women are weakened by malnutrition themselves or ignorant about how to feed their families, the consequences ripple outward to affect many other individuals. The importance of women in these countries is increasingly being appreciated, and many countries now offer development programs with women in mind.

Seven basic strategies are at the heart of women's programs:

- Removing barriers to financial credit so women can obtain loans for raw materials and equipment to enhance their role in food production
- Providing access to time-saving technologies—seed grinders, for example

In 1981, the World Health Assembly—comprising the health ministers of almost all countries—adopted the International Code of Marketing of Breastmilk Substitutes. The code stipulates that health facilities never be involved in the promotion of infant formula and that free samples should not be provided to new mothers. By 1997, only 17 countries have enacted laws that put them in compliance with the code. To improve this situation, WHO and UNICEF have begun a Baby-Friendly Hospital Initiative to help transform hospitals into centers that promote and support good infant feeding practices.

TABLE 16-7 *Women and Development: Fiction and Fact*

- **Men produce the world's food; women prepare it for the table.** In developing countries, where three-fourths of the world's people live, rural women account for more than half the food produced.

- **Women work to supplement the family's income.** Women are the sole breadwinners in one-fourth to one-third of the families in the world. The number of female-headed families is rapidly increasing.

- **When women receive the same education and training as men, they will receive equal pay.** So far, earning differentials persist even at equivalent levels of training. In professional fields, for example, comparisons of men's and women's salaries show a large gap between them even when samples are matched for training and experience.

- **Men are the heavy workers, and where food is short, they should have first priority.** As a rule, women work longer hours than men. Many carry triple work loads—in their household, labor force, and reproductive roles. Rural women often average an 18-hour day. Anemia resulting from a primary or secondary nutrient deficiency is a serious health problem for women in developing countries.

- **In modern societies, women have moved into all fields of work.** Relatively few women have entered occupations traditionally dominated by men. Most women remain highly segregated in low-paid jobs.

- **Women contribute a minor share of the world's economic product.** Women are a minority in the conventional measures of economic activity because these measures undercount women's paid labor and do not cover their unpaid labor. The value of women's work in the household alone, if given economic value, would add an estimated one-third to the world's gross national product.

Source: Reproduced with permission from *Women . . . A World Survey* by R. L. Sivard. Copyright © 1985 by World Priorities, Inc., Washington, D.C.

- Providing appropriate training to make women self-reliant
- Teaching management and marketing skills to help women avoid exploitation
- Making health and day care services available to provide a healthy environment for the women's children
- Forming women's support groups to foster strength through cooperative efforts[63]
- Providing information and technology to promote planned pregnancies

The recognition of women's needs by some development organizations is an encouraging trend in the efforts to contend with the world hunger crisis. The following examples from Sierra Leone and Ghana illustrate how women's development programs work:[64]

Balu Kamara is a farmer in Sierra Leone in West Africa, where farming is difficult, particularly for women. There women have little money and must take out loans to buy seed rice and to pay for the use of oxen. The price of rice is so low, though, that at the end of the growing season the women do not earn enough money to repay their loans. Yet, as the economy worsens, it is up to the women to carry the burdens; it is up to the women to stretch what resources are available to feed their families regardless of hardships.

Balu is the leader of the Farm Women's Club, a basket cooperative the women formed to make and sell baskets so they could pay their debts and continue farming. Finding time to weave baskets is difficult. Yet the women and their cooperative are succeeding. On the value of the Farm Women's Club, Balu says, "We have access to credit and a cash income. We have the opportunity to learn improved methods of

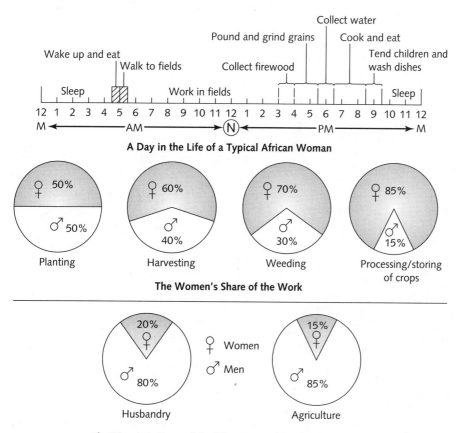

A Day in the Life of a Typical African Woman

The Women's Share of the Work

Planting — ♀ 50% / ♂ 50%

Harvesting — ♀ 60% / ♂ 40%

Weeding — ♀ 70% / ♂ 30%

Processing/storing of crops — ♀ 85% / ♂ 15%

The Women's Share of the Education and Training Programs

Husbandry — ♀ 20% / ♂ 80%

Agriculture — ♀ 15% / ♂ 85%

♀ Women   ♂ Men

FIGURE 16-8 *Women and Development*

Note that the women do the work for their families, while the men receive the training to produce goods and crops for export.

Source: Adapted form T. Flynn, Women in development, *Seeds* (Sprouts edition), May 1988.

agriculture and marketing and to increase our belief in ourselves and ease our families through the hungry season."

Gari (processed cassava) is becoming increasingly popular in Ghana because of the shortage of many other food items and because, once prepared, it is easy to cook. But it is very time-consuming to prepare gari—peeling and grating the fresh cassava, fermenting it over several days, squeezing the water from the fermented cassava, and, finally roasting it over a wood fire.

Gari, or processed cassava, is a bland-tasting food served in a form like porridge.

To help village women in the Volta Region increase their income through gari processing, an improved technology was introduced with the help of the National Council on Women and Development. The process involves a special mechanical grater, a pressing machine to squeeze the water from the grated cassava and a large enamel pan for roasting. This pan holds ten times the volume of the traditional cassava pot. The system was developed locally, with advice from the women themselves contributing to the success of the project.

Before, the women produced 50 gari bags every week. Now they are able to produce 5000 to 6000 bags a week. However, this increased output of gari can only be maintained with a higher yield of cassava in the area. Therefore, a male cassava growers' association has been formed to step up cassava production, and a tractor has been acquired by the women's cooperative so as to put more land under cassava cultivation.

The Vitamin A Field Support Project (VITAL)

Eye damage caused by vitamin A deficiency (VAD) affects an estimated 13 million preschool children worldwide.[65] Millions more children consume inadequate amounts of vitamin A–rich foods and are thus at risk for VAD. Manifestations of VAD range from mild xerophthalmia (night blindness and/or Bitot's spots) to dryness of the conjunctiva and cornea and, in severe cases, to melting of the cornea and blindness. In Asia alone, 200,000 children are blinded each year, with two-thirds of these children dying soon afterward. Recent epidemiological research has identified a relationship between marginal VAD, documented by reduced levels of circulating serum retinol, and higher mortality and morbidity rates from infectious diseases in children.[66] Researchers meeting at the International Conference on Nutrition in 1992 confirmed that even mild VAD significantly increases the death rate among children aged 6 months to 6 years. In particular, VAD significantly increases the severity and risk for diarrheal disease, measles, and pneumonia—the three main health threats facing children in the developing world.[67] Evidence from Africa suggests that vitamin A supplements can substantially reduce mortality and complications among children with measles, presumably by protecting epithelial tissue and ensuring the proper maintenance and functioning of the immune system.[68]

The most common factor contributing to the magnitude of VAD worldwide is the chronic inadequate dietary intake of vitamin A. Other contributors include poor nutrition status of mothers during pregnancy and lactation, low prevalence of breastfeeding, delayed or inappropriate introduction of supplementary foods, high incidence of infection (e.g., diarrhea, acute respiratory infection, measles), low levels of maternal education, drought, civil strife, poverty, and ecologic deprivation in some regions resulting in limited availability of vitamin A–rich foods.[69] For example, the production and consumption of vitamin A–rich foods in Africa (dark green, leafy vegetables, orange-colored fruits and tubers, and red palm oil) are influenced strongly by seasonal trends and cultural practices.[70]

VITAL is a VAD program administered by the U.S. Agency for International Development (USAID). Like other programs designed to eradicate and prevent VAD in developing countries, VITAL focuses on three approaches:

- **Dietary diversification**. VITAL strategies include stimulating the production and consumption of vitamin A–rich foods through agricultural production, home gardening, food preservation, nutrition education, and social marketing. For example, VITAL promotes the consumption of papayas, an excellent source of vitamin A, by pregnant women in the South Pacific. In some regions, foods containing vitamin A—especially vegetables and fruits—are readily available but underutilized by vulnerable groups, particularly weaning-age children and pregnant and lactating women, due to traditional customs and beliefs. Consequently, dietary diversification programs in these areas focus on intensive nutrition education and social marketing campaigns to foster necessary community understanding, motivation, and participation.

  Home gardens play a critical role in alleviating VAD in many communities. They provide a regular and secure supply of household food.

  VITAL also sponsors several projects aimed at increasing vitamin A consumption by improving food processing techniques. Solar drying has been introduced as the appropriate technology for the preservation of mangoes, papayas, sweet potatoes, pumpkins, green leaves, and other vitamin A–rich foods in several countries. When solar dried, these foods retain both flavor and carotene content and can alleviate seasonal variation in the availability of food. For example, the MANGOCOM Project in Senegal attempts to improve the vitamin A intake of weaning-age children by promoting dried mangoes, produced by women's cottage industries, as finger foods and fruit purees for toddlers.[71]

- **Food fortification**. Fortifying food is generally a large-scale undertaking and will only be effective if the target groups can buy and will consume the fortified product. In several developed countries, commonly consumed foods (e.g., margarine and milk) have been successfully fortified with vitamin A. Sugar is fortified with the vitamin in several regions of Central America, notably through VITAL projects in Guatemala, El Salvador, and Honduras.[72] Pilot trials with vitamin A–fortified rice are underway in Brazil. In both Indonesia and the Philippines, monosodium glutamate (MSG) was fortified with vitamin A. Program evaluations in both countries showed improvements in vitamin A status in the target populations. However, MSG fortification programs have since been halted due to unresolved technological problems, the questionable safety of MSG, and the lack of political will to support the fortification efforts. A pilot program in the Philippines is currently testing vitamin A–fortified margarine as an alternative means of improving the vitamin A status of children.

- **Distribution of vitamin A supplements**. Most commonly, VAD intervention programs periodically distribute vitamin A supplements in the form of high-dose capsules or oral dispensers. UNICEF typically donates the vitamin A capsules and helps organize the distribution efforts.* Often, supplements are delivered in conjunction with ongoing local health services, primary health care programs (e.g., maternal-child health projects), or national vaccination campaigns.

Vitamin A supplementation programs to date have reported a number of operational obstacles: low priority given to the distribution of the supplements by primary health care workers, a lack of community demand for vitamin A, and a lack of awareness among policy makers of the critical nature of vitamin A nutrition status. However, similar obstacles have been overcome by immunization programs, which now reach 80 percent of targeted children worldwide.[73] For this reason, WHO and UNICEF have suggested that vitamin A supplementation be integrated into existing immunization programs: "The provision of vitamin A supplementation through immunization services could dramatically expand the coverage of children in late infancy, giving a boost to vitamin A status before the critical period of weaning." Since immunization programs primarily target infants under 12 months of age, similar coverage would need to be provided by other means to children from one to six years of age.

Supplementation is considered a temporary measure for the control and prevention of VAD. More permanent solutions include cultivating vitamin A–rich foods, fortifying foods with vitamin A, promoting improved food habits, eliminating poverty, and improving sanitation worldwide (see Table 16-8).[74] Therefore, countries pursuing the high coverage achieved by integrating vitamin A supplement distribution into immu-

TABLE 16-8 *Examples of Cost-Effective Interventions to Reduce Micronutrient Malnutrition*

Intervention	Examples
Food/Menu-based	Modification in quantity and/or quality and diversity of menus Improved bioavailability through modified preservation/preparation procedures Household food-to-food fortification
Fortification	Sugar, salt, and other condiments Oils and margarine Cereals and flours
Supplementation	Periodic high doses Frequent low doses
Public health measures	Breastfeeding promotion Immunizations Parasite control Sanitation/safe water
Social and economic developmental measures	Female literacy Family spacing Income generation

Source: Adapted from B. A. Underwood, Micronutrient malnutrition, *Nutrition Today* 33 (1998): 125; and B. A. Underwood, From research to global reality: The micronutrient story, *Journal of Nutrition* 128 (1998): 145–51.

nization programs are encouraged to also allocate resources to these alternative efforts.

Fortunately, the problem of VAD is not insurmountable. The numerous VITAL projects worldwide have demonstrated three principles critical to successful VAD intervention efforts. First, successful interventions include preliminary formative research (e.g., focus group interviews), target audience segmentation, pretesting, and evaluation in program planning. Secondly, support by policy makers and participation by local community members is critical to sustaining the program. Lastly, multiple channels of communication are recommended (e.g., mass media, traditional forms of media, and personal communications).

In the final analysis, individual countries will need to choose the most practical and cost-effective mix of VAD interventions based on local customs, resources, and needs. Numerous international agencies, including WHO, UNICEF, FAO, the World Bank, and USAID have made the commitment to support country efforts to meet the 1990 World Summit for Children's goal of virtually eliminating VAD by the year 2000.

---

* The recommended dosing schedule for a vitamin A capsule distribution program is 200,000 IU given twice a year. The dose for children less than one year of age or those who are significantly underweight is 100,000 IU twice a year. (A. Gadomski and C. Kjolhede, *Vitamin A Deficiency and Childhood Morbidity and Mortality* [Baltimore, MD: Johns Hopkins University Publications, 1988].)

International Nutrition Programs

Nutrition programs in developing countries vary considerably. A sample job-description for a nutrition specialist working with FAO is shown in Figure 16-9. International nutrition programs include both large-scale and small-scale operations, may be supported with private or public funds, and may focus on emergency relief or long-term development.[75] In developing countries, emphasis has generally been placed on four types of nutrition interventions:

1. Breastfeeding promotion programs with guidance on preparing appropriate weaning foods
2. Nutrition education programs typically focusing either on infant and child feeding guidelines and practices and child survival activities or on the incorporation of nutrition education into primary school curriculums and teacher training programs
3. Food fortification and/or the distribution of nutrient supplements (e.g., vitamin A capsules) and the identification of local food sources of problem nutrients
4. Special feeding programs designed to provide particularly vulnerable groups with nutritious supplemental foods[76]

In many countries, there is mounting evidence of grass roots progress in improving agricultural, water, education, and health services, especially for children.[77] Experiences in Sierra Leone and Nepal are encouraging examples.

In Sierra Leone, a food product was developed from rice, sesame (benniseed), and peanuts that were hand pounded and cooked to make a flour meal. The local children not only found it tasty, but whereas they had been malnourished before, they thrived when this product supplemented their diets. The village women formed a cooperative to reduce the drudgery of preparing the food and rotated the work on a weekly or monthly basis.[78] The government also established a manufacturing plan to produce and market the mixture at subsidized prices. The success of the venture lies in the involvement of the local people in identifying the problem and devising a solution that meets their needs.

A similar success story is told in Nepal. A supplementary food made from soybeans, corn, and wheat, mixed in a 2:1:1 proportion, yielded a concentrated "superflour" of high biological value suitable for infants and children. A nutrition rehabilitation center tested this superflour by giving undernourished children and their mothers two cereal-based meals a day, and giving the children three additional small meals of superflour porridge daily. Within 10 days, the undernourished children had gained weight, lost their edema, and recovered their appetites and social alertness. The mothers, who saw the remarkable recoveries of their children, were motivated to learn how to make the tasty supplementary food and incorporate it into their local foodstuffs and customs.

These two examples offer hope, but the real issue of poverty remains to be addressed. One-shot intervention programs offering nutrition education, food distribution, food fortification, and the like are not enough. It is difficult to describe the misery a mother feels when she has received education about nutrition but cannot grow or purchase the foods her family needs. She now knows *why* her child is sick and dying but is unable to *apply* her new knowledge.

Position Title: Nutrition Officer (Nutrition Information)
Responsible To: Senior Officer (Nutrition Assessment)
Level Grade: P–4

Responsibilities:
Under the general direction of the Chief of the Nutrition Planning, Assessment and Evaluation Service, Food Policy and Nutrition Division, and the supervision of the Senior Officer (Nutrition Assessment), this specialist is responsible for the development of activities aimed at improving information necessary for assessing and analyzing nutrition status of populations at global, national, and local levels, especially in the drought-prone African countries.

Job Duties:
In accordance with departmental policies and procedures, this specialist carries out the following duties and responsibilities:

* Assists member countries in the design and implementation of data collection and analysis activities to meet the needs of policy makers and planners concerned with food and nutrition issues;
* In cooperation with other organization units, strengthens the skills and technical capacity of member countries, particularly within the agricultural sector, to assess, analyze, and monitor the food and nutrition situation of their own population;
* Provides data and background information for Expert Committees and Councils;
* Provides consultations on energy and nutrient requirements, dietary assessment methodology, and foodways;
* Prepares relevant reports for the organization and its committees; and
* Performs other related duties.

Qualifications and Experience—Essential
The specialist should have the following educational qualifications and experience:

* A university degree in Nutrition or Biological Science with postgraduate qualification in Nutrition.
* Seven years of progressively responsible professional experience, including the planning and implementation of nutrition data collection, processing, and analysis.
* Experience in the management and analysis of nutrition-related data using advanced statistical applications on microcomputers.

Qualifications and Experience—Desirable
* Experience in international work and research.
* Relevant work experience in Africa.
* Experience in the organization of seminars, training courses, or conferences.

Skills
* Working knowledge of English and French (level C).
* Demonstrated ability to write related technical reports. Familiarity with project design techniques.
* Courtesy, tact, and ability to establish and maintain effective work relationships with people of different national and cultural backgrounds.
* Experience using word-processing equipment and software.

Date:

Revisions:

FIGURE 16-9 *Sample Job Description for a Nutrition Specialist with the Food and Agriculture Organization*

Source: Adapted form Vacancy Announcement (No. 130-ESN), 1992, of the Food and Agriculture Organization of the United Nations. Used with permission.

Looking Ahead: The Global Challenges

In 1977, the thirtieth annual World Health Assembly, convened in Alma Ata, decided unanimously that the main social target of member governments and WHO itself should be "the attainment by all citizens of the world by the year 2000 of a level of health that will permit them to lead a socially and economically productive life." This goal is now referred to simply as "Health for All by the Year 2000."[79] The assembly hoped to make primary health care services accessible and affordable for all people in every country of the world. Today, however, over a billion and a half of the world's people continue to live without access to basic health care, and some 40,000 children die each day from the effects of malnutrition and related diseases—an estimated 100 million deaths in the 1990s.

In the developing world, the health care crisis revolves around the daily struggle for survival and the growing disparity between the haves and the have-nots. For the poor, the struggle for safe water, adequate nutrition, and access to basic health care leaves no energy or resources for other concerns—and most have little hope of winning the battle. For most of us, living in the developed world, health care reform means we are assured of meeting our own and our family's ongoing health care needs, including access to the latest miracle drug or the availability of a bone marrow transplant. But as one physician from the hospital ship M/V *Anastasis* remarked on visiting Ghana in West Africa: "We are foolhardy to believe that we can be a healthy society when the world around us is languishing with diseases and poverty that we could alleviate."[80]

In addition, we face many new challenges not even thought of in 1977 when the goal of "Health for All by the Year 2000" was set.[81] Before we can attain this goal, we must deal with these issues, among others:

- The HIV/AIDS pandemic. An estimated 8.4 million cases of AIDS have occurred since the start of the epidemic and over 22 million people, including 800,000 children, are today infected with the virus. In addition, by the year 2000, there will be 10 million uninfected orphans whose parents have died of AIDS.[82]
- The recent upsurge in cases of tuberculosis in the United States among the urban poor, homeless persons, migrant workers, immigrants and refugees from countries with high tuberculosis rates, intravenous drug users, residents of correctional facilities, nursing home residents, and those infected with HIV.[83]
- The trend toward urbanization.[84] Urbanization has been a factor in the world-wide decline in breastfeeding, the increased consumption of empty-calorie foods, and the outbreak of cholera in several Latin American countries since 1990 due to contamination of the urban water supply. As nations continue the rural-to-urban transition, the incidence of chronic diarrhea from polluted water and foodborne contamination remains a major public health problem in urban slums.
- The growing number of elderly persons in the developed world. This demographic change requires fundamental health care reform, as discussed earlier in Chapter 15, to ensure the maintenance of quality living and functional ability in the later years.
- The upward surge in global poverty. The proportion of the world's population that is poor, approximately 1.2 billion people, has increased over the last decade, halting a 40-year downward trend.[85]
- Rapid population growth. The earth's population will nearly double within the next 40 years—most of these people will be born into poor families in developing countries. Agricultural production will need to increase to feed these people. The number of nonagricultural jobs must also increase to support those not working in agriculture.

- Destruction of the global environment. The earth's capacity to sustain life is being impaired by many complex interrelated developments, including overconsumption, industrial pollution, overgrazing, and deforestation. This destruction of natural resources threatens the health and well-being of both today's people and future generations as well.

Personal Action: Opportunity Knocks

The problems addressed in this chapter may appear to be so great that they can be approached only through worldwide political decisions. Indeed, the members of the International Conference on Nutrition stressed that intensive worldwide efforts were needed to overcome hunger and malnutrition and to foster self-reliant development. To this end, many individuals and groups are working to improve the future well-being of the world and its people through a number of national and international organizations. Check out the resources available on-line from the organizations listed in the box on the following page.

Individuals can help change the world through their personal choices.[86] Our choices have an impact on the way the rest of the world's people live and die. As mentioned in Chapter 12, our nation, with 6 percent of the world's population, consumes about 40 percent of the world's food and energy resources. The world food problem depends partly on the demands we place on the world's finite natural resources. In a sense, we contribute to the world food problem. People in affluent nations have the freedom and means to choose their lifestyles; people in poor nations do not. We can find ways to reduce our consumption of the world's resources by using only what is required.

It is ironic that whereas other societies cannot secure enough clean water for people to drink, our society produces bottles of soda that contain one calorie of artificial sweetener in 12 ounces of water that cost 800 calories to produce. In the United States, $5 billion is spent annually to lower calorie consumption, while 500 million people in the rest of the world can rarely find an adequate number of calories to consume.[87] Thus, choosing a diet at the level of necessity, rather than excess, would reduce the resource demands made by our industrial agriculture. In fact, those

Internet Resources

Bread for the World Institute
    Seeks to inform, educate, nurture, and motivate concerned
    citizens for action on policies that affect hungry people.
www.bread.org

CARE
    Provides famine and disaster victims with emergency
    assistance, improves health care, helps subsistence farmers
    and small business owners produce more goods, and
    addresses population and environmental concerns.
www.care.org

Catholic Charities USA
    Network of independent social service organizations;
    promotes public policies and strategies that address human
    needs and social injustices.
www.catholiccharitiesusa.org

Food and Agriculture Organization
www.fao.org

Freedom from Hunger
    Fights chronic hunger through "Credit with Education"
    program that combines village banking with basic health,
    nutrition, family planning, and microenterprise
    management education.
www.freefromhunger.org

The International Fund for Agricultural Development (IFAD)
    Specialized international financial agency of the United
    Nations with the mandate to provide the rural poor of the
    developing world with cost-effective ways of overcoming
    hunger, poverty, and malnutrition.
www.ifad.org

Oxfam America
    Fights global poverty and hunger by working in
    partnership with grassroots organizations promoting
    sustainable development.
www.oxfamamerica.org

Pan American Health Organization
www.paho.org

UNICEF USA
    Works for the survival, protection, and development of
    children worldwide through education, advocacy, and
    fund-raising.
www.unicefusa.org

The United Nations Development Programme (UNDP)
    Works in 175 countries and territories to build national
    capacities for sustainable human development.
www.undp.org

World Health Organization
www.who.org

who study the future are convinced that the hope of the world lies in everyone's adopting a simple lifestyle. As one such person put it, "the widespread simplification of life is vital to the well-being of the entire human family."[88] Personal lifestyles do matter, for a society is nothing more than the sum of its individuals. As we go, so goes our world.

Community Learning Activity

You probably cannot solve the world food problem single-handedly, but the place to start is with yourself and your community. Phrases such as "think globally, act locally" reinforce this message—start where you are.

Activity 1

This activity asks you to consider what changes you could make in your personal life that might affect the problem of world hunger and the issues associated with it that were discussed in this chapter. We ask that you consider the following list of suggestions and then identify other strategies to add to this list:

- Examine your diet to become more conscious of your dependence on others.
- Consider your lifestyle, and make intentional choices as you use resources.
- Grow some of your own food. (Try community gardens in the city.)
- As you buy, think about who benefits.

Activity 2

Access and review the Web site for one or more of the organizations working to end world hunger (see the box on page 524). Learn more about their activities, and share a profile of an organization with your class.

References

1. G. G. Graham, Starvation in the modern world, *New England Journal of Medicine* 328 (1993): 1058–61.
2. Food and Agriculture Organization of the United Nations, *The Sixth World Food Survey* (Rome: FAO Statistical Analysis Service, 1996).
3. UNICEF, *The State of the World's Children 1998* (New York: Oxford University Press, 1998).
4. United Nations International Conference on Nutrition, *World Declaration on Nutrition* (Rome: FAO/WHO Joint Secretariat for the Conference, 1992), p. 1.
5. L. R. Brown, *Who Will Feed China? Wake Up Call For a Small Planet* (New York: Norton, 1995).
6. A. Durning, Life on the brink, *World Watch* 3 (1990): 22–30.
7. Ibid., p. 25.
8. FAO, *Sixth World Food Survey.*
9. Bread for the World Institute, *Hunger 1998: Hunger in a Global Economy* (Silver Spring, MD: Bread for the World Institute, 1997).
10. B. Hartman and J. Boyce, *Quiet Violence: View from a Bangladesh Village* (San Francisco: Institute for Food and Development Policy, 1983).
11. P. Uvin, The state of world hunger, *Nutrition Reviews* 52 (1994): 151–61.

12. Bread for the World Institute, *Hunger 1998: Hunger in a Global Economy* (Silver Spring, MD: Bread for the World Institute, 1997).
13. Community Nutrition Institute, Malnutrition accounts for over half of child deaths worldwide, says UNICEF, *Nutrition Week* 28 (1998): 1–2.
14. This estimate and the margin comment are from UNICEF, *The State of the World's Children, 1993* (London: Oxford University Press, 1993), pp. 1–3, 57.
15. B. A. Carlson and T. M. Wardlaw, A global, regional, and country assessment of child malnutrition, *UNICEF Staff Working Papers,* Number 7 (New York: UNICEF, 1990), pp. 1–30; and Bread for the World Institute, *Hunger 1997: What Governments Can Do* (Silver Springs, MD: Bread for the World Institute, 1996).
16. Population Reference Bureau, *1997 World Population Data Sheet* (Washington, DC: Population Reference Bureau, 1997).
17. M. Cameron and Y. Hofvander, *Manual on Feeding Infants and Young Children,* 3rd ed. (New York: Protein Advisory Board of the United Nations, 1989), pp. 11–13.
18. C. D. Williams, N. Baumslag, and D. B. Jelliffe, *Mother and Child Health: Delivering the Services,* 3rd ed. (New York: Oxford University Press, 1994).

19. UNICEF, *The State of the World's Children 1998*.

20. L. Robertson, Breastfeeding practices in maternity wards in Swaziland, *Journal of Nutrition Education* 23 (1991): 284–87.

21. UNICEF, *The State of the World's Children 1998*.

22. M. W. Rosegrant, *Water Resources in the Twenty-First Century*, Discussion Paper 20 (Washington, D.C.: International Food Policy Research Institute, 1997).

23. J. P. Habicht, J. DaVanzo, and W. P. Butz, Mother's milk and sewage: Their interactive effects on infant mortality, *Pediatrics* 81 (1988): 456–61.

24. Dr. Carol Dyer's findings related to social and cultural beliefs about food in India are from A. Berg, *The Nutrition Factor* (Washington, D.C.: Brookings Institute, 1973), p. 46.

25. World Bank, *World Development Report 1991* (New York: Oxford University Press, 1991), Table 21.

26. Y. W. Bradshaw and coauthors, Borrowing against the future: Children and third world indebtedness, *Social Forces* 71 (1993): 629–56.

27. UNICEF, *State of the World's Children* (London: Oxford University Press, 1989), p. 82.

28. Berg, *The Nutrition Factor*, p. 46.

29. Independent Commission on International Issues, *North-South: A Program for Survival* (Cambridge, MA: MIT Press, 1980), pp. 49–50.

30. Bread for the World Institute, *Hunger 1998*.

31. U.S. National Committee for World Food Day, *World Food Summit: Promises and Prospects*, Fourteenth Annual World Food Day Teleconference materials, 1997.

32. S. Lewis, Realism and vision in Africa, *Development: A Journal of the Society for International Development* 1 (1988): 46.

33. G. Kent, Food trade: The poor feed the rich, *Food and Nutrition Bulletin* 4 (1982): 25–33.

34. F. M. Lappe and J. Collins, *Food First: Beyond the Myth of Scarcity* (Boston: Houghton Mifflin, 1978), p. 15; Bread for the World Institute, *Hunger 1996: Countries in Crisis* (Silver Spring, MD: Bread for the World Institute, 1995).

35. Interreligious Taskforce, on U.S. Food Policy, *Identifying a Food Policy Agenda for the 1990s: A Working Paper* (Washington, D.C.: Interreligious Taskforce on U.S. Food Policy, 1989), pp. 1–30.

36. Brown, *Who Will Feed China?*

37. J. Kocher, Not too many but too little, in J. D. Gussow, *The Feeding Web: Issues in Nutritional Ecology* (Palo Alto, CA: Bull, 1978), pp. 81–83.

38. Bread for the World Institute, *Hunger 1997: What Governments Can Do* (Silver Spring, MD: Bread for the World Institute, 1996).

39. International Food Policy Research Institute, *Feeding the World, Preventing Poverty, and Protecting the Earth: A 2020 Vision* (Washington, D.C.: IFPRI, 1996).

40. Food and Agriculture Organization, *World Food Summit Papers* (Rome: FAO, 1996).

41. Bread for the World Institute, *Hunger 1998*.

42. E. O'Kelly, Appropriate technology for women, *Development Forum*, June 1984, p. 2.

43. G. Gardner, *Shrinking Fields: Cropland Loss in a World of Eight Billion* (Washington, D.C.: Worldwatch Institute, 1996).

44. L. Brown, *State of the World Report, 1990* (Washington, D.C.: Worldwatch Institute, 1990).

45. Position of the American Dietetic Association: World Hunger, *Journal of the American Dietetic Association* 95 (1995): 1160–62.

46. *Oxfam America News*, Fall 1992, p. 8.

47. The discussion of Sri Lankan development is adapted from M. Boyle, New heartbeat for an ancient people, *Seeds*, December 1984, pp. 14–18.

48. Mobilization for nutrition: Results from Iringa, *Mothers and Children: Bulletin on Infant Feeding and Maternal Nutrition* 8 (1989): 1–3; Improving child survival and nutrition, *Evaluation Report: Joint WHO/UNICEF Nutrition Support Program in Iringa, Tanzania* (United Republic of Tanzania: WHO/UNICEF, 1989).

49. *Evaluation Report: Joint WHO/UNICEF Nutrition Support Program*.

50. Durning, Life on the brink, p. 29.

51. World Declaration on Nutrition, *Nutrition Reviews* 51 (1993): 41–43

52. L. Miring'U and C. R. Mumaw, Needs assessment for in-service training for community nutrition educators in the Kiambu district in Kenya, *Journal of Nutrition Education* 25 (1993): 70–73.

53. L. Brown, *State of the World, 1997*; (Washington, D.C.: Worldwatch Institute, 1997); U.S. National Committee for World Food Day, 1997.

54. FAO, *The Sixth World Food Survey*; and United Nations Department of Public Information, *UN Chronicle* 33 (4) (1996): 24–28.

55. S. Lewis, Food security, environment, poverty, and the world's children, *Journal of Nutrition Education* 24 (1992): 3S–5S.

56. UNICEF, *The State of the World's Children 1998*.

57. Williams et al., *Mother and Child Health*.

58. P. Pellet, The role of food mixtures in combating childhood malnutrition, in *Nutrition in the Community*, ed. D. McLaren (New York: Wiley, 1978), pp. 185–202; Graham, Starvation in the modern world, p. 1060.

59. UNICEF, *The State of the World's Children, 1998*.

60. Lewis, Food security.

61. Ibid, p. 5S.

62. The discussion on progress towards meeting the Child Summit goals is adapted from: Bread for the World Institute, *Hunger 1995: Causes of World Hunger* (Silver Spring, MD: Bread for the World Institute, 1994), pp. 98–100.

63. Oxfam America, *Facts for Action: Women Creating a New World*, no. 3 (Boston: Oxfam America, 1991), pp. 2–3.

64. Trade and Development Program, *Exploring the Linkages: Trade Policies, Third World Development, and U.S. Agriculture* (Washington, D.C.: Bread for the World Institute, 1989), p. 23, as adapted from M. Carr, *Blacksmith, Baker, Roofing Sheet Maker* (London: Intermediate Technology Publications, 1984); Bread for the World Institute, *Women in Development* (Washington, D.C.: Bread for the World Institute, 1988).

65. Vitamin A deficiency in Asia, *VITAL NEWS* 3(3) (1992):

1–11; The State of the World's Children 1998: A UNICEF Report, *Nutrition Reviews* 56 (1998): 115–123.

66. K. P. West and coauthors, Efficacy of vitamin A in reducing preschool child mortality in Nepal, *Lancet* 338 (1991): 67–71.

67. State of the World's Children, 1998.

68. Vitamin A deficiency in Africa, *VITAL NEWS* 3(2) (1992): 1–10.

69. Vitamin A deficiency in Asia, p. 2

70. Ibid., p. 4.

71. J. Rankins, MANGOCOM: A nutrition social marketing module for field use, *Journal of Nutrition Education* 24 (1992): 192–94.

72. Vitamin A deficiency in Latin America and the Caribbean, *VITAL NEWS* 3(1) (1992): 8.

73. The discussion of vitamin A supplementation and immunization programs was adapted from Linking vitamin A activities to primary health care programs, *VITAL NEWS* 2(1) (1991): 1–7.

74. M. G. Herrera and coauthors, Vitamin A supplementation and child survival, *Lancet* 340 (1992): 267–71; and B. A. Underwood and P. Arthur, The contribution of vitamin A to public health, *The FASEB Journal* 10 (1996): 1040–48.

75. K. R. Nelson, R. M. Jenkins, and S. K. Nelson, A third world supplemental feeding project: Expectations and realities—A dichotomy, *Nutrition Today* 24 (1989): 19–26; E. Velempini and K. D. Travers, Accessibility of nutritious African foods for an adequate diet in Bulawayo, Zimbabwe, *Journal of Nutrition Education* 29 (1997): 120–27; S. Dalton, An education and research opportunity in Nepal: Dietetics in a developing country, *Topics in Clinical Nutrition* 11 (1996): 39–46; and W. Fawzi and coauthors, A prospective study of malnutrition in relation to child mortality in the Sudan, *American Journal of Clinical Nutrition* (1997): 1062.

76. The list of types of programs was adapted from G. M. Wardlaw, Hunger and undernutrition in the world, *Nutri-News* (St. Louis, MO: Mosby–Year Book, 1990), p. 14.

77. L. Brown, *State of the World*, 1997.

78. *National Conference on Primary Health Care* (Kathmandu: Ministry of Health, Health Services Coordination Committee, World Health Organization, and UNICEF, 1977), pp. 9, 25, as cited by M. E. Frantz, Nutrition problems and programs in Nepal, *Hunger Notes* 2 (1980): 5–8.

79. L. C. Chen, Primary health care in developing countries: Overcoming operational, technical, and social barriers, *Lancet*, 2 (1986): 1260–65; P. F. Basch, *Textbook of International Health* (New York: Oxford University Press, 1990), pp. 200–5.

80. C. Aroney-Sine, Health care crisis: The global challenge, *Seeds* 15 (1993): 9–11.

81. L. R. Brown, *Vital Signs 1993: The Trends That Are Shaping Our Future* (New York: Norton, 1993); L. R. Brown, A decade of discontinuity, *World Watch* (July–August 1993): 19–26.

82. M. Thuriaux and S. Cherney, AIDS affects us all, *World Health,* January–February, 1997, pp. 20–21.

83. Report of the Special Initiative on AIDS of the American Public Health Association, *Tuberculosis and HIV Disease* (Washington, D.C.: American Public Health Association, 1992), pp. 1–5; and R. Bayer, N. Neveloff, and S. Landesman, The dual epidemics of tuberculosis and AIDS: Ethical and policy issues in screening and treatment, *American Journal of Public Health* 83 (1993): 649–54.

84. Brown, *State of the World, 1997.*

85. Bread for the World Institute, *Hunger 1997.*

86. The case for optimism, in E. Cornish, *The Study of the Future: An Introduction to the Art and Science of Understanding and Shaping Tomorrow's World* (Washington, D.C.: World Future Society, 1977), pp. 34–37.

87. *1990 State of the World Report.*

88. D. Elgin, *Voluntary Simplicity: Toward a Way of Life That Is Outwardly Simple, Inwardly Rich* (New York: Morrow, 1981), p. 25.

Appendixes

Appendix A

Healthy People 2000 Objectives in the Nutrition Priority Area

Number	Nutrition Objective	Target Population	Baseline	Most Recent Figure*	2000 Target
	Risk Reduction Objectives				
2.5	Reduce dietary fat and saturated fat intake	People aged 2 years and older Average % of calories from total fat	34%[†]	33%	30%
		Average % of calories from saturated fat	12%	11%	10%
2.6	Increase daily intake of fruits, vegetables, and grains	People aged 2 years and older Average number servings of fruits/veg.	4.1	5.0	5.0
		Average number servings of grains	5.8	6.7	6.0
2.7	Increase adoption of sound weight loss practices among overweight people	Overweight people aged 12+ years	NA[‡]	NA	NA
		Overweight men	25%[§]	17%	50%
		Overweight women	30%	19%	50%
2.8	Increase calcium intake	People aged 11–24 years	20%	17%	50%
2.9	Decrease salt and sodium intake	Adults who rarely or never use salt at table	60%	58%	80%
2.10	Reduce prevalence of iron deficiency	Children 1–2 years	9%	9%	3%
		Women aged 20–44 years	5%	8%	3%
2.11	Increase the proportion of women who breastfeed in the early postpartum period	New mothers	54%	59%	75%
2.12	Increase the proportion of caretakers who use feeding practices to prevent bottle tooth decay	Parents and caretakers	55%	NA	75%
2.13	Increase the proportion of people who read food labels	People aged 18+ years	74%	75%	85%
2.25	Reduce the prevalence of high blood cholesterol levels (> 6.21 mmol/L)[‖]	People aged 20–74 years	27%	19%	20%
2.26	Increase the number of people whose high blood pressure is under control	People aged 18–74 years with high blood pressure	11%	29%	50%

Continued

Number	Nutrition Objective	Target Population	Baseline	Most Recent Figure*	2000 Target
	Risk Reduction Objectives—continued				
2.27	Reduce the mean serum cholesterol concentration to no more than 5.17 mmol/L	People aged 20–74 years	5.51 mmol/L	5.25 mmol/L	5.17 mmol/L
	Services and Protection Objectives				
2.14	Achieve informative nutrition labeling on virtually all processed foods		60%	96%	100%
2.15	Increase the number of products that are reduced in fat and saturated fat		2,500	5,618#	5,000
2.16	Increase proportion of large chain restaurants offering at least one low-fat, low-calorie item on the menu		70%	75%**	90%
2.17	Increase the proportion of nutritious school and child care food services (lunches and breakfasts)		NA	NA	90%
2.18	Increase receipt of home food services by people in need		7%	NA	80%
2.19	Increase nutrition education in schools		60%††	69%	75%
2.20	Increase the number of worksites with 50+ employees that offer nutrition education or weight/management programs		17%	31%‡‡	50%
2.21	Increase the proportion of primary care providers who provide nutrition assessment and counseling and/or referral to qualified nutritionists or dietitians		NA	19%§§	75%

*Most recent figure based on survey data published in 1994 or 1995, unless otherwise noted.
†Data for total fat and saturated fat were obtained from the Continuing Survey of Food Intakes by Individuals, 1989–1991.
‡NA = Data not available.
§Overweight data for 1993.
‖For blood total cholesterol concentrations, to convert mmol/L to mg/dL, divide by 0.02586. Thus, 6.21 mmol/L = 240 mg/dL.
#Based on 1991 data.
**Based on 1990 data.
††Expressed as the proportion of states requiring nutrition education in schools.
‡‡For nutrition education programs, the data were obtained for 1992.
§§The figure reported is for family physicians.
Source: National Center for Health Statistics, *Healthy People 2000 Review, 1997* (Hyattsville, MD: Public Health Service, 1997). Data obtained from the document file located on the NCHS homepage at www.cdc.gov/nchswww.

Appendix B

Organization of Government

Appendix B-1 Organization of the U.S. Government

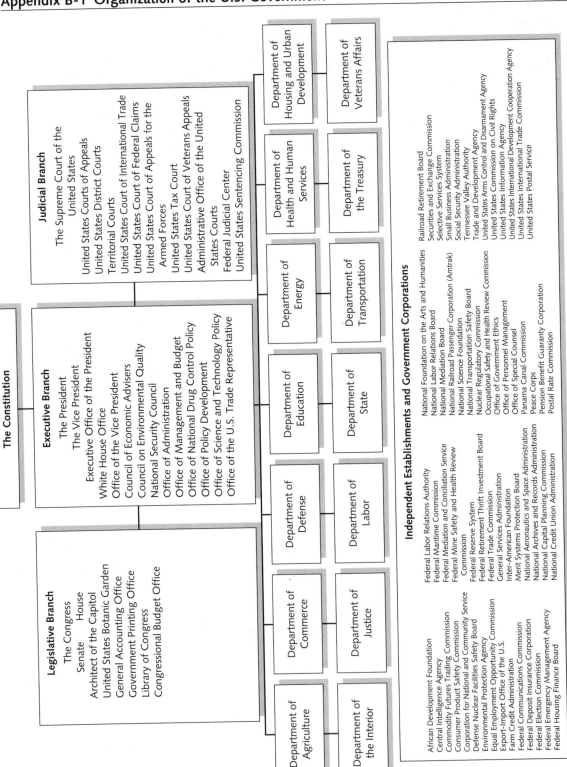

Source: Office of the Federal Register, National Archives and Records Administration, *The United States Government Manual 1997/1998* (Washington, D.C.: U.S. Government Printing Office, 1997), p. 22.

Appendix B-2 Organization of a Hypothetical State Health Agency

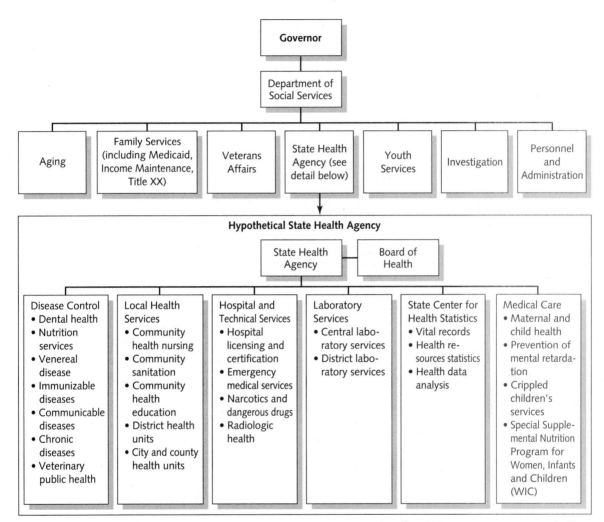

Source: G. Pickett and J. J. Hanlon, *Public Health: Administration and Practice* (St. Louis: Times Mirror/Mosby College Publishing, 1990), p. 109. (From National Public Health Program Reporting System.) Used with permission.

Appendix B-3 Organization Chart for a Hypothetical Local Health Department

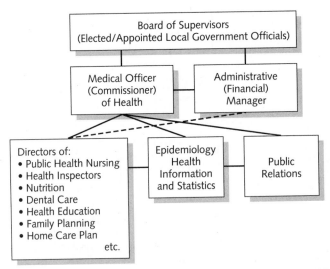

Source: J. M. Last, *Public Health and Human Ecology* (Stamford, CT: Appleton & Lange, 1998), p. 323. Used with permission.

Appendix B-4 Congressional Committees and Subcommittees, 105th Congress

There are four types of committees in Congress. *Standing* committees have the power to pass legislation within a certain policy arena. *Conference* committees exist temporarily when the two chambers need to reconcile differences between two versions of the same bill. *Select* or *special* committees may be either temporary or permanent and are struck to allow members to investigate a special problem; they are not empowered to pass legislation, only make recommendations for legislation. *Joint* committees likewise do not pass legislation; they are struck to address special topics of interest to Congress as a whole.*

The following are standing committees (and some of their subcommittees) of the 105th Congress with duties related to some aspect of health, food, or nutrition:

Senate—Standing Committees†

- Committee on Agriculture, Nutrition, and Forestry
  - Subcommittee on Marketing, Inspection, and Product Promotion
  - Subcommittee on Research, Nutrition, and General Legislation
- Committee on Appropriations
  - Subcommittee on Agriculture, Rural Development, and Related Agencies
  - Subcommittee on Labor, Health and Human Services, and Education
  - Subcommittee on Veterans Affairs, Housing and Urban Development, and Independent Agencies
- Committee on the Budget
- Committee on Finance
  - Subcommittee on Health Care
  - Subcommittee on Social Security and Family Policy
- Committee on Indian Affairs
- Committee on Labor and Human Resources
  - Subcommittee on Aging
  - Subcommittee on Children and Families
  - Subcommittee on Employment and Training
  - Subcommittee on Public Health and Safety
- Committee on Rules and Administration
- Committee on Veterans Affairs

Senate—Special Committees

- National Bipartisan Commission on the Future of Medicare
- Special Committee on Aging
- U.S. Bipartisan Commission on Comprehensive Health Care‡

House of Representatives—Standing Committees§

- Committee on Agriculture
  - Subcommittee on Department Operations, Nutrition, and Foreign Agriculture
  - Subcommittee on Livestock, Dairy, and Poultry
- Committee on Appropriations
  - Subcommittee on Agriculture, Rural Development, Food and Drug Administration, and Related Agencies
  - Subcommittee on Labor, Health and Human Services, and Education
  - Subcommittee on Veterans Affairs, Housing and Urban Development, and Independent Agencies
- Committee on the Budget
- Committee on Commerce
  - Subcommittee on Health and the Environment
- Committee on Education and the Workforce
  - Subcommittee on Early Childhood, Youth, and Families
- Committee on Resources
  - Subcommittee on Native American and Insular Affairs
- Committee on Rules
- Committee on Veterans Affairs
- Committee on Ways and Means
  - Subcommittee on Health
  - Subcommittee on Human Resources
  - Subcommittee on Social Security

*Source: W. Lyons, J. M. Scheb II, and L. E. Richardson, Jr., in *American Government: Politics and Political Culture* (Minneapolis: West, 1995), p. 356.

†Source: Legislative Activities: Committee and Subcommittee Membership at www.senate.gov/activities/cmte-mem.html.

‡Also known as the Pepper Commission.

§Source: The U.S. House of Representatives Committee Office Web Service at www.house.gov/CommitteeWWW.html.

Appendix C

Complementary and Alternative Medicine

Appendix C-1 Glossary of Complementary and Alternative Medicine Terminology

Additional information about the following therapies can be found on the Office of Alternative Medicine's Web site at altmed.od.nih.gov (click on "What is CAM?" and then on "Fields of Practice").

Acupuncture: a technique that involves piercing the skin with long thin needles at specific anatomical sites to stimulate, disperse, and regulate the flow of *chi*, or vital energy, and relieve pain or illness. Acupuncture sometimes uses heat, pressure, friction, suction, or electromagnetic energy to stimulate the points.

Aromatherapy: a technique that uses essential oils (the volatile oils distilled from plants) to enhance physical, psychological, and spiritual health. The oils are inhaled, massaged into the skin in diluted form, or used in baths.

Ayurveda (EYE-your-VAY-dah): a traditional Hindu system of improving health practiced in India for more than 5,000 years. In this approach, illness is a state of imbalance among the body's systems and can be detected through procedures involving reading the pulse and checking the tongue. It uses herbs, diet, meditation, massage, and yoga to stimulate the body to make its own natural drugs.

Bioelectromagnetic applications: the use of electrical energy, magnetic energy, or both, to stimulate bone repair, wound healing, and tissue regeneration.

Biofeedback: the use of special devices to convey information about heart rate, blood pressure, skin temperature, muscle relaxation, and the like, to enable a person to learn how to consciously control these medically important functions.

Chelation therapy: the use of ethylenediamine tetraacetic acid (EDTA) to bind with metallic ions, thus healing the body by removing toxic metals.

Chinese (Oriental) medicine: practitioners are trained to use a variety of ancient and modern therapeutic methods, including acupuncture, herbal medicine, massage, moxibustion (heat therapy), and nutrition and lifestyle counseling, to treat a broad range of acute and chronic diseases.

Chiropractic: a manual healing method of manipulating or adjusting vertebrae to relieve musculoskeletal pain suspected of causing problems with internal organs.

Guided imagery: a technique that guides clients to achieve a desired physical, emotional, or spiritual state by visualizing themselves in that state.

Herbal medicine: the use of plants to treat disease or improve health; also known as *botanical medicine* or *phytotherapy.*

Homeopathic medicine: a practice based on the theory that "like cures like"—that is, that substances that cause symptoms in healthy people can cure those symptoms when given in very dilute amounts.

Hypnotherapy: a technique that uses hypnosis and the power of suggestion to improve health behaviors, relieve pain, and heal.

Iridology: the study of changes in the iris of the eye and their relationships to disease.

Massage therapy: a healing method in which the therapist manually kneads muscles to reduce tension, increase blood circulation, improve joint mobility, and promote healing of injuries.

Meditation: a self-directed technique of relaxing the body and calming the mind.

Naturopathic medicine: a system that integrates traditional medicine with botanical medicine, clinical nutrition, homeopathy, acupuncture, hydrotherapy, and manipulative therapy.

Orthomolecular medicine: the use of large doses of vitamins to treat chronic disease.

Reflexology: manipulation of specific areas on the feet and hands that correspond to a particular organ or zone in the upper body.

Therapeutic touch: a technique that uses graceful, sweeping movements of the hands, a few inches from the body, to scan the patient's energy flow, replenish it when necessary, release congestion or obstruction, and generally restore order and balance.

Appendix C-2 Complementary and Alternative Medicine Research Centers and World Wide Web Sites

The Office of Alternative Medicine of the National Institutes of Health provides funding to several research centers that are designed to evaluate alternative treatments for chronic health conditions such as HIV/AIDS, cancer, addictions, asthma, stroke, and pain, among others. The research centers are listed below.*

Center for Addiction and Alternative Medicine Research University of Minnesota Medical School and Hennepin County Medical Center, Minneapolis, MN	www.winternet.com/~caamr
Complementary and Alternative Medicine Program at Stanford Stanford University, Palo Alto, CA	scrdp.stanford.edu/camps.html
Center for Complementary and Alternative Medicine Research in Asthma, Allergy and Immunology University of California at Davis, Davis, CA	www-camra.ucdavis.edu
Center for Alternative Medicine Research in Cancer University of Texas-Houston Health Science Center, Houston, TX	www.sph.uth.tmc.edu/utcam
Consortial Center for Chiropractic Research Palmer Center for Chiropractic Research, Davenport, IA	none
Center for Alternative Medicine Research Beth Israel Deaconess Medical Center and Harvard Medical School, Boston, MA	www.bidmc.harvard.edu/medicine/camr/index.html
Bastyr University AIDS Research Center Bastyr University, Bothell, WA	www.bastyr.edu/research/recruit.html
Center for the Study of Complementary and Alternative Therapies University of Virginia Health Sciences Center, Charlottesville, VA	www.med.Virginia.EDU/nursing/centers/alt-ther.html
Center for Alternative Medicine Pain Research and Evaluation University of Maryland School of Medicine, Baltimore, MD	207.123.250.14†
Center for Research in Complementary and Alternative Medicine for Stroke and Neurological Disorders Kessler Medical Rehabilitation Research and Education Corporation, West Orange, NJ, and New Jersey Medical School, Newark, NJ	www.umdnj.edu/altmdweb/web.html
Center for Complementary and Alternative Medicine Research in Women's Health Columbia University, New York, NY	cpmcnet.columbia.edu/dept/rosenthal

*Inclusion of a treatment or resource in this list does not imply endorsement by the Office of Alternative Medicine, the National Institutes of Health, or the U.S. Public Health Service.
†Site was under reconstruction in August, 1998.
Source: National Institutes of Health, Office of Alternative Medicine, Web site at altmed.od.nih.gov.

Appendix D

Comparison of National Health Care Systems

Country	Access	Costs
Canada	Canada provides health care for all its citizens through a national health insurance program operated by the provincial governments. Primary care providers are evenly distributed and specialists are concentrated in "centers of excellence."	Canada spends about 8.6% of GNP (gross national product) on health care. Doctors are paid on a fee-for-service basis with rates negotiated annually with the provincial government. Hospitals have global annual budgets set by the provincial government.
Germany*	Germany provides all its citizens with ready access to health care through a century-old system of social insurance that represents a middle ground in the spectrum of approaches Western countries have adopted to protect their populations from the economic consequences of illness.	Germany spends an average of 8.5% of GNP on health care. A formal price-fixing mechanism exists for physician fees and hospital rates. Fee increases are negotiated by not-for-profit insurance organizations and regional "doctors" payment organizations.
Japan	Japan provides access to health care for all its workers and their families through an employer-mandated health insurance program. All other citizens are covered through a national health insurance program.	Japan spends 6.7% of GNP on health care. The Health Ministry sets rates for private and public providers on a fee-for-service basis. It operates a utilization review system. Providers can, in some cases, charge more than the government rate, but this overcharge must be paid out-of-pocket.
Sweden	Sweden provides access to health care for all its citizens through an almost completely government-operated national health service program.	Sweden spends 9% of GNP on health care. 95% of doctors are employed by county councils that operate the health system locally. Hospitals are run by either local or area agencies with budgets based on the medical needs of the area.
United States	Access to health care is not a right in the U.S. Medicaid covers 17 million low-income people. Medicare covers 36 million seniors and 2 million people with disabilities. Private insurance, usually employment based, covers 147 million people. 41 million Americans are uninsured.	The U.S. spends over 14% of GNP on health care. Most providers are private with little rate setting (Medicaid, Medicare for hospital services, and a few private insurance companies). Hospital care consumes about half of health costs, with doctor services consuming another quarter.

Country	Finance	Benefits
Canada	The Canadian federal and provincial governments share the financing of health costs. The federal share comes primarily from income taxes, as does the provincial share.	The same benefits are provided as in Sweden (Sweden has the most comprehensive benefits package and is used as the standard for comparison) except that eye care is not covered and dental care is limited.
Germany*	90% of citizens receive care coverage by a system of "sickness funds" into which employees and employers are mandated by law to make matching contributions based on a fixed percentage of income (average 12.8%). Most other citizens purchase comprehensive private insurance.	All of Germany's citizens have access to a comprehensive set of benefits as in Sweden (comprehensive dental coverage is included; preventive services include visits to spas) with free choice of physicians and hospitals and virtually immediate access to all services.
Japan	About 60% of health costs in Japan come from the employment-based health program that is financed through an 8–9% payroll tax split by the employee-employer. 30% comes from local taxes, with the final 10% coming from direct patient payment.	The same benefits are provided as in Sweden except paramedical treatment, eye care, and psychiatric care are not covered and dental care is limited.
Sweden	60% of the financing for Sweden's health system is from a proportional wage tax of 13.5%. 35% of costs are covered by general federal revenues. The final 4% is paid for through direct patient fees.	Sweden covers hospital services (tests, lab work, nonelective surgery), physician services, preventive care, home care and nursing home care, prescription drugs, dental care, eye care, paramedical services, and psychiatric care.
United States	Private insurance pays 31% of health costs. Medicare, financed through a 3.5% payroll tax and premiums, pays 17% of health costs. Medicaid pays 10% and is funded jointly by federal and state taxes. Other government programs pay 14%. Direct patient payments pay 25%. Private sources pay the remaining 3%.	Private plans vary but few compare to Sweden. Medicare does not cover dental, eye and preventive care, or paramedical services and limits psychiatric, drug, and long-term care. Medicaid can have fuller coverage, but few states offer all options.

*Germany refers to the former West Germany. A formidable challenge facing the unified Germany is the transformation of East Germany's national health service into the market-oriented social insurance model of West Germany.

Source: Adapted from Comparison of national health care systems, *Nursing and Health Care* 13 (1992): 202–3. Reprinted by permission from the advocacy group, Citizen Action, 1120 19th Street, NW, Washington, D.C. 20036.

Medical Nutrition Therapy Protocols

HYPERLIPIDEMIA
Medical Nutrition Therapy Protocol

Setting: Ambulatory Care (Adult 18+ years old)
Number of sessions: 3

No. of interventions	Length of contact	Time between interventions	Cost/charge
1	60 minutes	3-4 weeks	
2	30 minutes	3-4 weeks	
3	30 minutes	as prescribed by PCP; recheck lab in 3 months	

Expected Outcomes of Medical Nutrition Therapy

Outcome assessment factors	Base-line	Evaluation of Intervention		Expected outcome	Ideal/goal value
	Intervention				
	1	2	3		
Clinical Outcomes					
• Biochemical parameters (measure < 30 days prior to nutrition session) Lipid profile (blood chol, trig, LDL-C, HDL-C)	✔		✔	Chol ↓ 20% Trig ↓ or no change LDL-C ↓ HDL-C ↑ or no change Ratio TC/HDL ↓ or no change	Chol < 200 mg/dL Fasting trig < 250 mg/dL LDL-C < 130 mg/dL (non CHD) LDL-C < 100 mg/dL (w/CHD) HDL-C > 35 mg/dL Ratio TC/HDL < 4.5
• Anthropometrics Weight, height, & BMI	✔	✔	✔	↓, ↑, or maintain as appropriate	Within reasonable body weight
• Clinical signs and symptoms	✔		✔	As appropriate: ↓ in retinal deposit ↓ shortness of breath ↓ in angina	
Behavioral Outcomes*					MNT Goal
• Food/meal planning	✔	✔	✔	• Limits foods ↑ in chol, total fat, & saturated fat • Uses monounsaturated fat as preferred fat	Fat and cholesterol consumed follow nutrition prescription, eg, < 20% total fat, 10% MUSF
• Food label reading			✔	• Accurately reads food label	
• Knowledge of soluble fiber			✔	• Increases intake of foods ↑ in soluble fiber	
• Recipe modification			✔	• Modifies recipes to ↓ total fat/saturated fat	
• Food preparation			✔	• Uses low-fat cooking techniques	
• Dining out		✔	✔	• Selects appropriately from restaurant menu	
• Simple sugar and alcohol intake	✔	✔	✔	• Limits per nutrition perscrip-tion, if applicable	
• Exercise pattern	✔	✔	✔	• Participates in aerobic activity 3x/wk, 45-min sessions	
• Smoking	✔	✔	✔	• Verbalizes importance of smoking cessation	
• Potential food/drug interactions	✔	✔	✔	• Verbalizes potential food/drug interaction	

*Session in which behavorial topics are covered may vary according to patient's readiness, skills, resources, and need for lifestyle changes.

HYPERLIPIDEMIA
Medical Nutritional Therapy Protocol

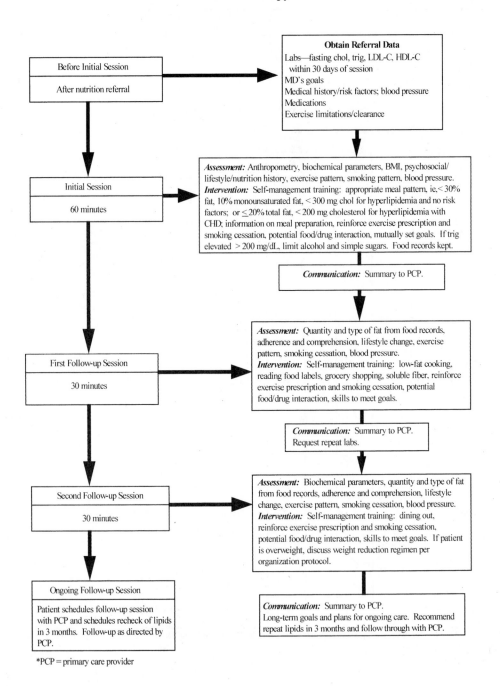

Before Initial Session

After nutrition referral

Obtain Referral Data

Labs—fasting chol, trig, LDL-C, HDL-C
  within 30 days of session
MD's goals
Medical history/risk factors; blood pressure
Medications
Exercise limitations/clearance

Initial Session

60 minutes

Assessment: Anthropometry, biochemical parameters, BMI, psychosocial/lifestyle/nutrition history, exercise pattern, smoking pattern, blood pressure.
Intervention: Self-management training: appropriate meal pattern, ie, < 30% fat, 10% monounsaturated fat, < 300 mg chol for hyperlipidemia and no risk factors; or ≤ 20% total fat, < 200 mg cholesterol for hyperlipidemia with CHD; information on meal preparation, reinforce exercise prescription and smoking cessation, potential food/drug interaction, mutually set goals. If trig elevated > 200 mg/dL, limit alcohol and simple sugars. Food records kept.

Communication: Summary to PCP.

First Follow-up Session

30 minutes

Assessment: Quantity and type of fat from food records, adherence and comprehension, lifestyle change, exercise pattern, smoking cessation, blood pressure.
Intervention: Self-management training: low-fat cooking, reading food labels, grocery shopping, soluble fiber, reinforce exercise prescription and smoking cessation, potential food/drug interaction, skills to meet goals.

Communication: Summary to PCP.
Request repeat labs.

Second Follow-up Session

30 minutes

Assessment: Biochemical parameters, quantity and type of fat from food records, adherence and comprehension, lifestyle change, exercise pattern, smoking cessation, blood pressure.
Intervention: Self-management training: dining out, reinforce exercise prescription and smoking cessation, potential food/drug interaction, skills to meet goals. If patient is overweight, discuss weight reduction regimen per organization protocol.

Ongoing Follow-up Session

Patient schedules follow-up session with PCP and schedules recheck of lipids in 3 months. Follow-up as directed by PCP.

Communication: Summary to PCP.
Long-term goals and plans for ongoing care. Recommend repeat lipids in 3 months and follow through with PCP.

*PCP = primary care provider

NUTRITION PROGRESS NOTES
Hyperlipidemia
Other diagnosis: ___HTN, CAD___

Patient's Name: _John Doe_
Medical Record #: _XX_
DOB: _9/4/35_ Gender: _M_
Ethnic background (optional): _Cau_
Referring physician: _Hart_

Outcomes of Medical Nutrition Therapy (MNT)

Expected outcome	Intervention provided to meet goal (Intervention = self-management training plus patient verbalizes/demonstrates)			Goal reached (Check indicates goal reached)		
Date / Session	5/10/95 1 (60 min)	6/7/95 2 (30 min)	7/5/95 3 (30 min)	5/10/95 1	6/7/95 2	7/5/95 3
Clinical Outcomes				Value		Value
Chol ↓ 20%				246 mg/dL		✔ 193 mg/dL
Trig ↓ or no change				345 mg/dL		✔ 142 mg/dL
LDL-C ↓				135 mg/dL		134 mg/dL
HDL-C ↑ or no change				42 mg/dL		34 mg/dL
Ratio TC/HDL ↓ or no change				5.9		5.7
Blood pressure				134/90	110/78	120/78
Patient to lose _20_ lb in _3_ months Weight:				182 lb	170 lb	163 lb
Height (initial visit): _69"_ BMI: _27_						
As appropriate: ↓ retinal deposit ↓ shortness of breath ↓ in angina, ↓ Na						
MNT Goal				2400 cal	1200 cal	✔ 1650 cal
• Fat and cholesterol consumed follows nutrition prescription	✔	✔	✔	170 g fat	45 g fat	✔ 50 g fat
				40 % fat	34 % fat	✔ 27 % fat
Behavioral Outcomes				5 gm Na	4 gm Na	3 gm Na
• Limits foods ↑ in chol, total fat, & saturated fat ↓ Na	✔	✔	✔			✔
• Uses allowed amount of monounsaturated fat		✔	✔	0 svg/day	0 svg/day	0 svg/day
• Accurately reads food label			✔			✔
• Increases intake of foods ↑ in soluble fiber			✔	0 svg/day	1 svg/day	1 svg/day
• Modifies recipes to ↓ total fat/saturated fat ↓ Na		✔	✔			∅ NA
• Uses low-fat cooking techniques ↓ Na		✔	✔			∅ NA
• Selects appropriately from restaurant menu			✔			✔
• Limits alcohol, if necessary	✔	✔	✔	NA svg/day	NA svg/day	NA svg/day
• Limits simple carbohydrates, if necessary	✔	✔	✔	5-6 d	2 wk	2 wk
• Participates in aerobic activity 3x/wk, 45-minute sessions	✔	✔	✔	5 x/wk / 15-20 min	6 x/wk / 20 min	5 x/wk / 40 min
• Verbalizes importance of smoking cessation	✔			NA ppd	NA ppd	NA ppd
• Verbalizes potential food/drug interaction						
Drug _Coumadin_	✔		✔	3 mg dose	3 mg dose	3 mg dose
Atenolol	✔		✔	100 mg dose	100 mg dose	100 mg dose
Vasotec	✔		✔	20 mg dose	10 mg dose	10 mg dose
Overall Compliance Potential						
• Comprehension				E Ⓖ P	E Ⓖ P	Ⓔ G P
• Receptivity				Ⓔ G P	Ⓔ G P	Ⓔ G P
• Adherence				E Ⓖ P	E Ⓖ P	Ⓔ G P

Comments 5/10/95 Eager to make changes. ↑ saturated fat intake, contributes ↑ 7 Chol. will need cooperation

of family to make changes.

R7C 1 month   Anne Brown, MS, RD   **RD**

Signature/Date Session 1

Comments 6/7/95 Feels great, pleased that Vasotec dose was reduced in half and BP still WNL.

R7C 1 month   Anne Brown, MS, RD   **RD**

Signature/Date Session 2

Comments 7/5/95 Excellent progress with fat modification; needs additional work with ↓ Na changes. Improved

clinical outcomes - will request 2 more visits.

R7C 1 month   Anne Brown, MS, RD   **RD**

Signature/Date Session 3

Appendix F

National Nutrition Monitoring
and Related Research Program

TABLE F-1 Sources of Data from the Five Component Areas of the National Nutrition Monitoring and Related Research Program*

Component Area and Survey or Study	Sponsoring Agency (Department)	Date	Population	Data Collected
Nutritional Status and Nutrition-Related Health Measurements				
Survey of Army Female Basic Trainees	ARIEM (DOD)	1993	Volunteers from females entering basic training at Fort Jackson, SC	Nutrition knowledge and beliefs, eating habits, and food attitudes; 7-day dietary intake; bone mineral mass, body fat, and biochemical analyses of blood
Nutritional and Physiological Assessment of the Special Forces Assessment and Selection Course	ARIEM (DOD)	1993	Special Operations Forces male soldiers	Body fat and bone mineral mass; biochemical analyses on blood drawn prior to conditions of deficient energy intakes
Ranger School Nutrition Intervention Study	ARIEM (DOD)	1992	Special Operations Forces male soldiers	Body fat and bone mineral mass; biochemical analyses on blood drawn prior to conditions of deficient energy intakes
Pregnancy Nutrition Surveillance System (PNSS)	NCCDPHP, CDC (HHS)	1992	Convenience population of low-income, high-risk pregnant women	Demographic information; pregravid-weight status, maternal weight gain during pregnancy, anemia (hemoglobin, hematocrit), pregnancy behavioral risk factors (smoking and drinking), birthweight, breastfeeding, and formula-feeding data
Pediatric Nutrition Surveillance System (PedNSS), see Chapter 4				
National Vital Registration System—Natality Statistics	NCHS, CDC (HHS)	1991	All live births for total U.S. population	Infant birthweight, gestational age, Apgar score, anemia, fetal alcohol syndrome, hyaline membrane disease, congenital anomalies; maternal weight gain during pregnancy, alcohol and tobacco use, and anemia, diabetes, and hypertension during pregnancy

Nutritional Status and Nutrition-Related Health Measurements—continued

Component Area and Survey or Study	Sponsoring Agency (Department)	Date	Population	Data Collected
Longitudinal Followup to the 1988 National Maternal and Infant Health Survey (NMIHS)	NCHS, CDC (HHS)	1991	Mothers of 3-year-olds who participated in the 1988 NMIHS, pediatricians, and hospitals	Use of vitamin and mineral supplements, WIC participation, and growth and hematological measurements from birth to 3 years of age
Assessment of Nutritional Status and Immune Function During the Ranger Training Course	ARIEM (DOD)	1991	Special Operations Forces male soldiers	Body fat and bone mineral mass; biochemical analyses of blood drawn prior to conditions of deficient energy intakes
Survey of Heights and Weights of American Indian School Children	IHS and CDC (HHS)	1990–91	American Indian schoolchildren 5–18 years of age	Height, weight, and body mass index
National Ambulatory Medical Care Survey (NAMCS)	NCHS, CDC (HHS)	1989–91	Visits to office-based physicians and hospital emergency and outpatient departments of nonfederal, short-stay general and specialty hospitals	Patients' symptoms and demographic characteristics, physicians' diagnoses, drugs prescribed, and referrals. Nutrition-related information is collected, including physician-reported hypertension, hypercholesterolemia, and obesity and whether the physician ordered or recommended counseling services for diet, exercise, cholesterol reduction, and weight reduction.
Fourth National Health and Nutrition Examination Survey (NHANES IV); see Chapter 4				
Third National Health and Nutrition Examination Survey (NHANES III); see Chapter 4				
Hispanic Health and Nutrition Examination Survey (HHANES); see Chapter 4				

Continued

Component Area and Survey or Study	Sponsoring Agency (Department)	Date	Population	Data Collected
Nutritional Status and Nutrition-Related Health Measurements—continued				
Second National Health and Nutrition Examination Survey (NHANES II); see Chapter 4				
First National Health and Nutrition Examination Survey (NHANES I); see Chapter 4				
First National Health Examination Survey (NHES I)	NCHS (HHS)	1960–62	Civilian, noninstitutionalized population of the conterminous United States; 18–79 years of age	Socioeconomic and demographic information, biochemical analyses of blood, physical examination, and body measurements
Food and Nutrient Consumption				
Survey of Army Female Basic Trainees	ARIEM (DOD)	1993	Volunteers from females entering basic training at Fort Jackson, SC	Nutrition knowledge and beliefs, eating habits, and food attitudes; 7-day dietary intake; bone mineral mass, body fat, and biochemical analyses of blood
School Nutrition Dietary Assessment Study (SNDA)	FCS (USDA)	1992	325 nationally representative schools in the 48 conterminous states and the District of Columbia and children and adolescents in grades 1–12 who attend those schools	For schools: lists of all foods served as part of a USDA meal (or all foods served if the school did not participate in the USDA meal programs), by meal and day of the week; complete descriptions of foods, recipes, and labels for prepared items; estimates of quantity served; à la carte food items; and food and beverage items in vending machines. For individuals: foods consumed and 24-hour recall by students (grades 3–12) or students and parents (grades 1–2).
Consumer Expenditure Survey (CES)	BLS (DOL)	1980–92	Civilian, noninstitutionalized population and a portion of the institutionalized population in the United States	No direct nutrition-related indicators collected. Weekly food expenditures collected at a detailed item level in the Diary Survey. Food Stamp Program participation data collected in the Interview Survey.

Continued

Food and Nutrient Consumption—continued

Component Area and Survey or Study	Sponsoring Agency (Department)	Date	Population	Data Collected
5 a Day for Better Health Baseline Survey; see Chapter 4				
Longitudinal Followup to the 1988 National Maternal and Infant Health Survey (NMIHS)	NCHS, CDC (HHS)	1991	Mothers of 3-year-olds who participated in the 1988 NMIHS, pediatricians, and hospitals	Use of vitamin and mineral supplements, WIC participation, and growth and hematological measurements from birth to 3 years of age
Continuing Survey of Food Intakes by Individuals (CSFII); see Chapter 4				
Strong Heart Dietary Survey	IHS and CDC (HHS)	1989–91	American Indian adults 45–74 years of age residing in South Dakota, Arizona, and Oklahoma areas	Food intake by 24-hour recall and quantitative food frequency
Third National Health and Nutrition Examination Survey (NHANES III)	NCHS, CDC (HHS)	1988–94	Civilian, noninstitutionalized population 2 months of age and older. Oversampling of non-Hispanic blacks and Mexican Americans, children <6 years of age, and adults aged ≥ 60 years	Dietary intake (one 24-hour recall and food frequency), socioeconomic and demographic information, biochemical analyses of blood and urine, physical examination, body measurements, blood pressure measurements, bone densitometry, dietary and health behaviors, and health conditions. Two additional 24-hour recalls for participants 50 years of age and older.
Total Diet Study (TDS); see Chapter 4				
Nationwide Food Consumption Survey (NFCS); see Chapter 4				

Component Area and Survey or Study	Sponsoring Agency (Department)	Date	Population	Data Collected
Food and Nutrient Consumption—continued				
National Health Interview Survey on Vitamin and Mineral Supplements	NCHS and FDA (HHS)	1986	Civilian, noninstitutionalized children 2–6 years of age and adults 18 years of age and older in the United States	Prevalence of use; sociodemographic characteristics of the users; intakes of 24 nutrients from supplements (12 vitamins and 12 minerals); potency, form, and the units used to declare potency; specific chemical compounds for mineral supplements; number of supplements taken, duration of use, and whether the supplement was prescribed
Continuing Survey of Food Intakes by Individuals (CSFII); see Chapter 4				
Hispanic Health and Nutrition Examination Survey (HHANES)	NCHS (HHS)	1982–84	Civilian, noninstitutionalized Mexican Americans in five southwestern states, Cuban Americans in Dade County, FL, and Puerto Ricans in metropolitan New York City; 6 months–74 years of age	Dietary intake (one 24-hour recall), food frequency, socioeconomic and demographic information, dietary and health behaviors, biochemical analyses of blood and urine, physical examination, body measurements, and health conditions
Vitamin and Mineral Supplement Use Survey; see Chapter 4				
Second National Health and Nutrition Examination Survey (NHANES II)	NCHS (HHS)	1976–80	Civilian, noninstitutionalized population of the United States; 6 months–74 years of age	Dietary intake (one 24-hour recall), food frequency, socioeconomic and demographic information, biochemical analyses of blood and urine, physical examination, and body measurements
Nationwide Food Consumption Survey (NFCS); see Chapter 4				
First National Health and Nutrition Examination Survey (NHANES I)	NCHS (HHS)	1971–74	Civilian, noninstitutionalized population of the conterminous United States, 1–74 years of age	Dietary intake (one 24-hour recall), food frequency, socioeconomic and demographic information, biochemical analyses of blood and urine, physical examination, and body measurements

Component Area and Survey or Study	Sponsoring Agency (Department)	Date	Population	Data Collected
Food and Nutrient Consumption—continued				
First National Health Examination Survey (NHES I)	NCHS (HHS)	1960–62	Civilian, noninstitutionalized population of the conterminous United States; 18–79 years of age	Socioeconomic and demographic information, biochemical analyses of blood, physical examination, and body measurements
Knowledge, Attitudes, and Behavior Assessments				
Survey of Army Female Basic Trainees	ARIEM (DOD)	1993	Volunteers from females entering basic training at Fort Jackson, SC	Nutrition knowledge and beliefs, eating habits, and food attitudes; 7-day dietary intake; bone mineral mass, body fat, and biochemical analyses of blood
Behavioral Risk Factor Surveillance System (BRFSS); see Chapter 4				
5 a Day for Better Health Baseline Survey	NCI (HHS)	1991	Adults 18 years of age and older in the United States	Demographic information; fruit and vegetable intake; knowledge, attitudes, and behaviors regarding fruit and vegetable intake
Weight Loss Practices Survey (WLPS); see Chapter 4				
Youth Risk Behavior Survey (YRBS); see Chapter 4				
Nutrition Label Format Studies	FDA (HHS)	1990, 1991	Primary food shoppers 18 years of age and older	Demographic information; objective performance measures and preference measures for the various formats for revised nutrition labels; frequency of food label reading; health status of household members with respect to heart disease, diabetes, high blood pressure, stroke, and cancer; household members' dieting practices with respect to weight control and intake of sodium, cholesterol, and fat
Diet and Health Knowledge Survey (DHKS); see Chapter 4				

Continued

Knowledge, Attitudes, and Behavior Assessments—continued

Component Area and Survey or Study	Sponsoring Agency (Department)	Date	Population	Data Collected
Continuing Survey of Food Intakes by Individuals (CSFII)	HNIS (USDA)	1989–91	Individuals in households in the 48 conterminous states. The survey was composed of two separate samples: households with incomes at any level (basic sample) and households with incomes ≤130% of the poverty thresholds (low-income sample).	One-day and 3-day food and nutrient intakes by individuals of all ages, names and times of eating occasions, and sources of food obtained and eaten away from home. Data collected over 3 consecutive days by use of a 1-day recall and a 2-day record. Intakes are available for 28 nutrients and food components.
Third National Health and Nutrition Examination Survey (NHANES III)	NCHS, CDC (HHS)	1988–94	Civilian, noninstitutionalized population 2 months of age and older. Oversampling of non-Hispanic blacks and Mexican Americans, children <6 years of age, and adults aged ≥60 years	Dietary intake (one 24-hour recall and food frequency), socioeconomic and demographic information, biochemical analyses of blood and urine, physical examination, body measurements, blood pressure measurements, bone densitometry, dietary and health behaviors, and health conditions. Two additional 24-hour recalls for participants 50 years of age and older.
Health and Diet Survey (HDS)	FDA and NHLBI[+] (HHS)	1983, 1986, 1988, 1990	Civilian, noninstitutionalized adults 18 years of age and older in the 48 conterminous states	Demographic information; data on awareness, beliefs, attitudes, knowledge, and reported behaviors regarding food, nutrition, and health; self-reported height, weight, health history, and health status by telephone interview
Nationwide Food Consumption Survey (NFCS)	HNIS (USDA)	1987–88	Households in the 48 conterminous states and individuals residing in those households. The survey was composed of two samples: a basic sample of all households and a low-income sample of households with incomes ≤130% of the poverty threshold.	For households; quantity (pounds), money value (dollars), and nutritive value of food used. For individuals: 1-day and 3-day food and nutrient intakes by individuals of all ages, names and times of eating occasions, and sources of food obtained and eaten away from home. Data collected over 3 consecutive days using a 1-day recall and a 2-day record. Intakes are available for 28 nutrients and food components.
National Survey of Family Growth (NSFG)	NCHS, CDC (HHS)	1982, 1988	Females 15–44 years of age	Demographic information; birthweight, breastfeeding, and prenatal care

Component Area and Survey or Study	Sponsoring Agency (Department)	Date	Population	Data Collected
Food Composition and Nutrient Data Bases				
National Nutrient Data Bank (NNDB); see Chapter 4				
USDA Nutrient Data Base for Standard Reference; see Chapter 4				
USDA Survey Nutrient Data Base; see Chapter 4				
Food Label and Package Survey (FLAPS); see Chapter 4				
Food Supply Determinations				
U.S. Food Supply Series; see Chapter 4				

*Within each component area, entries are listed in reverse chronological order. Some surveys and studies are listed more than once because their data are used in more than one component area. ARIEM, Army Research Institute of Environmental Medicine; DOD, Department of Defense; NCCDPHP, National Center for Chronic Disease Prevention and Health Promotion; CDC, Centers for Disease Control and Prevention; HHS, Department of Health and Human Services; NCHS, National Center for Health Statistics; WIC, Special Supplemental Nutrition Program for Women, Infants, and Children; IHS, Indian Health Service; FCS, Food and Consumer Service; USDA, U.S. Department of Agriculture; BLS, Bureau of Labor Statistics; DOL, Department of Labor; NCI, National Cancer Institute; HNIS, Human Nutrition Information Service; FDA, Food and Drug Administration; ARS, Agricultural Research Service; NHLBI, National Heart, Lung, and Blood Institute; CNPP, Center for Nutrition Policy and Promotion; ERS, Economic Research Service; NA, not applicable.

†Cosponsored with NHLBI in 1983, 1986, and 1990.

Source: Federation of American Societies for Experimental Biology, Life Sciences Research Office, prepared for the Interagency Board for Nutrition Monitoring and Related Research, *Third Report on Nutrition Monitoring in the United States*, Volume 1 (Washington, D.C.: U.S. Government Printing Office, 1995), pp. 7–15.

TABLE F-2 *Selected Blood Assessments for Adults Aged 20+ Years in the Third National Health and Nutrition Examination Survey (NHANES III)*

Whole Blood

CBC/RDW*
Platelets
Lead
Protoporphyrin
Red blood cell folate
Glycated hemoglobin

Serum

Nutrients/proteins	**Biochemistry profile**
Iron	Total carbon dioxide
Total iron binding capacity	Blood urea nitrogen
Ferritin	Total bilirubin
Folate	Alkaline phosphatase
Apolipoprotein A_1, B	Total cholesterol
Total cholesterol	AST (SGOT)§
HDL cholesterol†	ALT (SGPT)‖
Triglycerides	LDH#
Lp(a)‡	GGT**
C-reactive protein	Total protein
Rheumatoid factor	Albumin
Vitamin A (retinol)	Creatinine
Carotenoids	Glucose
Retinyl esters	Calcium
Vitamin E	Chloride
Vitamin B_{12}	Uric acid
Methyl malonic acid	Phosphorus
Homocysteine	Sodium
Vitamin C	Potassium
Vitamin D (25-hydroxyvitamin D_3)	
Total/ionized calcium	
Selenium	
Thyroxine (T_4)	

*CBC/RDW = Complete blood count/red cell distribution width; includes hematocrit, hemoglobin, red and white cell counts, mean corpuscular volume, mean corpuscular hemoglobin, and mean corpuscular hemoglobin concentration.
†HDL = High density lipoprotein.
‡Lp(a) = Lipoprotein(a).
§AST (SGOT) = Aspartate aminotransferase (serum glutamic oxaloacetic transaminase).
‖ALT (SGPT) = Alanine aminotransferase (serum glutamic pyruvic transaminase).
#LDH = Lactic dehydrogenase.
**GGT = Gamma glutamyl transferase.

Source: Centers for Disease Control and Prevention, National Center for Health Statistics, Plan and operation of the Third National Health and Nutrition Examination Survey, 1988–94, *Vital and Health Statistics,* Series 1, No. 32 (1994), pp. 48–49.

Appendix G

Canadian Dietary Guidelines and Nutrition Recommendations from WHO

Appendix G-1 Canada's Guidelines for Healthy Eating

Canada's Guidelines for Healthy Eating* were developed by the Communications/Implementation Committee as the key nutrition messages to be communicated to healthy Canadians over two years of age. The guidelines encourage people to:

- Enjoy a VARIETY of foods.
- Emphasize cereals, breads, other grain products, vegetables and fruits.
- Choose low-fat dairy products, lean meats, and foods prepared with little or no fat.
- Achieve and maintain a healthy body weight by enjoying regular physical activity and healthy eating.
- Limit salt, alcohol and caffeine.

Nutrition Recommendations for Canadians†

The Nutrition Recommendations for Canadians examines the relationships linking nutrition and disease. The intent is to make recommendations that will supply enough nutrients, while reducing the risk of chronic disease. The Scientific Review Committee adopted the following key statements as the Nutrition Recommendations for Canadians.

- The Canadian diet should provide energy consistent with the maintenance of body weight within the recommended range.
- The Canadian diet should include essential nutrients in amounts specified in the RNI (see Appendix G-2).

- The Canadian diet should include no more than 30 percent of energy as fat and no more than 10 percent as saturated fat.
- The Canadian diet should provide 55 percent of energy as carbohydrates from a variety of sources.
- The sodium content of the Canadian diet should be reduced.
- The Canadian diet should include no more than 5 percent of total energy as alcohol, or two drinks daily, whichever is less.
- The Canadian diet should contain no more caffeine than the equivalent of four cups of regular coffee per day.
- Community water supplies containing less than 1 milligram per liter should be fluoridated to that level.

*Health Canada, Communications/Implementation Committee, *Action Towards Healthy Eating … Canada's Guidelines for Healthy Eating and Recommended Strategies for Implementation*, Cat. No. H39-166/1990E (Ottawa, ON: Minister of Supply and Services Canada, 1990), p. 5.

†Adapted from Health Canada, Communications/ Implementation Committee, *Action Towards Healthy Eating … Canada's Guildelines for Healthy Eating and Recommended Strategies for Implementation*, Cat. No. H39-166/1990E (Ottawa, ON: Minister of Supply and Services Canada, 1990), p. 21.

Appendix G-2 Canada's Recommended Nutrient Intakes (RNI)

TABLE G2-1 *Recommended Nutrient Intakes for Canadians, 1990*

| | | | | Fat-Soluble Vitamins | | |
Age	Sex	Weight (kg)	Protein (g/day)*	Vitamin A (RE/day)[†]	Vitamin D (µg/day)[‡]	Vitamin E (mg/day)[§]
Infants (months)						
0–4	Both	6	12[#]	400	10	3
5–12	Both	9	12	400	10	3
Children and adults (years)						
1	Both	11	13	400	10	3
2–3	Both	14	16	400	5	4
4–6	Both	18	19	500	5	5
7–9	M	25	26	700	2.5	7
	F	25	26	700	2.5	6
10–12	M	34	34	800	2.5	8
	F	36	36	800	5	7
13–15	M	50	49	900	5	9
	F	48	46	800	5	7
16–18	M	62	58	1000	5	10
	F	53	47	800	2.5	7
19–24	M	71	61	1000	2.5	10
	F	58	50	800	2.5	7
25–49	M	74	64	1000	2.5	9
	F	59	51	800	2.5	6
50–74	M	73	63	1000	5	7
	F	63	54	800	5	6
75+	M	69	59	1000	5	6
	F	64	55	800	5	5
Pregnancy (additional amount needed)						
1st trimester			5	0	2.5	2
2nd trimester			20	0	2.5	2
3rd trimester			24	0	2.5	2
Lactation (additional amount needed)			22	400	2.5	3

Note: Recommended intakes of energy and certain nutrients are not listed in this table because of the nature of the variables upon which they are based. The figures for energy are estimates of average requirements for expected patterns of activity. For nutrients not shown, the following amounts are recommended based on at least 2000 kcalories per day and body weights as given: thiamin, 0.4 milligrams per 1000 kcalories (0.48 milligrams/5000 kilojoules); riboflavin, 0.5 milligrams per 1000 kcalories (0.6 milligrams/5000 kilojoules); niacin, 7.2 niacin equivalents per 1000 kcalories (8.6 niacin equivalents/5000 kilojoules); vitamin B_6, 15 micrograms, as pyridoxine, per gram of protein. Recommended intakes during periods of growth are taken as appropriate for individuals representative of the midpoint in each age group. All recommended intakes are designed to cover individual variations in essentially all of a healthy population subsisting upon a variety of common foods available in Canada.

*The primary units are expressed per kilogram of body weight. The figures shown here are examples.
[†]One retinol equivalent (RE) corresponds to the biological activity of 1 microgram of retinol, 6 micrograms of beta-carotene, or 12 micrograms of other carotenes.
[‡]Expressed as cholecalciferol or ergocalciferol.
[§]Expressed as δ-α-tocopherol equivalents, relative to whch ß- and γ-tocopherol and α-tocotrienol have activities of 0.5, 0.1, and 0.3, respectively.

TABLE G2-1 *Recommended Nutrient Intakes for Canadians, 1990—Continued*

Water-Soluble Vitamins			Minerals					
Vitamin C (mg/day)[II]	Folate (µg/day)	Vitamin B$_{12}$ (µg/day)	Calcium (mg/day)	Phosphorus (mg/day)	Magnesium (mg/day)	Iron (mg/day)	Iodine (µg/day)	Zinc (mg/day)
20	25	0.3	250	150	20	0.3[**]	30	2[††]
20	40	0.4	400	200	32	7	40	3
20	40	0.5	500	300	40	6	55	4
20	50	0.6	550	350	50	6	65	4
25	70	0.8	600	400	65	8	85	5
25	90	1.0	700	500	100	8	110	7
25	90	1.0	700	500	100	8	95	7
25	120	1.0	900	700	130	8	125	9
25	130	1.0	1100	800	135	8	110	9
30	175	1.0	1100	900	185	10	160	12
30	170	1.0	1000	850	180	13	160	9
40	220	1.0	900	1000	230	10	160	12
30	190	1.0	700	850	200	12	160	9
40	220	1.0	800	1000	240	9	160	12
30	180	1.0	700	850	200	13	160	9
40	230	1.0	800	1000	250	9	160	12
30	185	1.0	700	850	200	13[‡‡]	160	9
40	230	1.0	800	1000	250	9	160	12
30	195	1.0	800	850	210	8	160	9
40	215	1.0	800	1000	230	9	160	12
30	200	1.0	800	850	210	8	160	9
0	200	0.2	500	200	15	0	25	6
10	200	0.2	500	200	45	5	25	6
10	200	0.2	500	200	45	10	25	6
25	100	0.2	500	200	65	0	50	6

[II]Cigarette smokers should increase intake by 50 percent.

[#]The assumption is made that the protein is from breast milk or is of the same biological value as that of breast milk, and that between 3 and 9 months, adjustment for the quality of the protein is made.

[**]Based on the assumption that breast milk is the source of iron.

[††]Based on the assumption that breast milk is the source of zinc.

[‡‡]After menopause, the recommended intake is 8 milligrams per day.

Source: Health Canada, *Nutrition Recommendations: The Report of the Scientific Review Committee* (Ottawa: Canadian Government Publishing Centre, 1990), Table 20, p. 204. Reproduced with the permission of the Minister of Public Works and Government Services Canada, 1998.

TABLE G2-2 *Average Energy Requirements for Canadians*

Age	Sex	Average Height (cm)	Average Weight (kg)	Requirements* (kcal/kg†)	(MJ/kg)†	(kcal/day)	(MJ/day)	(kcal/cm)	(MJ/cm)
Infants (months)									
0–2	Both	55	4.5	120–100	0.50–0.42	500	2.0	9	0.04
3–5	Both	63	7.0	100–95	0.42–0.40	700	2.8	11	0.05
6–8	Both	69	8.5	95–97	0.40–0.41	800	3.4	11.5	0.05
9–11	Both	73	9.5	97–99	0.41	950	3.8	12.5	0.05
Children and adults (years)									
1	Both	82	11	101	0.42	1100	4.8	13.5	0.06
2–3	Both	95	14	94	0.39	1300	5.6	13.5	0.06
4–6	Both	107	18	100	0.42	1800	7.6	17	0.07
7–9	M	126	25	88	0.37	2200	9.2	17.5	0.07
	F	125	25	76	0.32	1900	8.0	15	0.06
10–12	M	141	34	73	0.30	2500	10.4	17.5	0.07
	F	143	36	61	0.25	2200	9.2	15.5	0.06
13–15	M	159	50	57	0.24	2800	12.0	17.5	0.07
	F	157	48	46	0.19	2200	9.2	14	0.06
16–18	M	172	62	51	0.21	3200	13.2	18.5	0.08
	F	160	53	40	0.17	2100	8.8	13	0.05
19–24	M	175	71	42	0.18	3000	12.6		
	F	160	58	36	0.15	2100	8.8		
25–49	M	172	74	36	0.15	2700	11.3		
	F	160	59	32	0.13	1900	8.0		
50–74	M	170	73	31	0.13	2300	9.7		
	F	158	63	29	0.12	1800	7.6		
75+	M	168	69	29	0.12	2000	8.4		
	F	155	64	23	0.10	1500	6.3		

*Requirements can be expected to vary within a range of ±30 percent.
†First and last figures are averages at the beginning and end of the three-month period.

Source: Health Canada, *Nutrition Recommendations: The Report of the Scientific Review Committee* (Ottawa: Canadian Government Publishing Centre, 1990). Tables 5 and 6, pp. 25, 27. Reproduced with the permission of the Minister of Public Works and Government Services Canada, 1998.

Appendix G-3 Canada's Food Guide to Healthy Eating

The 1992 *Canada's Food Guide to Healthy Eating* gives consumers detailed information for selecting foods to meet *Canada's Guidelines for Healthy Eating* (1990). The *Food Guide* was designed to meet the nutritional needs of all Canadians four years of age and older and takes a total diet approach, rather than emphasizing a single food, meal, or day's meals and snacks.

The rainbow side of the Food Guide shows the four food groups with their revised names and pictorial examples of foods in each group. Key statements direct consumers about selecting foods generally from all the groups, and more specifically within each group. The bar side shows the number of servings recommended for each group, using a range of servings instead of a single minimum number. Other notable changes include the number of servings for some food groups and the size of servings for some foods.

Health and Welfare
Canada

Santé et Bien-être social
Canada

Enjoy a variety
of foods from each
group every day.

Choose lower-
fat foods
more often.

Grain Products
Choose whole grain
and enriched
products more
often.

Vegetables & Fruit
Choose dark green and
orange vegetables and
orange fruit more often.

Milk Products
Choose lower-fat
milk products more
often.

Meat & Alternatives
Choose leaner meats,
poultry and fish, as well
as dried peas, beans and
lentils more often.

Different People Need Different Amounts of Food

The amount of food you need every day from the 4 food groups and other foods depends on your age, body size, activity level, whether you are male or female and if you are pregnant or breast-feeding. That's why the Food Guide gives a lower and higher number of servings for each food group. For example, young children can choose the lower number of servings, while male teenagers can go to the higher number. Most other people can choose servings somewhere in between.

Grain Products
5–12
SERVINGS PER DAY

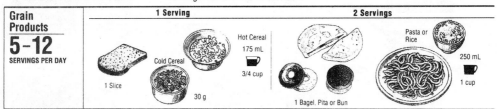

1 Serving

1 Slice

Cold Cereal — 30 g

Hot Cereal
175 mL
3/4 cup

2 Servings

1 Bagel, Pita or Bun

Pasta or Rice
250 mL
1 cup

Vegetables & Fruit
5–10
SERVINGS PER DAY

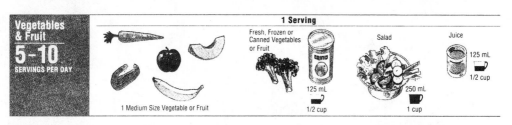

1 Serving

1 Medium Size Vegetable or Fruit

Fresh, Frozen or Canned Vegetables or Fruit
125 mL
1/2 cup

Salad
250 mL
1 cup

Juice
125 mL
1/2 cup

Milk Products
SERVINGS PER DAY
Children 4–9 years: 2–3
Youth 10–16 years: 3–4
Adults: 2–4
Pregnant & Breast-feeding
Women: 3–4

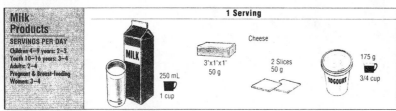

1 Serving

MILK
250 mL
1 cup

Cheese
3"x1"x1"
50 g

2 Slices
50 g

YOGOURT
175 g
3/4 cup

Other Foods

Taste and enjoyment can also come from other foods and beverages that are not part of the 4 food groups. Some of these foods are higher in fat or Calories, so use these foods in moderation.

Meat & Alternatives
2–3
SERVINGS PER DAY

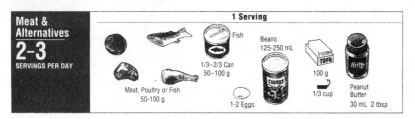

1 Serving

Meat, Poultry or Fish
50-100 g

Fish
1/3–2/3 Can
50–100 g

1-2 Eggs

Beans
125-250 mL
1/3 cup

TOFU
100 g

Peanut Butter
30 mL 2 tbsp

Enjoy eating well, being active and feeling good about yourself. That's VITALITÉ

Appendix G-4 Nutrition Recommendations from WHO

Like the Committee on Diet and Health in the United States, the World Health Organization (WHO) has also assessed the relationships between diet and the development of chronic diseases.* Its recommendations are expressed in average daily ranges that represent the lower and upper limits.

- Total energy: sufficient to support normal growth, physical activity, and body weight (body mass index = 20 to 22)

- Total fat: 15 to 30 percent of total energy
  Saturated fatty acids: 0 to 10 percent total energy
  Polyunsaturated fatty acids: 3 to 7 percent total energy
  Dietary cholesterol: 0 to 300 milligrams per day
- Total carbohydrate: 55 to 75 percent total energy
  Complex carbohydrates: 50 to 75 percent total energy
  Dietary fiber: 27 to 40 grams per day
  Refined sugars: 0 to 10 percent total energy
- Protein: 10 to 15 percent total energy
- Salt: upper limit of 6 grams/day (no lower limit set)

*Diet, Nutrition, and the Prevention of Chronic Diseases. Report of a WHO Study Group, WHO Technical Report Series, No. 797 (Geneva: World Health Organization, 1990), p. 108.

Appendix H

Community Nutrition Resources: Clearinghouses, Information Centers, and Data Archives

Clearinghouses and Information Centers

National Clearinghouse for Bilingual Education
The George Washington University Center for
the Study of Language & Education
2011 Eye Street, NW, Suite 200
Washington, D.C. 20006
Voice: (202) 467-0867
Fax: (202) 467-4283
Email: askncbe@ncbe.gwu.edu
Web site: www.ncbe.gwu.edu

National Clearinghouse for Alcohol and Drug
Information (NCADI)
P.O. Box 2345
Rockville, MD 20847-2345
Voice: (301) 468-2600
Fax: (301) 468-6433
Web site: www.health.org

National Heart, Lung, and Blood Institute (NHLBI)
NHLBI Information Center
P.O. Box 30105
Bethesda, MD 20824-0105
Fax: (301) 251-1223
Email: NHLBIIC@dgsys.com

National Clearinghouse on Families & Youth
P.O. Box 13505
Silver Spring, MD 20911-3505
Voice: (301) 608-8098
Fax: (301) 608-8721
Web site: www.ncfy.com

National Maternal and Child Health Clearinghouse
2070 Chain Bridge Road, Suite 450
Vienna, VA 22182-2536
Voice: (703) 356-1964
Fax: (703) 821-2098
Email: nmchc@circsol.com
Web site: www.circsol.com/mch

Office of Alternative Medicine (OAM) Clearinghouse
P.O. Box 8218
Silver Spring, MD 20907-8218
Toll Free Number: 1-888-644-6226
Fax: (301) 495-4957
Web site: altmed.od.nih.gov/oam/clearinghouse

Office of Cancer Communications
National Cancer Institute
Public Inquiries Section
Building #31, Room #10A-16
Bethesda, MD 20892-2580
Toll Free Number: 1-800-4-CANCER
Voice: (301) 496-5583
Fax: (301) 402-5874

Superintendent of Documents
U.S. Government Printing Office
To order by mail:
P.O. Box 371954
Pittsburgh, PA 15250-7954
Voice: (202) 512-1800
Fax: (202) 512-2250
Web site: www.gpo.gov

Data Archives

Information about population survey data, including data related to nutrition, can be obtained from both the government and private sectors. Several organizations publish catalogs of the data they collect and/or store. Each organization sets it own priorities, terms for accessing the data, and charge for obtaining the data and supporting documentation. A few of these organizations are listed here.

Centers for Disease Control and Prevention (CDC)
    Wonder Database
Web site: wonder.cdc.gov

Combined Health Information Data Base (CHID)*
Web site: chid.nih.gov

Louis Harris Data Center
Contact: David Sheaves
CB #3355, Manning Hall
University of North Carolina
Chapel Hill, NC 27599-3355
Voice: (919) 966-3348
Email: david_sheaves@unc.edu

Institute for Research in Social Science
CB #3355, Manning Hall
University of North Carolina
Chapel Hill, NC 27599-3355
Voice: (919) 962-3061
Web site: www.unc.edu/depts/irss/guide.htm

Inter-university Consortium for Political and Social
    Research
Institute for Social Research
University of Michigan
426 Thompson
Ann Arbor, MI 48104-2321
Voice: (734) 764-8363
Fax: (734) 647-4575
Email: ISR@mail.isr.umich.edu
Web site: www.isr.umich.edu

National Center for Health Statistics
Catalog of Electronic Products
Web site: www.cdc.gov/nchswww

National Technical Information Service (NTIS)
Contact: Document Library
5285 Port Royal Road
Springfield, VA 22161
Voice: (703) 487-4650
Publication orders: 1-800-553-6847

*A free bibliographic database of more than 101,000 entries, including teaching guides, audiotapes, videotapes, booklets, fact sheets, newsletters, journal articles, book chapters, and posters, combining the resources of the National Institutes of Health, Centers for Disease Control and Prevention, and other agencies of the Public Health Service.

Appendix I

Healthy Communities 2000: Model Standards—Guidelines for Community Attainment of the Year 2000 National Health Objectives*

Healthy Communities 2000: Model Standards puts the objectives of Healthy People 2000 into practice and encourages communities to establish achievable community health targets. It covers the priority areas and age groups used in Healthy People 2000 and includes all of the national objectives. Community leaders can adapt the national targets according to local needs and can establish objectives based on their own situations using the fill-in-the-blank approach used in Healthy Communities 2000: Model Standards (see Table I-1). By its direct use of the Healthy People 2000 goals and objectives, Model Standards provides a framework for communities to work toward their own priorities while pursuing national health objectives. Model Standards has been used successfully in large urban communities, in sparsely populated rural communities, and by city, county, district, and state health agencies.

Model Standards Principles

The Healthy Communities 2000 document serves as a guidebook and a process for planning community public health services, as the following principles for using Model Standards illustrate.

- **Emphasis on health outcomes.** Making progress in attacking major health problems depends on establishing an understandable set of health status objectives that are measurable and realistic and are accompanied by local process objectives for their achievement.
- **Flexibility.** Model Standards is a flexible planning tool using a "fill-in-the-blank" approach that allows communities to establish and quantify objectives and develop strategies based on their own situations.

- **Focus upon the entire community.** Cooperation among major community groups and organizations creates the foundation for communities to establish and achieve the goals and objectives suggested by Healthy Communities 2000: Model Standards.
- **A government presence at the local level (AGPALL).** Every locale and population should be served by a unit of government that takes a leadership role in assuring the public's health. Assuring that vital services are provided in all communities is an indispensable role of government. Government assures services by encouraging actions by other entities, requiring such actions by regulation, or providing services directly.
- **The importance of negotiation.** Negotiation is the principal way to maintain local flexibility and promote agreement among agencies and individuals who have an interest in and responsibility to protect the public's health.
- **Standards and guidelines.** "Standards" implies uniform objectives to assure equity and social justice, and "guidelines" emphasizes local discretion for decision making.
- **Accessibility of services.** Healthy Communities 2000 is designed to help communities tailor special population targets to assure services for those most in need.
- **Emphasis on programs.** Model Standards focuses on programs rather than professional practice because professional practice standards are usually set by the specialty practice organization.

Steps for Putting Model Standards to Use

A series of 11 steps have been developed to assist local health agencies in implementing Model Standards:[†]

*Adapted from Healthy Communities 2000: Model Standards, 3rd ed. (Washington, D.C.: American Public Health Association, 1991). Reprinted with permission.
[†]These steps were adapted from The Guide to Implementing Model Standards: Eleven Steps Toward a Healthy Community (Washington, D.C.: American Public Health Association, 1993), pp. 1–22.

1. **Assess and determine the role of one's health agency.** The local health department develops a mission statement and a long-range vision that provides employees and the community with a clear description of the agency's role and serves as a guide for the steps that follow.

2. **Assess the lead agency's organizational capacity.** The director and the staff of the agency should assess the organization's readiness to exercise leadership. Such an assessment can be accomplished by reviewing the department's structure to determine if it has the skills, community support, and staff capacity to lead the community.

3. **Develop an agency plan to build the necessary organizational capacity.** The agency should develop a plan to build on its internal strengths, overcome its weaknesses, and enhance its organizational effectiveness for carrying out community-wide efforts.

4. **Assess the community's organizational and power structures.** Each local health department should work with key community agencies, community leaders, interest groups, and community members. The agency should conduct an assessment of the community's organizational and power structures on either a formal or an informal basis.

5. **Organize the community to build a stronger constituency and establish a partnership for public health.** The local health agency should convene community groups to assess health needs, address health problems, and assist in the coordination of responsibilities.

6. **Assess the health needs and available community resources.** A community assessment provides the information needed to identify a community's most critical health problems (see this textbook's Chapter 5).

7. **Determine local priorities.** Establishing priorities should involve major health agencies, community organizations, and key interest groups and individuals. The community assessment aids in determining local priorities.

8. **Select outcome and process objectives that are compatible with local priorities and the *Healthy People 2000* objectives.** *Healthy Communities 2000: Model Standards* provides an array of goals and objectives (see Table I-1) to establish measurable health status (outcome) objectives. After establishing these objectives, a community coalition can develop process objectives for achieving them.

9. **Develop community-wide intervention strategies.** Developing community-wide interventions provides the means to achieve selected community goals and objectives. Responsibilities should then be assigned so that activities can be distributed and coordinated among agencies and organizations.

10. **Develop and implement a plan of action.** Establishing goals, objectives, and community-wide intervention strategies is an important step, but success depends on developing and executing a plan of action that implements intervention activities and services.

11. **Monitor and evaluate the effort on a continuing basis.** The achievement of improved health status will attest to the effectiveness of community efforts. In the short term, achievement of local process objectives will show movement toward improved health status, if effective interventions have been selected.

Two planning tools have been developed to assist in the performance of these 11 steps. The *Assessment Protocol for Excellence in Public Health* (APEXPH) enhances the capacity of the public health agency to address assessment, policy development, and quality assurance functions. APEXPH is presented in a workbook that local health departments can use to:

- Assess and improve their organizational capacity.
- Assess the health status of the community.
- Involve the community in improving public health.

The *Planned Approach to Community Health* (PATCH) is a program designed to help communities plan, implement, and evaluate health promotion and education programs that are directed at preventing and controlling chronic diseases. PATCH is a community health promotion methodology that increases a community's capacity to organize and mobilize members, collect and use local area data, set health priorities, select and implement appropriate interventions, and perform process and impact evaluation.

Local health agencies and communities can use *Healthy Communities 2000: Model Standards* and the complementary planning processes such as APEXPH and PATCH to translate national health objectives into community health action plans responsive to community needs.

Model Standards Nutrition Goals and Objectives

Model Standards has established the following goal for nutrition: Community residents will achieve optimal nutrition status that will reduce premature death and disability. The *Model Standards* health status objectives for the nutrition goal provide specific indicators focusing on nutrition-related disorders, deaths from coronary heart disease, cancer deaths, prevalence of overweight, low birthweight, weight gain during pregnancy, and growth retardation among low-income children. These are listed in Table I-1.

The *Model Standards* risk reduction objectives focus on dietary fat and saturated fat intake, consumption of complex carbohydrates and fiber-containing foods, practices to attain appropriate body weight, consumption of calcium-rich foods, salt and sodium intake, iron deficiency, breast-feeding, prevention of baby bottle tooth decay, and food labels. Specific objectives and indicators are provided in each category.

The *Model Standards* nutrition section also includes services and protection objectives. These focus on the development of a comprehensive nutrition plan, community nutrition education, nutrition services for at-risk populations, nutrition labeling, the availability of processed foods reduced in fat and saturated fat, the provision of nutrition information in grocery stores, healthful food choices in restaurants and foodservice operations, adequate delivery of home-delivered meals for older adults in need, nutrition education in schools and at worksites, nutrition assessment, counseling, and services as a provision of primary care services, breastfeeding promotion, and nutrition monitoring.

TABLE I-1 Model Standards *Goal for Nutrition*

Goal: Community residents will achieve optimal nutrition status that will reduce premature death and disability.

Focus	Objective	Indicator
Health Status Objectives		
Nutrition-related disorders	1. By _____ the prevalence of _____ nutrition-related disorders will be reduced to _____ among target population.*	Prevalence of specific nutrition-related disorders
Deaths from coronary heart disease	2. By _____ (2000) reduce coronary heart disease deaths to no more than _____ (100) per 100,000 people. (Age-adjusted baseline: 135 per 100,000 in 1987)	Coronary heart disease death rate

Special population target:

Coronary Deaths (per 100,000)	1987 Baseline	2000 Target
a. Blacks	163	_____ (115)
b. Other	_____	_____

Focus	Objective	Indicator
Cancer deaths	3. By _____ (2000) reverse the rise in cancer deaths to achieve a rate of no more than _____ (130) per 100,000 people. (Age-adjusted baseline: 133 per 100,000 in 1987) (*Note:* In its publications, the National Cancer Institute age-adjusts cancer death rates to the 1970 U.S. population. Using the 1970 standard, the equivalent baseline and target values for this objective would be 171 and 175 per 100,000, respectively.)	Nutritionally related cancer deaths (i.e., breast, colorectal)
Prevalence of overweight	4. By _____ (2000) reduce overweight to a prevalence of no more than _____ (20) % among people aged 20 and older and no more than 15% among adolescents aged 12 through 19. (Baseline: 26% for people aged 20 through 74 in 1976–80, 24% for men and 27% for women; 15% for adolescents aged 12 through 19 in 1976–80)	Percent overweight

Focus	Objective	Indicator

Health Status Objectives—Continued

Special population targets:

Overweight Prevalence	1976–80 Baseline[†]	2000 Target
a. Low-income women aged 20 and older	37%	_____ (25%)
b. Black women aged 20 and older	44%	_____ (30%)
c. Hispanic women aged 20 and older		_____ (25%)
• Mexican-American women	39%[‡]	
• Cuban women	34%[‡]	
• Puerto Rican women	37%[‡]	
d. American Indians/ Alaska Natives	29–75%[§]	_____ (30%)
e. People with disabilities	36%[‖]	_____ (25%)
f. Women with high blood pressure	50%	_____ (41%)
g. Men with high blood pressure	39%	_____ (35%)
h. Other	_____	_____

(*Note:* For people aged 20 and older, overweight is defined as body mass index (BMI) equal to or greater than 27.8 for men and 27.3 for women. For adolescents, overweight is defined as BMI equal to or greater than 23 for males aged 12 through 14, 24.3 for males aged 15 through 17, 25.8 for males aged 18 through 19, 23.4 for females aged 12 through 14, 24.8 for females aged 15 through 17, and 25.7 for females aged 18 through 19. The values for adolescents are the age- and gender-specific 85th percentile values of the 1976–80 National Health and Nutrition Examination Survey (NHANES II), corrected for sample variation. BMI is calculated by dividing weight in kilograms by the square of height in meters. The cutpoints used to define overweight approximate the 120% of desirable body weight definition used in the 1990 objectives.)

Focus	Objective	Indicator
Low birthweight	5. Reduce low birthweight to an incidence of no more than _____ % of live births. (*Model standards note:* The community may wish to develop special population targets, for example, by age, race, sex, income, handicapping conditions, etc., for community relevant subpopulations.)	Incidence of low and very low birthweights
Weight gain during pregnancy	6. Increase to at least _____ % the proportion of mothers who achieve the minimum recommended weight gain during their pregnancies. (*Model standards note:* All pregnancy weight gain should be adjusted for weight status prior to pregnancy. Recommended weight gain is defined as recommended in the 1990 report by the National Academy of Science, *Nutrition during Pregnancy.*)	Percent achieving appropriate weight gain

Continued

TABLE I-1 Model Standards *Goal for Nutrition—Continued*

Focus	Objective	Indicator

Health Status Objectives—Continued

Growth retardation among low-income children

7. By _____ (2000) reduce growth retardation among low-income children aged 5 and younger to less than _____ (10)% (Baseline: Up to 16% among low-income children in 1988, depending on age and race/ethnicity)

Indicator: Prevalence of growth retardation

Special population targets:

Prevalence of Short Stature	1988 Baseline	2000 Target
a. Low-income black children < age 1	15%	_____ (10%)
b. Low-income Hispanic children < age 1	13%	_____ (10%)
c. Low-income Hispanic children aged 1	16%	_____ (10%)
d. Low-income Asian/Pacific Islander children aged 1	14%	_____ (10%)
e. Low-income Asian/Pacific Islander children aged 2–4	16%	_____ (10%)
f. Other	_____	_____

(*Note:* Growth retardation is defined as height for age below the fifth percentile of children in the National Center for Health Statistics' reference population.)

*Insert specific nutrition-related disorder, e.g., obesity, anemia, retarded growth, elevated serum cholesterol, coronary artery disease, colon cancer, hypertension, and osteoporosis.

†Baseline for people aged 20–74.

‡1982–84 baseline for Hispanics aged 20–74.

§1984–88 estimates for different tribes.

‖1985 baseline for people aged 20–74 who report any limitation in activity due to chronic conditions.

Appendix J

Tips for Designing a Web Site

Web sites have moved into the mainstream, what with every Aunt Bess and Uncle Harry putting their own homepages on the Internet. Government agencies, companies, and nonprofit organizations have their respective Web sites, too. As a community nutritionist, you may have an opportunity to help develop your organization's Web site, or you may want to design and post your own homepage on the World Wide Web. In either event, there are basic questions to ask and issues to consider when designing a Web site.[1]

Basic Questions to Ask

Answer as many of the following questions as possible before designing a Web site. The more time you spend thinking about what you want—and don't want—the easier the design process will be.

Who Will Develop the Web Site? If you are helping design a Web site for your organization, the chances are good that a Webmaster has been hired to develop, test, update, and maintain the Web site. In this case, your role may be one of choosing the site's content. If it is your own Web site, you must decide whether to develop it yourself or hire a Web page designer to do it for you.

What Is the Purpose of the Web Site? Your site may be designed to sell a product or service, present information to a certain audience, or provide a collection of links to other Web sites. Perhaps it will do all three of these things. Specify the purpose of your site right from the beginning to help control costs and design time.

Who Is the Intended Audience? Your site may be designed to reach clients, customers, people who already know something about the subject matter or people who are unfamiliar with the topic. Specifying the intended audience helps determine how much background information must be provided and the terminology that must be explained. When answering this question, consider the typical user of your Web site and the type of problem he or she is trying to solve. In other words, think about why this user accessed your site in the first place.

Will Your Web Site Be Browser-Specific? Browsers are software programs that link to, interpret, and present information on the Internet. Two popular browsers today are Internet Explorer and Netscape Navigator. When you design your Web site, you have the option of coding it for one particular browser. It can also be designed for text only or frames (that is, special sections of the Web page that show graphics and text).[2] Making a decision in this area means thinking about the type of browser users are likely to have. Given that you probably won't have hard data on browser type, your best option is to design the Web site for a wide range of viewing capabilities.

How Many Links Will Your Site Provide? Links are the embedded codes that, when clicked on with a mouse, transport the user to another site on the World Wide Web. They can enhance or seriously detract from the Web site's presentation. When planning links, write about the subject as if there were no links in the text, and choose meaningful words or phrases for the link.

How Long Should the Web Site Pages Be? One frustrating aspect of Web browsing is scrolling up and down long pages. A rule of thumb is to keep page lengths to one window. This translates to about 1½ screenfuls of text. And remember, short pages are easier to maintain than long ones.

What Kind of Graphics Should Appear on the Web Site?
One of the great pleasures of Web browsing is the graphics—everything from Cow's Caught in the Web[3] to Sammy Salad-in-a-Bag.[4] There are issues to consider, however, when choosing graphics for your Web site. First, use only those that are essential to the Web site's purpose. Too many images clutter the page and increase the download time. Second, avoid large graphics—their long download time can annoy users. A rule of thumb here is to keep the total size of all images used on a page to less than 30K.

Other Issues to Consider

The number one consideration in Web site design is ease of use. If a Web site is cluttered, disorganized, or takes several minutes to download, users may seek information elsewhere—and you will have lost an opportunity to get your message across. To prevent clutter, use graphics wisely. To help keep the site organized, use document and chapter headings and put a title heading on each page.

The second consideration is quality. A Web site should be as presentable as an educational brochure, journal article, or textbook. Check spelling. Write well. Test every link. Update the site's pages often and date the pages. Avoid the "blink" feature. Many people are annoyed by a word or phrase that blinks on and off constantly while they are trying to read and understand text.

Finally, pay attention to netiquette. Don't publish registered trademarks or copyrighted material without permission. Don't publish a link to someone else's Web site without permission. Take time to respond to the people who send you queries about the information on your Web site. Provide good customer service to ensure that users keep coming back to your Web site.

A Word About Content

The Sun Microsystem style guide on the Internet summarizes the content issue nicely: "Publish things that solve people's problems."[5] Work to make your Web site useful to visitors. A good Web site enjoys frequent visits by satisified users.

References

1. Sun on the Net, sponsored by Sun Microsystems at www.sun.com; and TLC Systems at www.tlc-systems.com.
2. N. Estabrook, *Teach Yourself the Internet in 24 Hours* (Indianapolis: Sams, 1997).
3. Cows Caught in the Web, sponsored by Brandon University at www.mcs.brandonu.ca/~ennsnr/Cows/.
4. Dole 5-A-Day at www.dole5aday.com.
5. Sun on the Net.

Appendix K

The SMOG Readability Formula*

To calculate the SMOG reading grade level, begin with the entire written work that is being assessed, and follow these four steps:

1. Count off 10 consecutive sentences near the beginning, in the middle, and near the end of the text.
2. From this sample of 30 sentences, circle all of the words containing three or more syllables (polysyllabic), including repetitions of the same word, and total the number of words circled.
3. Estimate the square root of the total number of polysyllabic words counted. This is done by finding the nearest perfect square, and taking its square root.
4. Finally, add a constant of three to the square root. This number gives the SMOG grade, or the reading grade level that a person must have reached if he or she is to fully understand the text being assessed.

A few additional guidelines will help to clarify these directions:

- A sentence is defined as a string of words punctuated with a period (.), an exclamation point (!), or a question mark (?).
- Hyphenated words are considered as one word.
- Numbers that are written out should also be considered, and if in numeric form in the text, they should be pronounced to determine if they are polysyllabic.
- Proper nouns, if polysyllabic, should be counted, too.

- Abbreviations should be read as unabbreviated to determine if they are polysyllabic.

Not all pamphlets, fact sheets, or other printed materials contain 30 sentences. To test a text that has fewer than 30 sentences:

1. Count all of the polysyllabic words in the text.
2. Count the number of sentences.
3. Find the average number of polysyllabic words per sentence as follows:

$$\text{Average} = \frac{\text{Total number of polysyllabic words}}{\text{Total number of sentences}}$$

4. Multiply that average by the number of sentences *short of 30*.
5. Add that figure to the total number of polysyllabic words.
6. Find the square root and add the constant of 3.

Perhaps the quickest way to administer the SMOG grading test is by using the SMOG conversion table. Simply count the number of polysyllabic words in your chain of 30 sentences and look up the approximate grade level on the chart shown on page 575.

An example of how to use the SMOG Readability Formula is provided on the next page. The SMOG Conversion Table is shown on page 575.

*Source: U.S. Department of Health and Human Services, Public Health Service, *Pretesting in Health Communications*, NIH Pub. No. 84-1493, (Bethesda, MD: National Cancer Institute, 1984), pp. 43–45.

In Controlling Cancer—You Make a Difference

The key is ACTION. You can help protect yourself against cancer. Act promptly to:

Prevent some cancers through simple changes in lifestyle.

Find out about early detection tests in your home.

Gain peace of mind through regular medical checkups.

Cancers You Should Know About

Lung Cancer is the number one cancer among men, both in the number of new cases each year (79,000) and deaths (70,500). Rapidly increasing rates are due mainly to cigarette smoking. By not smoking, you can largely prevent lung cancer. The risk is reduced by smoking less, and by using lower tar and nicotine brands. But quitting altogether is by far the most effective safeguard. The American Cancer Society offers Quit Smoking Clinics and self-help materials.

Colorectal Cancer is second in cancer deaths (25,100) and third in new cases (49,000). When it is found early, chances of cure are good. A regular general physical usually includes a digital examination of the rectum and a guaiac slide test of a stool specimen to check for invisible blood. Now there are also Do-It-Yourself Guaiac Slides for home use. Ask your doctor about them. After you reach the age of 40, your regular check-up may include a "Procto," in which the rectum and part of the colon are inspected through a hollow, lighted tube.

Prostate Cancer is second in the number of new cases each year (57,000) and third in deaths (20,600). It occurs mainly in men over 60. A regular rectal exam of the prostate by your doctor is the best protection.

A Check-Up Pays Off

Be sure to have a regular, general physical including an oral exam. It is your best guarantee of good health.

How Cancer Works

If we know something about how cancer works, we can act more effectively to protect ourselves against the disease. Here are the basics:

1. Cancer spreads; time counts—Cancer is uncontrolled growth of abnormal cells. It begins small and if unchecked, spreads. If detected in an early, local stage, the chances for cure are best.

2. Risk increases with age—This is not a reason to worry, but a signal to have more regular thorough physical check-ups. Your doctor or clinic can advise you on what tests to get and how often they should be performed.

3. What you can do—Don't smoke and you will sharply reduce your chances of getting lung cancer. Avoid too much sun, a major cause of skin cancer. Learn cancer's Seven Warning Signals, listed on the back of this leaflet, and see your doctor promptly if they persist. Pain usually is a late symptom of cancer; don't wait for it.

Unproven Remedies

Beware of unproven cancer remedies. They may sound appealing, but they are usually worthless. Relying on them can delay good treatment until it is too late. Check with your doctor or the American Cancer Society.

More Information

For more information of any kind about cancer—free of cost—contact your local Unit of the American Cancer Society.

Know Cancer's Seven Warning Signals

1. Change in bowel or bladder habits.
2. A sore that does not heal.
3. Unusual bleeding or discharge.
4. Thickening or lump in breast or elsewhere.
5. Indigestion or difficulty in swallowing.
6. Obvious change in wart or mole.
7. Nagging cough or hoarseness.

If you have a warning signal, see your doctor.

(This pamphlet is from the American Cancer Society.)

We have calculated the reading grade level for this example. Compare your results to ours, then check both with the SMOG Conversion Table:

Readability Test Calculations

Total number of polysyllabic words	= 38
Nearest perfect square	= 36
Square root	= 6
Constant	= 3
SMOG reading grade level	= 9

SMOG Conversion Table*

Total Polysyllabic Word Counts	Approximate Grade Level (+1.5 Grades)
0–2	4
3–6	5
7–12	6
13–20	7
21–30	8
31–42	9
43–56	10
57–72	11
73–90	12
91–110	13
111–132	14
133–156	15
157–182	16
183–210	17
211–240	18

*Developed by Harold C. McGraw, Office of Educational Research, Baltimore County Schools, Towson, Maryland.

Appendix L

Nutrition Assessment and Screening

Appendix L-1 Assessment of Children's Growth Status

Several indices for assessing growth are derived from a child's height, weight, and age. The actual values for height, weight, and age are plotted on a nomogram or growth chart, which makes it possible to compare a particular child's height and weight for age with a national standard. The anthropometric reference data used in the United States for assessing physical growth are based on a large, nationally representative sample of children from birth to 18 years of age. The percentile curves (for example, 5th, 10th, 25th, 50th, 75th, 90th, and 95th) for this reference population are displayed on growth charts developed by the National Center for Health Statistics (NCHS). The physical growth charts for girls and boys are shown in Figures L1-1 through L1-4. A discussion of indices derived from growth measurements is given here.

Height-for-Age. Low height-for-age is defined as a height-for-age value below the 5th percentile or the −1.65 Z-score of the NCHS-CDC height reference.[1] (Refer to Figure L1-5 for a description of Z-scores.) In other words, low height-for-age reflects a child's failure to achieve for a given age group distribution a height that conforms to standards established for a well-nourished, healthy population of children.[2] Low height-for-age is sometimes referred to as growth stunting, which may be the consequence of poor nutrition, a high frequency of infections, or both.[3] Stunting refers to a slowing of skeletal growth and stature, the "end result of a reduced rate of linear growth."[4]

Although shortness in an individual child may be a normal reflection of the child's genetic heritage, a high prevalence rate of growth stunting reflects poor socioeconomic conditions. In some developing countries, the prevalence rate is as high as 60 to 70 percent. In developed countries, the rate is about 2 to 5 percent.[5] The prevalence of stunting is highest during the second or third year of life.[6]

Weight-for-Height. Low weight-for-height, or thinness, is defined as a weight-for-height value below the 5th percentile.[7] This indicator is a sensitive index of current nutritional status and is often associated with recent severe disease. Between the ages of 1 and 10 years, the indicator is relatively independent of age. It is relatively independent of ethnic groups, particularly among children aged 1 to 5 years.[8]

Weight-for-height differentiates between nutritional stunting, when a child's weight may be appropriate for her height, and wasting, when weight is very low for height owing to reductions in both tissue and fat mass. The prevalence of low weight-for-height is usually less than 5 percent except during periods of famine, war, or other extreme conditions. A prevalence rate greater than 5 percent indicates the presence of serious nutritional problems.

At the other end of the spectrum, a high weight-for-height (a value greater than the 95th percentile) correlates well with obesity. It typically indicates excess food consumption, low activity levels, or both.

Weight-for-Age. Weight-for-age is a composite index of height-for-age and weight-for-height. In children aged 6 months to about 7 years, weight-for-age is an indicator of acute malnutrition. It is widely used to assess protein-energy malnutrition and overnutrition. However, one limitation of using weight-for-age as an indicator of protein-energy malnutrition is that it does not consider height differences. For example, a child who has low weight-for-age may be genetically short with proportionally low height and low weight, rather than too thin with a low weight for height. As a result, the prevalence of protein-energy malnutrition in small children may be overstated if only this indicator is used.[9]

Weight-for-age is most useful in clinical settings where repeated measurements of the indicator are used to evaluate children who are not gaining weight.[10]

Velocity Growth Curves. The nomograms for assessing growth are based on cross-sectional data, meaning that measurements were made once on many children of different ages. Sometimes, however, it is useful to determine how a child's growth is changing over time. In this case, cross-sectional data are inappropriate, and longitudinal data—that is, those data obtained on the same children who are measured at specific ages over a period of many years—are needed. The velocity growth curves for boys and girls aged 2 to 18 years are shown in Figure L1-6.[11]

Weight/Height Ratios. Weight/height ratios measure body weight corrected for height. The underlying assumption of this calculation is that the ratios are correlated with obesity. For this reason, these ratios are often called obesity or body mass indices. Several such indices exist. The two ratios most commonly used in large-scale population and field studies are Quetelet's index and the ponderal index:

$$\text{Quetelet's index} = \frac{\text{Weight in kg}}{\text{Height in m}^2}$$

$$\text{Ponderal index} = \frac{\text{Height}}{\sqrt[3]{\text{Weight}}}$$

Quetelet's index is sometimes referred to as the body mass index (BMI). The question of whether BMI is an appropriate index for measuring body composition in children and adolescents has not been fully resolved. Some researchers believe that BMI is a good measure of body size for growing children when the child's biological age, which measures the year of peak height velocity, rather than his chronological age is computed.[12] Others maintain that Quetelet's index is appropriate only for the adult population. Canada, for example, does not consider Quetelet's index a valid index for individuals under the age of 20 or over 65 years or for women who are pregnant or lactating.[13] Part of the problem with measuring body fat in children is the lack of reference data for this population subgroup.

References

1. NCHS-CDC = National Center for Health Statistics-Centers for Disease Control and Prevention.
2. C. A. Miller and coauthors, *Monitoring Children's Health: Key Indicators* (Washington, D.C.: American Public Health Association, 1989), pp. 70–77.
3. R. Yip and coauthors, Pediatric Nutrition Surveillance System—United States, 1980–1991, *Morbidity and Mortality Weekly Report* 41/No. SS-7 (November 27, 1992): 1–24.
4. As cited in R. S. Gibson, *Principles of Nutritional Assessment* (New York: Oxford, 1990), p. 175.
5. Yip and coauthors, Pediatric Nutrition Surveillance System.
6. Gibson, *Principles of Nutritional Assessment*, pp. 175–76.
7. Yip and coauthors, Pediatric Nutrition Surveillance System.
8. Gibson, *Principles of Nutritional Assessment*, pp. 173–74.
9. Ibid., pp. 172–73.
10. Yip and coauthors, Pediatric Nutrition Surveillance System.
11. F. Falkner, Measurement of health, in *Measurement in Health Promotion and Protection*, ed. T. Abelin, Z. J. Brzezinski, and V. D. L. Carstairs (Copenhagen: World Health Organization, 1987), pp. 109–22.
12. V. A. Casey and coauthors, Body mass index from childhood to middle age: A 50-y follow-up, *American Journal of Clinical Nutrition* 56 (1992): 14–18.
13. Health and Welfare Canada, *Canadian Guidelines for Healthy Weights* (Ottawa: Health Services and Promotion Branch, Health and Welfare Canada, 1988).

FIGURE L1-1 *Girls: 2 to 18 Years Physical Growth NCHS Percentiles—Height and Weight for Age**

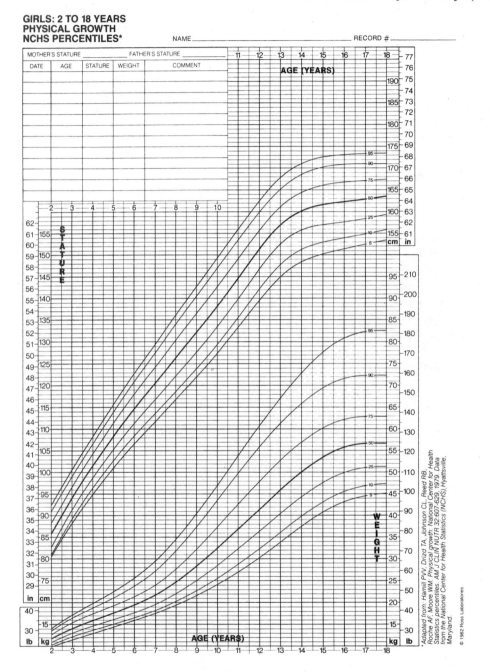

*Check the National Center for Health Statistics' Web site at www.cdc.gov/nchswww for revised growth charts.

FIGURE L1-2 *Boys: 2 to 18 Years Physical Growth NCHS Percentiles—Height and Weight for Age**

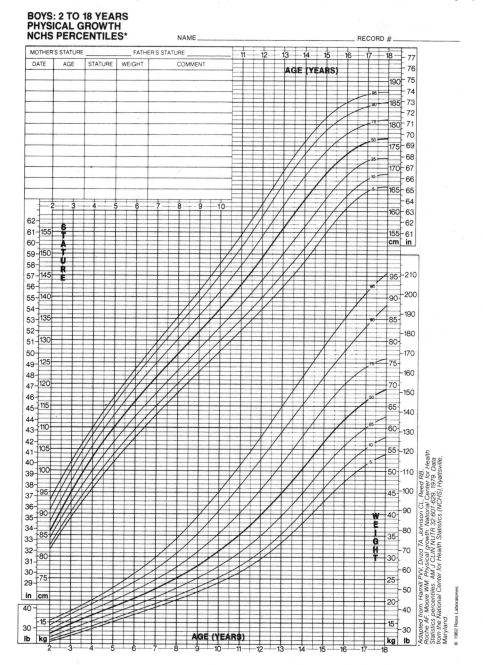

*Check the National Center for Health Statistics' Web site at www.cdc.gov/nchswww for revised growth charts.

FIGURE L1-3 *Girls: Prepubescent Physical Growth NCHS Percentiles—Weight for Height*

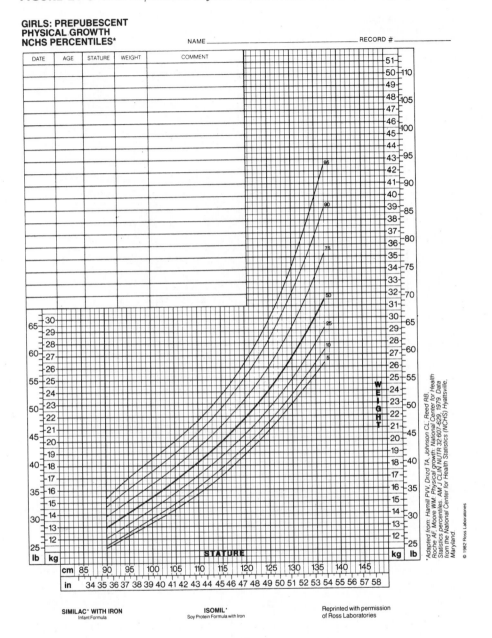

FIGURE L1-4 Boys: Prepubescent Physical Growth NCHS Percentiles—Weight for Height*

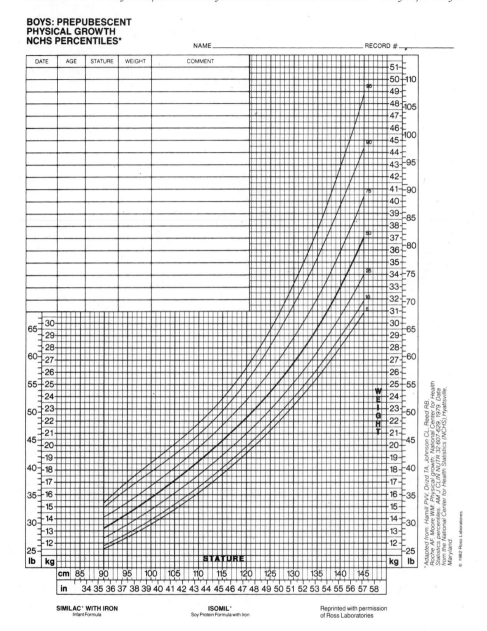

*Check the National Center for Health Statistics' Web site at www.cdc.gov/nchswww for revised growth charts.

FIGURE L1-5

Understanding Z-Scores

Source: R. Yip and coauthors, Pediatric Nutrition Surveillance System—United States, 1980–1991, *Morbidity and Mortality Weekly Report* 41/No. SS–7 (November 27, 1992): 8.

Growth indices can be expressed in percentile values or as standard deviation values or Z-scores. The Z-score of each growth index for each child is calculated by the following formula:

$$Z = \frac{\text{Observed value} - \text{Reference mean value}}{\text{Reference standard deviation value}}$$

The relationship of Z-scores to percentile values is shown on the distribution curve in the figure. Note that a Z-score of 0.0 corresponds to the 50th percentile on the distribution curve.

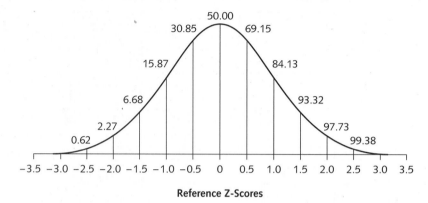

FIGURE L1-6 *Percentile Velocity Growth Curves for Boys and Girls, Aged 2 to 18 Years*

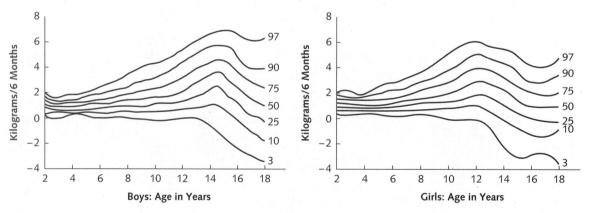

Source: Reproduced with permission from Abelin, T. et al. *Measurement in Health Promotion and Protection,* WHO Regional Office for Europe, 1987.

Appendix L-2 Nutritional Side Effects of Drugs Commonly Used by the Elderly

Drug	Nutritional Side Effects
• **Drugs influencing appetite and food intake:**	
Antipsychotics and sedatives	Somnolence; disinterest in food
Digoxin	Marked anorexia; nausea; vomiting; weakness
Cancer chemotherapies	Nausea; vomiting; aversion to food
• **Drugs affecting absorption of nutrients:**	
Laxatives and cathartics	Malabsorption of fat-soluble vitamins; fluid and electrolyte loss
Corticosteroids	Decreased vitamin D activity
Anticonvulsants	Decreased vitamin D activity
H_2 receptor blockers	Vitamin B_{12} malabsorption
Aluminum or magnesium antacids	Phosphate depletion
Cholestyramine	Decreased absorption of lipids, folate, iron, vitamin B_{12}, and fat-soluble vitamins; gastrointestinal side effects
• **Drugs affecting metabolism of nutrients:**	
Isoniazid, hydralazine, L-dopa	Vitamin B_6 antagonists
Salicylates	Iron loss secondary to gastrointestinal bleeding
Anticoagulants	Vitamin K antagonists
• **Drugs affecting excretion of nutrients:**	
Thiazides	Loss of potassium, sodium, magnesium, and zinc
Furosemide	Loss of potassium, magnesium, chloride, sodium, and water
Spironolactone	Potassium sparing; hyperkalemia; fluid and electrolyte changes

Source: J. E. Kerstetter, B. A. Holthausen, and P. A. Fitz, Malnutrition in the institutionalized older adult. Copyright The American Dietetic Association. Reprinted by permission from *Journal of the American Dietetic Association* 92 (1992): 1113.

FIGURE L3-1

*Nutrition Screening
Initiative Level I Screen*

Level I Screen

Body Weight

Measure height to the nearest inch and weight to the nearest pound. Record the values below and mark them on the Body Mass Index (BMI) scale to the right. Then use a straight edge (ruler) to connect the two points and circle the spot where this straight line crosses the center line (body mass index). Record the number below.

Healthy older adults should have a BMI between 22 and 27.

Height (in): _____

Weight (lbs): _____

Body Mass Index: _____

(number from center column)

Please make a check by any statements that are true for the individual:

- ⭕ Has lost or gained 10 pounds (or more) in the past 6 months.
- ⭕ Body mass index <22
- ⭕ Body mass index >27

For the remaining sections, please ask the individual which of the statements (if any) is true for him or her and place a check by each that applies.

Eating Habits

- ⭕ Does not have enough food to eat each day
- ⭕ Usually eats alone
- ⭕ Does not eat anything on one or more days each month
- ⭕ Has poor appetite
- ⭕ Is on a special diet
- ⭕ Eats vegetables two or fewer times daily
- ⭕ Eats milk or milk products once or not at all daily
- ⭕ Eats fruit or drinks fruit juice once or not at all daily
- ⭕ Eats breads, cereals, pasta, rice, or other grains five or fewer times daily
- ⭕ Has difficulty chewing or swallowing
- ⭕ Has more than one alcoholic drink per day (if woman); more than two drinks per day (if man)
- ⭕ Has pain in mouth, teeth, or gums

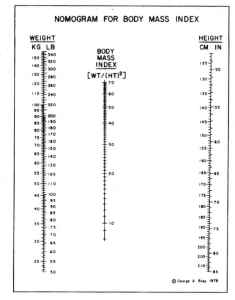

Living Environment

- ⭕ Lives on an income of less than $6000 per year (per individual in the household)
- ⭕ Lives alone
- ⭕ Is housebound
- ⭕ Is concerned about home security
- ⭕ Lives in a home with inadequate heating or cooling
- ⭕ Does not have a stove and/or refrigerator
- ⭕ Is unable or prefers not to spend money on food (<$25-30 per person spent on food each week)

Functional Status

Usually or always needs assistance with

(check each that apply):

- ⭕ Bathing
- ⭕ Dressing
- ⭕ Grooming
- ⭕ Toileting
- ⭕ Eating
- ⭕ Walking or moving about
- ⭕ Traveling (outside the home)
- ⭕ Preparing food
- ⭕ Shopping for food or other necessities

Level II Screen

Complete the following screen by interviewing the patient directly and/or by referring to the patient chart. If you do not routinely perform all of the described tests or ask all of the listed questions, please consider including them but do not be concerned if the entire screen is not completed. Please try to conduct a minimal screen on as many older patients as possible, and please try to collect serial measurements, which are extremely valuable in monitoring nutritional status. Please refer to the manual for additional information.

Anthropometrics

Measure height to the nearest inch and weight to the nearest pound. Record the values below and mark them on the Body Mass Index (BMI) scale to the right. Then use a straight edge (paper, ruler) to connect the two points and circle the spot where this straight line crosses the center line (body mass index). Record the number below; healthy older adults should have a BMI between 22 and 27; check the appropriate box to flag an abnormally high or low value.

Height (in): _____

Weight (lbs): _____

Body Mass Index (weight/height): _____

Please place a check by any statement regarding BMI and recent weight loss that is true for the patient.
- Body mass index <22
- Body mass index >27
- Has lost or gained 10 pounds (or more) of body weight in the past 6 months

Record the measurement of mid-arm circumference to the nearest 0.1 centimeter and of triceps skinfold to the nearest 2 millimeters.

Mid-Arm Circumference (cm): _____

Triceps Skinfold (mm): _____

Mid-Arm Muscle Circumference (cm): _____

NOMOGRAM FOR BODY MASS INDEX

© George A. Bray 1978

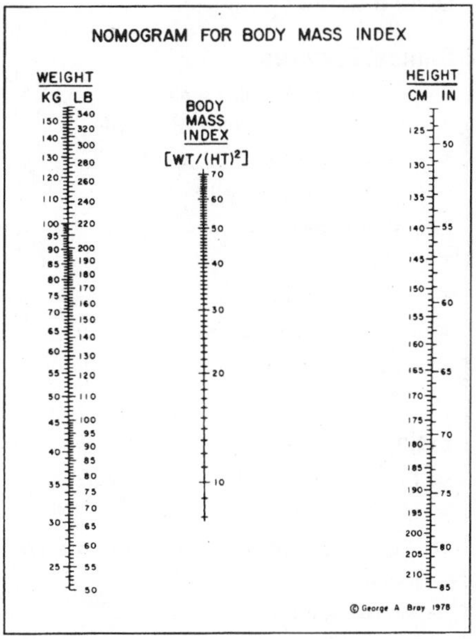

Refer to the table and check any abnormal values:
- Mid-arm muscle circumference <10th percentile
- Triceps skinfold <10th percentile
- Triceps skinfold >95th percentile

Note: mid-arm circumference (cm) - {0.314 x triceps skinfold (mm)}= mid-arm muscle circumference (cm)

For the remaining sections, please place a check by any statements that are true for the patient.

Laboratory Data
- Serum albumin below 3.5 g/dl
- Serum cholesterol below 160 mg/dl
- Serum cholesterol above 240 mg/dl

Drug Use
- Three or more prescription drugs, OTC medications, and/or vitamin/mineral supplements daily

Continued

FIGURE L3-2
Nutrition Screening Initiative Level II Screen

FIGURE L3-2

Nutrition Screening Initiative Level II Screen—continued

Level II Screen

Clinical Features

Presence of (check each that apply):
- Problems with mouth, teeth, or gums
- Difficulty chewing
- Difficulty swallowing
- Angular stomatitis
- Glossitis
- History of bone pain
- History of bone fractures
- Skin changes (dry, loose, nonspecific lesions, edema)

Eating Habits

- Does not have enough food to eat each day
- Usually eats alone
- Does not eat anything on one or more days each month
- Has poor appetite
- Is on a special diet
- Eats vegetables two or fewer times daily
- Eats milk or milk products once or not at all daily
- Eats fruit or drinks fruit juice once or not at all daily
- Eats bread, cereals, pasta, rice, or other grains five or fewer times daily
- Has more than one alcoholic drink per day (if woman); more than two drinks per day (if man)

Living Environment

- Lives on an income of less than $6000 per year (per individual in the household)
- Lives alone
- Is housebound
- Is concerned about home security
- Lives in a home with inadequate heating or cooling
- Does not have a stove and/or refrigerator
- Is unable or prefers not to spend money on food (<$25-30 per person spent on food each week)

Functional Status

Usually or always needs assistance with (check all that apply):
- Bathing
- Dressing
- Grooming
- Toileting
- Eating
- Walking or moving about
- Traveling (outside the home)
- Preparing food
- Shopping for food or other necessities

Mental/Cognitive Status

- Clinical evidence of impairment, e.g. Folstein <26
- Clinical evidence of depressive illness, e.g. Beck Depression Inventory>15, Geriatric Depression Scale>5

Source: Nutrition Screening Initiative, *Keeping Older Americans Healthy at Home* (Washington, D.C.: Nutrition Screening Initiative, 1996). Reprinted with permission by the Nutrition Screening Initiative, a project of the American Academy of Family Physicians, The American Dietetic Association, and the National Council on Aging, Inc. and funded in part by a grant from Ross Laboratories, a division of Abbott Laboratories.

TABLE L3-1 *Mini Mental Status Exam*

Orientation:

- Ask for the date (What is the year/season/date/day/month?). (1 point each)
- Where are we (state) (county) (town) (hospital) (floor)? (1 point each)

Registration:

- Name 3 unrelated objects, (e.g., ball, tree, flag) clearly and slowly. Then ask the patient to say all 3 after you have said them. (Give 1 point for each correct answer. Then repeat them until the patient learns all 3. Count trials and record.)

Attention and calculation trial:

- Serial 7s: Ask patient to begin with 100 and count backward by 7. (1 point for each correct answer. Stop after 5 answers.) Alternatively, spell "world" backwards. (The score is the number of letters in the correct order.)

Recall trial:

- Ask the patient to recall the 3 objects repeated above. (Give 1 point for each correct answer.)

Language trial:

- Show the patient a wristwatch and ask him what it is. Repeat for a pencil. (Score 0–2.)
- Repeat the following "No ifs, ands, or buts." (1 point)
- Follow a 3-stage command: "Take a paper in your right hand, fold it in half, and put it on the floor." (3 points)
- On a blank piece of paper, print the sentence "Close your eyes." Ask patient to read it and do what it says. (1 point)
- Write a sentence. (1 point)
- Copy design: draw intersecting pentagons; ask the patient to copy it exactly as it is. All 10 angles must be present and two must intersect. (1 point)

Assessment:

• ASSESS level of consciousness along a continuum.			
Alert	Drowsy	Stupor	Coma

Source: Reprinted from M. F. Folstein, S. E. Folstein, and P. McHugh. Mini-mental state: A practical method for grading the cognitive state of patients for the clinician, *Journal of Psychiatric Research* 12, Copyright 1975, with kind permission from Elsevier Science Ltd, The Boulevard, Langford Lane, Kidlington, OX5 1GB, UK. Some professional education and training are needed to administer and interpret this instrument.

TABLE L3-2 *Geriatric Depression Scale*

Choose the best answer ("yes" or "no") for how you felt this past week.[†]

* 1. Are you basically satisfied with your life?
   2. Have you dropped many of your activities and interests?
   3. Do you feel that your life is empty?
   4. Do you often get bored?
* 5. Are you hopeful about the future?
   6. Are you bothered by thoughts you can't get out of your head?
* 7. Are you in good spirits most of the time?
   8. Are you afraid that something bad is going to happen to you?
* 9. Do you feel happy most of the time?
  10. Do you often feel helpless?
  11. Do you often get restless and fidgety?
  12. Do you prefer to stay at home, rather than going out and doing new things?
  13. Do you frequently worry about the future?
  14. Do you feel you have more problems with memory than most?
*15. Do you think it is wonderful to be alive now?
  16. Do you often feel downhearted and blue?
  17. Do you feel pretty worthless the way you are now?
  18. Do you worry a lot about the past?
*19. Do you find life very exciting?
  20. Is it hard for you to get started on new projects?
*21. Do you feel full of energy?
  22. Do you feel that your situation is hopeless?
  23. Do you think that most people are better off than you are?
  24. Do you frequently get upset over little things?
  25. Do you frequently feel like crying?
  26. Do you have trouble concentrating?
*27. Do you enjoy getting up in the morning?
  28. Do you prefer to avoid social gatherings?
*29. Is it easy for you to make decisions?
*30. Is your mind as clear as it used to be?

Norms:		
Normal	5 ± 4
Mildly depressed	15 ± 6
Very depressed	23 ± 5

---

*Appropriate (nondepressed) answers = yes, all others = no.
[†]This instrument is intended for professional administration and interpretation. Any significant score (depressed range) indicates probable need for professional advice.

Source: Reprinted from T. Yesavage, T. Brink, and coauthors. Development and validation of a geriatric screening scale: A preliminary report, *Journal of Psychiatric Research* 17, Copyright 1983, with kind permission from Elsevier Science Ltd, The Boulevard, Langford Lane, Kidlington OX5 1GB UK.

Appendix M

Acronyms

AAA Area Agency on Aging
AAP American Academy of Pediatrics
ADA American Dietetic Association
ADL Activities of daily living
AFDC Aid to Families with Dependent Children
AHA American Heart Association
AHP Accountable health partnership or plan
AI Adequate Intake
AIDS Acquired immunodeficiency syndrome
BMI Body mass index
BRFSS Behavioral Risk Factor and Surveillance System
CACFP Child and Adult Care Food Program
CAM Complementary and Alternative Medicine
CARE Cooperative for American Relief Everywhere
CAT Computerized axial tomography
CATCH Child and Adolescent Trial for Cardiovascular Health
CCHIP Community Childhood Hunger Identification Project
CDC Centers for Disease Control and Prevention
CDR Crude death rate
CFDA Catalog of Federal Domestic Assistance
CFR *Code of Federal Regulations*
CHAMPUS Civilian Health and Medical Program of the Uniformed Services
CHD Coronary heart disease
CSFII Continuing Survey of Food Intakes by Individuals
CSFP Commodity Supplemental Food Program
CSREES Cooperative State Research, Education, and Education Service
CSSS Coordinated State Surveillance Survey
CVD Cardiovascular disease
DHHS Department of Health and Human Services
DHKS Diet and Health Knowledge Survey
DNA Deoxyribonucleic acid
DRG Diagnosis-related group
DRI Dietary Reference Intake
DTR Dietitian technician registered
EAR Estimated Average Requirement

EFNEP Expanded Food and Nutrition Education Program
ENP Elderly Nutrition Program
EPSDT Early Periodic Screening, Diagnosis, and Treatment
FASEB Federation of American Societies for Experimental Biology
FAO Food and Agriculture Organization
FAS Fetal alcohol syndrome
FDA Food and Drug Administration
FD&C Act Food, Drug, and Cosmetic Act
FDPIR Food Distribution Program on Indian Reservations
FLAPS Food Label and Package Survey
FMNP WIC Farmers' Market Nutrition Program
FNS Food and Nutrition Service (USDA)
FRAC Food Research and Action Center
FSP Food Stamp Program
GAO General Accounting Office
GASP Growing as a Single Parent (program)
GDP Gross domestic product
GIGO Garbage in/garbage out
GNP Gross national product
GOBI Growth charts, oral rehydration therapy, breast milk, and immunization
GPO Government Printing Office
GRAS Generally Recognized as Safe
GTT Glucose tolerance test
HACCP Hazard Analysis and Critical Control Point
Hb Hemoglobin
HCFA Health Care Financing Administration
Hct Hematocrit
HDL-C High-density lipoprotein cholesterol
HDS Health and Diet Study
HHANES Hispanic Health and Nutrition Examination Survey
HHH Harvest Health at Home—Eating for the Second Fifty Years
HIPC Health insurance purchasing cooperative
HIV Human immunodeficiency virus
HMO Health maintenance organization

HNIS   Human Nutrition Information Service (USDA)
HR   House of Representatives
HRA   Health risk appraisal
IADL   Instrumental activities of daily living
ICD-9-CM   *International Classification of Diseases—Clinical Manifestations*
ICN   International Conference on Nutrition
ICPSR   Inter-university Consortium for Political and Social Research
IDDM   Insulin-dependent diabetes mellitus
IMR   Infant mortality rate
INCAP   Institute of Nutrition of Central America and Panama
IOM/NAS   Institute of Medicine/National Academy of Sciences
IOP   Improvement of Organizational Performance
IPA   Independent practice association
JCAHO   Joint Commission on Accreditation of Healthcare Organizations
LBW   Low birthweight
LD   Licensed dietitian
LDL-C   Low-density lipoprotein cholesterol
LEAN   Lowfat Eating for America Now (Project LEAN)
MCH   Maternal and child health
MCV   Mean corpuscular volume
MIC   Maternal and infant care
MMR   Maternal mortality rate
MMWR   *Morbidity and Mortality Weekly Report*
MNT   Medical Nutrition Therapy
MRFIT   Multiple Risk Factor Intervention Trial
NAS/NRC   National Academy of Sciences/National Research Council
NCEP   National Cholesterol Education Program
NCES   Nutrition Counseling and Education Services
NCHS   National Center for Health Statistics
NCI   National Cancer Institute
NET   Nutrition Education and Training (program)
NFCS   Nationwide Food Consumption Survey
NHANES   National Health and Nutrition Examination Survey
NHANES-MA   National Health and Nutrition Examination Survey—Mexican American
NHEFS   NHANES I Epidemiologic Followup Study
NHI   National health insurance
NHIS   National Health Interview Survey
NHLBI   National Heart, Lung, and Blood Institute
NIA   National Institute on Aging
NIDDM   Non-insulin-dependent diabetes mellitus
NIEO   New International Economic Order

NIH   National Institutes of Health
NLEA   Nutrition Labeling and Education Act
NMN   N-methyl nicotinamide
NNMRRP   National Nutrition Monitoring and Related Research Program
NPE   Nutrition Program for the Elderly
NSI   Nutrition Screening Initiative
NSLP   National School Lunch Program
OAA   Older Americans Act
OMB   Office of Management and Budget
ORT   Oral rehydration therapy
PAC   Political Action Committee
PAHO   Pan American Health Organization
PedNSS   Pediatric Nutrition Surveillance System
PEM   Protein-energy malnutrition
PHHP   Pawtucket Heart Health Program
PIH   Pregnancy-induced hypertension
PKU   Phenylketonuria
PL   Public law
PPO   Preferred provider organization
PPS   Prospective payment system
PREGNSS   Pregnancy Nutrition Surveillance System
PSA   Public service announcement
QA   Quality assurance
RD   Registered dietitian
RDA   Recommended Dietary Allowances
RNA   Ribonucleic acid
RNI   Recommended Nutrient Intakes (Canadian)
S   Senate
SCORE   Screening, counseling, and referral event
SCT   Social Cognitive Theory
SFSP   Summer Food Service Program
SIDS   Sudden Infant Death Syndrome
SMP   Special Milk Program
SNAP   Supplemental Nutrition Assistance Program
SSA   Social Security Administration
SSI   Supplemental Security Income
SWOT   Strengths, weaknesses, opportunities, and threats
TANF   Temporary Assistance for Needy Families
TDS   Total Diet Study
TEFAP   Emergency Food Assistance Program
TIBC   Total iron-binding capacity
TQM   Total Quality Management
U5MR   Under-5 mortality rate
UL   Tolerable Upper Intake Level
UN   United Nations
UNICEF   United Nations Children's Fund
USAID   U.S. Agency for International Development
USDA   U.S. Department of Agriculture

VA   Veterans Affairs
VAD   Vitamin A deficiency
VITAL   Vitamin A Field Support Project
VMSIS   Vitamin/Mineral Supplement Intake Survey

WHO   World Health Organization
WIC   Special Supplemental Nutrition Program for Women, Infants, and Children
WLPS   Weight Loss Practices Survey

Appendix N

Directory of Internet Addresses

Internet Addresses

Administration for Children and Families   www.acf.dhhs.gov

Administration on Aging   www.aoa.dhhs.gov

Adolescence Directory On-Line   education.indiana.edu/cas/adol/adol.html

Agency for Health Care Policy and Research   www.ahcpr.gov

Agricultural Research Service   www.ars.usda.gov

Agriculture and Agri-Food Canada's Electronic Information Service   aceis.agr.ca/newintre.html

Alzheimer's Association   www.alz.org

American Association of Health Plans   www.aahp.org

American Association of Retired Persons (AARP)   www.aarp.org

American Diabetes Association   www.diabetes.org

American Dietetic Association   www.eatright.org

American Geriatrics Society   www.americangeriatrics.org

American Heart Association   www.americanheart.org

American Institute for Cancer Research   www.aicr.org

American Public Health Association   www.apha.org

Arbor Nutrition Guide   arborcom.com

Asian Diet Pyramid   www.news.cornell.edu/science/Dec95/st.asian.pyramid.html

Bastyr University AIDS Research Center   www.bastyr.edu/research/recruit.html

Bibliography on Evaluating Internet Resources   refserver.lib.vt.edu/libinst/critTHINK.HTM

Bread for the World Institute   www.bread.org

BRFSS Summary Prevalence Report   www.cdc.gov/nccdphp/brfss

Canada's Food Guide to Healthy Eating   www.hc-sc.gc.ca

Canada's National Plan of Action for Nutrition   www.hc-sc.gc.ca

Canadian Food Inspection Agency   www.cfia-acia.agr.ca

Canadian Public Health Association   www.cpha.ca

CancerNet   cancernet.nci.nih.gov

CARE   www.care.org

Catalog of Federal Domestic Assistance Programs   **family.info.gov/cfda/index.htm**

Catholic Charities USA   **www.catholiccharitiesusa.org**

Census Bureau   **www.census.gov**

Center for Addiction and Alternative Medicine Research   **www.winternet.com/~caamr**

Center for Alternative Medicine Pain and Research and Evaluation   **207.123.250.14**

Center for Alternative Medicine Research (Harvard)   **www.bidmc.harvard.edu/medicine/camr/index.html**

Center for Alternative Medicine Research in Cancer   **www.sph.uth.tmc.edu/utcam**

Center for Complementary and Alternative Medicine Research in Asthma, Allergy and Immunology
www-camra.ucdavis.edu

Center for Complementary and Alternative Medicine Research in Women's Health
cpmcnet.columbia.edu/dept/rosenthal

Center for Food Safety and Applied Nutrition   **vm.cfsan.fda.gov/list.html**

Center for Nutrition Policy and Promotion   **www.usda.gov/cnpp**

Center for Research in Complementary and Alternative Medicine for Stroke and Neurological Disorders
www.umdnj.edu/altmdweb/web.html

Center for the Study of Complementary and Alternative Therapies   **www.med.Virginia.EDU/nursing/centers/alt-ther.html**

Centers for Disease Control and Prevention   **www.cdc.gov**

Centers for Disease Control and Prevention Wonder Database   **wonder.cdc.gov**

CHID On-line   **chid.nih.gov**

Code of Federal Regulations   **www.access.gpo.gov/nara/cfr/index.html**

Complementary and Alternative Medicine Program   **scrdp.stanford.edu/camps.html**

Congressional Hunger Center   **www.hungercenter.org/chc**

Consumer Information Center   **www.pueblo.gsa.gov**

Continuing Survey of Food Intakes by Individuals, 1996, and 1996 Diet and Health Knowledge Survey
www.barc.usda.gov/bhnrc/foodsurvey/home.htm

Cooperative State Research, Education, and Extension Service   **www.reeusda.gov/csrees.htm**

Cooperative State Research, Education, and Extension Service Grant and Funding Opportunities
www.reeusda.gov/new/funding.htm

Cornucopia of Disability Information (CODI)   **codi.buffalo.edu**

CRISP (Computer Retrieval of Information on Scientific Projects)   **www-commons.dcrt.nih.gov**

Cultural Diversity—Eating in America   **www.ag.ohio-state.edu/~ohioline/lines/food.html**

Democratic Party Headquarters   **www.democrats.org/party**

Department of Health and Human Services   **www.hhs.gov**

Department of Health and Human Services Gateway Partnership Funding Page   **www.odphp.osophs.dhhs.gov/partner/funding.htm**

Department of Health and Human Services Racial and Ethnic Disparities in Health   **raceandhealth.hhs.gov**

Dietary Assessment Calibration/Validation Register   **www-dacv.ims.nci.nih.gov**

Dietary Guidelines for Americans, History   **www.nal.usda.gov/fnic/Dietary/12dietapp1.htm**

Dietetics Online®   **www.dietetics.com**

Dietitians of Canada   www.dietitians.ca

Discussion Paper on Domestic Food Security   www.fas.usda.gov/icd/summit/discussi.html

Economic Research Service   www.econ.ag.gov

ERS Research on Food Assistance and Nutrition   www.econ.ag.gov/Briefing/foodasst/contracts.htm

Evaluating Internet Research Sources   www.sccu.edu/faculty/R_Harris/evalu8it.htm

Evaluating Quality on the Net   www.tiac.net/users/hope/findqual.html

Evaluating Web Resources   www.science.widener.edu/~withers/webeval.htm

Evaluating World Wide Web Information   thorplus.lib.purdue.edu/research/classes/gs175/3gs175/evaluation.html

Federal Information Exchange   nscp.fie.com

Federal Register   www.access.gpo.gov/su_docs/aces/aces140.html

Federal Trade Commission's Consumer Response Center   www.ftc.gov/bcp/conline/fraud.htm

FedStats   www.fedstats.gov

FedWorld   www.fedworld.gov

5 a Day for Better Health Program   dccps.nci.nih.gov/5aday

5 a Day Week Community Intervention Kit   www.dcpc.nci.nih.gov/5aday/week98/CommunityKit98.html

Food and Agriculture Organization   www.fao.org

Food and Drug Administration   www.fda.gov

Food and Nutrition Information Center   www.nal.usda.gov/fnic

Food and Nutrition Service   www.usda.gov/fcs

Food Composition Resource List for Professionals   www.nal.usda.gov/fnic/pubs/bibs/gen/97fdcomp.htm

Food Research and Action Center   www.frac.org

Food Safety and Inspection Service   www.fsis.usda.gov

Foodchain   www.foodchain.org

Foundation Center   www.fdncenter.org/2index.html

Freedom from Hunger   www.freefromhunger.org

Government of Canada   canada.gc.ca

Government Printing Office (GPO) Access   www.access.gpo.gov

Government Resources—General Disability Information   mirconnect.com/feds/general.html

Grants Database (Oryx Press)   www.oryxpress.com/grants.htm

Grantsmanship Center   www.tgci.com

GrantsNet   www.os.dhhs.gov/progorg/grantsnet

Grateful Med   www.nlm.nih.gov

Guide to Community Preventive Services   web.health.gov/communityguide

Head Start Bureau   www.acf.dhhs.gov/programs/hsb

Health Canada   www.hc-sc.gc.ca

Health Care Financing Administration   www.hcfa.gov

Health Insurance Association of America   www.hiaa.org

Health Pages   www.thehealthpages.com

Health Resources and Services Administration   www.hrsa.dhhs.gov

Healthfinder   www.healthfinder.gov

Healthy People 2000   www.odphp.osophs.dhhs.gov/pubs/hp2000

Healthy People 2010   web.health.gov/healthypeople

Hispanic Customer Service Home Page   www.dhhs.gov/about/heo/hispanic.html

How to Critically Analyze Information Sources   www.library.cornell.edu/okuref/research/skill26.htm

Human Genome Education Project   www.nhgri.nih.gov/HGP

HungerWeb   www.brown.edu/Departments/World_Hunger_Program/index.html

Indian Health Service   www.tucson.ihs.gov

Infant Feeding Action Coalition (INFACT) Canada   www.infactcanada.ca

Institute for Research in Social Science   www.unc.edu/depts/irss/guide.htm

Institute for Social Research   www.isr.umich.edu

InteliHealth   www.intelihealth.com

Inter*Action*   www.interaction.org

International Food Information Council   ificinfo.health.org

International Fund for Agricultural Development (IFAD)   www.ifad.org

International Network of Food Data Systems (INFOODS)   www.crop.cri.nz/foodinfo/infoods/infoods.htm

Jean Mayer USDA, Human Nutrition Research Center on Aging   www.hnrc.tufts.edu

Joint Commission on Accreditation of Healthcare Organizations (JCAHO)   www.jcaho.org

Kids Count   www.aecf.org/aeckids.htm

Kids Food Cyberclub   www.kidsfood.org

KidsHealth   www.kidshealth.org

La Leche League   www.lalecheleague.org

Latin American Diet Pyramid   www.oldwayspt.org/html/p_latin.htm

Library of Congress State and Local Government Information   lcweb.loc.gov/global/state/stategov.html

Mayo Clinic Health Oasis: Children's Health Center   www.mayohealth.org/mayo/common/htm/pregpg.htm

Mediterranian Diet Pyramid   www.oldwayspt.org/html/p_med.htm

MEDLINE   www.nlm.nih.gov

Milton's Web Site for Evaluating Information Found on the Internet   milton.mse.jhu.edu:8001/research/education/net.html

Miscellaneous Food Guide Pyramids   www.kde.state.ky.us/bmss/odss/dscn/intpyr.htm

National Academy of Sciences, Institute of Medicine   www2.nas.edu/iom

National Agricultural Statistics Service   www.usda.gov/nass

National Association of State Information Resource Executives State Search   www.nasire.org/ss/index.html

National Cancer Institute   www.nci.nih.gov

National Center for Children in Poverty   cpmcnet.columbia.edu:80/dept/nccp

National Center for Chronic Disease Prevention and Health Promotion   www.cdc.gov/nccdphp

National Center for Health Statistics   www.cdc.gov/nchswww

National Center for HIV, STD, and TB Prevention, Division of HIV/AIDS Prevention (CDC)
    www.cdc.gov/nchstp/hiv_aids/dhap.htm

National Clearinghouse for Alcohol and Drug Information   www.health.org

National Clearinghouse for Bilingual Education   www.ncbe.gwu.edu

National Clearinghouse on Families & Youth   www.ncfy.com

National Committee for Quality Assurance (NCQA)   www.ncqa.org

National Council Against Health Fraud   www.ncahf.org

National Council on Disability   www.ncd.gov

National Evaluation of the Elderly Nutrition Program, 1993–1995   www.aoa.dhhs.gov/aoa/pages/nutreval.html

National Fraud Information Center   www.fraud.org

National Health Information Center   nhic-nt.health.org

National Heart, Lung, and Blood Institute   www.nhlbi.nih.gov/nhlbi/nhlbi.htm

National Information Center for Children and Youth with Disabilities   www.nichcy.org

National Institute on Aging   www.nih.gov/nia

National Institute on Aging's Alzheimer's Disease Education and Referral Center   www.alzheimers.org

National Institutes of Health   www.nih.gov

National Maternal and Child Health Clearinghouse   www.circsol.com/mch

National Osteoporosis Foundation   www.nof.org

National Policy and Resource Center on Nutrition and Aging,   www.aoa.dhhs.gov/aoa/dir/186.html

National Science Foundation   www.nsf.gov

NIH Consumer Health Information   www.nih.gov/health/consumer/conicd.htm

North American Menopause Society   www.menopause.org

Notices of Funding Availability   ocd.usda.gov/nofa.htm

Nutrition Expedition   Fscn.che.umn.edu/nutrexp/default.html

Nutrition on the Web—for teens   hyperion.advanced.org/10991

Office of Alternative Medicine   altmed.od.nih.gov

Office of Alternative Medicine Clearinghouse   altmed.od.nih.gov/oam/clearinghouse

Office of Management and Budget   www1.whitehouse.gov/WH/EOP/OMB/html/ombhome-plain.html

Oregon State University Extension Home Economics   www.osu.orst.edu/dept/ehe/diverse.html

Oxfam America   www.oxfamamerica.org

Pan American Health Organization   www.paho.org

Produce for Better Health Foundation   www.5aday.com

Promotion of Mother's Milk, Inc. (ProMom)   www.promom.org

Puerto Rican Food Guide Pyramid   www.hispanichealth.com/pyramid.htm

Pyramid Servings Data: Results from USDA's 1995 and 1996 Continuing Survey of Food Intakes by Individuals   www.barc.usda.gov/bhnrc/foodsurvey/home.htm

Quackwatch   www.quackwatch.com

Republican National Committee   www.rnc.org

Second Harvest   www.secondharvest.org

Shape Up America BMI Chart   www.shapeup.org/sua/bmi/chart.htm

Share Our Strength   www.foggy.com/SOS/involved.html

Slack's Pediatric Internet Directory   **www.slackinc.com/child/pednet-x.htm**

Social Security Administration   **www.ssa.gov**

State and Local Governments   **lcweb.loc.gov/global/state/stategov.html**

Substance Abuse and Mental Health Services Administration   **www.samhsa.gov**

Superintendent of Documents   **www.gpo.gov**

Thinking Critically About World Wide Web Resources
    www.library.ucla.edu/libraries/college/instruct/critical.htm

Third National Health and Nutrition Examination Survey (NHANES III) Public-Use Data Files
    www.cdc.gov/nchswww/products/catalogs/subject/nhanes3/nhanes3.htm#descriptional

Third Report on Nutrition Monitoring in the United States   **www.nalusda.gov/fnic/usda/thirdreport.html**

Thomas Legislative Information on the Internet   **thomas.loc.gov**

Trends in Food and Nutrient Intakes by Adults: NFCS 1977–78, CSFII 1989–91, and CSFII 1994–95
    www.barc.usda.gov/bhnrc/foodsurvey/home.htm

Tufts Nutrition Navigator   **navigator.tufts.edu**

United Nations Development Programme (UNDP)   **www.undp.org/index5.html**

UNICEF   **www.unicef.org**

UNICEF USA   **www.unicefusa.org**

United Way   **www.unitedway.org**

U.S. Action Plan on Food Security   **www.fas.usda.gov/icd/summit/framewor.html**

U.S. Department of Agriculture   **www.usda.gov**

U.S. Food and Drug Administration, Center for Food Safety and Applied Nutrition: Information for Women Who Are
    Pregnant   **vm.cfsan.fda.gov/~dms/wh-preg.html**

U.S. House of Representatives   **www.house.gov**

U.S. Senate   **www.senate.gov**

U.S. State and Local Government Gateway   **www.health.gov/statelocal**

USDA Food and Nutrition Research Briefs   **www.nal.usda.gov/fnic/usda/fnrb**

USDA Nutrient Database for Standard Reference, Release 12   **www.nal.usda.gov/fnic/foodcomp**

USDA Food Guide and Other Pyramids   **www.nal.usda.gov/fnic/Fpyr/pyramid.html**

USDA's School Meals Initiative for Healthy Children/Team Nutrition   **schoolmeals.nal.usda.gov:8001**

Vegetarian Food Guide Pyramid   **www.eatright.org/adap1197.html**

Web of Culture   **www.webofculture.com**

Weight-Control Information Network   **www.niddk.nih.gov/health/nutrit/win.htm**

White House   **www.whitehouse.gov**

WIC Program   **www.usda.gov/fcs/wic.htm**

WIC Program Database of Nutrition Education Materials   **www.nal.usda.gov/fnic**

Women's Health Initiative   **www.nih.gov**

Working with Culturally Diverse Audiences   **www.osu.orst.edu/dept/ehe/diverse.html**

World Alliance for Breastfeeding Action   **www.elogica.com.br/waba**

World Health Organization   **www.who.org**

Youth Risk Behavior Surveillance System   **www.cdc.gov/nccdphp/youthris.htm**

Index

Photo Credits

Internet Resources

Professional Organizations

American Dietetic Association	www.eatright.org
American Diabetes Association	www.diabetes.org
American Heart Association	www.americanheart.org
American Public Health Association	www.apha.org
Canadian Public Health Association	www.cpha.ca
Dietitians of Canada	www.dietitians.ca

Health Organizations

Food and Agriculture Organization	www.fao.org
Pan American Health Organization	www.paho.org
World Health Organization	www.who.org

Government Agencies and Offices

Administration on Aging	www.aoa.dhhs.gov
Agency for Health Care Policy and Research	www.ahcpr.gov
Center for Food Safety and Applied Nutrition	vm.cfsan.fda.gov/list.html
Centers for Disease Control and Prevention	www.cdc.gov
Canadian Food Inspection Agency	www.cfia-acia.agr.ca
Consumer Information Center	www.pueblo.gsa.gov
Department of Health and Human Services	www.hhs.gov
Federal Register	www.access.gpo.gov/su_docs/aces/aces140.html
Food and Drug Administration	www.fda.gov
Food and Nutrition Service	www.usda.gov/fcs
Food Safety and Inspection Service	www.fsis.usda.gov
Government Printing Office Access	www.access.gpo.gov
Health Canada	www.hc-sc.gc.ca
Health Care Financing Administration	www.hcfa.gov
National Cancer Institute	www.nci.nih.gov
National Institutes of Health	www.nih.gov
Office of Alternative Medicine	altmed.od.nih.gov
Thomas Legislative Information	thomas.loc.gov
United States Department of Agriculture	www.usda.gov
USDA's School Meals Initiative/Team Nutrition	schoolmeals.nal.usda.gov:8001
U.S. House of Representatives	www.house.gov
U.S. Senate	www.senate.gov

Health Promotion/Consumer Health Sites

American Association of Retired Persons	www.aarp.org
CancerNet	cancernet.nci.nih.gov
Healthfinder	www.healthfinder.gov
Healthy People 2000	www.odphp.osophs.dhhs.gov/pubs/hp2000
Healthy People 2010	web.health.gov/healthypeople
InteliHealth	www.intelihealth.com
International Food Information Council	ificinfo.health.org

Useful Databases

Arbor Nutrition Guide	arborcom.com
Combined Health Information Database	chid.nih.gov
FedStats	www.fedstats.gov
FedWorld	www.fedworld.gov
Food and Nutrition Information Center	www.nal.usda.gov/fnic
MEDLINE	www.nlm.nih.gov
National Center for Health Statistics	www.cdc.gov/nchswww
U.S. Census Bureau	www.census.gov